bnas

W9-CRB-746

Modern Chromatographic Analysis of Vitamins

Modern
Chromatographic
Analysis of Vitamins

Third Edition, Revised and Expanded

edited by

André P. De Leenheer
Willy E. Lambert
Jan F. Van Bocxlaer
University of Ghent
Ghent, Belgium

MARCEL DEKKER, INC.   NEW YORK • BASEL

SEP /be
CHEM

Chemistry Library

ISBN: 0-8247-0316-2

This book is printed on acid-free paper.

Headquarters
Marcel Dekker, Inc.
270 Madison Avenue, New York, NY 10016
tel: 212-696-9000; fax: 212-685-4540

Eastern Hemisphere Distribution
Marcel Dekker AG
Hutgasse 4, Postfach 812, CH-4001 Basel, Switzerland
tel: 41-61-261-8482; fax: 41-61-261-8896

World Wide Web
http://www.dekker.com

The publisher offers discounts on this book when ordered in bulk quantities. For more information, write to Special Sales/Professional Marketing at the headquarters address above.

Copyright © 2000 by Marcel Dekker, Inc. All Rights Reserved.

Neither this book nor any part may be reproduced or transmitted in any form or by any means, electronic or mechanical, including photocopying, microfilming, and recording, or by any information storage and retrieval system, without permission in writing from the publisher.

Current printing (last digit):
10 9 8 7 6 5 4 3 2 1

PRINTED IN THE UNITED STATES OF AMERICA

QP771
M63
2000
CHEM

Preface

This edition of *Modern Chromatographic Analysis of Vitamins* is the third of its kind. The first edition was published in 1985 and the second in 1992. The last decade has again witnessed an enormous proliferation of chromatographic techniques for the determination of vitamins in different matrices. It is exactly this progress that is covered in the third edition of this book.

The work covers both the fat-soluble (Chapters 1–4) and the water-soluble (Chapters 5–13) vitamins, with emphasis on state-of-the-art chromatography, sample preparation, and final measurement. Following present analytical evolution, sections on recent techniques such as capillary electrophoresis and mass spectrometry have been added or expanded. Information on metabolism and biochemical function has also been revised to incorporate current knowledge.

The contributors include many of the best-known and most knowledgeable workers in the field of vitamin analysis throughout the world and are all recognized experts on their topics.

The book aims to be a valuable source of information for scientists with a high degree of expertise in chromatographic analysis of vitamins. Nevertheless, most chapters contain considerable introductory and background material, making this book also appropriate for the relatively inexperienced researcher in this field.

We gratefully acknowledge all of the contributors, whose collaboration let us realize this third volume, which provides broad and up-to-date information on vitamins and their analysis.

André P. De Leenheer
Willy E. Lambert
Jan F. Van Bocxlaer

Contents

Contributors

Arun B. Barua, Ph.D.   Department of Biochemistry, Biophysics, and Molecular Biology, Iowa State University, Ames, Iowa

Jiří Davídek, Ph.D., Dr.Sc.   Department of Food Chemistry and Analysis, Institute of Chemical Technology, Prague, Czech Republic

André P. De Leenheer, Ph.D.   Laboratory of Toxicology, University of Ghent, Ghent, Belgium

Eva D'Haese, Ph.D.   Department of Pharmaceutical Microbiology, University of Ghent, Ghent, Belgium

Yoshiko Egi, Ph.D.   Department of Nutrition, Suzugamine Women's College, Hiroshima, Japan

Günter Fauler, Ph.D.   Department of Analytical Biochemistry and Mass Spectrometry, University Children's Hospital, University of Graz, Graz, Austria

Paul Michael Finglas, B.Sc.   Diet Health and Consumer Science Division, Institute of Food Research, Colney, Norwich, Norfolk, England

Harold C. Furr, Ph.D.   Department of Nutritional Sciences, University of Connecticut, Storrs, Connecticut

Glenville Jones, Ph.D.   Departments of Biochemistry and Medicine, Queen's University, Kingston, Ontario, Canada

Takashi Kawasaki, M.D., Ph.D.   Department of Biochemistry, Hiroshima University School of Medicine, Hiroshima, Japan

Willy E. Lambert, Ph.D. Laboratory of Toxicology, University of Ghent, Ghent, Belgium

Hans Jörg Leis, Ph.D. Department of Analytical Biochemistry and Spectrometry, University Children's Hospital, University of Graz, Graz, Austria

Jan Lindemans, Ph.D. Department of Clinical Chemistry, University Hospital Rotterdam, Rotterdam, The Netherlands

Hugh L. J. Makin, P.L.D., D.Sc., F.R.S.C., F.R.C.Path Department of Clinical Biochemistry, St. Bartholomew's and the Royal London School of Medicine and Dentistry, Whitechapel Campus, London, England

Wolfgang Muntean, M.D. Department of Pediatrics, University Children's Hospital, University of Graz, Graz, Austria

Hans J. Nelis, Ph.D. Department of Pharmaceutical Microbiology, University of Ghent, Ghent, Belgium

Peter Nielsen, M.D., Ph.D. Institute of Medical Biochemistry, University Hospital Eppendorf, Hamburg, Germany

Kristiina Nyyssönen, Ph.D. Research Institute of Public Health, University of Kuopio, Kuopio, Finland

James Allen Olson, Ph.D. Department of Biochemistry, Biophysics, and Molecular Biology, Iowa State University, Ames, Iowa

Markku T. Parviainen, Ph.D., Eur.Clin.Chem. Department of Clinical Chemistry, Kuopio University Hospital, Kuopio, Finland

Olivier Ploux, Ph.D. Department of Bioorganic Chemistry, University of Pierre and Marie Curie, Paris, France

Jukka T. Salonen, M.D., Ph.D., M.Sc.P.H. Research Institute of Public Health, University of Kuopio, Kuopio, Finland

Katsumi Shibata, Ph.D. Department of Life Style Studies, The University of Shiga Prefecture, Hikone, Shiga, Japan

Hiroshi Taguchi, Ph.D. Department of Bioresources, Mie University, Tsu, Mie, Japan

Johan B. Ubbink, D.Sc. Department of Chemical Pathology, University of Pretoria, Pretoria, South Africa

Liisa Vahteristo, Ph.D. Department of Applied Chemistry and Microbiology, University of Helsinki, Helsinki, Finland

Jan F. Van Bocxlaer, Ph.D. Laboratory of Medical Biochemistry and Clinical Analysis, University of Ghent, Ghent, Belgium

Richard B. van Breemen, Ph.D. Department of Medicinal Chemistry and Pharmacognosy, University of Illinois at Chicago, Chicago, Illinois

Jan Velíšek, Ph.D., D.Sc. Department of Food Chemistry and Analysis, Institute of Chemical Technology, Prague, Czech Republic

Karen Vermis Department of Pharmaceutical Microbiology, University of Ghent, Ghent, Belgium

Modern Chromatographic Analysis of Vitamins

1
Vitamin A and Carotenoids

Arun B. Barua and James Allen Olson
Iowa State University, Ames, Iowa

Harold C. Furr
University of Connecticut, Storrs, Connecticut

Richard B. van Breemen
University of Illinois at Chicago, Chicago, Illinois

I. CHEMISTRY

A. Retinoids

The structures and the IUPAC-approved numbering system of some naturally occurring compounds in the vitamin A group are depicted in Figure 1. A group of synthetic retinoids is shown in Figure 2. Approximately 2500 retinoids have thus far been synthesized and characterized (1,2). Although some of the more recently synthesized retinoids, such as 2.i and 2.j, bear little chemical resemblance to retinol (1.a), they all possess a highly conjugated double-bond system, roughly similar physical dimensions, and similar types of biological activities. Inasmuch as some naturally occurring compounds, such as 1.c and 1.d, are also used as therapeutic agents, the distinction between naturally occurring and synthetic retinoids is no longer helpful relative to their usage in society. Thus, the general term "retinoids" embraces a large number of sometimes distantly related chemical compounds.

The so-called parent compound, all-*trans*-retinol (1.a), is an isoprenoid with five conjugated double bonds, absorption maximum at 325 nm in hexane or ethanol, and a single oxygen function. In acid, retinol rapidly dehydrates to anhydroretinol (1.h). In the presence of light, retinol and its derivatives, such as 1.b,

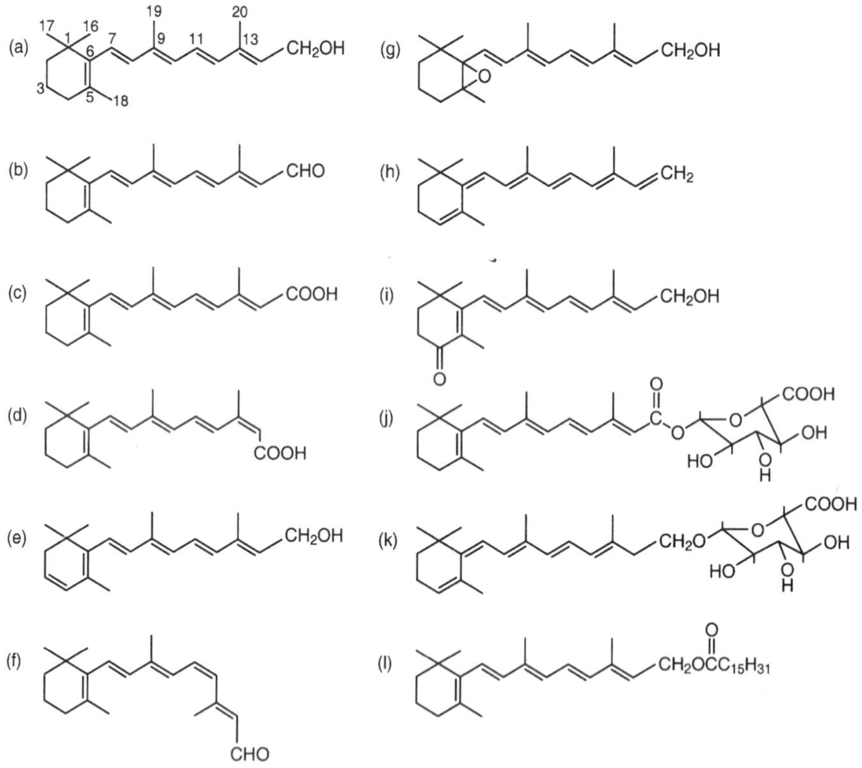

Figure 1 Some naturally occuring retinoids: (a) all-*trans*-retinol; (b) all-*trans*-retinal; (c) all-*trans*-retinoic acid; (d) 13-*cis*-retinoic acid; (e) all-*trans*-3,4-didehydroretinol (vitamin A$_2$ alcohol); (f) 11-*cis*-retinal; (g) all-*trans*-5,6-epoxy retinol; (h) all-*trans*-anhydroretinol; (i) all-*trans*-4-oxoretinol; (j) all-*trans*-retinoyl β-glucuronide; (k) all-*trans*-retinyl β-glucuronide; (l) all-*trans*-retinyl palmitate.

1.c, and 1.l, isomerize to a mixture of isomers, primarily the 13-*cis* and 9-*cis* forms. Isomerization is much enhanced by iodine. The nature of the isomer mixture markedly depends on the solvent used.

All retinoids are sensitive to oxidation and peroxidation, particularly in the presence of transition group metals. A very large assortment of products results, including those with shortened carbon chains, ring oxidations, and hydroxymethyl groups. Thus, in their isolation and analysis, care must be taken to minimize the formation of artifactual products.

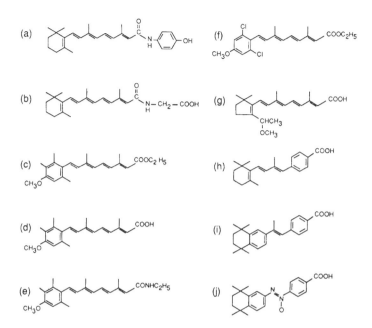

Figure 2 Some synthetic all-*trans*-retinoids: (a) 4-hydroxyphenylretinamide (4-HPR), also called N-retinoyl-4-aminophenol or fenretinide; (b) N-retinoyl glycine; (c) trimethyl-methoxyphenol analog of ethyl retinoate, trivially called "etretinate"; (d) trimethoxyphenol analog of retinoic acid, trivially called "acitretin"; (e) trimethylmethoxyphenol analog of ethyl retinamide, trivially called "motretinide"; (f) methylmethoxydichlorophenol analog of ethyl retinamide, trivially called "dichloroetretinate"; (g) dimethylmethoxyethyl-cyclopentenyl analog of retinoic acid; (h) an aryl triene analog of retinoic acid; (i) tetrahydrotetramethylnaphthalenylpropenylbenzoic acid, abbreviated TTNPB and trivially termed an "arotinoid"; (j) tetrahydrotetramethylnaphthalenyl-azobenzoic acid, abbreviated TTNAOB and trivially called "Az90."

Because of its conjugated double-bond system, vitamin A and other retinoids can be characterized and quantitated by their UV absorption spectra and often by their fluorescence. The light absorbances of selected retinoids are given in Table 1 (3). Proton and ^{13}C nuclear magnetic resonance (NMR) spectra are also useful in characterizing *cis-trans* isomers, and mass spectrometry (MS) is valuable in assessing molecular weights and structures (1).

Formulas and molecular weights of some naturally occurring retinoids are: retinol ($C_{20}H_{30}O$, 286.44); retinal ($C_{20}H_{28}O$, 284.42); retinoic acid ($C_{20}H_{28}O_2$, 300.42); retinyl palmitate ($C_{36}H_{60}O_2$, 524.88); and retinoyl β-glucuronide ($C_{26}H_{36}O_8$, 476.54). At room temperature, retinol and its esters fluoresce strongly

Table 1 Light Absorbances of Selected Retinoids[a]

Retinoid	Solvent	λ_{max}	ϵ	$E_{1cm}^{1\%}$
All-*trans* retinol	Ethanol	325	52770	1845
	Hexane	325	51770	1810
13-*cis* retinol	Ethanol	328	48305	1689
11-*cis* retinol	Ethanol	319	34890	1220
	Hexane	318	34320	1200
9-*cis* retinol	Ethanol	323	42300	1477
11,13-di-*cis* retinol	Ethanol	311	29240	1024
9,13-di-*cis* retinol	Ethanol	324	39500	1379
All-*trans* retinyl acetate	Ethanol	325	51180	1560
	Hexane	325	52150	1590
All-*trans* retinyl palmitate	Ethanol	325	49260	940
All-*trans* retinal	Ethanol	383	42880	1510
	Hexane	368	48000	1690
13-*cis* retinal	Ethanol	375	35500	1250
	Hexane	363	38770	1365
11-*cis* retinal	Ethanol	380	24935	878
	Hexane	365	26360	928
9-*cis* retinal	Ethanol	373	36100	1270
11,13-di-*cis* retinal	Ethanol	373	19880	700
9,13-di-*cis* retinal	Ethanol	368	32380	1140
All-*trans* retinal oxime (*syn*)	Hexane	357	55600	1850
(*anti*)	Hexane	361	51700	1723
11-*cis* retinal oxime (*syn*)	Hexane	347	35900	1197
(*anti*)	Hexane	351	30000	1000
All-*trans* retinoic acid	Ethanol	350	45300	1510
13-*cis* retinoic acid	Ethanol	354	39750	1325
9-*cis* retinoic acid	Ethanol	345	36900	1230
11,13-di-*cis* retinoic acid	Ethanol	346	25890	863
9,13-di-*cis* retinoic acid	Ethanol	346	34500	1150
all-*trans* Methyl retinoate	Ethanol	354	44340	1415
13-*cis* Methyl retinoate	Ethanol	359	38310	1220
All-*trans* retinoyl β-glucuronide	Methanol	360	50700	1065
13-*cis* retinoyl β-glucuronide	Methanol	369	38310	805
9-*cis* retinoyl β-glucuronide	Methanol	353	36900	775
All-*trans* retinyl β-glucuronide	Methanol	325	44950	973
α-Retinol	Ethanol	311	47190	1650
α-Retinal	Ethanol	368	48800	1720
α-Retinoic acid	Ethanol	340	33000	1100
All-*trans* 3,4-didehydroretinol	Ethanol	350	41320	1455
13-*cis* 3,4-didehydroretinol	Ethanol	352	39080	1376
9-*cis* 3,4-didehydroretinol	Ethanol	348	32460	1143
9,13-di-*cis* 3,4-didehydroretinol	Ethanol	350	29950	1030

Table 1 Continued

Retinoid	Solvent	λ_{max}	ϵ	$E_{1cm}^{1\%}$
all-*trans* 3,4-didehydroretinal	Ethanol	401	41450	1470
All-*trans* 3,4-didehydroretinoic acid	Ethanol	370	41570	1395
13-*cis* 3,4-didehydroretinoic acid	Ethanol	372	38740	1300
9-*cis* 3,4-didehydroretinoic acid	Ethanol	369	36950	1240
9,13-di-*cis* 3,4-didehydroretinoic acid	Ethanol	366	32990	1107
5,6-Epoxyretinol	Ethanol	310	73140	2422
5,6-Epoxyretinal	Ethanol	365	45330	1511
5,6-Epoxyretinoic acid	Ethanol	338	45280	1442
	Hexane	338	48860	1556
5,8-Epoxyretinol	Ethanol	278	53390	1768
5,8-Epoxyretinal	Ethanol	331	43800	1460
5,8-Epoxyretinoic acid	Ethanol	298	39470	1257
	Hexane	306	45590	1452
14-Hydroxy-4,14-retroretinol	Ethanol(?)	348	53960	1785
All-*trans* 4-oxoretinoic acid	Ethanol	360	58220	1854
	Hexane	350	54010	1720
13-*cis* 4-oxoretinoic acid	Ethanol	361	39000	1242
All-*trans* 4-oxoretinoyl β-glucuronide	Methanol	364	58220	
13-*cis* 4-oxoretinoyl β-glucuronide	Methanol	367	39000	
9-*cis* 4-oxoretinoyl β-glucuronide	Methanol	356		
Anhydroretinol	Ethanol	371	97820	3650
Anhydrovitamin A$_2$	Ethanol	370	79270	2980
Axerophthene	Hexane	326	49950	1850
Retinyl methyl ether	2-Propanol	328	49800	1660
All-*trans* N-(4-hydroxyphenyl)retinamide	Methanol	362	47900	1225
All-*trans* acitretin (TMMP-retinoic acid)	Ethanol	361	41400	1270
13-*cis* acitretin	Ethanol	361	40450	1241
TMMP-retinol	Ethanol	325	49800	1596

[a] Where the geometric isomer is not indicated, the preparation is assumed to be predominantly all-*trans*. For many compounds, molar extinction coefficients have been calculated from $E_{1cm}^{1\%}$ values in the original references or vice versa.
Source: Ref. 3.

at 480 to 490 nm when excited at 325 to 345 nm, whereas retinal and retinoic acid do not. Fluorescence is enhanced in hexane and dioxane and depressed in ethanol and other alcohols. Because other polyenes, such as phytofluene, also fluoresce strongly under these conditions, care must be taken in differentiating between vitamin A fluorescence and that due to other compounds.

In addition to all-*trans*-vitamin A, all 15 of its possible *cis* isomers have been characterized. In general, the introduction of increasing numbers of *cis*

bonds lowers both the wavelength of peak absorption and the molecular extinction coefficient relative to the all-*trans* isomer.

B. Carotenoids

Although the carotenoids, the highly colored plant pigments, have been of interest
to biologists and chemists since the early 1800s, their chemical characteristics
as a class were first well defined in the 1920s. A major treatise on carotenoids
appeared in 1971 (4) and another recently (5–7).

   All-*trans*-β-carotene (Fig. 3) is generally considered as a class prototype
(8). β-Carotene is a symmetrical molecule of 40 carbon atoms, consists of 8
isoprene units, has 11 conjugated double bonds, and possesses two β-ionone rings
at the ends of the molecule. Other common carotenoids possess hydroxy, alde-
hydic, oxo, or epoxy substituents; may be acylic or contain different ring systems;
and may contain variable numbers of conjugated and isolated double bonds (Fig.
3). The numbering system is similar to that for retinoids in that the gem-dimethyl
carbon in the ring is denoted C-1, the first carbon atom in the central chain C-

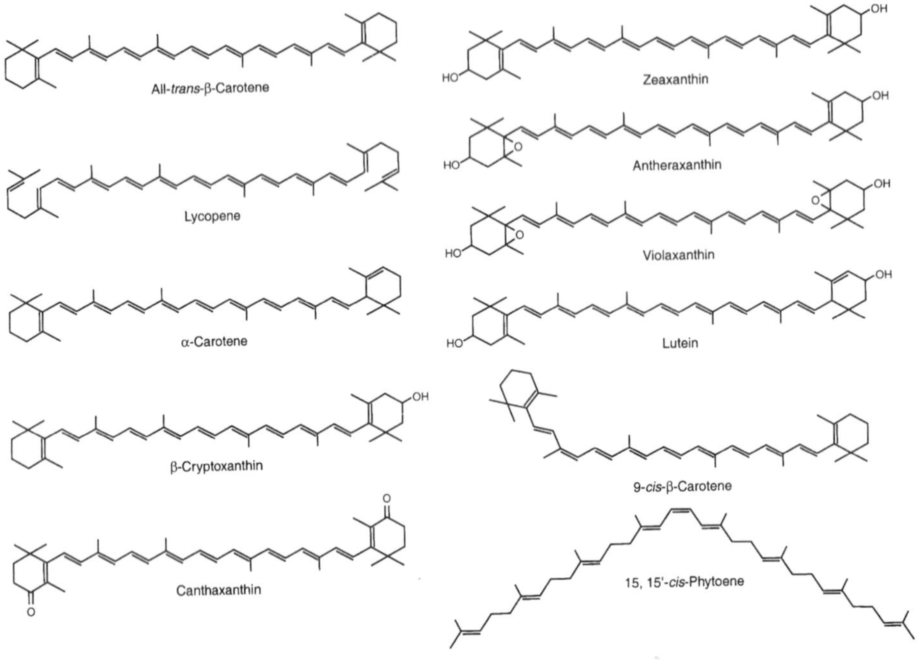

Figure 3   Some common carotenoids found in nature. (From Ref. 8.)

7, and the *two* carbons at the point of symmetry C-15 and C-15'. Thus, violaxanthin (Fig. 3) is 3,3'-dihydroxy-5,6:5',6'-diepoxy-β-carotene. Carotenoids can also contain additional isoprene chains (homocarotenoids), less than 40 carbon atoms (apocarotenoids), and allenic and acetylenic bonds. Hydroxycarotenoids (xanthophylls) can form ester linkages with fatty acids and glycosidic linkages with sugars.

In the most recent compilation, 563 distinct carotenoids are listed (9). In addition to the most common all-*trans* (all-*E*) form, many *cis* isomers of each carotenoid are possible. In fact, many of the latter have also been specifically characterized. Carotenoids are very widely distributed in nature; indeed, it is rare to find a genus without some carotenoid-containing species.

Because of their highly conjugated double-bond system, carotenoids show characteristic ultraviolet and visible absorption spectra (10). For most carotenoids, three peaks, or two peaks and a shoulder, absorb in the range of 400 to 500 nm. Absorption maxima, values of $E^{1\%}_{1cm}$, and molecular extinction values for common carotenoids in petroleum ether (10,11) are given in Table 2. Both the wavelength maximum and $E^{1\%}_{1cm}$ are significantly affected by the solvent used. Thus, for all-*trans*-β-carotene, the wavelength maximum and $E^{1\%}_{1cm}$ are 453 nm and 2592 in petroleum ether, 453 nm and 2620 in ethanol, 465 nm and 2337 in

Table 2 Light Absorbances of Selected Carotenoids[a]

Carotenoid	Solvent	Absorption maxima (nm)	$E^{1\%}_{1cm}$	ε
Antheraxanthin	Ethanol	446	2350	137200
Astaxanthin	Ethanol	472	2135	124000
Canthaxanthin	Light petroleum	466	2200	124000
β-apo-8'-Carotenal	Light petroleum	457	2640	109800
α-Carotene	Light petroleum	422, *444*, 474	2800	150000
β-Carotene	Light petroleum	425, *453*, 479	2592	139000
β-Cryptoxanthin	Light petroleum	425, *452*, 479	2386	132000
Lutein	Ethanol	421, *445*, 475	2550	145000
Lycopene	Light petroleum	444, *472*, 502	3450	
Neoxanthin	Ethanol	416, *439*, 467	2243	135000
Phytoene	Light petroleum	275, *285*, 296	1250	68100
15-*cis* Phytoene	Light petroleum	286	757	41200
Phytofluene	Light petroleum	331, *348*, 367	1350	73300
Violaxanthin	Ethanol	420, *443*, 470	2550	153000
Zeaxanthin	Light petroleum	426, *452*, 479	2348	133400

$E^{1\%}_{1cm}$ and ε values are given for the major wavelength, which is italized when more than one peak is present.
[a] all-*trans* Isomers except as noted.
Source: Refs. 10 and 29.

benzene, 465 nm and 2396 in chloroform, and 484 nm and 2008 in carbon disulfide. The *cis* isomers not only absorb less strongly than the all-*trans* isomer, but also show a so-called *cis* peak of absorbance at 330 to 340 nm.

Carotenoids have also been characterized by nuclear magnetic resonance (NMR), infrared (IR), and Raman spectroscopy as well as by x-ray analysis and mass spectrometry (6). Because of their low volatility and relatively high molecular weight, the C-40 carotenoids have not as yet been separated by gas chromatography. Currently, carotenoids are primarily isolated by liquid chromatography on solid supports.

Carotenoids are unstable in the presence of light and oxygen (1,2,4). The central chain of conjugated double bonds is oxidatively cleaved chemically at various points, giving rise to a family of apocarotenoids. Most carotenoids, but not vitamin A, also serve as singlet oxygen quenchers (12). In essence, singlet oxygen, which is an electronically excited and highly reactive form of oxygen, interacts with the highly conjugated, ground-state carotenoid to give triplet states of both molecules. The triplet state of oxygen is its less active ground state, whereas the triplet carotenoid returns to the ground state by the emission of thermal energy (12). Carotenoids can also serve as antioxidants and free-radical-quenching agents. Carotenoids interact rapidly with free radicals or with oxygen, thereby inhibiting the propagation step of lipid peroxidation. Carotenoids serve this function best at low oxygen tensions; indeed, carotenoids can be pro-oxidants in 100% oxygen (12). These chemical properties of carotenoids, as noted below, are inherently related to their biological functions and actions.

II. BIOCHEMISTRY

A. Retinoids

Vitamin A functions primarily in vision, cell differentiation, and pattern formation during embryogenesis. Vitamin A consequently influences many physiologic processes, including growth, reproduction, and the immune response (13–15).

The diet usually provides both preformed vitamin A from animal products and provitamin A carotenoids from vegetables and fruits. Approximately 50 of the 600 known carotenoids can be oxidatively converted into retinal in mammals, primarily by central cleavage (16), but also in part by eccentric cleavage (16). Retinyl esters, the major dietary form of vitamin A from animal products, are hydrolyzed in the intestinal lumen in the presence of pancreatic esterases and conjugated bile salts (13–15). Retinal produced from carotenoid cleavage is reduced to retinol, which is esterified and incorporated into chylomicra. Chylomicron remnants are taken up primarily by parenchymal cells of the liver but also by other tissues. Retinyl esters are hydrolyzed to retinol, which can (1) combine with the apo form of retinol-binding protein (RBP) to form holo-RBP, (2) be transferred to the stellate cells and be re-esterified, (3) be reesterified in the parenchymal cell, (4) be oxidized through retinal to retinoic acid, or (5) be conjugated with UDP-glucuronic acid to form retinyl β-glucuronide. In a similar way, retinoic acid forms retinoyl β-glucuronide. Although the mode of transferring retinol between parenchymal and stellate cells is still not clear, RBP has been suggested as a possible intercellular carrier within liver (13–15).

Vitamin A is inactivated biologically in mammals either by oxidation at the C-4 position to yield hydroxy and oxoderivatives or by oxidative shortening of the conjugated carbon chain. 4-Oxoretinol and 4-oxoretinoic acid, however, show activity in some biological systems. The methyl groups may also be hydroxylated.

Within cells, vitamin A is largely bound to specific carrier proteins, termed cellular retinoid-binding proteins (13–15). At least two such proteins bind retinol; another two bind retinoic acid, and another, uniquely in the eye, binds retinal. An additional binding protein for retinol and retinal, termed interphotoreceptor (interstitial) retinoid binding protein (IRBP), is found in the interphotoreceptor matrix between the retinal pigment epithelium and the outer segments of rod cells. In addition to the all-*trans* form, specific *cis* isomers are bound by some of these cellular binding proteins, but not by others. The binding proteins serve as transport and protective agents for retinoids and, in some cases, enhance their enzymatic transformations.

Another set of binding proteins, specific for retinoic acid, is found in the nucleus of cells (13–15). Because they influence genome expression, they have been called retinoic acid receptors. Two families of these exist, the RAR family, which binds both all-*trans* and 9-*cis* retinoic acid, and the RXR family, which binds only 9-*cis* retinoic acid. These retinoic acid receptors belong to a superfamily of receptors, which also includes those for some steroids, tri-iodothyronine, and 1α,25-hydroxycholecalciferol (13–15).

In the plasma, vitamin A is transported from its major depots in the liver to tissues in several forms, primarily as a 1:1 molar complex of all-*trans*-retinol with RBP. Low concentrations of all-*trans*- and 13-*cis*-retinoic acid, probably bound to albumin, and retinyl and retinoyl β-glucuronides, are also present. Holo-RBP also interacts strongly with transthyretin in the plasma. All of these forms of vitamin A can be taken up by various tissue cells. Several tissues besides the parenchymal cells of the liver can also synthesize RBP, as evidenced by the presence of mRNA for RBP within such cells. Thus, the extensive recycling of retinol between the liver and peripheral tissue cells may well occur as complexes with RBP. Another possibility is that retinyl ester, which is synthesized in essentially all cells of the body, might be carried back to the liver in lipoproteins.

As already noted, vitamin A functions in vision, cellular differentiation, and embryonic development (13–15). In the visual process (17), holo-RBP of the plasma is taken up by the retinal pigment epithelium (RPE) by a receptor-mediated process. All-*trans*-retinol is isomerized to the 11-*cis* form in the RPE by a concerted reaction involving ester formation and hydrolysis. The 11-*cis* isomer, like the all-*trans* form, can be esterified by transacylation from endogenous membrane phospholipids or can be ferried to the rod outer segment on IRBP.

Within the RPE or outer segment, 11-*cis*-retinol is oxidized to 11-*cis*-retinal, which combines in the outer segment with opsin to yield rhodopsin. Light destabilizes rhodopsin by isomerizing the 11-*cis* chromophore through a transoid state to the all-*trans* isomer. Concomitantly, the protein passes through several conformational states to metarhodopsin II, which interacts with transducin, a G-protein with three subunits. A series of molecular events follows, including GTP binding, phosphodiesterase activation, a decrease in the cytosolic concentration of cGMP, a blocking of sodium ion uptake, and resultant hyperpolarization of the rod cell membrane (17). Needless to say, these events can be reversed. The three types of cone cells presumably have similar mechanisms for transducing impulses of light into membrane potentials.

In vitamin A deficiency, mucus-secreting epithelial cells in many tissues are replaced by squamous cells (13–15). Although this process of ''keratinization'' is the most prominent cellular manifestation of vitamin A deficiency, more subtle changes occur in most cells of the body. In the absence of vitamin A, keratinocytes produce a different pattern of keratins. Low concentrations of retinoic acid rapidly induce transglutaminase in macrophages, granulocytic differentiation in promyelocytic leukemia (HL-60) cells, and the differentiation of embryonic teratocarcinoma (F-9) cells. Retinoic acid and, to a lesser extent, other naturally occurring retinoids produce a variety of cellular effects in other cell lines as well. Many synthetic retinoids show similar actions. Although these effects cannot be placed in a cohesive, all-embracing pattern, retinoids clearly influence gene expression in many cells (13–15).

The current leading hypothesis is that nuclear RAR and RXR play a direct role in this process. The retinoid receptors can be activated by physicochemical binding of free retinoic acid to RAR and RXR. Alternatively, covalent forms, such as retinoyl derivatives of RAR and RXR, might also exist. Interestingly, retinyl and retinoyl β-glucuronide stimulate the differentiations of HL-60 cells well without evident conversion to retinol and retinoic acid, respectively. Retinoic acid has also been implicated as a morphogen in embryonic development (18). The adverse effects of vitamin A deficiency on reproduction, growth, and the immune response, in all likelihood, are an expression of perturbations in the process of cellular differentiation.

High doses of vitamin A and other retinoids also cause serious toxic manifestations: namely, teratogenicity, chronic toxicity, and acute hypervitaminosis (13–15). Some individuals may also show a genetic sensitivity to vitamin A—

termed vitamin A intolerance—at intakes not much above those normally ingested.

B. Carotenoids

Carotenoids are primarily synthesized in plants and micro-organisms (19,20). A
major function of carotenoids clearly relates to the absorbance of light in the
process of photosynthesis (12,19,20). Carotenoids serve as ancillary pigments in
the antenna systems of photosynthetic organisms. Thus, the energy in the excited
states of the carotenoids is ultimately transferred to the chlorophyll molecule at
the reaction center of the antenna system (12,20).

A related function of carotenoids in photosynthetic organisms is photoprotection (12,20). Mutant photosynthetic organisms lacking carotenoids are killed
by exposure to light and oxygen. Similarly, added carotenoids protect organisms
from mutations induced by photo-oxidation.

As *cis-trans* isomerization is directly involved in the visual function of
vitamin A in animals and in energy transduction in bacteria via the bacteriorho-
dopsin system, so also might the *cis-trans* isomerization of carotenoids play physiological roles in other organisms. For example, Nelis et al. (21) found a high
concentration of *cis*-canthaxanthin in the female reproductive system of the brine
shrimp *Artemia* and a progressive decrease in the *cis/trans* ratio during nauplii
development. Thus, the relation between isomeric forms of carotenoids and their
biological actions deserves further attention.

In mammals, a small group of carotenoids, approximately 50 of the 600
known compounds, serve as precursors of vitamin A (13–16). Carotenoids in
vertebrates seem to be oxidatively cleaved primarily at the central 15,15' double
bond to yield two molecules of retinal (16), but eccentric cleavage to yield β-
apo-carotenals also seems to occur. Fish and birds can also form vitamin A from
hydroxy- and oxocarotenoids. In plants and micro-organisms, eccentric cleavage
is common.

Carotenoid cleavage products with physiologic roles in plants and micro-
organisms include abscisic acid, β-ionone, trisporic acid, and crocetin (22). Carotenoids undergo a variety of other biological transformations, including epoxidation (primarily at the 5,6 and 5',6' positions but also elsewhere in the molecule),
reduction of oxo to hydroxy analogs and of epoxy to dihydrohydroxy compounds,
esterification of hydroxy carotenoids, glycosylation of both hydroxy and carboxy
compounds, and oxidation of methyl groups. Some of these metabolic reactions
relate to physiologic functions and others seem to be degradative in nature.

In mammals, carotenoids, freed from foodstuffs by proteolytic digestion of
proteins in the stomach and small intestine, are solubilized by conjugated bile
salts in the small intestine and are absorbed with other lipids (16,23,24). Dietary

carotenoids are primarily converted into vitamin A in the intestinal mucosa. A significant portion of absorbed carotenoids in humans, chickens, and cows, but not in rodents, is also incorporated per se into chylomicra, whose remnants are cleared from the plasma primarily by the liver (16,23,24).

Besides chylomicra, carotenoids are mainly associated in plasma with low-density lipoproteins (LDL) and high-density lipoproteins (HDL). The major carotenoids in human plasma are lutein, zeaxanthin, lycopene, β-cryptoxanthin, α-carotene, and β-carotene. β-Carotene usually comprises 10% to 25% of the total human plasma carotenoids (16,23,24).

Carotenoid patterns in the tissues of humans and monkeys are usually similar to those in plasma (16,23,24). Of various tissues, adipose tissue and liver seem to be major depositories of carotenoids in humans. Presumably carotenoids are both deposited in tissues and released from them into the plasma. Outside of the cleavage of several carotenoids into vitamin A, little is known of the manner in which other carotenoids, particularly those that are nutritionally inactive, are metabolized.

Carotenoids also play a role in the coloration of living things, the most dramatic and pleasing example being the plumage of birds. Specific coloration may well serve as an attractant in sexual processes as well as providing protection from predators. In many cases, specific carotenoproteins are formed at given sites in these organisms (12). The ligands in such complexes are usually oxocarotenoids. Hydrocarbon carotenoids are usually loosely associated with lipoproteins in the plasma and tissues of animals, although a β-carotene binding protein of rat liver has recently been reported (16).

Apart from defined physiological functions, carotenoids have several biological actions (16,25). Whereas functions can be considered as essential physiological roles, actions can be defined as physiological or pharmacological responses, either beneficial or adverse, to treatment of an organism with carotenoids (16,25). In some cases, very little carotenoid is needed to invoke a response; in other cases, very large amounts are required.

Biological actions of carotenoids include enhancement of the immune response, a reduction of photoinduced or chemically induced neoplasm, decreased mutagenesis and sister-chromatid exchange, reduced cellular transformation, inhibited micronuclei formation in epithelial cells, and photoprotection of patients with erythropoietic protoporphyria (16,25).

Finally, carotenoids have been associated with a reduced incidence of some chronic diseases (26,27). These associations, although not inferring a cause-and-effect relationship, have nonetheless been seriously considered in a public health context. For example, the dietary intake of carotenoids, but not of vitamin A, is inversely associated in epidemiological surveys with the risk of lung cancer. Together with many other singlet oxygen- and free-radical-quenching compounds

found in fruits and vegetables, carotenoids may thereby be contributing to a reduced oxidative stress on cells of the lung.

III. SAMPLE EXTRACTION, HANDLING, AND STORAGE OF RETINOIDS AND CAROTENOIDS

Retinoids and carotenoids are susceptible to geometric isomerization in the presence of oxygen, light, and heat; therefore, precautions should be taken to avoid
these destructive factors during extraction and handling. Previous editions of this
book also provide detailed information on sample handling and extraction
(28,29). The reader is also referred to other detailed descriptions of the analysis,
handling, and storage of retinoids (3,30) and carotenoids (5).

Although the all-*trans* (all-*E*) forms predominate in most biological tissues,
isomerization is catalyzed by light and heat. Even if total retinoid or carotenoid
content rather than quantitation of specific isomers is desired, it is best to avoid
artifactual isomerization because the molar extinction coefficients of the *cis* isomers (Z isomers) are usually lower than that of the all-*trans* isomer (Table 1);
significant *cis*-isomerization in samples will lead to underestimates of retinoid
or carotenoid content. It is possible to keep solutions in amber glassware and to
work under subdued lighting or under dim red lights, but more convenient to
outfit the laboratory with yellow (F40 Gold) fluorescent lights (31).

There has been considerable concern about the stability of retinoids and
carotenoids in stored biological samples. Craft et al. (32) found that retinol, α-
tocopherol, and carotenoids were stable when stored frozen at $-70°C$ for at least
28 months or at $-20°C$ for 5 months, but carotenoids were less stable than retinol
and tocopherol on prolonged frozen storage. Flushing oxygen from samples with
nitrogen gas before freezing had no apparent beneficial effect on sample stability.
Similarly, Edmonds and Nierenberg (33) found no significant effects of storage
at $-70°C$ for 5 years on serum concentrations of retinol, α-tocopherol, or β-
carotene. Comstock et al. (34,35) concluded also that storage at $-70°C$ maintains
retinol, α-tocopherol, and carotenoids for years, but that storage at $-40°C$ leads
to marked losses of β-carotene. Ito et al. found that retinol and α-tocopherol
concentrations were stable in plasma stored refrigerated for 25 days, but that
carotenoid concentrations began to fall after 21 days of refrigerated storage (36).
Hankinson et al. found that whole-blood samples could be stored overnight for
48 h at 4°C with loss of about 5% to 7% per day of retinol and β-carotene (considerably less than the subject-to-subject variation), suggesting that overnight cou-
rier shipment should not be a major problem for survey samples (37).

Regarding exposure of plasma samples to light and air, Gross et al. (38)
concluded that limited exposure (as in routine clinical blood handling) resulted

in no loss of carotenoids or of α-tocopherol. Other tissues (liver, kidney) also should be collected quickly and either frozen intact, homogenized, and frozen, or homogenized and lyophilized; Peng et al. (39) found that the buccal cell content of carotenoids and α-tocopherol did not change during storage at −80°C for 8 months. Barreto-Lins et al. (40) noted that retinol decomposes rapidly in solvent extracts even at ice temperatures. Some authors recommend that an antioxidant (for example, ascorbic acid or butylated hydroxytoluene; BHT) be added during sample extraction (41,42), although Craft et al. (32) found no advantage in this practice. In this regard, pyrogallol and ascorbic acid are poorly soluble in organic solvents, whereas butylated hydroxyanisole (BHA) and BHT are readily soluble; BHA's light absorption maximum (290 nm in methanol) approaches that of retinol, while BHT (absorption maximum 275 nm in methanol) is less well separated from retinol chromatographically and thus is more likely to interfere in HPLC analyses (H. C. Furr, unpublished observations).

To summarize, we recommend (1) that biological samples be stored at −70°C if storage for more than a few days is required; (2) that samples not be extracted until shortly before chromatographic analysis and that overnight storage of extracts be avoided; (3) that an appropriate antioxidant such as BHA be added at an early stage of the extraction process; and (4) that samples be protected from excessive heat and, as much as possible, from white light.

To take advantage of the fact that retinol is much more stable when complexed with plasma retinol-binding protein, its transport protein, Oliver et al. (43) allowed 200-μL serum aliquots to dry onto ethanol-washed, dry filter paper. The retinol was redissolved for HPLC analysis by suspending the paper in phosphate buffer, followed by shaking with ethanol, and then extracted with hexane. Shi et al. (44) similarly shipped and stored whole blood or serum samples dried on filter paper, with subsequent analysis by capillary electrophoresis. This procedure allows simple, inexpensive shipment of samples (e.g., population screening surveys) over great distances.

Vitamin A compounds and carotenoids are generally lipophilic—i.e., much more soluble in organic solvents than in aqueous systems. The solubility of retinol and retinal in water is approximately 0.1 μM (45), whereas their solubility in many organic solvents is on the order of 5 mM. Thus, most vitamin A compounds and carotenoids can be extracted from biological matrices with organic solvents. In analysis of soft tissues (e.g., liver) the tissue can be ground with anhydrous sodium sulfate (to rupture tissues and to remove water) under an appropriate solvent (e.g., dichloromethane, chloroform:methanol, diethyl ether, or hexane:2-propanol) in a mortar, and the resulting suspension filtered (46,47). Blanco et al. found that some pharmaceutical preparations of vitamin A may be extracted directly with hexane, but that some require pretreatment (48). For extraction of fibrous tissues, homogenization may be necessary before solvent extraction (49). Ito et al. lyophilized tissue homogenates before solvent extraction (50); the lyophilized homogenate may be extracted with dichloromethane: methanol and methanol (50), chloroform and methanol (51), diethyl ether (52), or methanol and hexane (53). The mixture hexane: 2-propanol has been recommended for extraction of lipids from biological tissues as being less toxic than chlorinated solvents (54). The simple Bligh-Dyer solvent extraction process with methanol: chloroform (49) was used to extract human cheek cell samples for retinol and α-tocopherol analysis (55).

In the analysis of carotenoids, sometimes it is necessary to pretreat the samples before extraction with organic solvents (56). Details of extraction of carotenoids can be found in a recent overview (5). To prevent the denaturing activity of some of the enzymes present in plant materials such as paprika (57) or green peas (58), the samples are blanched in water. It is also necessary to add antioxidants such as BHA, hydroquinone, or BHT during the extraction process to prevent lipid oxidation activity and pigment oxidation. Strong acids can bleach carotenoids, resulting in total decoloration of the pigment solution; can catalyze dehydration of retinol to anhydroretinol; can isomerize 5,6-epoxides to 5,8-epoxides; and can hydrolyze glucuronide conjugates. Therefore, addition of acid neutralizers such as magnesium carbonate is recommended during extraction when acids are released during extraction (59). Traces of HCl present in chloroform used as a solvent must be removed before use. Solvents should be free from peroxides as well as acids, since retinoids and carotenoids are susceptible to peroxidation. The principal extraction solvents for carotenoids are acetone, light petroleum, n-hexane, diethyl ether, and tetrahydrofuran, used alone or in binary mixtures. The Association of Official Agricultural Chemists (AOAC) recommends using acetone: n-hexane (6:4). Acetone solubilizes carotenoids but is miscible with water, thus giving efficient carotenoid extraction. Extraction is continued until no more color can also be extracted. After extraction, depending on the amount of water present, the extract can be evaporated to dryness, or the carotenoids can be transferred to hexane by adding hexane and water to the extract. Solvents from extracts should be removed under reduced pressure at temperatures $<40°C$ to reduce isomerization. When samples, such as green leaves, contain chlorophyll, saponification of the extract with alkali helps removal of chlorophyll. Other compounds, such as oils and fats, which often cause difficulty during chromatographic analysis, can also be removed by saponification. However, saponification may result in destruction of certain carotenoids and should thus be conducted under the mildest possible conditions. The presence of acetone during saponification may result in aldol condensations giving rise to artifacts (60). The effectiveness of the saponification step in the quantitative determination of carotenoids and provitamins A in several fruits and vegetables has been assessed (61). It is often useful to perform enzymatic hydrolysis instead of alkaline saponification to isolate unstable carotenoids (56,62,63).

Blood plasma or serum is often first treated with an equal volume of methanol or acetonitrile to denature plasma proteins, and then retinoids and carotenoids are extracted with hexane or ethyl acetate (two or three extractions for completeness). Diethyl ether is also used for extractions (64–66), but it has the disadvantages of high vapor pressure and flammability. An internal standard and/or an antioxidant may be dissolved in the organic solvent used to precipitate proteins. The choice of methanol or acetonitrile is a matter of preference: acetonitrile gives a rubbery pellet of precipitated protein that is more difficult to resuspend for subsequent extractions. Hexane is the most popular solvent for the extraction of carotenoids because it does not carry water from serum and is easy to evaporate under a stream of nitrogen or argon. A mixture of ethyl acetate and hexane can extract more hydrocarbon carotenoids such as α- and β-carotene and lycopene than does hexane alone (67,68). Addition of a trace of acetic acid to the sample improves extraction of acidic compounds such as retinoic acid and retinoid glucuronides but avoids the possible problems associated with stronger acids (68).

The organic extracts of plasma or other tissues are usually concentrated before chromatographic analysis. For small volumes (<5 mL), it is convenient to evaporate volatile solvents under a gentle stream of inert gas, such as argon or nitrogen (commercial nitrogen may contain traces of oxygen which can destroy retinoids or carotenoids). The residual lipid is then redissolved in a known volume of suitable solvent. Ideally this solvent should be of low volatility (to minimize volume changes by evaporation), capable of dissolving large amounts of lipid, and fully compatible with the chromatographic technique to be used.

Alternatively, it is possible to inject the solvent extract directly for HPLC analysis. For good quantitation this method requires an internal standard or very careful measurement of liquid volumes. McClean et al. (69) and Nierenberg and Lester (70) used acetonitrile to denature plasma proteins and then solubilized retinol with ethyl acetate:1-butanol (1:1) for direct injection; Siddiqui et al. precipitated proteins from serum by addition of acetonitrile, centrifuged, and then injected the supernatant (71). Retinoids and carotenoids can be efficiently extracted from one volume of serum or plasma with three volumes of 2-propanol:dichloromethane (2:1) followed by centrifugation; an aliquot can be directly analyzed by HPLC (72,73). The risks of sample loss and degradation during processing can thus be avoided.

In an analysis of retinal production from cleavage of β-carotene by intestinal homogenates, acetonitrile was simply mixed with the reaction mixture and centrifuged, and an aliquot of the supernatant was injected directly for reversed-phase HPLC analysis (74).

Solid-phase extraction techniques can be used to purify retinol from human plasma, employing first a C18 cartridge to remove aqueous impurities, followed by an aminopropyl cartridge to remove lipid contaminants (75). Although retinol recovery was only ~50%, the retinol was of high purity, suitable for analysis of isotope ratios by mass spectrometry.

Conventional methods of extracting solid biological tissues are labor-intensive and time-consuming. Precipitation of solids from liquid solvents, with dissolution by high-pressure CO_2 as an antisolvent to create supersaturation, is a potentially attractive crystallization process. Supercritical fluid extraction (SCE), which is widely used for extractions of lipids such as pesticides from a variety of matrices on an industrial scale, is slowly gaining attention. Burri et al. applied supercritical fluid extraction techniques to extraction of retinyl esters from calf liver. They obtained recoveries of 108% compared with liquid-liquid (Bligh-Dyer) extraction, with relative standard deviations of 8% (76). They also applied this technique to extraction of β-carotene from beadlets, and obtained recoveries ≥90%. Design and control of a process to extract β-carotene with supercritical CO_2 has been described (77,78). The gas antisolvent (GAS) process has been used to separate and purify β-carotene from mixtures containing carotene oxidation products (79); it seems that this extraction technique deserves more attention.

Saponification has often been used to break up tissue matrices and to eliminate triglycerides in biological tissues. Several specific procedures have been used in which the concentration of alkali, temperature, and length of time vary. As examples, Vahlquist used ethanolic KOH at 80°C for 20 min to release retinoids and carotenoids from human skin biopsy samples (80). Rettenmaier and Schuep recommend methanolic KOH to hydrolyze liver samples (30 min at reflux), with ascorbic acid present as antioxidant (81). An automated, microprocessor-controlled saponification system for analysis of vitamin A (and vitamins D and E) in milk has been described (82); the milk sample is saponified in a knotted flow reactor, then neutralized and trapped on a C18 cartridge before analysis on a C18 HPLC column. Other authors have described an automated system to transesterify fats with methanolic potassium methylate before analysis of cholesterol and tocopherols (83); although this system was not applied to vitamin A analysis, it would seem appropriate as an alternative to saponification for samples with high lipid content. Sommerburg et al. compared enzymatic lipolysis (lipase plus cholesterol esterase) with saponification for digestion of plasma, and concluded that enzymatic hydrolysis is preferable (84).

Retinoic acid and its analogs are generally ionizable and hence require a different sort of extraction from biological matrices, as described below.

A. Quantitation of Standards

Because of the chemical instability of retinoids and carotenoids, standards are frequently contaminated with degradation products. Furthermore, retinoids and carotenoids are usually present in very small quantities (picomoles or nanomoles) in biological samples. Thus solutions of retinoid and carotenoid standards are invariably quantitated (and their purity assessed) by absorbance spectroscopy. Full absorption spectra should be scanned and HPLC should be performed to confirm integrity of standards. Molar extinction coefficients and $E_{1cm}^{1\%}$ values for a number of retinoids and carotenoids are presented in Tables 1 and 2, and spectroscopic values for many additional carotenoids are given by Britton (10); example calculations of determining concentration from absorbance data are given by Barua and Furr (3). The use of molar units is encouraged for quantitation, to avoid ambiguities. The effects of different solvents on relative solubility, stability, and light absorbance of lutein, zeaxanthin, lycopene, and β-carotene have been studied (85,86).

Blanco et al. used multiple linear regression analysis of light absorption at selected wavelengths to determine concentrations of vitamin A esters, vitamin D, and vitamin E in pharmaceutical preparations; the novel aspect of their application was use of first-derivative spectra in addition to the absorption spectra (48).

B. Internal Standards

When chosen and used properly, internal standards help to correct quantitative analyses for losses of sample during sample workup. It should be remembered, however, that the internal standard is added during the analyte extraction, and so can not completely replicate the extraction from the original sample matrix. The ideal internal standard should be as chemically similar to the analyte as possible, but must be resolved by the analytical method. Radioactive forms of retinoids have been used as internal standards by, among others, Cullum and Zile (53) and Beeman and Kronmiller (87). Deuterated retinoic acid has been used as internal standard in GC/MS analysis of endogenous retinoic acid (88,89). A sulfonic retinoid has been used as internal standard for HPLC analyses (90). The recent flood of synthetic aromatic retinoid analogs, produced as ligands for nuclear retinoic receptor proteins, provide a diverse variety of potential internal standards for retinoid extractions and analyses (91,92).

Retinyl acetate is sometimes used as internal standard for retinol analysis; it is not ideal because it is an ester, not a free alcohol (and is hydrolyzed by saponification processes), but superior in principle to the use of tocol or other nonretinoid forms for retinoid quantitation. Substituted retinal oximes have been used as internal standards (93). 15-Methylretinol (94) is conceptually an excellent internal standard for retinol analysis, because of its very similar chemical structure and properties.

For analysis of carotenoids, 8′-apo-β-carotenal and C45-β-carotene (95), ethyl β-apo-8′-carotenoate and 3R-8′-apo-β-caroten-3,8′-diol (65), β-apo-10′-carotenal oxime (96), β-apo-carotenyl decanoate (68), and echinenone (97) have been used as internal standards.

IV. THIN-LAYER CHROMATOGRAPHY OF RETINOIDS AND CAROTENOIDS

Although thin-layer chromatography (TLC) and open-column chromatography have played important roles historically in analysis of retinoids and carotenoids, both methods have been almost completely supplanted by HPLC. Paper chromatography has never been employed extensively for retinoid or carotenoid analysis.

TLC is still useful for rapid assessment of HPLC methods, because it provides rapid analyses with different mobile phases in much less time than required for reequilibration of HPLC columns. TLC is also practical for rapid qualitative analysis of reaction products from synthetic or semisynthetic reactions and for confirming purity of concentrated solutions, in which impurities might escape detection by HPLC. TLC is in fact preferable to HPLC for analysis of radiochemical impurities (degradation products) in radiolabeled retinoids and carotenoids, since in HPLC some impurities might not be eluted and detected; radioactivity on thin-layer plates may be detected by autoradiography (exposure of the plate to x-ray film (98)) or by cutting the TLC plate into short sections which can be counted individually in vials by liquid scintillation.

Short (7.5 to 8 cm), commercially available plates are convenient for either adsorption (silica or alumina) or reversed-phase (C8 or C18) chromatography, using mobile phases as appropriate for adsorption or reversed-phase HPLC. On reversed-phase TLC plates, as in reversed-phase HPLC, methanol or acetonitrile gives adequate mobility and resolution of retinol and retinyl acetate; methanol:water or acetonitrile:water mixtures are suitable for polar retinoids, whereas mixtures of acetonitrile with less polar solvents (e.g., dichloroethane) give good resolution of retinyl esters or carotenes. For adsorption chromatography, mobile phases of hexane:ethyl acetate or hexane:acetone (4:1, v:v) are useful for preliminary analyses. The use of TLC for analysis of retinoids has been reviewed by Lambert et al. (28), and tables of R_f values (relative mobility) of representative retinoids on silica and alumina are given by John et al. (99), Varma et al. (100) and Fung et al. (101). Thin-layer chromatography of carotenoids has been reviewed recently by Schiedt (102).

Retinol, retinyl esters, and retinyl ethers and the carotene phytofluene display bright yellow-green fluorescence on TLC plates or open columns when illuminated with a long-wavelength (366-nm) mercury lamp, and anhydroretinol appears red. Other retinoids and carotenoids are not fluorescent under these conditions, and must be identified by other methods such as absorption of iodine vapor (formation of brown spots when placed in a chamber containing iodine crystals) or quenching of the background fluorescence of fluorescent TLC plates (illuminated by a mercury lamp at 254 nm).

In a novel application, carotenoids were separated on commercially available silica Chromarods; β-carotene and other nonsaponfiable lipids were chromatographed with a nonpolar mobile phase (light petroleum:chloroform:acetone, 89.5:10:0.5), and a more polar mobile phase (light petroleum:chloroform: 2-propanol, 50:40:10) was then used to resolve the xanthophylls canthaxanthin, lutein, violaxanthin, and neoxanthin (103). Quantitation was obtained by flame ionization detection, using methyl tetracosanoate as internal standard; the working range was 0.6 to 5 μg β-carotene. The Chromarods themselves can be cleaned and reused.

A TLC method for analysis of bixin, lycopene, canthaxanthin, and β-apo-8′-carotenal was recently presented (104) for rapid screening of food color adulteration. Another method that was quickly supplanted by HPLC was that of liquid-gel partition chromatography, using stationary phases such as Sephadex LH-20 or hydroxyalkoxypropyl Sephadex (50). This method, which is capable of resolving retinyl esters, retinol, retinal, and retinoic acid, was applied to studying vitamin A metabolism, but it is slower and gives less resolution than HPLC on sturdy silica stationary phases.

V. HIGH-PERFORMANCE LIQUID CHROMATOGRAPHY

High-performance liquid chromatography (HPLC) has become the choice of most investigators for the analysis of retinoids and carotenoids in serum and tissues. The advantages of HPLC over other conventional methods of analysis such as open-column and thin-layer chromatography include shorter analysis times, greater resolution, ease of quantitation, and lower limits of detection; HPLC equipment is now widely available. With the increasing use of more sensitive photodiode array detectors for HPLC, it is now possible not only to separate retinoids and carotenoids but also to identify the compound in each peak by its absorption spectrum. Both normal-phase (adsorption) and reversed-phase (partition) HPLC have been widely used, although reversed-phase HPLC has become more popular. Detailed descriptions of these two types of HPLC can be found in earlier editions of this book (28,29). A review on the limitations of HPLC and supercritical fluid chromatography of carotenoids (105) and detailed descriptions of HPLC equipment and analysis of carotenoids by HPLC have been provided by Pfander et al. (106,107). Both normal (straight-phase, adsorption) HPLC and reversed-phase (RP) HPLC can be used for the separation of carotenoids and retinoids. Normal bonded-phase HPLC has been increasingly used in LC-MS and LC-MS-MS techniques in recent years. However, reversed-phase HPLC is by far the most popular mode.

A. Retinoids

1. Retinol

HPLC is particularly well suited for analysis of all-*trans* retinol; retinol (from extracts of plasma or from saponification of other tissues) is easily resolved from other tissue components by either reversed-phase or adsorption chromatography, and is readily quantitated using absorbance detectors (108–110). Resolution of geometric isomers of retinol is discussed below, with retinal separations. Octadecylsilane (C18) columns with mobile phases of methanol: water or acetonitrile: water (typically 98:2 or 95:5 to obtain a capacity factor k' in the optimum range, i.e., $1 < k' < 10$) are very suitable for analysis of retinol from biological extracts. α-Tocopherol is readily extracted and chromatographed under the same conditions that are suitable for retinol, with detection at a compromise wavelength (e.g., 300 nm) or with wavelength-switching (325 nm for retinol, 295 nm for tocopherols), or with fluorescence detection for either of these analytes (e.g., λ_{ex} 325, λ_{em} 450 nm for retinol; λ_{ex} 295, λ_{em} 390 nm for vitamin E). Likewise, many researchers have performed simultaneous extraction and HPLC analysis of retinol and carotenoids from serum and other biological tissues. Rather than giving an exhaustive catalog of all publications presenting such analyses, some representative examples will be cited in this chapter.

Limits of detection for absorbance detectors (325 nm) with conventional (5 μm particle size) C18 columns and methanol: water mobile phases are typically 0.35 pmol (0.1 ng) at a 5:1 signal:noise ratio. Even low serum retinol concentrations as found in vitamin A deficiency (0.35 to 0.7 μM, i.e., 10 to 20 μg/dL) require sample volumes of only 1 μL! Nonetheless, fluorescence detection can give even lower limits of detection (0.07 pmol, 20 pg in tear fluid (111); 5-μL sample sizes have been used for routine plasma assays (112). Electrochemical detection has also been used for simultaneous analysis of retinol and tocopherol (113,114). Microbore columns and smaller packing particle sizes could give improved limits of detection (115,116) but require low-dispersion fittings and detector cells. The requirements for plasma retinol quantitation are not so stringent that use of these techniques has become popular.

Solvent extraction is not obligatory for analysis of serum retinol. Retinol bound to plasma retinol-binding protein (RBP) in plasma or serum can be directly detected by fluorescence in size exclusion chromatography (117,118) or protein reversed-phase HPLC (119). Sensitivity and specificity are good because retinol fluorescence is enhanced some 14-fold by its binding to RBP (120), so the procedure requires only 20 μL or less of serum or plasma. This method has the advantages of allowing direct injection of serum or plasma (no extraction step) and of requiring only very small samples (potentially, fingerprick blood samples).

Capillary zonal electrophoresis has also been applied to the rapid quantitative analysis of serum retinol-RBP complex, requiring <5 min for analysis of 8 to 10 nL of sample (injected electrokinetically). Again, fluorescence of the retinol-RBP complex was used for detection (excitation at 325 nm by a He-Cd laser, emission at 465 nm) to achieve a limit of detection of 10 fg retinol at 5:1 signal: noise ratio (121). This method was applied also to quantitation of retinol in whole blood spots dried on filter paper, requiring only 2 drops of blood and providing a limit of detection of serum retinol of 10 nM (0.3 µg/dL) (44).

Open-column reversed-phase chromatography has also been employed for analysis of retinol from plasma extracts, either in micro mode (50-µm diameter C18 particles, column dimensions 0.5 × 5 cm) or macro mode (column dimensions 1.5 × 6 cm), using elution with methanol:water (90:10) (122). Limit of detection (fluorometric determination in collected fractions) was 0.044 µM (1.25 µg retinol/dL) for the microcolumn method, with 0.1 mL serum used. The major fluorescent interference in serum (phytofluene) was retained on the column under these elution conditions.

2. Retinyl Esters

Esters of retinol with long-chain fatty acids are a major storage form of vitamin A, especially in liver but also in other tissues. In some cases, it may be convenient to saponify the sample and assay total vitamin A as retinol; in other cases, more information can be gained by analyzing the intact retinyl esters. Because all retinyl esters and retinol have identical molar absorptivity in a given solvent (123), retinyl palmitate can be used as a single quantitative standard for retinyl ester analysis if peak area (but not peak height) is used.

The major problems involved in analysis of retinyl esters are: (1) the difficulty in either resolving individual retinyl esters or eluting all retinyl esters as a single peak; (2) separating geometric isomers of retinyl esters; (3) obtaining good elution behavior and quantitation for free retinol and for retinyl esters in a single run. Reversed-phase chromatography in general is efficient at separating homologs which differ by chain length, and this is true for separation of retinyl esters. The most difficult separation, in practice, is that of retinyl oleate from retinyl palmitate; in general, mobile phases based on methanol (with or without modifiers of tetrahydrofuran or toluene or chlorinated hydrocarbons) separate retinyl laurate, myristate, palmitate, and stearate well but do not separate retinyl oleate from palmitate. De Ruyter and De Leenheer cleverly used silver nitrate as mobile phase modifier to separate this pair (124), but this approach is reportedly awkward in use. However, acetonitrile-based mobile phases (with modifiers of dichloromethane, dichloroethane, or chloroform) readily separate that "critical pair" as well as most other naturally occurring retinyl esters (125). Although methanol:water mixtures have been used to elute retinyl esters on reversed-phase HPLC, elution times are unacceptably long; nonaqueous mobile phases give equally good resolution with much shorter elution times.

Adsorption ("normal-phase") chromatography, on the other hand, is well suited to separation of geometric isomers but much less efficient at discriminating by chain length. However, Biesalski and Weiser achieved separation of 11-*cis*, 13-*cis*, and all-*trans* isomers of retinyl palmitate, stearate, and oleate by adsorption HPLC on a 3-μm silica column (126). In those adsorption HPLC methods in which all the retinyl esters are eluted as a single very early peak (127), there is increased risk of coeluting and misidentifying other, nonretinoid, compounds.

High-pressure gel permeation chromatography (μStyragel columns eluted with dichloromethane or tetrahydrofuran) has been used for preliminary purification of retinyl palmitate in foodstuffs (128,129) but has not gained wide usage.

3. Geometric Isomers of Retinal and Retinol

Retinal (retinaldehyde) occurs in significant quantities only in ocular tissue of mammals. Inasmuch as 11-*cis* and all-*trans* isomers of retinal and retinol play important roles in the visual cycle, the separation of geometric isomers of both retinoids is discussed here. All retinoids (and carotenoids) are isomerized by heat and light, so samples should be handled with extreme precautions if information on the original isomer distribution is desired. Because the *cis*-isomers generally have lower molecular extinction coefficients than the all-*trans* forms (Table 1), the presence of large amounts of *cis*-isomers may result in underestimation of retinoid (or carotenoid) quantities.

Retinal is not always well resolved from retinol in reversed-phase HPLC systems; resolution is affected both by the stationary phase and by the mobile phase composition. Curley et al. found that methanol:water was superior to acetonitrile:water for resolution of retinol from retinal on an end-capped C18 column, whereas acetonitrile:water was superior for separating retinal from retinol on a non-end-capped column (130); this difference presumably is due to hydrogen bonding between the analytes and the mobile phase or stationary phase. Plots of log (k′) versus carbon number are linear and parallel for reversed-phase separation of alcohol versus aldehyde/ketone analogs of apo-retinoids and apo-carotenoids using methanol:water mobile phases, but nonparallel for acetonitrile:water mobile phases (H.C. Furr, unpublished observations). Thus, separation of retinol and retinal is more dependent on the specific column and mobile phase than are most retinoid separations.

MacCrehan and Schonberger used reversed-phase HPLC to separate geometric isomers of retinol (with electrochemical detection), and found methanol:water mobile phases to be superior to acetonitrile:water for resolution (113,114); addition of *n*-butanol reduced elution times without sacrificing resolution. The order of elution from this reversed-phase system (either Vydac 201TP or Zorbax ODS columns) was di-*cis* < 11-*cis* < 9-*cis* < 13-*cis* < 7-*cis* < all-*trans*.

In general, however, best resolution of geometric isomers of retinoids is achieved by adsorption chromatography, usually on silica columns. Thorough reviews of preparation and HPLC separation of isomers have been given by Groenendijk et al. (131) and Bridges (132). Stancher and Zonta used 2-propanol: hexane and dioxane:hexane to separate retinal isomers, and 1-octanol:hexane to separate retinol isomers (133). Bruening et al. used methyl *t*-butyl ether with 1,1,2-trichloroethane (Freon 113) to achieve very good separation of 14 retinal isomers (134).

Noll found that the elution order of retinol isomers is dependent on the mobile phase: in hexane (or heptane) modified with methyl *t*-butyl ether, the order is 13-*cis* < 11-*cis* < 9-*cis* < all-*trans*-retinol, whereas in hexane modified with dioxane, 11-*cis* elutes before 13-*cis*-retinol (in contradiction to peak identifications in some older literature) (135); peaks were identified by absorption spectroscopy and ^{1}H-NMR.

Retinal oximes are convenient to prepare and are useful for quantitative extraction of retinal from ocular tissue (although there is a disadvantage in that two conformers, the *syn* and *anti*, are formed for each geometric isomer of retinal). Tsukida et al. (136) achieved resolution of the isomers (*syn* and *anti* conformers) of 7-*cis*, 9-*cis*, 11-*cis*, 13-*cis*, and all-*trans*-retinal oxime using a mobile phase of 2-propanol:diethyl ether:benzene. Landers and Olson used solvent optimization techniques to achieve separation of retinol and retinal oxime isomers on silica columns (137). The order of elution was: retinyl esters (unresolved) < 13-*cis*-retinal < 11-*cis*-retinal < 9-*cis*-retinal < 7-*cis*-retinal < all-*trans*-retinal < *syn* 11-*cis*-retinal oxime < *syn* all-*trans*-retinal oxime < *syn* 9-*cis*-retinal oxime + *syn* 13-*cis*-retinal oxime < *anti* 13-*cis*-retinal oxime < *anti* 11-*cis*-retinal oxime < 11-*cis*-retinol < 13-*cis*-retinol < *anti* 9-*cis*-retinal oxime < 9,11- *dicis*-retinol + 9,13-*dicis*-retinol < *anti* all-*trans*-retinal oxime < 9-*cis*-retinol < all-*trans*-retinol < *syn* 9,11-*dicis*-retinal oxime < *syn* 7-*cis*-retinal oxime < *syn* 9,13-*dicis*-retinal oxime < *anti* 9,13-*dicis*-retinal oxime < *anti* 9,11-*dicis*-retinal oxime < *anti* 7-*cis*-retinal oxime < 7-*cis*-retinol, using ethyl acetate:dioxane: 1-octanol:hexane as mobile phase; retention of retinol isomers relative to retinals and retinal oximes could be shifted by varying the proportion of 1-octanol (138).

Biesalski and Weiser separated all-*trans* from 13-*cis*, 11-*cis*, and 9-*cis* isomers of retinyl esters (retinyl palmitate, stearate, oleate, palmitoleate, and linoleate) by isocratic adsorption HPLC (126). Bridges et al. used step gradients to separate geometric isomers of retinyl esters, retinal, retinal oximes, and retinal (both vitamin A_1 and vitamin A_2 forms), also by adsorption ("normal-phase") HPLC (139,140).

4. Retinoic Acid and Synthetic Retinoids

Polar retinoids can also be analyzed by either reversed-phase or adsorption HPLC. Absorbance detection is invariably used, because these compounds are not fluorescent. Extraction of these compounds usually requires acidification, and chromatographic analysis requires use of mild acid or a salt such as ammonium acetate in the mobile phase to repress analyte ionization and peak spreading (52). Wyss has recently reviewed analysis of these compounds (141). The following listing is not exhaustive, but will serve to illustrate some of the methods used for extraction and analysis of these compounds.

Meyer et al. analyzed plasma retinol plus endogenous all-*trans* and 13-*cis* retinoic acid isocratically by normal-phase HPLC, using hexane:2-propanol: acetic acid as mobile phase (90). Extraction of the retinoic acids required acidification of the sample; however, too much acid can result in dehydration of retinol to anhydroretinol, and hydrolysis of endogenous retinoyl β-glucuronide (90). A synthetic retinoid sulfonic acid was used as internal standard. By using absorbance at 350 nm, limits of detection were 1.7 nM (0.5 µg/L) for retinoic acid isomers, 35 nM (10 µg/L) for retinol. 13-Demethyl retinoic acid has also been used as internal standard (142). Lanvers et al. used normal-phase HPLC with gradient elution (hexane:2-propanol:glacial acetic acid) to analyze 13-*cis* retinoic acid, 9-*cis* retinoic acid, all-*trans* retinoic acid, retinol, and the 4-oxo metabolites of the retinoic acid isomers (143). Plasma samples were treated with ethanol to denature proteins, and then were extracted with hexane after addition of saturated ammonium sulfate solution (pH 5).

Several older methods demonstrate the power of gradient elution to resolve classes of retinoids in a single chromatographic run; most often these procedures use reversed-phase HPLC, with mobile-phase gradients from methanol:water or acetonitrile:water (usually with ammonium acetate to improve chromatography of ionizable retinoids) to nonaqueous organic solvents (52,53,93,144,145). A procedure for the simultaneous analysis of very polar retinoids like 4-oxoretinoic acid and retinoyl β-glucuronide as well as nonpolar retinyl esters, along with bilirubin, carotenoids, and tocopherols, has been described recently (Fig. 4) (73).

Retinoic acid and metabolites (all-*trans*, 13-*cis*, 9-*cis*, and their 4-oxo metabolites) were analyzed by reversed-phase HPLC, using a gradient of aqueous acetic acid in methanol:acetonitrile:tetrahydrofuran (92). Plasma samples were extracted with diethyl ether-ethyl acetate after addition of phosphate buffer (pH 7); acitretin and 13-*cis* acitretin were used as internal standards. Dimitrova et al. extracted retinoid metabolites (all-*trans*, 9-*cis*, and 13-*cis*-retinoic acid; all-*trans*-4-oxoretinoic acid; and all-*trans*-5,6-epoxyretinoic acid) and retinol from mouse plasma or homogenized tissues by adding acetonitrile to precipitate proteins, then centrifuging and injecting the supernatant directly (91). A C18 column with an isocratic mobile phase of acetonitrile:water:methanol:*n*-butyl alcohol:ammonium acetate:acetic acid was used. Lower limits of detection were on the order of 80 to 170 nM (25 to 50 ng/mL) for retinoic acid and its metabolites, using absorbance detection; an anthracenyl retinoid was used as internal standard. A similar extraction procedure was used by Guiso et al. to analyze retinoic acid and its metabolites in human plasma; proteins were precipitated with acetonitrile,

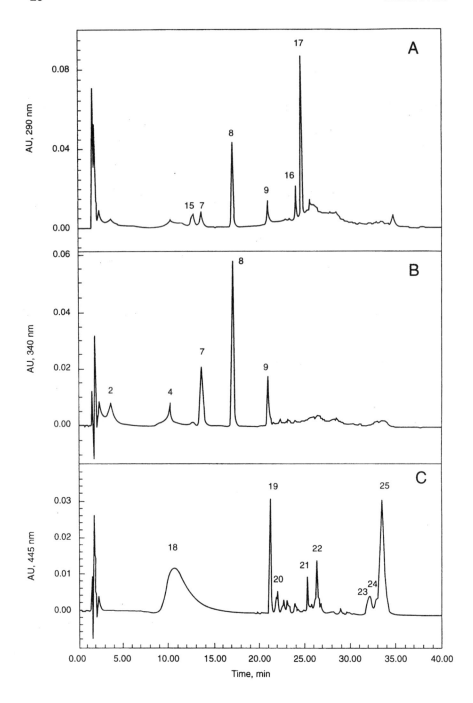

and an aliquot of the acetonitrile supernatant (after concentration) was analyzed by isocratic reversed-phase HPLC (146).

Solid-phase extraction cartridges have been used by several groups for retinoic acid extraction. Periquet et al. extracted neutral and basic lipids from alkaline plasma with hexane, then acidified the aqueous residue and extracted retinoic acid with hexane (147). This extract was applied to an aminopropyl cartridge, and the all-*trans* and 13-*cis*-retinoic acid were eluted with acidic ether for subsequent analysis by reversed-phase HPLC. Lefebvre et al. applied acidified plasma directly to octadecylsilane solid-phase extraction cartridges; after washing the cartridges with acetone:acetic acid, acidic retinoids (all-*trans* and 13-*cis*-retinoic acid and their 4-oxo metabolites) were eluted with acetonitrile and then analyzed by adsorption HPLC (mobile phase of hexane:dichloromethane:dioxane:acetic acid) (148). Van Wauwe et al. used C18 cartridges to extract and concentrate metabolites of 4-oxoretinoic acid, after initially precipitating plasma with acetonitrile (149). Urinary and biliary metabolites of the synthetic arotinoid Am-80 were extracted on SepPak C18 cartridges before HPLC analysis (150,151). Solid-phase extraction C18 columns were also used to extract retinoyl β-glucuronides directly from rat urine (152); acitretin β-glucuronide served as internal standard.

Vitamin A metabolites (retinol, all-*trans*-retinoic acid, all-*trans*-4-oxoretinoic acid, retinyl palmitate, and retinyl stearate) were extracted from embryonic tissues by first homogenizing the tissues in 2-propanol and then extracting the retinoids onto C2 (ethyl) cartridges (153). The retinoids were eluted from the cartridges in a Varian AASP module which was directly connected to the analytical HPLC column, using a mobile-phase gradient from aqueous ammonium acetate:methanol to methanol:2-propanol. Arafa et al. also used extraction cartridges with automated column switching for analysis of synthetic retinoids; tissue samples were homogenized with acetonitrile before being applied to C18 extraction cartridges, and were eluted directly onto the analytical Spherisorb ODS column with a gradient from aqueous ammonium acetate:methanol to methanol (154).

Alkaline hydrolysis was used to release retinoids from skin biopsy samples, with hexane extraction before and after adjusting the hydrolysate pH to 4 or 5

Figure 4 Reversed-phase HPLC elution profiles of tocopherols (panel A), retinoids (B), and carotenoids (C) present in human plasma (200 μL). Blood was collected 3 h after an oral dose of retinoic acid. The chromatogram was obtained by use of gradient elution (Table 3). Peak identification: 2, 4-oxo-retinoic acid; 4, retinoyl β-glucuronide; 7, retinoic acid; 8, retinol; 9, retinyl acetate (internal standard); 15, butylated hydroxytoluene; 16, γ-tocopherol; 17, α-tocopherol; 18, free bilirubin; 19, lutein; 20, zeaxanthin; 21, 2′,3′-anhydrolutein; 22, β-cryptoxanthin; 23, lycopene; 24, α-carotene; 25, β-carotene. (From Ref. 73.)

(155). These samples were analyzed by isocratic reversed-phase HPLC (ODS column) with acetonitrile:water:acetic acid as mobile phase.

Wyss pioneered use of automated column-switching techniques for analysis of retinoic acid and other polar retinoids (156–160). In general, a plasma sample is diluted with acetonitrile:ethanol (95:5) containing an appropriate internal standard; the supernatant is passed across a preconcentration column (such as Bondapak C18 Corasil, 14 mm × 4.6 mm ID), and the preconcentration column is rinsed with ammonium acetate:acetonitrile. The retinoids were eluted from the preconcentration column with a gradient of ammonium acetate:acetonitrile:acetic acid, and analyzed on a Spherisorb ODS 1 column by gradient elution (159).

The realization that 9-*cis* retinoic acid plays an important role in control of gene expression has stimulated the development of improved methods for its analysis. Cahnmann, after preparing 9-*cis* retinoic acid by photoisomerization, obtained complete resolution of 9-*cis* retinoic acid from all-*trans* retinoic acid by reversed-phase HPLC (Waters NovaPak C18 column, acetonitrile:aqueous ammonium acetate mobile phase) (161).

Although not strictly a method of retinoid chromatography, high-performance size-exclusion chromatography can be used to assess the binding affinities of synthetic retinoid ligands for the nuclear retinoid receptor proteins RAR and RXR (162,163).

5. Vitamin A_2 Forms

The 3,4-didehydro analogs of retinoids are referred to as vitamin A_2 forms, have been found in mammalian skin and some other tissues (80,164,165), and are used as visual pigments in some fish and amphibia (140,166,167). 3,4-Didehydroretinoic acid has similar activity to retinoic acid in control of gene expression in some systems (168). 3,4-Didehydroretinol is also used in an indirect assessment of vitamin A status, the Modified Relative Dose Response (MRDR) test (169). The dehydroretinoids are slightly more polar than their vitamin A_1 analogs, so the vitamin A_2 forms elute slightly earlier on reversed-phase HPLC, slightly later on adsorption chromatography; but in general the conditions suitable for tissue extraction and HPLC analysis of vitamin A_1 forms are suitable for the vitamin A_2 analogs. Because 3,4-didehydroretinol is much less fluorescent than retinol (170), absorbance detection is used (maximum absorbance at 340 nm; Table 1). Geometric isomers of retinol and dehydroretinol, and of retinal and dehydroretinal, were separated by adsorption HPLC on silica columns with mobile phases of hexane containing dioxane, 2-propanol, or 1-octanol (133,171).

B. Carotenoids

1. Normal-Phase HPLC Systems

Normal-phase (also called straight-phase, or adsorption) HPLC is particularly valuable for separation of geometric isomers. Adsorbents used for normal-phase

HPLC of carotenoids include silica, alumina, magnesia, and calcium hydroxide. A description of analytical conditions for the separation of carotenoids by normal-phase HPLC can be found in previous reviews (56,105). The separation of *cis-trans* isomers of α- and β-carotene by adsorption HPLC on a calcium hydroxide column with a solvent mixture of hexane/acetone (99.5:0.5, 99.2:0.8, or 99:1) has been described (172,173). The retention times, capacity factors, and chromatographic conditions for the elution of several polar and nonpolar carotenoids and retinoids on cyclodextrin-bonded phases (Cyclobond I, beta-CD, 5 μm) have been reported (174). A normal-phase system was described for the qualitative and analytical separation of individual carotenoids in a saponified extract of *Capsicum annum*: analytical separation was achieved on a Spherisorb 5 μm spherical particle column (0.46 × 25 cm), and semipreparative separation was done on a Waters 10-μm irregular particle column (0.78 × 30 cm), with a 30-min linear gradient of light petroleum/acetone 95:5 to 75:25, and a flow rate of 1 mL/min and 4 mL/min, respectively (175,176). Analytical separation on a Spherisorb (0.46 × 25 cm) 5-μm column, and semipreparative separation on a 10-μm Porasil (0.78 × 30 cm) column with light petroleum (bp 40–60°C) and ethanol (97:3) for 7 min, followed by a linear gradient to light petroleum and ethanol (85:15), was used to separate nonpolar β-carotene from chlorophyll, polar lutein, and very polar violaxanthin and neoxanthin present in the green leaves of *Citrus limon* (177).

By use of a Nucleosil 300-5 stationary phase and hexane/N-ethyldiisopropylamine (2000:1) as the mobile phase, all-*trans*, 9-*cis*, 13-*cis*, 9,13-*dicis*, and several other *cis* isomers of lycopene have been resolved well (178).

Normal-phase Spherisorb silica (3 μm) as well as ODS-1 and ODS-2 columns and Nucleosil C18 RP (3-μm) columns were used to resolve radiolabeled reaction products formed in specific carotenogenic enzymatic reactions, such as those catalyzed by β-carotene hydroxylase, lycopene cyclase, and phytoene synthase; isocratic elution with a variety of mobile phases was used (179).

2. Reversed-Phase HPLC Systems

Reversed-phase high-performance liquid chromatography (RP-HPLC) is more popular for analysis of carotenoids than is normal-phase HPLC because (1) retention is very little affected by small variations in the mobile-phase composition, and (2) the risk of artifact formation on passage through the column is minimal as solute-support interactions on non-polar-bonded phases only involve weak forces. A variety of stationary phases of various polarities are available, such as C18, C8, C4, C2, C1, phenyl, and cyano derivatives; the C18 phase is the most popular.

Extensive reviews on RP-HPLC of carotenoids have been published by Craft (97) and Pfander and Riesen (106). Another review on RP-HPLC of carotenoids evaluated sixty commercially available and five experimental LC columns for the separation and recovery of seven carotenoids using methanol- and acetonitrile-based solvents, either straight or modified with ethyl acetate or tetrahydrofuran. Methanol-based solvents provided better recoveries, and polymeric C18 phases were better than monomeric C18 phases for separation (180). The effects of mobile-phase modifiers, modifier concentration, and column temperature on the separation of seven carotenoids (lutein, zeaxanthin, β-cryptoxanthin, echinenone, α-carotene, β-carotene, and lycopene) on a polymeric C18 column were investigated (42). In RP-LC systems, the retention of nonpolar carotenoids is often attributed to solvophobic interactions, but other properties of the solute may provide an additional basis for separation (181). A variety of bonded phase parameters such as endcapping, phase chemistry, ligand length, and substrate parameters were studied for their effects on column retention and selectivity toward carotenoids. The result was a C30 column designed with the properties of high absolute retention, enhanced shape recognition of structured solutes, and moderate silanol activity (182); the C30 stationary phase was much superior to Vydac 201TP54 or Suplex PKB-100 stationary phases in resolving six common carotenoids and their geometric isomers. Another long-chain (C34) alkyl-bonded stationary phase was shown to exhibit increased selectivity for certain *cis*-isomers of α- and β-carotenes as compared to the C30 stationary phase (183).

Moreover, dramatic differences in the separation of carotenoids can be achieved by changes in temperature of the column (184). In a study of the temperature dependence of carotenoid elution on C18, C30, and C34 stationary phases, mostly linear van't Hoff plots were observed for all solutes on the C18 phase, but a variety of retention behaviors was observed on the C30 and C34 phases (185).

A comparison of LC methods for determination of *cis-trans* isomers of β-carotene was made on Vydac C18 201 TP and calcium hydroxide columns (186). The purity and relative distribution of β-carotene and its isomers in several commercially available products were evaluated by HPLC on several columns using a mobile phase of methanol/water (97:3) (187). Because carotenoids and chlorophylls are very sensitive to the nature of injection solvent, such as acetone, sample-solvent interaction may give rise to distorted and even false peaks (188). Because metal column frits may damage carotenoids, Teflon column frits should be used (189). Also artifacts may be produced on the column by reactions among the carotenoids, injection solvents, and mobile phase. Losses that occur during extraction and saponification can be reduced by use of suitable antioxidants (189).

3. Human Tissues

Serum. Many of the current HPLC procedures used for the analysis of carotenoids in plasma or serum are based on the RP isocratic procedure reported by Nelis and De Leenheer (190). When acetonitrile is used as a component of the mobile phase, it is necessary to include ammonium acetate as otherwise there is a tendency for loss of carotenoids on the column. A number of RP gradient procedures have also been reported. Some representative HPLC systems used for the analysis of carotenoids in human plasma or serum since the publication of the second edition of this book are listed in Table 3 (96,190–197), and a typical analysis is presented in Figure 5. Some of these methods allow simultaneous analysis of carotenoids, retinoids, and tocopherols. Some of the older procedures listed in the second edition of this book (29) have been used successfully in recent epidemiological surveys of carotenoids in human plasma and tissues. By use of HPLC, it has now been possible to separate and characterize at least 21 carotenoids in human plasma (64,65).

The metabolism of three carotenoid analogs was studied by HPLC on a Waters Resolve 5-μm C18 column using solvent mixtures of $CH_3CN:C_2H_4Cl_2$: $CH_3OH:H_2O$ of varying composition (198). Accumulation and clearance of capsanthin in blood plasma after the ingestion of paprika juice has been demonstrated in humans by HPLC on a Lichrosphere RP-18 5-μm column with a mobile phase of $CH_3CN/CH_3OH/CH_2Cl_2$/water (199). The appearance of bixin and norbixin in plasma of humans who consumed annato food color has been demonstrated by analysis of plasma by HPLC (200). The uptake of lycopene and its isomers from unprocessed and heat-processed tomato juice was studied by analyzing serum of human volunteers by HPLC on an end-capped 5-μm Merck C18 column (0.4 $\times$ 25 cm) with $CH_3OH/CH_3CN/CH_2Cl_2/H_2O$ (7:7:3:0.16) (201). To determine the bioavailability of lycopene from fresh tomato and tomato paste in humans, plasma was analyzed by step gradient HPLC on a 5-μm Suplex PKB 100 column (0.46 $\times$ 25 cm, Supelco) with $CH_3OH/CH_3CN/2$-propanol (54:44:2; 97%) and water (3%) for the first 10 min, and then with 100% of the former solvent for another 15 min (202).

Skin. By use of a solvent mixture of $CH_3CN/CH_2Cl_2/CH_3OH$ (70:20: 10) and a C18 5-μm column, the absorption of β-carotene in human serum after multiple doses of β-carotene was measured (203). To evaluate the rate of accumulation of β-carotene in the skin of these persons, remittance spectra (300 to 600 nm) of palms were taken by means of a spectrophotometer fitted with a 15-cm integrating sphere (203). Human skin specimens obtained from subjects undergoing facial surgery were extracted by a procedure which consisted of saponification of the samples, followed by extraction with organic solvents. The extract was subjected to HPLC on a Biosil-C18 5-μm column with CH_3CN/tetrahydrofuran/ CH_3OH/1% ammonium acetate (74/15/6/5) as the mobile phase (204). To study the effects of ingestion of a single large dose on the concentrations of β-carotene and lycopene in plasma and skin, the carotenoids were analyzed by RP gradient HPLC on a Pecosphere-3 (3 μm) C18 cartridge (205).

Macula. HPLC analysis was used to study the distribution of carotenoids in the macula (206). The extraction procedure and HPLC analysis of macular carotenoids obtained from human donor eyes on a 5-μm Spherisorb ODS-1 C18

Table 3 HPLC Systems Recently Used in the Analysis of Carotenoids in Human Plasma or Serum

Compounds	Stationary phase	Mode	Mobile phase	Reference
L, Z, Cr, Ly, αC, βC	Zorbax ODS, 5 μm	Isocratic	MeCN/DCM/MeOH (70:20:10)	190
A, Z, Cr, Ly, αC, βC, and 12 others	Regis spherical 5 μm, Microsorb 5 μm and Vydac 201 TP54 5 μm	Isocratic	Hex/DCM/MeOH/Disa (74.65:25:0.25:0.1)	64,65,217
		Gradient	MeCN/MeOH/DCM/Hex (85:10:2.5:2.5 to 45:10:22.5:22.5)	
L/Z, Cr, Ly, αC, βC	Waters Resolve 5 μm	Isocratic	MeCN/DCM/MeOH/PrOH (90:15:10:0.1)	68,72
L, Z, Cr, Ly, βC	Microsorb-MV 3 μm	Gradient	MeOH/H₂O (75:25) to MeOH/DCM (80:20)	73
L/Z, Cr, Ly, αC, βC	Zorbax ODS, 5 μm	Isocratic	MeCN/THF/MeOH/H₂O (65:25:6:4)	191
L, Z, Cr, isomers of Ly, αC and βC	Adsorbsphere HS, 3 μm	Gradient	MeCN/MeOH (85:15) to MeCN/MeOH/PrOH (59.5:10.5:30)	96,192
Cr and isomers of Ly, βC	Suplex PKB 100, 5 μm	Isocratic	MeCN/MeOH/PrOH (44:54:2) or MeCN/MeOH/PrOH/H2O (40:10:40:10)	195
L, Z, Cr, Ly, αC, βC	Waters Novapak C18, 4 μm	Gradient	MeCN/THF/MeOH/AA + 0.1% Tea (varying composition)	196
L/Z, Can, Cr, Ly, αC, βC and isomers	Merck RP-18, 5 μm	Isocratic	MeCN/MeOH/DCM/H2O (7:7:2:0.16)	194
L, Z, Cr, αC, βC and isomers	Bakerbond 5 μm	Gradient	MeCN/EtAc (90:10) to MeOH/EtAc (90:10)	197
L, Z, Cr, Ly, αC, βC	Spheri-5 RP	Isocratic	MeCN/DCM/MeOH (7:2:1)	193

Abbreviations: L, lutein; Z, zeaxanthin; Cr, cryptoxanthin; Cn, canthaxanthin; Ly, lycopene; αC, α-carotene; βC, β-carotene; MeCN, acetonitrile; MeOH, methanol; DCM, dichloromethane; THF, tetrahydrofuran; EtAc, ethyl acetate; Hex, hexane; PrOH, 2-propanol; Disa, N,N-diisopropylethylamine; Tea, triethylamine; AA, ammonium acetate.

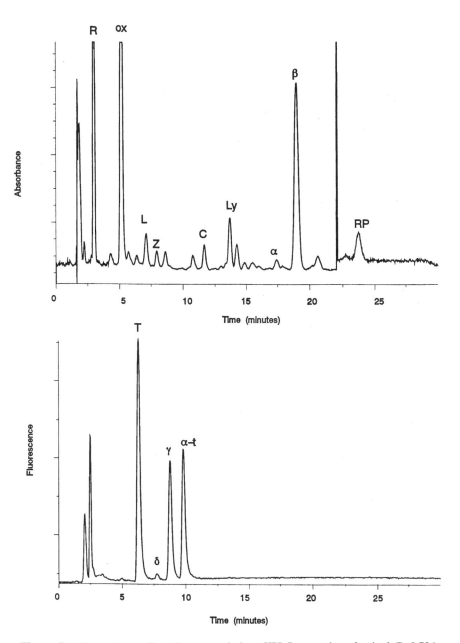

Figure 5 Chromatogram from the reversed-phase HPLC separation of retinol (R, 0.734 μg/mL), *trans*-β-apo-10′-carotenal oxime (ox, internal standard), lutein (L, 0.110 μg/mL), zeaxanthin (Z, 0.051 μg/mL), β-cryptoxanthin (C, 0.060 μg/mL), *trans*-lycopene (Ly, 0.124 μg/mL), α-carotene (α, 0.022 μg/mL), *trans*-β-carotene (β, 0.567 μg/mL), and retinyl palmitate (RP, spiked, 0.297 μg/mL) in human serum using absorbance detection with wavelength switching. Chromatogram of tocol (T, internal standard), δ-tocopherol (δ, 0.084 μg/mL), γ-tocopherol (γ, 1.94 μg/mL), and α-tocopherol (α-t, 5.69 μg/mL) using fluorescence detection. (From Ref. 197.)

column (0.46 × 25 cm) using a mobile phase of methanol (92%) and water/acetonitrile (3:1, 8%) has been described (207,208). The use of a 3-μm Ultracarb ODS (0.2 × 25 cm) (Phenomenex) column and a solvent mixture of acetonitrile/methanol (90:10) containing 0.1% triethylamine has been described for studying the distribution of lutein and zeaxanthin in human retina (209). 3′-Epilutein and 3′-hydroxy-beta, epsilon-caroten-3′-one have been identified by HPLC on a silica-based nitrile bonded column with hexane/CH_2Cl_2/CH_3OH/N,N-diisopropyl-ethylamine as the mobile phase, as have lutein and zeaxanthin in human and monkey retina (210). Carotenoids, retinoids, and tocopherols in human lenses were determined by HPLC on a Pecosphere-3 C18 cartridge column by use of a gradient of CH_3CN, THF, and H_2O solvent mixture (211). HPLC separation of carotenoids in human and monkey retina on an Adsorbsphere-HS 3-μm column using solvent mixtures of CH_3OH and CH_3CN has been reported (212).

Lung Fibroblasts. The uptake kinetics of β-carotene into rat small intestinal cells (hBRIE 380) in culture and human lung fibroblasts and the conversion of β-carotene to retinol and retinoic acid were studied by HPLC on a Beckman Ultrasphere ODS 5-μm (0.46 × 15 cm) column with a 12-min linear gradient of 100% methanol containing ammonium acetate (0.5%) to methanol/toluene (80:20) (213).

Buccal Cells. Cell pellets were first incubated at 37°C with a protease solution, and then extracted with hexane. The extract was analyzed by HPLC on two Novapak C18 columns (4 μm) connected in series, developed with a gradient mobile phase (39,214). β-Carotene in plasma, blood cells, and buccal mucosa cells was determined by HPLC analysis on a Vydac C18 (0.46 × 25 cm) with methanol/acetonitrile (95:5) containing 50 nM $NaClO_4$ (215).

Milk. Simultaneous analysis of retinol and carotenoids present in human milk was carried out on a YMC RP 5-μm ODS column (0.46 × 25 cm) by isocratic elution with tetrahydrofuran/methanol (90:10) (216). Thirty-four carotenoids, including 13 geometric isomers and eight metabolites, along with vitamins A and E in milk of lactating mothers, were separated by a combination of normal and RP-HPLC (217).

Other Human Tissues. Details of extraction procedures and isocratic as well as RP-HPLC analysis of carotenoids present in various human tissues have been summarized by Parker (218) and by Schmitz et al. (219). *Cis* and *trans* isomers of carotenoids may have different biological activities. Thus the isomeric composition of lycopene and β-carotene was determined in the serum of healthy volunteers and in seven human tissues obtained by autopsy soon after death, using RP-HPLC on a Merck 5-μm C18 end-capped column with a solvent mixture of CH_3OH/CH_3CN/CH_2Cl_2/H_2O) (7/7/2/0.16) and a photodiode array detector (220).

Schmitz et al. determined the concentrations of the carotenoids α-carotene, β-carotene, cryptoxanthin, lutein, and lycopene and of retinol in human liver, kidney, and lung tissues obtained from autopsy specimens by HPLC on a 5-μm ODS Supelco LC-18 or a Vydac C18 column, using a mobile phase of $CH_3OH/$ $CH_3CN/CHCl_3$ (221). The carotenoids present in the adrenal gland, testes, spleen, fat, heart, thyroid, ovary, kidney, liver, and pancreas, obtained as autopsy samples, were also analyzed by Kaplan et al. (222).

4. Other Vertebrate Tissues

Calf and Ferret Tissues. The comparative absorption and transport of five common carotenoids in calves was studied by HPLC of serum extracts on a Supelco LC-18 column using $CH_3CN/CH_3OH/CHCl_3$ (47:47:3) for lycopene and a Vydac 201TP5415 column with $CH_3CN/CH_3OH/H_2O$ (9:88:3) for α-carotene, β-carotene, lutein, and canthaxanthin (223). Poor et al. (224) used HPLC analysis to study the accumulation of α- and β-carotenes in serum and tissues of preruminant calves fed raw and steamed carrot slurries. Serum and various tissues of ferrets following a dose with β-carotene were extracted with a mixture of chloroform-methanol (2:1) followed by extraction with hexane; the extracts were analyzed by HPLC (225–229).

Rat Hepatoma Cells. *Cis-trans* conversion of β-carotene in FU-5 rat hepatoma cells was studied by HPLC analysis of cell extracts using a 5-μm Suplex PKB 100 column and a mobile phase of methanol/dichloromethane (97:3) (230).

Astaxanthin in Marine Organisms. The resolution of all-*trans* astaxanthin on a Pirkle covalent D-phenylglycine type column made in Japan has been described (231). A Pirkle covalent L-leucine column (5 μm) was used with a solvent mixture of hexane/THF to resolve the stereoisomers of all-*trans* astaxanthin (232). A combination of column chromatography, TLC and HPLC on a Spherisorb S3-CN column, and a mobile phase of 17% isopropyl acetate and 7% acetone in hexane was used to analyze the metabolites of astaxanthin in shrimp (233).

Birds. Details of extraction procedure and analysis of carotenoids from bird feathers have been described (234). The carotenoid pattern in several species of birds, palearctic *Carduelis, Serinus, Loxia*, and *Pinicola enucleator* was examined by HPLC on two Lichrocart Purospher RP-18 columns (0.4 $\times$ 25 cm) in series at 40°C with CH_3CN/CH_3OH (70:30) as the mobile phase (235).

5. Cleavage Products of Carotene

The cleavage products of β-carotene in intestinal preparations, mostly from the rat but also from other species, were analyzed on a Pecosphere-3 C18 cartridge column (0.46 $\times$ 8.3 cm) by RP isocratic elution with $CH_3CN/THF/H_2O$ (50:

20:30), then by a linear gradient to 50:44:6 of the same solvent mixture, and finally followed by isocratic elution with the latter solvent mixture (236–238). The enzymatic formation of retinal and its isomers from β-carotene and its isomers was analyzed by HPLC on a RP Vydac 218TPP54 column with a mobile phase of $CH_3CN/CH_3OH/THF$ (45:56:4) (239). Similarly, the formation of 9-cis-retinoic acid from 9-cis-β-carotene in ferret intestine was studied by HPLC on a Vydac TP201 column using a gradient of CH_3CN/CH_3OH (45:55) to $CH_3CN/CH_3OH/2$-propanol (240). Recently, the formation of retinoic acid from orally administered β-carotene in vitamin A–deficient rats has been demonstrated by RP gradient HPLC on a Rainin 3-μm Microsorb-MV column using a gradient of CH_3OH/H_2O (3:1) to CH_3OH/CH_2Cl_2 (4:1) (73). A simple procedure for assay of β-carotene 15,15′-dioxygenase activity by RP HPLC on a TSK ODS-80TS C18 RP column with a mobile phase of CH_3CN/H_2O has been described (74).

6.   Plants, Fruits, and Vegetables

The separation, identification, and quantification by HPLC of the major carotenoids in extracts of several vegetables, fruits, and some foods have been reported (95,241). A combination of isocratic and gradient HPLC was used to determine the effect of food preparation techniques such as boiling, steaming, or microwaving on the carotenoid contents of tomatoes and several green vegetables, e.g., broccoli, spinach, and green beans (241).

Carotenoid contents in fruits continue to be studied. β-Carotene and other hydrocarbon carotenoids, e.g., lycopene, phytofluene, phytoene, and α-carotene, for example, have been analyzed in extracts of red grapefruit by RP HPLC on an Analytichem C18 (0.46 × 25 cm) column by use of $CH_3CN/CH_2Cl_2/CH_3OH$ (65:25:10) (242). Similarly, gradient HPLC separation of chlorophyll and carotenoids in four kiwi fruit cultivars, using a photodiode array detector, has been described (243). Also, the contents of α-carotene, β-carotene, and cryptoxanthin in 14 varieties of Japanese and American persimmons were determined on saponified extracts by HPLC on a Zorbax 5-μm ODS column (0.46 × 25 cm) using a mobile phase of $CH_3CN/CH_2Cl_2/0.001\%$ triethylamine in CH_3OH (350:150:1) (244).

Many spices also contain carotenoids. The composition and changes in concentration of different carotenoids during the ripening process in different species of pepper were studied by use of a binary gradient of water/acetone and a C18 RP column (245–247). An isocratic RP HPLC separation of capsanthin and capsorubin in *Capsicum annum* paprika and oleoresin on a Merck LiChrospher 100 C18 5-μm (0.4 × 25 cm) column with $CH_3CN/2$-propanol/ethyl acetate (80:10:10) has been described (248). The carotenoids in saponified extracts of irradiated and ethylene oxide–treated red pepper were determined by HPLC on a Chromsil 10-μm C18 column (0.46 × 25 cm) using a gradient of acetone/water (100:38 to 100:5). Although these treatments did not result in significant changes in carotenoid composition, storage time did (249). The carotenoids in yellow pepper were separated by HPLC on Chromsil C18 6-μm columns, end-capped or not end-capped, using a gradient system of water/methanol (12:88) to methanol (100%) to CH₃OH/acetone (50:50). Fifty-five carotenoids including the very polar neoxanthin and violaxanthin and the least polar α- and β-carotenes were separated (250).

Eighteen pigments, including chlorophylls and carotenoids, were separated by RP ion-pair HPLC during lactic fermentation and the later preservation phase of green table olives (251). Gradient elution of H₂O/t-butylammonium acetate (ion-pairing agent)/CH₃OH (1:1:8) to CH₃OH/CH₃COCH₃ (1:1) was used for this purpose on a Spherisorb 5-μm ODS-2 column (0.4 × 25 cm) (251). The method of liquid-phase distribution was used for the selective separation of the pigments present in olive oil between N,N-dimethylformamide and hexane. Thus partition of oil in the solvent mixture resulted in retention of chlorophylls in the DMF layer, whereas repeated extraction with hexane resulted in recovery of carotenoids in hexane extracts (252).

Extracts of green leaves from a variety of specimens grown under the sun or shade were analyzed by RP gradient HPLC on a Dupont non-end-capped Zorbax 5-μm ODS column. A solvent mixture of CH₃CN/CH₃OH (85:15) was used for the first 14.5 min followed by a 2-min linear gradient to 100% methanol/ethyl acetate (68:32). HPLC on an end-capped column failed to provide separation of lutein and zeaxanthin (253).

A voltametric method for determination of β-carotene in brine and soya oil has been reported (254).

Several reports on the carotenoid composition in 17 species of lichens have appeared. Column and thin-layer chromatography on alumina or silica were used to separate the carotenoids (255–259).

7. Bacteria and Algae

The usefulness of HPLC and PDA in the analysis and identification of carotenoids present in several species of alga and bacteria has been described (260). Carotenoids in the green algae *Scenedesmus obliquus* were analyzed by HPLC (261). Column chromatography on DEAE-Toyopearl with acetone followed by another column chromatography on silica gel with hexane, followed by HPLC on a C18 cartridge with methanol, was used to isolate and purify a novel carotenoid ester, loroxanthin dodecenoate, from *Pyramimonas parkeae* and from a chlorarachniophycean alga (262). Braunixanthins 1 and 2 were separated and characterized as new carotenoids in the green microalga *Botryococcus braunii* by column chromatography on silica gel and ODS, followed by HPLC on a Develosil 60-3 column with hexane/diethyl ether (9:1) as solvents (263). The extraction procedure and RP-HPLC purification of carotenoids in extracts from *Rhodococcus rhodochrous* on a Radial-Pak Bondapak C18 cartridge using methanol (264) and in the photosynthetic bacterium *Erythrobacter longus* have been reported (265). HPLC analysis of extracts of several species of Heliobacteria on C18 μBondapak and Novapak columns with methanol as the mobile phase showed that the C30 carotenoid 4,4′-diaponeurosporene was the major carotenoid in all the species (266). Minor carotenoids were diapolycopene, lycopene, diapo-β-carotene, and β-carotene. In the green sulfur bacterium *Chlorobium tepidum*, 1′,2′-dihydro-β-carotene, 1′,2′-dihydrochlorobactene, and hydroxy-chlorobactene glucoside ester were identified by subjecting extracts to column chromatography first on a DEAE-Sepharose CL-6B column, and then to HPLC on C18 μBondapak and Novapak columns (267).

The separation by HPLC of seven zeaxanthin glycosides (glucopyranoside and rhamnopyranoside) from *Sulfolobus* on a Spherisorb 5-μm ODS-2 column using a gradient of methanol/water to methanol/ethyl acetate, and their characterization have been reported (268).

A new carotenoid, termed rhodobacterioxanthin, was isolated from *Rhodobacter capsulatus* and purified by column chromatography on silica gel followed by preparative HPLC on a Shim-pack ODS 15 column using acetonitrile/dichloromethane (269). β-Cryptoxanthin glucoside, zeaxanthin monoglucoside, and zeaxanthin diglucoside were found to be the major carotenoids in *Erwinia herbicola* and in *Escherchia coli* expressing *E. herbicola* genes by TLC on silica gel followed by HPLC analysis on a Nucleosil 5-μm C18 RP column using acetonitrile or acetonitrile/water (270). A monocyclic carotenoid glucoside ester, hydroxy-β-carotene glucoside ester, was identified as a major carotenoid in the green filamentous bacterium *Chloroflexus aurantiacus* strain J-10-fl (271). HPLC was carried out on a Bondapak C18 column with methanol as the mobile phase (271). Canthaxanthin has been identified as the major carotenoid in *Bradyrhizobium* strains by normal-phase chromatography on a Lichrosorb Si60 5-μm column using CH_2Cl_2/ethyl acetate, and by RP HPLC on a Hypersil 5-μm column using $CH_3CN/CH_3OH/CH_2Cl_2$ (272).

8. Other Species

By a combination of column and thin-layer chromatography and of spectrophotometry as many as 37 carotenoids in extracts of butterflies have been characterized. In most species lutein epoxide was found to be the predominant carotenoid. Specimens of the same species of butterfly collected in different months quite frequently contained different carotenoids (273). Saponified acetone extracts of the Japanese stick insect *Neophirasea japonica* were analyzed by a combination of TLC and HPLC on Hitachi gel columns with $CHCl_3/CH_3CN$ as the mobile phase (274). Carotenoids in poplar hawkmoth caterpillars were analyzed by TLC under alkaline conditions, by absorption spectroscopy and by mass spectrometry (275).

VI. GAS CHROMATOGRAPHY OF RETINOIDS AND CAROTENOIDS

Gas chromatography has been little used for the analysis of retinoids and carotenoids, because (1) these compounds are particularly labile to destruction in the presence of high temperatures and reactive sites in uncoated injection ports, (2) they isomerize readily at high temperatures, (3) they are not very volatile, and (4) liquid chromatography has proven fully adequate for almost all analytical needs. Early attempts at gas chromatography of carotenoids and retinoids have been reviewed previously (29,276). In brief, the most successful results with retinol have been achieved by chromatographing the trimethylsilyl ether derivative (277), or by using cold on-column injection for capillary gas chromatography of underivatized apo-retinoids, retinoids, and apo-carotenoids (278,279). By using the latter technique, retinol, retinal, retinyl acetate, methyl retinoate, and anhydroretinol, as well as retinyl esters up to retinyl dodecanoate, and the apo-carotenoid β-apo-12′-carotenal were chromatographed (279). The upper temperature limit of the chromatograph limited the range of compounds that could be analyzed. By linear extrapolation, it was estimated that β-carotene would have a Kovats Index of approximately 5000; it is not known at what operating temperature the thermal isomerization of these compounds will limit the usefulness of the technique.

Retinoic acid and other retinoid carboxylic acids, however, can be readily converted to derivatives that are suitable for gas chromatography. Many of these applications have used mass spectrometry for detection (see below). Diazomethane is used, at room temperature, to prepare the methyl esters without apparent isomerization. Pentafluorobenzyl esters of retinoic acid and its analogs have also been prepared for GC-MS (280) or HPLC-MS (281). Deuterated analogs of retinoic acid or other retinoid carboxylic acids have been used as internal standards, with mass spectrometric detection (88,282) (reviewed by Napoli [283] and by De Leenheer and Lambert [89]). The pentafluorobenzyl ester of a synthetic retinoid, Ro 13-7410, was analyzed by column switching: the peak of interest from a SE54 column was cut to an OV 240 column, with subsequent detection by negative ion chemical ionization mass spectrometry (280).

VII. MASS SPECTROMETRY

A. Retinoids

Gas chromatography–mass spectrometry (GC-MS) and high-performance liquid chromatography–mass spectrometry (LC-MS) have been applied successfully to the identification and quantification of retinoids. Because of the thermal instability and polarity of some retinoids and their derivatives, derivatization of retinoids is usually required, and even then, gas chromatography may be unsuitable. Although HPLC is ideal for the separation of thermally labile, polar retinoids, LC-MS technology has lagged behind GC-MS by several decades and has only recently become highly sensitive, robust, and widely available for retinoid analysis. Although GC-MS analyses of retinoids continue to be reported, the research focus has shifted to the application of LC-MS techniques such as electrospray, particle beam, and atmospheric-pressure chemical ionization. In addition to advances in LC-MS of retinoids, laser desorption time-of-flight mass spectrometry and fast atom bombardment mass spectrometry have been applied to the analysis of retinoids, and the use of tandem mass spectrometry is just beginning.

1. GC-MS

Electron impact ionization (EI) and chemical ionization (CI) mass spectrometry using a direct insertion probe continue to be used for molecular weight confirmation and identification of purified retinoids. Retinoid fragmentation patterns are useful for identification, especially when mass spectra of unknown compounds are compared to those of reference standards. For example, Buck et al. (284) used EI and CI mass spectrometry with a direct insertion probe to identify retinol as an essential growth factor for the culturing of human B cells, and Lakshman et al. (285,286) used EI mass spectrometry to identify retinal O-ethyloxime. Barua (287) reported the desorption chemical ionization mass spectra of retinoyl β-glucuronide after methylation with diazomethane and pertrimethylsilylation. Molecular ions were detected in very low abundance.

Unless retinoid samples are already purified, GC-MS with EI or CI is usually used instead of a direct insertion probe for more efficient and sensitive on-line separation and mass spectrometric analysis. Furr et al. (279,288) used GC-MS with EI ionization for the analysis of retinoids and apo-retinoids. Molecular ions were observed, and the base peaks of the mass spectra were typical fragment ions, such as $[M-18]^{+\bullet}$ for retinol, $[M-28]^{+\bullet}$ for retinal, and loss of $C_2H_4O_2$ for retinyl acetate. In a study by Hashimoto et al. (289), GC-MS was used to elucidate structures of rat metabolites of a synthetic retinoid, 4-[(5,6,7,8-tetrahydro-5,5,8,8-tetramethyl-2-naphthyl)carbamoyl]benzoic acid. A GC-MS method using positive ion EI for quantitative analysis of the methyl ester of retinoic acid was reported by De Leenheer and Lambert (89) using deuterated retinoic acid as an internal standard. The sensitivity of GC-MS for retinoid analysis may be enhanced by using negative ion chemical ionization (NCI). In their review of the literature, De Leenheer and Lambert (89) noted a 10-fold enhancement of sensitivity for GC-MS analysis of retinoids using NCI compared to EI. The limit of detection of NCI may be improved even further by using pentafluorobenzyl esters of retinoic acid and its analogs (280). The electronegativity of the fluorine atoms increases the efficiency of electron capture during ionization.

Besides being used as internal standards for quantitation, retinoids labeled with stable isotopes are frequently utilized as tracers during bioavailability, distribution, and metabolism studies in order to distinguish an administered retinoid dose from endogenous retinoids. For a human study of vitamin A plasma kinetics, von Reinersdorff et al. (290) administered [8,9,19-^{13}C]retinyl palmitate and measured retinol and [8,9,19-^{13}C]retinol using GC-MS with NCl. [2,3,6,7,10,11-^{13}C]all-*trans*-Retinol was used as an internal standard. Retinol was converted to its trimethylsilyl derivative prior to GC-MS analysis, and the fragment ions corresponding to loss of the trimethylsilyl group from labeled and unlabeled retinol were measured using selected ion monitoring. A similar stable isotope dilution mass spectrometry method was used by Clifford and coworkers (291) to assess vitamin A status. First, HPLC was used to separate retinol from retinyl esters in serum extracts. Then, the retinyl esters were hydrolyzed to retinol, and finally, both retinol samples were analyzed using GC-MS with positive ion EI to form molecular ions. Although molecular ions constituted <5% of the total retinol signal during GC-MS, EI was preferred to CI in this study because protonated retinol quantitatively dehydrates during CI. In a revision of this protocol, retinol was purified using HPLC, converted to its *tert*-butyldimethylsilyl ether, and then analyzed using GC-MS with EI and selected ion monitoring (292). Alternatively, a solid-phase extraction step was used instead of HPLC purification of retinol to simplify sample preparation and increase sample throughput to up to 32 samples per day per analyst (75). In these applications, selected ion monitoring was used to measure the abundance of the fragment ion formed by elimination of the *tert*-butyldimethylsilyl group. More recently, this group used GC-MS to measure the concentration of (d_4)-retinol in plasma following oral administration of (d_8)-β-carotene (293,294). Retinol was derivatized to its *tert*-butyldimethylsilyl ether prior to analysis. Tang and coworkers (295) used GC-MS with NCl for the quantification of the *tert*-butyldimethylsilyl derivatized deuterated retinol in human serum after administration of deuterated β-carotene.

In another approach using stable isotopes to measure retinoids in human serum, Parker and coworkers (296,297) studied the absorption and metabolism of ^{13}C-labeled β-carotene using GC-isotope ratio mass spectrometry. Retinol and retinyl esters were separated using HPLC, the retinyl esters were hydrolyzed to retinol, and each retinol sample was analyzed using GC-isotope ratio mass spectrometry. As the retinol peak eluted from the GC, it was combusted to carbon dioxide, and the $^{13}CO_2/^{12}CO_2$ ratio was measured using an isotope ratio mass spectrometer. In a similar way, lutein isolated from C_3 and C_4 plant sources was hydrogenated to perhydrocarotene which was subjected to GC-combustion-interfaced isotope ratio–mass spectrometry (GC-C-IR-MS) to determine natural abundance ^{13}C in lutein (298).

2. LC-MS

A wide variety of liquid chromatography–mass spectrometry (LC-MS) techniques have been reported for retinoid analysis including direct liquid introduction with CI (281,282,299–301), particle beam (302–304), thermospray (305–307), electrospray (308,309), and atmospheric pressure chemical ionization (APCI) (310). Because many of the early LC-MS applications to retinoids carried out hydrolysis of retinyl esters and then derivatization of retinoic acid and related retinoids, Wyss (141) predicted in a review of retinoid analysis that derivatization of retinoids would be necessary for all LC-MS techniques, even the anticipated application of atmospheric pressure chemical ionization (APCI). Recently, LC-MS analyses of retinoids have been carried out using APCI (310) and electrospray (308,309), and highly sensitive LC-MS analyses of retinoic acid and retinyl esters were achieved without hydrolysis or derivatization.

During direct liquid introduction LC-MS, the HPLC column eluate is pumped directly into the ion source of the mass spectrometer at μL/min flow rates. The maximum flow rate is determined by the capacity of the vacuum pumps serving the ion source of the mass spectrometer. Because of the high gas pressure produced by the evaporating solvent, CI is the usual ionization technique, and the solvent molecules may be used as the reagent gas. A microbore normal-phase HPLC-negative chemical ionization mass spectrometric method was developed by Fayer et al. (282) for the quantitative analysis of acitretin and 13-*cis*-acitretin over the range of 1 to 20 ng/mL in plasma. The limit of quantification was 1 ng/mL, but to achieve this level of sensitivity during NCI, acitretin and 13-*cis*-acitretin had to be derivatized to their pentafluorobenzyl esters. Molecular ions were not observed, and the base peaks of the mass spectra corresponded to fragment ions formed by elimination of the pentafluorobenzyl group. Huselton et al. (299) and Muindi et al. (300) used similar methods for the measurement of retinoic acid in human plasma. A variation of this microbore LC-MS method was used by Ranalder et al. (281) for the quantitation of all-*trans*- and 13-*cis*-retinoic acids and their 4-oxo metabolites in human plasma with a limit of detection of 1 pg and a limit of quantitation of 0.3 ng/mL. Finally, Rubio et al. (301) used normal-phase HPLC during direct inlet LC-MS with NCI to identify the *tert*-butyldimethylsilyl esters of acitretin and major metabolites during a study of the disposition of acitretin in humans. No molecular ions were observed, and abundant fragment ions were formed by elimination of the *tert*-butyldimethylsilyl groups.

In one of several spray techniques used as interfaces for LC-MS, the particle beam interface forms an aerosol containing clusters of sample molecules as the mobile phase evaporates in a heated, reduced-pressure chamber. The aerosol is separated from the lower-molecular-weight solvent molecules in a momentum separator, and then EI or CI ionization is carried out when the sample aerosol disintegrates as it strikes a heated metal surface in the ion source. An LC-MS method for the quantitation of all-*trans*-retinoic acid and 13-*cis*-retinoic acid in plasma was reported by Lehman and Franz (304), who used reversed-phase HPLC–particle beam mass spectrometry with selected ion monitoring. The retinoic acids were derivatized to their pentafluorobenzyl esters, which fragmented during NCl with methane reagent gas to form carboxylate anions of each retinoid. The limit of detection was 25 pg injected on-column. Bempong et al. (302) used normal-phase HPLC–particle beam mass spectrometry to separate and identify *cis/trans* isomers and degradation products of retinoic acid. EI was used for ionization, and an in-line UV detector provided additional retinoid characterization. Positive ion EI and positive and negative ion CI were compared by Careri et al. (303) during the particle beam LC-MS analysis of retinol, retinyl acetate and retinyl palmitate. EI produced the most abundant molecular ions of retinol and retinyl acetate, but none of these ionization techniques produced molecular ions of retinyl palmitate, which fragmented to eliminate palmitic acid. In an application of this technique, Andreoli et al. (311) used LC-MS with a particle beam interface to analyze vitamin A in infant formula.

During thermospray, which is another spray technique, the HPLC eluate is pumped through a heated capillary to produce a supersonic jet that expands into an evacuated chamber. As solvent evaporates, sample ions are formed from droplets that have a net charge, and ionization may be enhanced by using a heated filament to produce electron impact ionization or chemical ionization. Eckhoff et al. (305) used LC-MS with positive ion thermospray to characterize eight retinol metabolites, including glucuronides, that had been isolated from monkey plasma using a preliminary HPLC purification step. In similar studies, intact underivatized glucuronides of 13-*cis*- and all-*trans*-retinoic acid were characterized and identified by Kraft and coworkers (306) using LC-MS with positive ion thermospray; and acitretin and isoacitretin metabolites formed using perfused rat liver were characterized by Cotler et al. (307), using LC-MS with thermospray.

Solvent removal and ionization are combined during electrospray in which the HPLC eluate (at flow rates as high as 1 mL/min) is sprayed through a capillary electrode at high potential to form a fine mist of charged droplets at atmospheric pressure. As the solvent evaporates, gas-phase sample ions are formed. The first application of electrospray LC-MS to the analysis of retinoids was reported by van Breemen and Huang (308). Retinoic acid, retinol, retinal, and retinyl acetate were analyzed without derivatization using a C30 reversed-phase HPLC column. Retinoic acid formed abundant deprotonated molecules, $[M-H]^-$, with a limit of detection of 23 pg injected on-column. Positive ion electrospray produced an abundant protonated molecule for retinal, and a base peak of m/z 269 was observed for retinol and retinyl acetate, which corresponded to elimination of water or acetic acid, respectively, from their protonated molecules. The limits of detection of retinal, retinol, and retinyl acetate were 1.0 ng, 0.5 ng, and 10 ng, respectively. Protonated molecules were almost undetectable for intact retinol and retinyl acetate. In another study, by Shirley et al. (309), LC-MS with negative ion electrospray ionization was used for analysis of rat biliary metabolites of the synthetic retinoid, LGD1069, which is under investigation for treating cancer. Glucuronide, sulfate, and taurine conjugates were observed, and the structures of selected metabolites were confirmed using electrospray MS-MS.

Atmospheric pressure chemical ionization (APCI) uses a heated nebulizer to spray the HPLC mobile phase into an atmospheric pressure chamber. As the solvent evaporates, it is ionized by a corona discharge to form a reagent gas that, in turn, ionizes the analyte by chemical ionization. Van Breemen et al. (310) published the first application of atmospheric pressure chemical ionization for the LC-MS quantitative analysis of retinol and retinyl palmitate in human serum (Fig. 6). Retinyl acetate was used as an internal standard. Sample preparation consisted of hexane extraction of serum, which was followed by C30 reversed-phase HPLC separation with on-line positive ion APCI mass spectrometric quan-

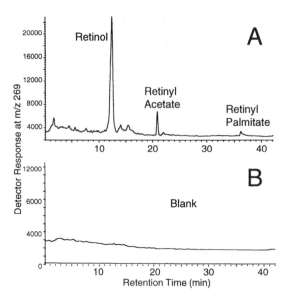

Figure 6   Positive ion atmospheric pressure chemical ionization (APCI) LC-MS analyses of (A) hexane extract of 200 μL of human serum, and (B) solvent blank showing no signals at m/z 269. The computer-reconstructed mass chromatogram at m/z 269 is shown from the full-scale data. The HPLC solvent system consisted of a 25 min linear gradient from 50:44.5:5:0.5 to 50:4.5:45:0.5 methanol/water/methyl-*tert*-butyl ether/acetic acid (v/v/v/v). (From Ref. 310.)

tification using selected ion monitoring. No sample derivatization was required. Although molecular ions or protonated molecules were detected in low abundance, the fragment ion of m/z 269, which corresponded to elimination of water, palmitic acid, or acetic acid from protonated retinol, retinyl palmitate, or retinyl acetate, respectively, was the base peak in each mass spectrum (Fig. 7). This APCI method showed a linear detector response over 4 orders of magnitude, and limits of detection of 34 fmol/µL for all-*trans*-retinol and 36 fmol/µL for all-*trans*-retinyl palmitate. The limit of quantitation was approximately 0.670 pmol for all-*trans*-retinol and 0.720 pmol for all-*trans*-retinyl palmitate injected in 20 µL on-column.

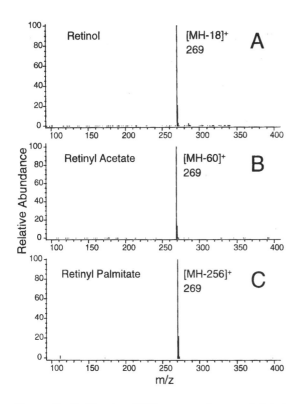

Figure 7 Positive ion APCI mass spectra recorded during the LC-MS analysis shown in Figure 6 including (A) all-*trans*-retinol at 16.1 min, (B) all-*trans*-retinyl acetate at 21.9 min, and (C) all-*trans*-retinyl palmitate at a retention time of 38.5 min. Note that all three retinoids fragmented during APCI to form a common base peak at m/z 269, which was used for selected ion monitoring during quantitative analysis. (From Ref. 310.)

3. New Ionization Methods and Tandem Mass Spectrometry

In addition to GC-MS and LC-MS analyses of retinoids, the mass spectrometric ionization techniques of laser desorption and fast atom bombardment (FAB) have been applied to retinoid analysis. For example, Wingerath et al. (312) used a nitrogen laser with a wavelength of 337 nm for the direct desorption/ionization mass spectrometry of retinyl esters that had been HPLC-purified from a rat liver extract. Matrix-assisted laser desorption ionization (MALDI) was unnecessary, and the addition of matrix was found to produce ions that interfered with retinoid analysis. Positive ion laser desorption mass spectra of retinyl esters showed molecular ions, $M^{+\cdot}$ of low abundance and base peaks of m/z 269, which were formed by elimination of the fatty acid. The only other significant fragment ion (slightly less abundant than the molecular ion) was formed by elimination of carbon dioxide from the molecular ion. The main advantage of laser desorption/ionization compared to other ionization techniques was that molecular ions of intact retinyl esters could be detected. FAB mass spectrometry has been used by Humphries and Curley (313), Salyers et al. (314), and Eldred and Lasky (315) for the analysis of retinoic acid, methyl retinoate, etretinate, retinoyl glucuronic acid, and an adduct of retinaldehyde with ethanolamine, respectively. In the study by Eldred and Lasky (315), FAB MS-MS with collision-induced dissociation was used to identify a Schiff base reaction product between retinaldehyde and ethanolamine, which form autofluorescent lysosomal residual bodies in retinal pigment epithelium. Another application of tandem mass spectrometry for retinoid analysis was reported by Shirley et al. (309), who used electrospray MS-MS for the identification of glucuronide, sulfate, and taurine conjugates of a synthetic retinoid.

B. Carotenoids

The high sensitivity and specificity of mass spectrometry is ideal for the identification and structural analysis of small quantities of carotenoids typically obtained from biological matrices, such as plants, animals, or human serum and tissue. Advances in soft ionization techniques, such as MALDI, FAB, electrospray, and atmospheric pressure chemical ionization (APCI), have facilitated the molecular weight determination of carotenoids by minimizing fragmentation that is typical of the classical ionization techniques of electron impact and chemical ionization. Once the molecular weight of a carotenoid has been established, collision-induced dissociation and tandem mass spectrometry may be used to augment fragmentation and obtain structurally significant fragment ions that may aid in the differentiation of structural isomers, such as differentiation of lutein from zeaxanthin, or α-carotene from β-carotene and lycopene.

1. New Ionization Methods and Tandem Mass Spectrometry

FAB is a matrix-mediated technique in which the analyte is dissolved in a liquid of low volatility such as glycerol or 3-nitrobenzyl alcohol, and then subjected to bombardment by a beam of energetic atoms or ions (usually xenon atoms at 3000 to 10,000 V or cesium ions at 10,000 to 20,000 V). The most effective matrix for carotenoid ionization is 3-nitrobenzyl alcohol (316,317), which facilitates the formation of molecular ion cations, $M^{+\bullet}$, for both xanthophylls and carotenes with a minimum of fragmentation. The observation of carotenoid molecular ions in FAB mass spectra is significant because most compounds form protonated or deprotonated molecules using this ionization technique. Although the production of abundant molecular ions is critical for molecular weight determination and accurate mass measurements, fragmentation would be helpful for structural analysis. Therefore, Caccamese and Garozzo (316) and van Breemen et al. (318) used high-energy collision-induced dissociation (CID) to generate structurally significant fragment ions from molecular ion precursors that had been produced during FAB ionization and then B/E linked scanning on a magnetic sector mass spectrometer to record the tandem mass spectrum of the product ions. In addition to providing information about carotenoid functional groups such as the presence of hydroxyl groups, esters, rings, or the extent of conjugation of the polyene chain, FAB with CID and tandem mass spectrometry provided unique fragment ions that differentiated isomeric carotenoids such as lutein and zeaxanthin, or lycopene, α-carotene, and β-carotene (318). Another advantage of tandem mass spectrometry following FAB ionization is that matrix ions and any other contaminating ions are eliminated, which simplifies interpretation of the mass spectrum. FAB ionization has been used in combination with LC-MS in a technique called continuous-flow FAB LC-MS, which is described below.

Sies' research group (319,320) reported the first application of MALDI time-of-flight mass spectrometry to the analysis of carotenoids including carotenoid esters isolated from fruit juice (Fig. 8). They dissolved purified carotenoids in acetone containing the matrix 2,5-dihydroxybenzoic acid, loaded each sample onto the MALDI probe, and used a pulsed nitrogen laser (λ = 337 nm) to obtain molecular ions, $M^{+\bullet}$. By using delayed extraction, molecular ions were observed as the base peaks of the MALDI mass spectra even for labile carotenoid mono- and diesters. Structurally significant fragment ions were observed owing to metastable fragmentation using post-source-decay analysis. Fragment ions were similar to those observed by van Breemen et al. (318) using FAB ionization followed by CID and tandem mass spectrometry.

EI and CI, which have been standard ionization methods for carotenoids for many years (321), continue to be used for carotenoid analysis. However, there have been some new developments in the application of negative ion electron

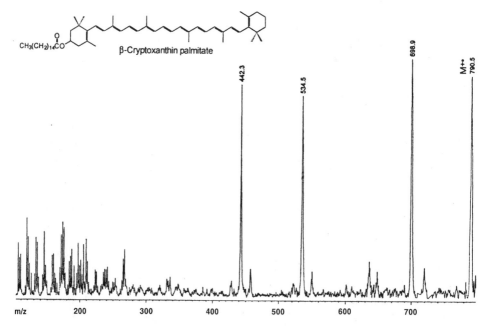

Figure 8 Positive ion matrix-assisted laser desorption (MALDI) time-of-flight mass spectrum of β-cryptoxanthin palmitate isolated from tangerine juice. Post source decay was used to enhance detection of structurally significant fragment ions such as loss of toluene (m/z 698), loss of palmitic acid (m/z 534), and loss of both toluene and palmitic acid (m/z 442). (From Ref. 320.)

capture chemical ionization (NCI). For example, McClure and Liebler (322) used negative ion chemical ionization to analyze oxidation products of β-carotene. Using NCI with a direct insertion probe, molecular anions, $M^{-\cdot}$, were obtained with little fragmentation. In order to obtain structurally significant fragment ions, McClure and Liebler (323) combined negative ion chemical ionization with high-energy CID and then recorded tandem mass spectra using B/E-linked scanning on a magnetic sector mass spectrometer. Fragment ions were observed that were complementary to the positive ion FAB tandem mass spectra reported by van Breemen et al. (318). Similar to charge-remote fragmentation, product ions were observed that were formed by cleavage of the polyene chain and permitted localization of branch points and double bonds within carotenoids.

2. LC-MS

Because carotenoids are thermally labile and are not compatible with gas chromatography or GC-MS, carotenoid separations are usually carried out by using reversed-phase HPLC. By combining HPLC separation and mass spectrometric detection (LC-MS), carotenoid analysis time is reduced, the sensitivity of the analysis is enhanced, and carotenoid handling and potential degradation are minimized (321,324). New developments in LC-MS technology, such as electrospray and atmospheric pressure ionization, are being applied to carotenoid analysis and are making LC-MS more robust and more accessible than ever before. Furthermore, LC-MS may be combined with on-line photodiode array absorbance spectroscopy and tandem mass spectrometry to obtain even more structural information (318,321). Five LC-MS techniques have been applied successfully to carotenoid analysis including moving belt, particle beam, continuous-flow FAB, electrospray, and atmospheric pressure CI.

In 1990, Taylor et al. (325) applied LC-MS to carotenoid analysis for the first time using a moving belt interface with EI. This study was also the first application of LC-MS-MS to carotenoid analysis. The most complex LC-MS system (and no longer in commercial production), the moving belt interface consists of two steps, solvent removal and then evaporation/ionization. First, the HPLC eluate is deposited on a belt that passes through a series of heated and evacuated chambers to evaporate the mobile phase. Then, the belt transports the solid analyte into the ion source of the mass spectrometer, where it is flash-evaporated and ionized by EI or CI. Using a moving belt system, Taylor et al. (324) observed the molecular ion of β-carotene in low abundance, whereas lutein formed dehydrated ions, $[M-H_2O]^{+\cdot}$, and many other fragment ions but no molecular ions.

Molecular ions of carotenoids are more abundant in particle beam mass spectra than in moving belt spectra, because sample molecules do not require volatilization from the belt (a step that can result in considerable pyrolysis). During particle beam LC-MS, the LC eluate is sprayed into a heated, near-atmospheric pressure chamber to evaporate the mobile phase, and the resulting sample aerosol is separated from the lower-molecular-weight solvent molecules in a momentum separator. Next, the sample aerosol enters the mass spectrometer ion source where the aggregates strike a heated metal surface and disintegrate; the resulting gas phase sample molecules are ionized by EI or CI. To minimize fragmentation and enhance sensitivity during particle beam LC-MS, NCI (negative ion chemical ionization) is usually used for carotenoid ionization. For example, Khachik et al. used particle beam LC-MS with NCI to analyze polar carotenoids extracted from human serum (64,326), human milk (217), and human and monkey retinas (210). They observed abundant molecular anions, $M^{-\cdot}$, and simple

fragmentation patterns such as loss of water from the molecular ion for xanthophylls like lutein. Careri et al. (327) reported molecular ions with little fragmentation for the xanthophylls lutein and zeaxanthin, and the carotenes β-carotene,
lycopene, and phytoene, which had been extracted from vegetables.

Fast Atom Bombardment.   Since FAB ionization requires that the analyte
be dissolved in a liquid matrix, this ionization technique may be readily interfaced
to HPLC in a system called continuous-flow FAB (CF-FAB). During CF-FAB,
the matrix (3-nitrobenzyl alcohol for carotenoids) is ideally mixed with the mobile phase and analyte in the ion source by coaxial-flow addition (328). The more
volatile mobile phase quickly evaporates, leaving behind a thin film of matrix
containing the analyte, which is exposed to the FAB beam. A disadvantage of
CF-FAB is that the upper flow limit is 10 μL, so that either a microbore HPLC
system must be used or the HPLC effluent must be split prior to entering the
mass spectrometer. Schmitz et al. (317) and van Breemen et al. (328) reported
the application of CF-FAB LC-MS and LC-MS-MS, respectively, to the analysis
of carotenoids extracted from a variety of fruits and vegetables. Abundant molecular ions, $M^{+\cdot}$, with little fragmentation were obtained for both carotenes and
xanthophylls, and the limits of detection ranged from 9 to 28 pmol injected on-
column. Structurally significant fragment ions were obtained by using CID during
LC-MS-MS.

Electrospray.   Unlike other LC-MS techniques that have been used for
the analysis of carotenoids, electrospray is both a solvent removal interface and
an ionization technique. During electrospray, the HPLC eluate is sprayed through
a capillary electrode at high potential (usually 2000 to 7000 V) to form a fine
mist of charged droplets at atmospheric pressure. As the charged droplets are
electrostatically attracted toward the opening of the mass spectrometer, they en-
counter a cross flow of heated nitrogen that increases solvent evaporation and
prevents most of the solvent molecules from entering the mass spectrometer.
Although ions produced by electrospray are usually preformed in solution by
acid/base reactions (i.e., $[M+nH]^{n+}$ or $[M-nH]^{n-}$), carotenoid ions are probably
formed by a field desorption mechanism at the surface of the droplet which appears to be enhanced by the presence of halogenated compounds such as hep-
tafluorobutanol (329). As a result of this unusual ionization process, electrospray
of carotenoids produces abundant molecular cations, $M^{+\cdot}$, with little fragmentation, and no molecular anions. Alternatively, van Berkel et al. (330) reported that
doubly charged carotenoid ions, M^{2+}, may be formed by solution phase oxidation
and then detected using electrospray. Because of the efficiency of this combined
ionization and desolvation, electrospray is compatible with a wide range of HPLC
flow rates (nL/min up to 1 mL/min) and with a variety of mobile phases including
the methanol/methyl-*tert*-butyl ether solvent system ideal for separations using

C30 carotenoid columns (329). However, a potential limitation of electrospray is the relatively narrow range of linearity of response it shows for carotenoid quantitation (331).

   Atmospheric Pressure Chemical Ionization. Most mass spectrometers equipped for electrospray ionization can be easily converted to atmospheric pressure CI (APCI). APCI uses a heated nebulizer instead of a strong electromagnetic field to facilitate solvent evaporation and obtain a fine spray of the HPLC mobile phase. Ionization takes place by spraying the analyte into a corona discharge, which produces reactive species from the solvent gas (a chemical reagent gas) that can ionize the analyte. Ions are then drawn into the aperture of the mass spectrometer. The first APCI LC-MS analyses of carotenoids were published by van Breemen et al. (321,331) and show molecular ions and protonated molecules in positive ion mode, and molecular ions and deprotonated molecules during negative ion analysis. Later, Liebler and McClure (332) used APCI ionization but not LC-MS to analyze β-carotene oxidation products. During APCI, the relative abundance of molecular ions and protonated or deprotonated molecules vary with the mobile-phase composition. For example, polar solvents such as alcohols lead to an increased abundance of protonated carotenoids (even protonated β-carotene), and nonpolar solvents such as methyl-*tert*-butyl ether facilitate the formation of molecular ions. Because van Breemen et al. (321,331) used a solvent system containing both methanol and methyl-*tert*-butyl ether during HPLC separation on a C30 column, a mixture of molecular ions and protonated molecules was observed in the carotenoid mass spectra. The main advantage of APCI compared to electrospray for carotenoid analysis is the higher linearity of detector response (>4 orders of magnitude of carotenoid concentration), which suggests that APCI LC-MS might become the preferred mass spectrometric technique for carotenoid quantitation (324,331). Disadvantages of APCI include the multiplicity of molecular ion species, which might lead to ambiguous molecular weight determinations, and abundant fragmentation, which tends to reduce the abundance of the molecular ions.

   Because the majority of LC-MS systems currently being installed are designed for electrospray and/or APCI, these interfaces will become the most widely available to carotenoid researchers during the next several years. Fortunately, electrospray and APCI provide high sensitivity for carotenoids (≤ low pmol level), and may be operated unattended by use of an autosampler. The superior linearity of detector response of APCI for carotenoids suggests that this LC-MS technique may become the standard for carotenoid quantitation. However, the soft ionization of electrospray, which produces molecular ions without fragmentation, greatly facilitates molecular weight confirmation and might be preferable to APCI for identification of carotenoids in mixtures and biological extracts, since APCI can produce abundant fragment ions.

Several research groups have been investigating carotenoid bioavailability using carotenoids labeled with stable isotopes. Although HPLC purification (293,295–297) and even ionization techniques such as APCI (295) are being used routinely in these studies, LC-MS has not been utilized for carotenoid bioavailability studies using stable isotope labeling. Instead, HPLC with absorbance detection is used for quantitation (293), or fractions containing labeled carotenoids are collected and analyzed off-line by mass spectrometry (295–297).

Future Prospects. As quadrupole and ion trap mass spectrometers are replaced by more efficient time-of-flight mass analyzers, the limits of detection of electrospray and APCI will be enhanced by at least 1 and perhaps 2 orders of magnitude. Electrospray time-of-flight mass spectrometers are already commercially available, although no carotenoid applications have been reported using these instruments. Another improvement being introduced into electrospray and APCI interfaces is the use of an orthogonal spray design. By spraying the HPLC effluent at a right angle to the entrance of the mass spectrometer, noise may be reduced 10- to 100-fold without sacrificing signal. MALDI time-of-flight mass spectrometry shows great potential for improving the sensitivity of carotenoid analysis, especially for carotenoid esters (319,320). Studies of carotenoids would benefit substantially by the availability of a sensitive, low-cost, bench top MALDI LC-MS instrument. Such instruments are under development. Finally, studies of carotenoid bioavailability using stable isotope labeling will be facilitated by improvements in the sensitivity of LC-MS, and major advances in this area are expected soon.

VIII. OTHER METHODS

This chapter focuses on the extraction and handling of retinoids and carotenoids, their separation by various chromatographic techniques, and their detection and quantitation, primarily by absorption spectrophotometry, fluorescence, and mass spectrometry. A variety of other methods exist for their identification and characterization, including circular dichroism (333), infrared spectroscopy (334), resonance Raman spectroscopy (335), NMR spectroscopy (336), and x-ray crystallography (337). Although some of these procedures require substantial amounts of a retinoid or a carotenoid in an essentially pure form for study, others, such as resonance Raman spectroscopy, are extremely sensitive and can be used to detect the localization of carotenoids in single cells (338,339).

In rigorously identifying a retinoid or carotenoid species, however, its chromatographic behavior and its absorption spectrum are not sufficient (340). Additional physiochemical data, e.g., mass spectra, NMR spectra, and information on chemical derivatives, are needed (340).

IX. RECENT DEVELOPMENTS AND FUTURE TRENDS

Although all chromatographic procedures can provide useful information about the structures, metabolism, and localization of retinoids and carotenoids in nature, HPLC has clearly emerged as the most useful way of separating and tentatively identifying these classes of compounds. In the last several years, the development of photodiode array detectors and appropriate software for their use has further enhanced the utility of HPLC. The development of new stationary phases, more uniform and smaller support materials, and new combinations of mobile phases have further broadened the use of HPLC.

Although absorption spectrophotometry and fluorescence emission spectroscopy remain important devices for detection and characterization of retinoids and/or carotenoids, technical advances in mass spectrometry are impressive. The development of new LC-MS methods for retinoid and carotenoid analysis utilizing electrospray and APCI have streamlined the sample preparation process by simplifying sample purification prior to chromatography and by eliminating the need for derivatization. Improvements in LC-MS design, such as the introduction of orthogonal spray interfaces and the use of time-of-flight mass spectrometers instead of scanning instruments, should increase the sensitivity of retinoid and carotenoid analyses by at least 2 orders of magnitude. Finally, the use of tandem mass spectrometry and LC-MS-MS for retinoid analysis has been minimal to date, but their application to retinoid studies should increase in the future. Development of benchtop MALDI time-of-flight LC-MS instruments will provide improved sensitivity for studies of carotenoid metabolism. These developments will allow use of stable isotope labeling to increase our understanding of carotenoid bioavailability and metabolism.

X. SUMMARY

In the first edition of *Modern Chromatographic Analysis of the Vitamins*, published in 1985, HPLC had already become the method of choice for the separation and quantitation of retinoids and carotenoids (28). Its dominance in this area of separation science increased up to the time of the second edition of this work, in 1992 (29), and has further increased since then. Together with improvements in instrumentation, stationary phases, and columns, the development of photodiode array detectors has added an important new dimension to the characterization of eluted compounds. Interfaces between mass spectrometry and either gas chromatography or liquid chromatography have also greatly improved. In addition, new sensitive methods for measuring these labile compounds by mass spectrometry have been devised. Significant further progress in these methodologies is anticipated in the next few years.

ACKNOWLEDGMENTS

The authors gratefully acknowledge support of their research studies on retinoids and carotenoids by the National Institutes of Health, U.S. Public Health Service DK-39733 (J.A.O. and A.B.B.), the U.S. Department of Agriculture 97-35200-0490 (J.A.O. and A.B.B.), CDFIN/ISU 96-34115-2835 (J.A.O.), National Cancer Institute CA70771 (R.B.B.), and a Hatch Project of the Storrs Agricultural Experiment Station (H.C.F.).

REFERENCES

1. F Frickel. Chemistry and physical properties of retinoids. In: MB Sporn, AB Roberts, DS Goodman, eds. The Retinoids. Vol. 1. Orlando, FL: Academic Press, 1984, pp 7–145.
2. MI Dawson, PD Hobbs. Synthetic chemistry of retinoids. In: MB Sporn, AB Roberts, DS Goodman, eds. The Retinoids. New York: Raven Press, 1994, pp 5–178.
3. AB Barua, HC Furr. Properties of retinoids: structure, handling, and preparation. In: CPF Redfern, ed. Retinoid Protocols. Totowa, NJ: Humana Press, 1998, pp 3–28.
4. O Isler, H Gutmann, U Solms. Carotenoids. Basel: Birkhauser Verlag, 1971, pp 1–932.
5. G Britton, S Liaaen-Jensen, H Pfander, eds. Carotenoids. 1A. Isolation and Analysis. Basel: Birkhauser Verlag, 1995, pp 1–317.
6. G Britton, S Liaaen-Jensen, H Pfander, eds. Carotenoids. 1B: Spectroscopy. Basel: Birkhauser Verlag, 1995, pp 1–354.
7. G Britton, S Liaaen-Jensen, H Pfander, eds. Carotenoids. Vol. 2: Synthesis. Basel: Birkhauser Verlag, 1996, pp 1–353.
8. JA Olson, NI Krinsky. Introduction: the colorful, fascinating world of the carotenoids: important physiologic modulators. FASEB J 9:1547–1550, 1995.
9. O Straub. Key to Carotenoids. Basel: Birkhauser Verlag, 1987, pp 1–276.
10. G Britton. UV/visible spectroscopy. In: G Britton, S Liaaen-Jensen, H Pfander, eds. Carotenoids. 1B: Spectroscopy. Basel: Birkhauser Verlag, 1995, pp, 13–62.
11. E De Ritter, AE Purcell. Carotenoid analytical methods. In: JC Bauernfeind, ed. Carotenoids as Colorants and Vitamin A Precursors. New York: Academic Press, 1981, pp 815–923.
12. G Britton. Carotenoids 1: structure and properties of carotenoids in relation to function. FASEB J 9:1551–1558, 1995.
13. JA Olson. Vitamin A, retinoids, and carotenoids. In: ME Shils, JA Olson, M Shike, eds. Modern Nutrition in Health and Disease. 8th ed. Philadelphia: Lea & Febiger, 1994, pp 287–307.
14. R Blomhoff, ed. Vitamin A in Health and Disease. New York: Marcel Dekker, 1994.

15. AC Ross. Vitamin A and retinoids. In: ME Shils, JA Olson, M Shike, AC Ross, eds. Modern Nutrition in Health and Disease. 9th ed. Baltimore: Williams & Wilkins, 1999, pp 305–327.

16. JA Olson. Carotenoids. In: ME Shils, JA Olson, M Shike, AC Ross, eds. Modern Nutrition in Health and Disease. 9th ed. Baltimore: Williams & Wilkins, 1999, pp 529–541.

17. JC Saari. Retinoids in photosensitive systems. In: MB Sporn, AB Roberts, DS Goodman, eds. The Retinoids. New York: Raven Press, 1994, pp 351–385.

18. C Hofmann, G Eichele. Retinoids in development. In: MB Sporn, AB Roberts, DS Goodman, eds. The Retinoids. 2nd ed. New York: Raven Press, 1994, pp 387–441.

19. GA Armstrong, JE Hearst. Carotenoids 2: genetics and molecular biology of carotenoid pigment biosynthesis. FASEB J 10:228–237, 1996.

20. B Demmig-Adams, AM Gilmore, WW Adams. Carotenoids 3: in vivo function of carotenoids in higher plants. FASEB J 10:403–412, 1996.

21. HJ Nelis, P Lavens, MM Van Steenberge, P Sorgeloos, GR Criel, AP De Leenheer. Qualitative and quantitative changes in the carotenoids during development of the brine shrimp *Artemia*. J Lipid Res 29:491–499, 1988.

22. JA Olson. Molecular actions of carotenoids. Ann NY Acad Sci 691:156–166, 1993.

23. RS Parker. Carotenoids 4: absorption, metabolism, and transport of carotenoids. FASEB J 10:542–551, 1996.

24. HC Furr, RM Clark. Intestinal absorption and tissue distribution of carotenoids. J Nutr Biochem 8:364–377, 1997.

25. NI Krinsky. The biological properties of carotenoids. Pure Appl Chem 66:1003–1010, 1994.

26. ST Mayne. Carotenoids 5: beta-carotene, carotenoids, and disease prevention in humans. FASEB J 10:690–701, 1996.

27. LM Canfield, NI Krinsky, JA Olson, eds. Carotenoids in Human Health. New York: New York: Academy of Sciences, 1993, pp 1–300.

28. WE Lambert, HJ Nelis, MGM De Ruyter, AP De Leenheer. Vitamin A: retinol, carotenoids, and related compounds. In: AP De Leenheer, WE Lambert, MGM De Ruyter, eds. Modern Chromatographic Analysis of the Vitamins. New York: Marcel Dekker, 1985, pp 1–72.

29. HC Furr, AB Barua, JA Olson. Retinoids and carotenoids. In: HJ Nelis, WE Lambert, AP DeLeenheer, eds. Modern Chromatographic Analysis of the Vitamins. 2nd ed. New York: Marcel Dekker, 1992, pp 1–71.

30. JL Napoli, RL Horst. Quantitative analyses of naturally occurring retinoids. In: CPF Redfern, ed. Retinoid Protocols. Totowa, NJ: Humana Press, 1998, pp 29–40.

31. GM Landers, JA Olson. Absence of isomerization of retinyl palmitate, retinol, and retinal in chlorinated and nonchlorinated solvents under gold light. J Assoc Off Anal Chem 69:50–55, 1986.

32. NE Craft, ED Brown, JC Smith Jr. Effects of storage and handling conditions on concentrations of individual carotenoids, retinol and tocopherol in plasma. Clin Chem 34:44–48, 1988.

33. BK Edmonds, DW Nierenberg. Serum concentrations of retinol, d-alpha-tocopherol

and beta-carotene: effects of storage at -70 degrees C for five years. J Chromatogr 614:169–174, 1993.

34. GW Comstock, AJ Alberg, KJ Helzlsouer. Reported effects of long-term freezer storage on concentrations of retinol, beta-carotene, and alpha-tocopherol in serum or plasma summarized. Clin Chem 39:1075–1078, 1993.

35. GW Comstock, EP Norkus, SC Hoffman, MW Xu, KJ Helzlsouer. Stability of ascorbic acid, carotenoids, retinol, and tocopherols in plasma stored at -70 degrees C for 4 years. Cancer Epidemiol Biomarkers Prev 4:505–507, 1995.

36. Y Ito, J Ochiai, R Sasaki, S Suzuki, Y Kusuhara, Y Morimitsu, M Otani, K Aoki. Serum concentrations of carotenoids, retinol, and alpha-tocopherol in healthy persons determined by high-performance liquid chromatography. Clin Chim Acta 194: 131–144, 1990.

37. SE Hankinson, SJ London, CG Chute, RL Barbieri, L Jones, LA Kaplan, FM Sacks, MJ Stampfer. Effect of transport conditions on the stability of biochemical markers in blood. Clin Chem 35:2313–2316, 1989.

38. MD Gross, CB Prouty, DR Jacobs Jr. Stability of carotenoids and alpha-tocopherol during blood collection and processing procedures. Clin Chem 41:943–944, 1995.

39. YS Peng, YM Peng, DL McGee, DS Alberts. Carotenoids, tocopherols, and retinoids in human buccal mucosal cells: intra- and interindividual variability and storage stability. Am J Clin Nutr 59:636–643, 1994.

40. MC Barreto-Lins, FACS Campos, MNA Azevedo, H Flores. A re-examination of the stability of retinol in blood and serum, and effects of a standardized meal. Clin Chem 34:2308–2310, 1988.

41. WJ Driskell, AD Lackey, JS Hewett, MM Bashor. Stability of vitamin A in frozen sera. Clin Chem 31:871–872, 1985.

42. NE Craft, SA Wise, JH Soares. Optimization of an isocratic high-performance liquid chromatographic separation of carotenoids. J Chromatogr 589:171–176, 1992.

43. RWA Oliver, EM Kafwembe, D Mwandu. Stability of vitamin A circulating complex in spots of dried serum samples adsorbed onto filter paper. Clin Chem 39: 1744–1745, 1993.

44. H Shi, Y Ma, JH Humphrey, NE Craft. Determination of vitamin A in dried human blood spots by high-performance capillary electrophoresis with laser-excited fluorescence detection. J Chromatogr B Biomed Appl 665:89–96, 1995.

45. EZ Szuts, FI Harosi. Solubility of retinoids in water. Arch Biochem Biophys 287: 297–304, 1991.

46. SR Ames, HA Risley, PL Harris. Simplified procedure for extraction and determination of vitamin A in liver. Anal Chem 26:1378–1381, 1954.

47. JA Olson. Liver vitamin A reserves of neonates, preschool children and adults dying of various causes in Salvador, Brazil. Arch Latinoam Nutr 29:521–545, 1979.

48. M Blanco, J Coello, H Iturriaga, S Maspoch, T Gomez-Cotin, S Alaoui-Ismaili, E Rovira. Simultaneous spectrophotometric determination of fat-soluble vitamins in multivitamin pharmaceutical preparations. Fresenius J Anal Chem 351:315–319, 1995.

49. EG Bligh, WJ Dyer. A rapid method of total lipid extraction and purification. Can J Biochem Physiol 37:911–917, 1959.

50. YL Ito, M Zile, H Ahrens, HF DeLuca. Liquid-gel partition chromatography of

vitamin A compounds; formation of retinoic acid from retinyl acetate in vivo. J Lipid Res 15:517–524, 1974.

51. S Zhang, G Tang, RM Russell, KA Mayzel, MJ Stampfer, WC Willett, DJ Hunter. Measurement of retinoids and carotenoids in breast adipose tissue and a comparison of concentrations in breast cancer cases and control subjects. Am J Clin Nutr 66: 626–632, 1997.

52. AB Roberts, MD Nichols, CA Frolik, DL Newton, MB Sporn. Assay of retinoids in biological samples by reverse-phase high-pressure liquid chromatography. Cancer Res 38:3327–3332, 1978.

53. ME Cullum, MH Zile. Quantitation of biological retinoids by high-pressure liquid chromatography: primary internal standardization using tritiated retinoids [published erratum appears in Anal Biochem 1988 Jan; 168(1):229]. Anal Biochem 153: 23–32, 1986.

54. NS Radin. Extraction of tissue lipids with a solvent of low toxicity. Methods Enzymol 72:5–7, 1981.

55. NR Badcock, DA O'Reilly, CB Pinnock. Liquid chromatographic determination of retinol and alpha-tocopherol in human buccal mucosal cells. J Chromatogr 382: 290–296, 1986.

56. E Lesellier, A Tchapla, C Marty, A Lebert. Analysis of carotenoids by high-performance liquid chromatography and supercritical fluid chromatography. J Chromatogr 633:9–23, 1993.

57. HG Daood, PA Biacs. Evidence for the presence of lipoxygenase and hydroperoxide-decomposing enzyme in red pepper seeds. Acta Aliment 15:307–318, 1986.

58. KS Rhee, BM Watts. Evaluation of lipid oxidation in plant tissues. J Food Sci 31: 664–668, 1966.

59. I Stewart. High performance liquid chromatographic determination of provitamin A in orange juice. J Assoc Off Anal Chem 60:132–136, 1977.

60. H Yokoyama, MJ White. Citrus carotenoids. II. The structure of citranaxanthin, a new carotenoid analog. J Org Chem 30:2481–2482, 1965.

61. M Kimura, DB Rodriguez-Amaya, HT Godoy. Assessment of the saponification step in the quantitative determination of carotenoids and provitamins A. Food Chem 35:187–195, 1990.

62. E Tee, C Lim. The analysis of carotenoids and retinoids: a review. Food Chem 41: 147–193, 1991.

63. G Scita. Stability of beta-carotene under different laboratory conditions. Methods Enzymol 213:175–185, 1992.

64. F Khachik, GR Beecher, MB Goli, WR Lusby, JC Smith Jr. Separation and identification of carotenoids and their oxidation products in the extracts of human plasma. Anal Chem 64:2111–2122, 1992.

65. F Khachik, GR Beecher, MB Goli, WR Lusby, CE Daitch. Separation and quantification of carotenoids in human plasma. Methods Enzymol 213:205–219, 1992.

66. F Khachik, G Englert, CE Daitch, GR Beecher, LH Tonucci, WR Lusby. Isolation and structural elucidation of the geometrical isomers of lutein and zeaxanthin in extracts from human plasma. J Chromatogr 582:153–166, 1992.

67. AB Barua, RO Batres, HC Furr, JA Olson. Analysis of carotenoids in human serum. J Micronutr Anal 5:291–302, 1989.

68. AB Barua, HC Furr, D Janick-Buckner, JA Olson. Simultaneous analysis of individual carotenoids, retinol, retinyl esters, and tocopherols in serum by isocratic nonaqueous reversed-phase HPLC. Food Chem 46:419–424, 1993.

69. SW McClean, ME Ruddel, EG Gross, JJ DeGiovanna, GL Peck. Liquid-chromatographic assay for retinol (vitamin A) and retinol analogs in therapeutic trials. Clin Chem 28:693–696, 1982.

70. DW Nierenberg, DC Lester. Determination of vitamins A and E in serum and plasma using a simplified clarification method and high-performance liquid chromatography. J Chromatogr 345:275–284, 1985.

71. FQ Siddiqui, F Malik, FR Fazli. Determination of serum retinol by reversed-phase high-performance liquid chromatography. J Chromatogr B Biomed Appl 666:342–346, 1995.

72. AB Barua, D Kostic, JA Olson. New, simplified procedures for the extraction and simultaneous high-performance liquid chromatographic analysis of retinol, tocopherols and carotenoids in human serum. J Chromatogr 617:257–264, 1993.

73. AB Barua, JA Olson. Reversed-phase gradient high-performance chromatography procedure for simultaneous analysis of very polar to nonpolar retinoids, carotenoids and tocopherols in animal and plant tissues. J Chromatogr B. Biomed Appl 707: 69–79, 1998.

74. A During, A Nagao, C Hoshino, J Terao. Assay of beta-carotene 15,15′-dioxygenase activity by reverse-phase high-pressure liquid chromatography. Anal Biochem 241:199–205, 1996.

75. SR Dueker, JM Lunetta, AD Jones, AJ Clifford. Solid-phase extraction protocol for isolating retinol-d4 and retinol from plasma for parallel processing for epidemiological studies. Clin Chem 39:2318–2322, 1993.

76. BJ Burri, TR Neidlinger, AO Lo, C Kwan, MR Wong. Supercritical fluid extraction and reversed-phase liquid chromatography methods for vitamin A and beta-carotene heterogeneous distribution of vitamin A in the liver. J Chromatogr A 762: 201–206, 1997.

77. F Favati, JW King, JP Friedrich, K Eskins. Supercritical CO_2 extraction of carotene and lutein from leaf protein concentrates. J Food Sci 53:1532–1536, 1988.

78. ML Cygnarowicz, WD Seider. Design and control of a process to extract beta-carotene with supercritical carbon dioxide. Biotechnol Prog 6:82–91, 1990.

79. CJ Chang, AD Randolph, NE Craft. Separation of beta-carotene mixtures precipitated from liquid solvents with high pressure CO_2. Biotechnol Prog 7:275–278, 1991.

80. A Vahlquist. Vitamin A in human skin: I. Detection and identification of retinoids in normal epidermis. J Invest Dermatol 79:89–93, 1982.

81. R Rettenmaier, W Schuep. Determination of vitamins A and E in liver tissue. Int J Vitam Nutr Res 62:312–317, 1992.

82. MM Delgado-Zamarreno, A Sanchez-Perez, MC Gomez-Perez, J Hernandez-Mendez. Directly coupled sample treatment-high-performance liquid chromatography for on-line automatic determination of liposoluble vitamins in milk. J Chromatogr A 694:399–406, 1995.

83. E Ballesteros, M Gallego, M Valcarcel. Gas chromatographic determination of cholesterol and tocopherols in edible oils and fats with automatic removal of interfering triglycerides. J Chromatogr A 719:221–227, 1996.

84. O Sommerburg, LY Zang, FGM van Kuijk. Simultaneous detection of carotenoids and vitamin E in human plasma. J Chromatogr B Biomed Appl 695:209–215, 1997.
85. NE Craft, JH Soares. Relative solubility, stability, and absorptivity of lutein and beta-carotene in organic solvents. J Agric Food Chem 40:431–434, 1992.
86. LY Zang, O Sommerburg, FJ van Kuijk. Absorbance changes of carotenoids in different solvents. Free Radic Biol Med 23:1086–1089, 1997.
87. CS Beeman, JE Kronmiller. Temporal distribution of endogenous retinoids in the embryonic mouse mandible. Arch Oral Biol 39:733–739, 1994.
88. JL Napoli, BC Pramanik, JB Williams, MI Dawson, PD Hobbs. Quantification of retinoic acid by gas–liquid chromatography–mass spectrometry: total versus all-*trans*-retinoic acid in human plasma. J Lipid Res 26:387–392, 1985.
89. AP De Leenheer, WE Lambert. Mass spectrometry of methyl ester of retinoic acid. Methods Enzymol 189:104–111, 1990.
90. E Meyer, WE Lambert, AP De Leenheer. Simultaneous determination of endogenous retinoic acid isomers and retinol in human plasma by isocratic normal-phase HPLC with ultraviolet detection. Clin Chem 40:48–57, 1994.
91. B Dimitrova, M Poyre, G Guiso, A Badiali, S Caccia. Isocratic reversed-phase liquid chromatography of all-*trans*-retinoic acid and its major metabolites in new potential supplementary test systems for developmental toxicology. J Chromatogr B Biomed Appl 681:153–160, 1996.
92. B Disdier, H Bun, J Catalin, A Durand. Simultaneous determination of all-*trans*-, 13-*cis*-, 9-*cis*-retinoic acid and their 4-oxo-metabolites in plasma by high-performance liquid chromatography. J Chromatogr B Biomed Appl 683:143–154, 1996.
93. FGM van Kuijk, GJ Handelman, EA Dratz. Rapid analysis of the major classes of retinoids by step gradient reversed-phase high-performance liquid chromatography using retinyl (O-ethyl) oxime derivatives. J Chromatogr 348:241–251, 1985.
94. MH Green, JB Green. Experimental and kinetic methods for studying vitamin A dynamics in vivo. Methods Enzymol 190:304–317, 1990.
95. F Khachik, GR Beecher, MG Goli. Separation, identification, and quantification of carotenoids in fruits, vegetables and human plasma by high performance liquid chromatography. Pure Appl Chem 63:71–80, 1991.
96. GJ Handelman, B Shen, NI Krinsky. High resolution analysis of carotenoids in human plasma by high-performance liquid chromatography. Methods Enzymol 213:336–346, 1992.
97. NE Craft. Carotenoid reversed-phase high-performance liquid chromatography methods: reference compendium. Methods Enzymol 213:185–205, 1992.
98. RM McKenzie, ML McGregor, EC Nelson. Purification of labelled retinoic acid and characterization of a major impurity. J Label Compds Radiopharm 15:265–278, 1978.
99. KV John, MR Lakshmanan, FB Jungalwala, HR Cama. Separation of vitamins A$_1$ and A$_2$ and allied compounds by thin-layer chromatography. J Chromatogr 18:53–56, 1965.
100. TNR Varma, T Panalaks, TK Murray. Thin layer chromatography of vitamin A and related compounds. Anal Chem 36:1864–1865, 1964.
101. YK Fung, RG Rahwan, RA Sams. Separation of vitamin A compounds by thin-layer chromatography. J Chromatogr 147:528–531, 1978.
102. K Schiedt. Chromatography: Part III. Thin-layer chromatography. In: G Britton,

S Liaaen-Jensen, H Pfander, eds. Carotenoids. 1A: Isolation and Analysis. Basel: Birkhauser Verlag, 1995, pp 131–144.

103. AJ Rosas Romero, JC Herrera, E Martinez De Aparicio, EA Molina Cuevas. Thin-layer chromatographic determination of beta-carotene, canthaxanthin, lutein, violaxanthin and neoxanthin on Chromarods. J Chromatogr A 667:361–366, 1994.

104. MI Minguez-Mosquera, D Hornero-Mendez, J Garrido-Fernandez. Detection of bixin, lycopene, canthaxanthin, and beta-apo-8′-carotenal in products derived from red pepper. J AOAC Int 78:491–496, 1995.

105. H Pfander, P Riesen, U Niggli. HPLC and SFC of carotenoids—scope and limitations. Pure Appl Chem 66:947–954, 1994.

106. H Pfander, R Riesen. Chromatography: Part IV. High-performance liquid chromatography. In: G Britton, S Liaaen-Jensen, H Pfander, eds. Carotenoids. 1A: Isolation and Analysis. Basel: Birkhauser Verlag, 1995, pp 145–190.

107. H Pfander, U Niggli. Chromatography: Part V. Supercritical-fluid chromatography. In: G Britton, S Liaaen-Jensen, H Pfander, eds. Carotenoids. 1A: Isolation and Analysis. Basel: Birkhauser Verlag, 1995, pp 191–198.

108. AP De Leenheer, VORC De Bevere, MGM De Ruyter, AE Claeys. Simultaneous determination of retinol and alpha-tocopherol in human serum by high-performance liquid chromatography. J Chromatogr 162:408–413, 1979.

109. GL Catignani, JG Bieri. Simultaneous determination of retinol and alpha-tocopherol in serum or plasma by liquid chromatography. Clin Chem 29:708–712, 1983.

110. GL Catignani. An HPLC method for the simultaneous determination of retinol and alpha-tocopherol in plasma or serum. Methods Enzymol 123:215–219, 1986.

111. AJ Speek, EJ van Agtmaal, S Saowakontha, WHP Schreurs, NJ van Haeringen. Fluorometric determination of retinol in human tear fluid using high-performance liquid chromatography. Current Eye Res 5:841–845, 1986.

112. AJ Speek, C Wongkham, N Limratana, S Saowakontha. Microdetermination of vitamin A in human plasma using high-performance liquid chromatography with fluorescence detection. J Chromatogr 382:284–289, 1986.

113. WA MacCrehan, E Schonberger. Reversed-phase high-performance liquid chromatographic separation and electrochemical detection of retinol and its isomers. J Chromatogr 417:65–78, 1987.

114. WA MacCrehan. Determination of retinol, alpha-tocopherol, and beta-carotene in serum by liquid chromatography. Methods Enzymol 189:172–181, 1990.

115. M Mulholland, RJ Dolphin. Analysis of fat-soluble vitamins using narrow-bore high-performance liquid chromatography with multichannel UV-VIS detection. J Chromatogr 350:285–291, 1985.

116. M Mulholland. Linking low dispersion liquid chromatography with diode-array detection for the sensitive and selective determination of vitamins A, D and E. Analyst 111:601–604, 1986.

117. HC Furr, JA Olson. A direct microassay for serum retinol (vitamin A alcohol) by using size-exclusion high-pressure liquid chromatography with fluorescence detection. Anal Biochem 171:360–365, 1988.

118. BJ Burri, MA Kutnink. Liquid-chromatographic assay for free and transthyretin-bound retinol-binding protein in serum from normal humans. Clin Chem 35:582–586, 1989.

119. BJ Burri, MA Kutnink, TR Neidlinger. Assay of human transthyretin-bound holo-retinol-binding protein with reversed-phase high-performance liquid chromatography. J Chromatogr 567:369–380, 1991.

120. H Muhilal, J Glover. The affinity of retinol and its analogues for retinol-binding protein. Biochem Soc Trans 3:744–746, 1975.

121. Y Ma, Z Wu, HC Furr, C Lammi-Keefe, NE Craft. Fast minimicroassay of serum retinol (vitamin A) by capillary zone electrophoresis with laser-excited fluorescence detection. J Chromatogr 616:31–37, 1993.

122. W Luo, AB al-Abdulaly, K Yoon, KL Simpson. Rapid determination of blood serum retinol by reverse phase open column chromatography. Int J Vitam Nutr Res 63:82–86, 1993.

123. AC Ross. Separation of long-chain fatty acid esters of retinol by high-performance liquid chromatography. Anal Biochem 115:324–330, 1981.

124. MGM De Ruyter, AP De Leenheer. Effect of silver ions on the reversed-phase high performance liquid chromatographic separation of retinyl esters. Anal Chem 51:43–46, 1979.

125. HC Furr. Reversed-phase high-performance liquid chromatography of retinyl esters. Methods Enzymol 189:85–94, 1990.

126. HK Biesalski, H Weiser. Sensitive analysis of retinyl esters by isocratic adsorption chromatography. J Clin Chem Clin Biochem 27:65–74, 1989.

127. DD Bankson, RM Russell, JA Sadowski. Determination of retinyl esters and retinol in serum or plasma by normal-phase liquid chromatography: method and applications. Clin Chem 32:35–40, 1986.

128. WO Landen. Application of gel permeation chromatography and nonaqueous reverse phase chromatography to high pressure liquid chromatographic determination of retinyl palmitate in fortified breakfast cereals. J Assoc Off Anal Chem 63:131–136, 1980.

129. WO Landen. Application of gel permeation chromatography and nonaqueous reverse phase chromatography to high performance liquid chromatographic determination of retinyl palmitate and alpha-tocopheryl acetate in infant formulas. J Assoc Off Anal Chem 65:810–816, 1982.

130. RW Curley, DL Carson, CN Ryzewski. Effect of end-capping of reversed-phase high-performance liquid chromatographic matrices on the analysis of vitamin A and its metabolites. J Chromatogr 370:188–193, 1986.

131. GWT Groenendijk, PAA Jansen, SL Bonting, FJM Daemen. Analysis of geometrically isomeric vitamin A compounds. Methods Enzymol 67(F):203–220, 1980.

132. CDB Bridges. High-performance liquid chromatography of retinoid isomers: an overview. Methods Enzymol 189:60–69, 1990.

133. B Stancher, F Zonta. High-performance liquid chromatography of the unsaponifiable from samples of marine and freshwater fish: fractionation and identification of retinol (vitamin A_1) and dehydroretinol (vitamin A_2) isomers. J Chromatogr 287:353–364, 1984.

134. RC Bruening, F Derguini, K Nakanishi. Rapid high-performance liquid chromatographic analysis of retinal mixtures. J Chromatogr 361:437–441, 1986.

135. GN Noll. High-performance liquid chromatographic analysis of retinal and retinol isomers. J Chromatogr A 721:247–259, 1996.

136. K Tsukida, M Ito, T Tanaka, I Yagi. High-performance liquid chromatographic and spectroscopic characterization of stereoisomeric retinal oximes. J Chromatogr 331:265–272, 1985.

137. GM Landers, JA Olson. Statistical solvent optimization for the separation of geometric isomers of retinol by high-performance liquid chromatography. J Chromatogr 291:51–57, 1984.

138. GM Landers, JA Olson. Rapid, simultaneous determination of isomers of retinal, retinal oxime and retinol by high-performance liquid chromatography. J Chromatogr 438:383–392, 1988.

139. CDB Bridges, SL Fong, RA Alvarez. Separation by programmed-gradient high-pressure liquid chromatography of vitamin A isomers, their esters, aldehydes, oximes and vitamin A_2: presence of retinyl ester in dark-adapted goldfish pigment epithelium. Vision Res 20:355–360, 1980.

140. ATC Tsin, RA Alvarez, SL Fong, CDB Bridges. Use of high-performance liquid chromatography in the analysis of retinyl and 3,4-didehydroretinyl compounds in tissue extracts of bullfrog tadpoles and goldfish. Vision Res 24:1835–1840, 1984.

141. R Wyss. Chromatographic and electrophoretic analysis of biomedically important retinoids. J Chromatogr B Biomed Appl 671:381–425, 1995.

142. E Meyer, WE Lambert, AP De Leenheer, JP Bersaques, AH Kint. Improved quantitation of 13-*cis*- and all-*trans*-acitretin in human plasma by normal-phase high-performance liquid chromatography. J Chromatogr 570:149–156, 1991.

143. C Lanvers, G Hempel, G Blaschke, J Boos. Simultaneous determination of all-*trans*-, 13-*cis*- and 9-*cis*-retinoic acid, their 4-oxo metabolites and all-*trans*-retinol in human plasma by high-performance liquid chromatography. J Chromatogr B Biomed Appl 685:233–240, 1996.

144. HC Furr, O Amedee-Manesme, JA Olson. Gradient reversed-phase high-performance liquid chromatographic separation of naturally occurring retinoids. J Chromatogr 309:299–307, 1984.

145. LR Chaudhary, EC Nelson. Separation of vitamin A and retinyl esters by reversed-phase high-performance liquid chromatography. J Chromatogr 294:466–470, 1984.

146. G Guiso, A Rambaldi, B Dimitrova, A Biondi, S Caccia. Determination of orally administered all-*trans*-retinoic acid in human plasma by high-performance liquid chromatography. J Chromatogr B Biomed Appl 656:239–244, 1994.

147. B Periquet, W Lambert, J Garcia, G Lecomte, AP De Leenheer, B Mazieres, JP Thouvenot, J Arlet. Increased concentrations of endogenous 13-*cis*- and all-*trans*-retinoic acids in diffuse idiopathic skeletal hyperostosis, as demonstrated by HPLC. Clin Chim Acta 203:57–65, 1991.

148. P Lefebvre, A Agadir, M Cornic, B Gourmel, B Hue, C Dreux, L Degos, C Chomienne. Simultaneous determination of all-*trans* and 13-*cis* retinoic acids and their 4-oxo metabolites by adsorption liquid chromatography after solid-phase extraction. J Chromatogr B Biomed Appl 666:55–61, 1995.

149. J Van Wauwe, MC Coene, W Cools, J Goossens, W Lauwers, L Le Jeune, C Van Hove, G Van Nyen. Liarozole fumarate inhibits the metabolism of 4-keto-all-*trans*-retinoic acid. Biochem Pharmacol 47:737–741, 1994.

150. K Mizojiri, H Okabe, K Sugeno, Y Esumi, M Takaichi, T Miyake, H Seki, A Inaba. Studies on the metabolism and disposition of the new retinoid 4-[(5,6,7,8-

tetrahydro-5,5,8,8-tetramethyl-2-naphthyl)carbamoyl] benzoic acid. First communication: absorption, distribution, metabolism and excretion after topical application and subcutaneous administration in rats. Arzneimittelforschung 47:59–69, 1997.

151. K Mizojiri, H Okabe, K Sugeno, A Misaki, M Ito, G Kominami, Y Esumi, M Takaichi, T Harada, H Seki, A Inaba. Studies on the metabolism and disposition of the new retinoid 4-[(5,6,7,8-tetrahydro-5,5,8,8-tetramethyl-2-naphthyl)carbamoyl] benzoic acid. 4th communication: absorption, metabolism, excretion and plasma protein binding in various animals and man. Arzneimittelforschung 47:259–269, 1997.

152. S Li, AB Barua, CA Huselton. Quantification of retinoyl-beta-glucuronides in rat urine by reversed-phase high-performance liquid chromatography with ultraviolet detection. J Chromatogr B Biomed Appl 683:155–162, 1996.

153. MD Collins, C Eckhoff, I Chahoud, G Bochert, H Nau. 4-Methylpyrazole partially ameliorated the teratogenicity of retinol and reduced the metabolic formation of all-*trans*-retinoic acid in the mouse. Arch Toxicol 66:652–659, 1992.

154. HM Arafa, FM Hamada, MM Elmazar, H Nau. Fully automated determination of selective retinoic acid receptor ligands in mouse plasma and tissue by reversed-phase liquid chromatography coupled on-line with solid-phase extraction. J Chromatogr A 729:125–136, 1996.

155. FG Larsen, C Vahlquist, E Andersson, H Törma, K Kragballe, A Vahlquist. Oral acitretin in psoriasis: drug and vitamin A concentrations in plasma, skin and adipose tissue. Acta Derm Venereol (Stockh) 72:84–88, 1992.

156. R Wyss, F Bucheli. Quantitative analysis of retinoids in biological fluids by high-performance liquid chromatography using column switching. J Chromatogr 424:303–314, 1988.

157. R Wyss, F Bucheli. Ultra-sensitive coupled-column liquid chromatographic determination of retinoids by direct injection of large plasma volumes and ultraviolet detection. J Pharm Biomed Anal 8:1033–1037, 1990.

158. R Wyss. Determination of retinoids in plasma by high-performance liquid chromatography and automated column switching. Methods Enzymol 189:146–155, 1990.

159. R Wyss, F Bucheli. Use of direct injection precolumn techniques for the high-performance liquid chromatographic determination of the retinoids acitretin and 13-cis-acitretin in plasma. J Chromatogr 593:55–62, 1992.

160. R Wyss, F Bucheli, B Hess. Determination of the arotinoid mofarotene in human, rat and dog plasma by high-performance liquid chromatography with automated column switching and ultraviolet detection. J Chromatogr A 729:315–322, 1996.

161. HJ Cahnmann. A fast photoisomerization method for the preparation of tritium-labeled 9-*cis*-retinoic acid of high specific activity. Anal Biochem 227:49–53, 1995.

162. MT Cavey, B Martin, I Carlavan, B Shroot. In vitro binding of retinoids to the nuclear retinoic acid receptor alpha. Anal Biochem 186:19–23, 1990.

163. B Martin, JM Bernardon, MT Cavey, B Bernard, I Carlavan, B Charpentier, WR Pilgrim, B Shroot, U Reichert. Selective synthetic ligands for human nuclear retinoic acid receptors. Skin Pharmacol 5:57–65, 1992.

164. A Vahlquist, JB Lee, G Michaëlsson, O Rollman. Vitamin A in human skin: II.

Concentrations of carotene, retinol and dehydroretinol in various components of normal skin. J Invest Dermatol 79:94–97, 1982.

165. E Andersson, C Bjorklind, H Torma, A Vahlquist. The metabolism of vitamin A to 3,4-didehydroretinol can be demonstrated in human keratinocytes, melanoma cells and HeLa cells, and is correlated to cellular retinoid-binding protein expression. Biochim Biophys Acta 1224:349–354, 1994.

166. T Seki, S Fujishita, M Ito, N Matsuoka, K Tsukida. Retinoid composition in the compound eyes of insects. Exp Biol 47:95–103, 1987.

167. I Provencio, ER Loew, RG Foster. Vitamin A_2-based visual pigments in fully terrestrial vertebrates. Vision Res 32:2201–2208, 1992.

168. H Torma, D Asselineau, E Andersson, B Martin, P Reiniche, P Chambon, B Shroot, M Darmon, A Vahlquist. Biologic activities of retinoic acid and 3,4-didehydroretinoic acid in human keratinocytes are similar and correlate with receptor affinities and transactivation properties. J Invest Dermatol 102:49–54, 1994.

169. SA Tanumihardjo, HC Furr, JW Erdman Jr, JA Olson. Use of the modified relative dose response (MRDR) assay in rats and its application to humans for the measurement of vitamin A status. Eur J Clin Nutr 44:219–224, 1990.

170. AT Tsin, HA Pedrozo-Fernandez, JM Gallas, JP Chambers. The fluorescence quantum yield of vitamin A_2. Life Sci 43:1379–1384, 1988.

171. F Zonta, B Stancher. High-performance liquid chromatography of retinals, retinols (vitamin A_1) and their dehydro homologues (vitamin A_2): improvements in resolution and spectroscopic characterization of the stereoisomers. J Chromatogr 301: 65–75, 1984.

172. CA O'Neil, SJ Schwartz. Chromatographic analysis of *cis/trans* carotenoid isomers. J Chromatogr 624:235–252, 1992.

173. A Pettersson, L Jonsson. Separation of *cis-trans* isomers of alpha- and beta-carotene by adsorption HPLC and identification with diode array detection. J Micronutr Anal 8:23–41, 1990.

174. AM Stalcup, HL Jin, DW Armstrong, P Mazur, F Derguini, K Nakanishi. Separation of carotenes on cyclodextrin-bonded phases. J Chromatogr 499:627–635, 1990.

175. L Almela, J Lopez-Roca, ME Candela, MD Alcazar. Separation and determination of individual carotenoids in a *Capsicum* cultivar by normal-phase high performance liquid chromatography. J Chromatogr 502:95–106, 1990.

176. L Almela, J Lopez-Roca, ME Candela, MD Alcazar. Carotenoid composition of new cultivars of red pepper for paprika. J Agric Food Chem 39:1606–1609, 1991.

177. L Almela, JA Fernandez-Lopez, JA Lopez-Roca. High-performance liquid chromatography-diode-array detection of photosynthetic pigments. J Chromatogr 607: 215–219, 1992.

178. U Hengartner, K Bernhard, K Meyer, G Englert, E Glinz. Synthesis, isolation, and NMR-spectroscopic characterization of fourteen (Z)-isomers of lycopene. Helv Chim Acta 75:1848–1865, 1992.

179. PD Fraser, M Albrecht, G Sandmann. Development of high-performance liquid chromatographic systems for the separation of radiolabelled carotenes and precursors formed in specific enzymatic reactions. J Chromatogr 645:265–272, 1993.

180. KS Epler, LC Sander, RG Ziegler, SA Wise, NE Craft. Evaluation of reversed-

phase liquid chromatographic columns for recovery and selectivity of selected carotenoids. J Chromatogr 595:89–101, 1992.

181. LC Sander, KE Sharpless, NE Craft, SA Wise. Development of engineered stationary phases for the separation of carotenoid isomers. Anal Chem 66:1667–1674, 1994.

182. C Emenhiser, LC Sander, SJ Schwartz. Capability of a polymeric C-30 stationary phase to resolve cis-trans carotenoid isomers in reversed-phase liquid chromatography. J Chromatogr A 707:205–216, 1995.

183. CM Bell, LC Sander, JC Fetzer, SA Wise. Synthesis and characterization of extended length alkyl stationary phases for liquid chromatography with application to the separation of carotenoid isomers. J Chromatogr A 753:37–45, 1996.

184. KJ Scott, DJ Hart. Further observations on problems associated with the analysis of carotenoids by HPLC. 2. Column temperature. Food Chem 47:403–405, 1993.

185. CM Bell, LC Sander, SA Wise. Temperature dependence of carotenoids on C18, C30 and C34 bonded stationary phases. J Chromatogr 757:29–39, 1997.

186. CA O'Neil, SJ Schwartz, GL Catignani. Comparison of liquid chromatographic methods for the determination of cis-trans isomers of beta-carotene. J Assoc Off Anal Chem 74:36–42, 1991.

187. NE Craft, LC Sander, HF Pierson. Separation and relative distribution of all-trans-beta-carotene and its cis isomers in beta-carotene preparations. J Micronutr Anal 8:209–221, 1990.

188. M Zapata, JL Garrido. Influence of injection conditions in reversed-phase high performance liquid chromatography of chlorophylls and carotenoids. Chromatographia 31:589–594, 1990.

189. KJ Scott. Observations on some of the problems associated with the analysis of carotenoids in foods by HPLC. Food Chem 45:357–364, 1992.

190. HJCF Nelis, AP De Leenheer. Isocratic nonaqueous reversed-phase liquid chromatography of carotenoids. Anal Chem 55:270–275, 1983.

191. LR Cantilena, DW Nierenberg. Simultaneous analysis of five carotenoids in human plasma by isocratic high performance liquid chromatography. J Micronutr Anal 6:127–145, 1989.

192. NI Krinsky, MD Russett, GJ Handelman, DM Snodderly. Structural and geometrical isomers of carotenoids in human plasma. J Nutr 120:1654–1662, 1990.

193. B Olmedilla, F Granado, E Rojas-Hidalgo, I Blanco. A rapid separation of ten carotenoids, three retinoids, alpha-tocopherol and tocopherol acetate by high performance liquid chromatography and its application to serum and vegetable samples. J Liquid Chrom 13:1455–1483, 1990.

194. DW Nierenberg, SL Nann. A method for determining concentrations of retinol, tocopherol, and five carotenoids in human plasma and tissue samples. Am J Clin Nutr 56:417–426, 1992.

195. W Stahl, AR Sundquist, M Hanusch, W Schwarz, H Sies. Separation of beta-carotene and lycopene geometrical isomers in biological samples. Clin Chem 39:810–814, 1993.

196. YM Peng, YS Peng, Y Lin, T Moon, M Baier. Micronutrient concentrations in paired skin and plasma of patients with actinic keratoses: effect of prolonged retinol supplementation. Cancer Epidemiol Biomarkers Prev 2:145–150, 1993.

197. KS Epler, RG Ziegler, NE Craft. Liquid chromatographic method for the determination of carotenoids, retinoids and tocopherols in human serum and in food. J Chromatogr 619:37–48, 1993.

198. S Zeng, HC Furr, JA Olson. Human metabolism of carotenoid analogs and apocarotenoids. Methods Enzymol 214:137–147, 1993.

199. S Oshima, H Sakamoto, Y Ishiguro, J Terao. Accumulation and clearance of capsanthin in blood plasma after the ingestion of paprika juice in men. J Nutr 127: 1475–1479, 1997.

200. LW Levy, E Regalado, S Navarrete, RH Watkins. Bixin and norbixin in human plasma: determination and study of the absorption of a single dose of Annatto food color. Analyst 122:977–980, 1997.

201. W Stahl, H Sies. Uptake of lycopene and its geometrical isomers is greater from heat-processed than from unprocessed tomato juice in humans. J Nutr 122:2161–2166, 1992.

202. C Gartner, W Stahl, H Sies. Lycopene is more bioavailable from tomato paste than from fresh tomatoes. Am J Clin Nutr 66:116–122, 1997.

203. MR Prince, JK Frisoli. Beta-carotene accumulation in serum and skin. Am J Clin Nutr 57:175–181, 1993.

204. AJ Culling-Berglund, SA Newcomb, M Gagne, WS Morfitt, TP Davis. A sensitive and specific procedure for the analysis of beta-carotene in human skin. J Micronutr Anal 5:139–148, 1989.

205. JD Ribaya-Mercado, M Garmyn, BA Gilchrest, RM Russell. Skin lycopene is destroyed preferentially over beta-carotene during ultraviolet irradiation in humans. J Nutr 125:1854–1859, 1995.

206. GJ Handelman, DM Snodderly, NI Krinsky, MD Russett, AJ Adler. Biological control of primate macular pigment. Biochemical and densitometric studies. Invest Ophthalmol Vis Sci 32:257–267, 1991.

207. RA Bone, JT Landrum. Distribution of macular pigment components, zeaxanthin and lutein, in human retina. Methods Enzymol 213:360–366, 1992.

208. RA Bone, JT Landrum, GW Hime, A Cains, J Zamor. Stereochemistry of the human macular carotenoids. Invest Ophthalmol Vis Sci 34:2033–2040, 1993.

209. RA Bone, JT Landrum, LM Friedes, CM Gomez, MD Kilburn, E Menendez, I Vidal, W Wang. Distribution of lutein and zeaxanthin stereoisomers in the human retina. Exp Eye Res 64:211–218, 1997.

210. F Khachik, PS Bernstein, DL Garland. Identification of lutein and zeaxanthin oxidation products in human and monkey retinas. Invest Ophthalmol Vis Sci 38:1802–1811, 1997.

211. KJ Yeum, A Taylor, G Tang, RM Russell. Measurement of carotenoids, retinoids, and tocopherols in human lenses. Invest Ophthalmol Vis Sci 36:2756–2761, 1995.

212. GJ Handelman, DM Snodderly, AJ Adler, MD Russett, EA Dratz. Measurement of carotenoids in human and monkey retinas. Methods Enzymol 213:220–230, 1992.

213. G Scita, GW Aponte, G Wolf. Uptake and cleavage of beta-carotene by cultures of rat small intestinal cells and human lung fibroblasts. J Nutr Biochem 3:118–123, 1992.

214. YS Peng, YM Peng. Simultaneous liquid chromatographic determination of carotenoids, retinoids, and tocopherols in human buccal mucosal cells. Cancer Epidemiol Biomarkers Prev 1:375–382, 1992.

215. T Murata, H Tamai, T Morinobu, M Manago, A Takenaka, H Takenaka, M Mino. Determination of beta-carotene in plasma, blood cells and buccal mucosa by electrochemical detection. Lipids 27:840–843, 1992.

216. AR Giuliano, EM Neilson, BE Kelly, LM Canfield. Simultaneous quantitation and separation of carotenoids and retinol in human milk by high-performance liquid chromatography. Methods Enzymol 213:391–399, 1992.

217. F Khachik, CJ Spangler, JC Smith Jr, LM Canfield, A Steck, H Pfander. Identification, quantification, and relative concentrations of carotenoids and their metabolites in human milk and serum. Anal Chem 69:1873–1881, 1997.

218. RS Parker. Analysis of carotenoids in human plasma and tissues. Methods Enzymol 214:86–93, 1993.

219. HH Schmitz, CL Poor, ET Gugger, JW Erdman Jr. Analysis of carotenoids in human and animal tissues. Methods Enzymol 214:102–116, 1993.

220. W Stahl, W Schwarz, AR Sundquist, H Sies. cis-trans Isomers of lycopene and beta-carotene in human serum and tissues. Arch Biochem Biophys 294:173–177, 1992.

221. HH Schmitz, CL Poor, RB Wellman, JW Erdman. Concentrations of selected carotenoids and vitamin A in human liver, kidney and lung tissue. J Nutr 121:1613–1621, 1991.

222. LA Kaplan, JM Lau, EA Stein. Carotenoid composition, concentrations, and relationships in various human organs. Clin Physiol Biochem 8:1–10, 1990.

223. TL Bierer, NR Merchen, JW Erdman Jr. Comparative absorption and transport of five common carotenoids in preruminant calves. J Nutr 125:1569–1577, 1995.

224. CL Poor, TL Bierer, NR Merchen, GC Fahey, JW Erdman. The accumulation of alpha- and beta-carotene in serum and tissues of preruminant calves fed raw and steamed carrot slurries. J Nutr 123:1296–1304, 1993.

225. JD Ribaya-Mercado, SC Holmgren, JG Fox, RM Russell. Dietary beta-carotene absorption and metabolism in ferrets and rats. J Nutr 119:665–668, 1989.

226. JD Ribaya-Mercado, JG Fox, WD Rosenblad, MC Blanco, RM Russell. Beta-carotene, retinol and retinyl ester concentrations in serum and selected tissues of ferrets fed beta-carotene. J Nutr 122:1898–1903, 1992.

227. WS White, KM Peck, EA Ulman, JW Erdman Jr. The ferret as a model for evaluation of the bioavailabilities of all-trans-beta-carotene and its isomers. J Nutr 123:1129–1139, 1993.

228. WS White, KM Peck, EA Ulman, JW Erdman Jr. Evaluation of the bioavailability of natural and synthetic forms of beta-carotenes in a ferret model. Ann NY Acad Sci 691:229–231, 1993.

229. JR Zhou, ET Gugger, JW Erdman Jr. The crystalline form of carotenes and the food matrix in carrot root decrease the relative bioavailability of beta- and alpha-carotene in the ferret model. J Am Coll Nutr 15:84–91, 1996.

230. M Oarada, W Stahl, H Sies. Cellular levels of all-trans-beta-carotene under the influence of 9-cis-beta-carotene in FU-5 rat hepatoma cells. Biol Chem Hoppe Seyler 374:1075–1081, 1993.

231. T Maoka, T Komori, T Matsuno. Direct diastereomeric resolution of carotenoids. J Chromatogr 318:122–124, 1985.

232. SA Turujman. Rapid direct resolution of the stereoisomers of all-*trans*-astaxanthin on a Pirkle covalent L-leucine column. J Chromatogr 631:197–199, 1993.

233. K Schiedt, S Bischof, E Glinz. Metabolism of carotenoids and in vivo racemization of (3S, 3′S)-astaxanthin in the crustacean *Penaeus*. Methods Enzymol 214:148–168, 1993.

234. J Hudon, AH Brush. Identification of carotenoid pigments in birds. Methods Enzymol 213:312–321, 1992.

235. R Stradi, G Celentano, K Schiedt. Carotenoids in birds' plumage. Comp Biochem Physiol 1:131–143, 1994.

236. XD Wang, GW Tang, JG Fox, NI Krinsky, RM Russell. Enzymatic conversion of beta-carotene into beta-apo-carotenals and retinoids by human, monkey, ferret, and rat tissues. Arch Biochem Biophys 285:8–16, 1991.

237. GW Tang, XD Wang, RM Russell, NI Krinsky. Characterization of beta-apo-13-carotenone and beta-apo-14′-carotenal as enzymatic products of the excentric cleavage of beta-carotene. Biochemistry 30:9829–9834, 1991.

238. XD Wang, NI Krinsky, GW Tang, RM Russell. Retinoic acid can be produced from excentric cleavage of beta-carotene in human intestinal mucosa. Arch Biochem Biophys 293:298–304, 1992.

239. A Nagao, JA Olson. Enzymatic formation of 9-*cis*, 13-*cis*, and all-*trans* retinals from isomers of beta-carotene. FASEB J 8:968–973, 1994.

240. X Hebuterne, XD Wang, EJ Johnson, NI Krinsky, RM Russell. Intestinal absorption and metabolism of 9-*cis*-beta-carotene in vivo: biosynthesis of 9-*cis*-retinoic acid. J Lipid Res 36:1264–1273, 1995.

241. F Khachik, MB Goli, GR Beecher, J Holden, WR Lusby, MD Tenorio, MR Barrera. Effect of food preparation on qualitative and quantitative distribution of major carotenoid constituents of tomatoes and several green vegetables. J Agric Food Chem 40:390–398, 1992.

242. RS Rouseff, GD Sadler, TJ Putnam, JE Davis. Determination of beta-carotene and other hydrocarbon carotenoids in red grapefruit cultivars. J Agric Food Chem 40:47–51, 1992.

243. MP Cano. HPLC separation of chlorophyll and carotenoid pigments of four kiwi fruit cultivars. J Agric Food Chem 39:1786–1791, 1991.

244. A Homnava, J Payne, P Koehler, R Eitenmiller. Provitamin A (alpha-carotene, beta-carotene and beta-cryptoxanthin) and ascorbic acid content of Japanese and American persimmons. J Food Qual 13:85–95, 1990.

245. MI Minguez-Mosquera, D Hornero-Mendez. Separation and quantification of the carotenoid pigments in red peppers (*Capsicum annuum* L.), paprika, and oleoresin by reversed-phase HPLC. J Agric Food Chem 41:1616–1620, 1993.

246. MI Minguez-Mosquera, D Hornero-Mendez. Formation and transformation of pigments during the fruit ripening of *Capsicum annum* Cv Bola and Agridulce. J Agric Food Chem 42:38–44, 1994.

247. MI Minguez-Mosquera, D Hornero-Mendez. Changes in carotenoid esterification during the fruit ripening of *Capsicum annuum* Cv Bola. J Agric Food Chem 42:640–644, 1994.

248. M Weissenberg, I Schaeffler, E Menagem, M Barzilai, A Levy. Isocratic non-aqueous reversed-phase high-performance liquid chromatographic separation of capsanthin and capsorubin in red peppers (*Capsicum annum* L.), paprika and oleoresin. J Chromatogr A 757:89–95, 1997.

249. G Zachariev, I Kiss, J Szabolcs, G Toth, P Molnar, Z Matus. HPLC analysis of carotenoids in irradiated and ethylene oxide treated red pepper. Acta Aliment 20: 115–122, 1991.

250. Z Matus, J Deli, J Szabolcs. Carotenoid composition of yellow pepper during ripening: isolation of beta-cryptoxanthin 5,6-epoxide. J Agric Food Chem 39:1907–1914, 1991.

251. MI Minguez-Mosquera, B Gandul-Rojas, A Montano-Asquerino, J Garrido-Fernandez. Determination of chlorophylls and carotenoids by high performance liquid chromatography during olive lactic fermentation. J Chromatogr 585:259–266, 1991.

252. MI Minguez-Mosquera, B Gandul-Rojas, ML Gallardo-Guerrero. Rapid method of quantitation of chlorophylls and carotenoids in virgin olive oil by high-performance liquid chromatography. J Agric Food Chem 40:60–63, 1991.

253. SS Thayer, O Bjorkman. Leaf xanthophyll content and composition in sun and shade determined by HPLC. Photosynth Res 23:331–343, 1990.

254. BV Pfund, AM Bond, TC Hughes. Simple voltammetric method for the determination of beta-carotene in brine and soya oil samples at mercury and glassy carbon electrodes. Analyst 117:857–861, 1992.

255. B Czeczuga, IF Skirina, CB Maximov, LS Stepanenko. Investigations on carotenoids in lichens. XIX. Carotenoids in lichens of the tundra of Kamchatka region (Far East). Phyton 29:7–13, 1989.

256. B Czecuzuga, S Caccamese, MV Passadore. Investigations on carotenoids in lichens. XX. Carotenoids in lichens from various Italian environments. Phyton 29: 15–22, 1989.

257. B Czeczuga. Investigations on carotenoids in lichens. XXII. Lichens from the upper Tracja valley (Bulgaria). Acta Soc Botan Polon 57:447–456, 1988.

258. B Czeczuga, M Olech. Investigations on carotenoids in lichens. XXIV. Further studies of carotenoids in lichens of the Antarctica. Ser Cient INACH 39:91–96, 1989.

259. B Czeczuga, M Olech. Investigations on carotenoids in lichens. XXV. Studies of carotenoids in lichens from Spitsbergen. Phyton 30:235–245, 1990.

260. AP De Leenheer, HJ Nelis. Profiling and quantitation of carotenoids by high-performance liquid chromatography and photodiode array detection. Methods Enzymol 213:251–265, 1992.

261. NI Bishop, T Urbig, H Senger. Complete separation of the beta, epsilon- and beta, beta-carotenoid biosynthetic pathways by a unique mutation of the lycopene cyclase in the green alga, *Scenedesmus obliquus*. FEBS Lett 367:158–162, 1995.

262. T Sasa, S Takaichi, N Hatakeyama, MM Watanabe. A novel carotenoid ester, loroxanthin dodecenoate, from *Pyramimonas parkeae* (Prasinophyceae) and a chlorarachniophycean alga. Plant Cell Physiol 33:921–925, 1992.

263. S Okada, I Tonegawa, H Matsuda, M Murakami, K Yamaguchi. Braunixanthins 1 and 2, new carotenoids from the green microalga *Botryococcus brauni*. Tetrahedron 53:11307–11316, 1997.

264. S Takaichi, J Ishundu. Carotenoid glycoside ester from *Rhodococcus rhodochrous*. Methods Enzymol 213:366–374, 1992.

265. S Takaichi, K Shimada. Characterization of carotenoids in photosynthetic bacteria. Methods Enzymol 213:374–385, 1992.

266. S Takaichi, K Inoue, M Akaike, M Kobayashi, H Oh-oka, MT Madigan. The major carotenoid in all known species of heliobacteria is the C30 carotenoid 4,4'-diapo-neurosporene, not neurosporene. Arch Microbiol 168:277–281, 1997.

267. S Takaichi, Z Wang, M Umetsu, T Nozawa, K Shimada, MT Madigan. New carotenoids from the thermophilic green sulfur bacterium *Chlorobium tepidum*: 1',2'-dihydro-beta-carotene, 1',2'-dihydrochlorobactene and hydroxychlorobactene glucoside ester, and the carotenoid composition of different strains. Arch Microbiol 168:270–276, 1997.

268. DR Kull, H Pfander. Isolation and structure elucidation of carotenoid glycosides from the thermoacidophilic Archaea *Sulfolobus shibatae*. J Nat Prod 60:371–374, 1997.

269. T Maoka, K Mochida, Y Okuda, Y Ito, T Fujiwara. A novel purple carotenoid, rhodobacterioxanthin, from *Rhodobacter capsulatus*. Chem Pharm Bull 45:1225–1227, 1997.

270. BS Hundle, P Beyer, H Kleinig, G Englert, JE Hearst. Carotenoids of *Erwinia herbicola* and an *Escherichia coli* HB101 strain carrying the *Erwinia herbicola* carotenoid gene cluster. Photochem Photobiol 54:89–93, 1991.

271. S Takaichi, K Tsuji, K Matsuura, K Shimada. A monocyclic carotenoid glucoside ester is a major carotenoid in the green filamentous bacterium *Chloroflexus aurantiacus*. Plant Cell Physiol 36:773–778, 1995.

272. J Lorquin, F Moluba, BL Dreyfus. Identification of the carotenoid pigment canthaxanthin from photosynthetic *Bradyrhizobium* strains. Appl Environ Microbiol 63:1151–1154, 1997.

273. B Czeczuga. Investigations on carotenoids in insects. X. Changes in the carotenoids in butterflies (Lepidoptera). Folia Biol (Krakow) 38:5–12, 1990.

274. T Matsuno, T Maoka, Y Toriiminami. Carotenoids in the Japanese stick insect *Neophirasea japonica*. Comp Biochem Physiol [B] 95B:583–587, 1990.

275. J Grayson, M Edmunds, EH Evans, G Britton. Carotenoids and coloration of poplar hawkmoth caterpillar (*Lathoe populi*). Biol J Linnean Soc 42:457–465, 1991.

276. HC Furr, AB Barua, JA Olson. Analytical methods. In: MB Sporn, AB Roberts, DS Goodman, eds. The Retinoids. 2nd ed. New York: Raven Press, 1994, pp 179–209.

277. M Vecchi, W Vetter, W Walther, SF Jermstad, GW Schutt. Gas-chromatographische und massenspektrometrische Untersuchung der Trimethylsilylather von Vitamin A und einigen seiner Isomeren. Helv Chim Acta 50:1243–1248, 1967.

278. CR Smidt, AD Jones, AJ Clifford. Gas chromatography of retinol and alpha-tocopherol without derivatization. J Chromatogr 434:21–29, 1988.

279. HC Furr, AJ Clifford, AD Jones. Analysis of apocarotenoids and retinoids by capillary gas chromatography–mass spectrometry. Methods Enzymol 213:281–290, 1992.

280. HJ Egger, UB Ranalder, EU Koelle, M Klaus. Determination of the new retinoic

acid Ro 13-7410 in plasma by two-dimensional gas chromatography–negative ion chemical ionization mass spectrometry with selected-ion monitoring. Biomed Environ Mass Spec 18:453–463, 1989.

281. UB Ranalder, BB Lausecker, C Huselton. Micro liquid chromatography-mass spectrometry with direct liquid introduction used for separation and quantitation of all-*trans*- and 13-*cis*-retinoic acids and their 4-oxo metabolites in human plasma. J Chromatogr 617:129–135, 1993.

282. BE Fayer, CA Huselton, WA Garland, DJ Liberato. Quantification of acitretin in human plasma by microbore liquid chromatography–negative chemical ionization mass spectrometry. J Chromatogr 568:135–144, 1991.

283. JL Napoli. Quantification of physiological levels of retinoic acid. Methods Enzymol 123:112–124, 1986.

284. J Buck, G Ritter, L Dannecker, V Katta, SL Cohen, BT Chait, U Hammerling. Retinol is essential for growth of activated human B cells. J Exp Med 171:1613–1624, 1990.

285. MR Lakshman, I Mychkovsky, M Attlesey. Enzymatic conversion of all-*trans*-beta-carotene to retinal by a cytosolic enzyme from rabbit and rat intestinal mucosa. Proc Natl Acad Sci USA 86:9124–9128, 1989.

286. MR Lakshman, C Okoh. Enzymatic conversion of all-*trans*-beta-carotene to retinal. Methods Enzymol 214:256–269, 1993.

287. AB Barua. Analysis of water-soluble compounds: glucuronides. Methods Enzymol 189:136–145, 1990.

288. HC Furr, SH Zeng, AJ Clifford, JA Olson. Capillary gas chromatography of retinoids (vitamin A compounds) and apo-retinoids: determination of Kovats retention indices. J Chromatogr 527:406–413, 1990.

289. S Hashimoto, M Mizobuchi, T Kuroda, H Okabe, K Mizojiri, S Takahashi, J Kikuchi, Y Terui. Biotransformation of a new synthetic retinoid, 4-[5,6,7,8-tetrahydro-5,5,8,8-tetramethyl-2-naphthyl)carbamoyl]benzoic acid (Am-80), in the rat. Structure elucidation of the metabolites by mass and NMR spectrometry. Xenobiotica 24:1177–1193, 1994.

290. D von Reinersdorff, E Bush, DJ Liberato. Plasma kinetics of vitamin A in humans after a single oral dose of [8,9,19-13C]retinyl palmitate. J Lipid Res 37:1875–1885, 1996.

291. AJ Clifford, AD Jones, HC Furr. Stable isotope dilution mass spectrometry to assess vitamin A status. Methods Enzymol 189:94–104, 1990.

292. GJ Handelman, MJ Haskell, AD Jones, AJ Clifford. An improved protocol for determining ratios of retinol-d4 to retinol isolated from human plasma. Anal Chem 65:2024–2028, 1993.

293. SR Dueker, AD Jones, GM Smith, AJ Clifford. Stable isotope methods for the study of beta-carotene-d8 metabolism in humans utilizing tandem mass spectrometry and high-performance liquid chromatography. Anal Chem 66:4177–4185, 1994.

294. JA Novotny, SR Dueker, LA Zech, AJ Clifford. Compartmental analysis of the dynamics of beta-carotene metabolism in an adult volunteer. J Lipid Res 36:1825–1838, 1995.

295. G Tang, BA Andrien, GG Dolnikowski, RM Russell. Atmospheric pressure chemical ionization and electron capture negative chemical ionization mass spectrometry

in studying beta-carotene conversion to retinol in humans. Methods Enzymol 282: 140–154, 1997.

296. CS You, RS Parker, KJ Goodman, JE Swanson, TN Corso. Evidence of *cis-trans* isomerization of 9-*cis*-beta-carotene during absorption in humans. Am J Clin Nutr 64:177–183, 1996.

297. RS Parker, JT Brenna, JE Swanson, KJ Goodman, B Marmor. Assessing metabolism of beta-[^{13}C] carotene using high-precision isotope ratio mass spectrometry. Methods Enzymol 282:130–140, 1997.

298. Y Liang, WS White, L Yao, RE Serfass. Use of high precision gas isotope ratio mass spectrometry to determine ^{13}C in lutein isolated from C3 and C4 plant sources. J Chromatogr 800:51–58, 1998.

299. CA Huselton, BE Fayer, WA Garland, DJ Liberato. Quantification of endogenous retinoic acid in human plasma by liquid chromatography/mass spectrometry. In: MA Brown, ed. Liquid Chromatography/Mass Spectrometry. Applications in Agricultural, Pharmaceutical, and Environmental Chemistry. Washington: American Chemical Society, 1990, pp 166–178.

300. JRF Muindi, SR Frankel, C Huselton, F DeGrazia, WA Garland, CW Young, RP Warrell Jr. Clinical pharmacology of oral all-*trans* retinoic acid in patients with acute promyelocytic leukemia. Cancer Res 52:2138–2142, 1992.

301. F Rubio, BK Jensen, L Henderson, WA Garland, A Szuna, C Town. Disposition of [C-14] acitretin in humans following oral administration. Drug Metab Dispos 22:211–215, 1994.

302. DK Bempong, IL Honigberg, NM Meltzer. Normal phase LC-MS determination of retinoic acid degradation products. J Pharm Biomed Anal 13:285–291, 1995.

303. M Careri, MT Lugari, A Mangia, P Manini, S Spagnoli. Identification of vitamin A, vitamin D and vitamin E by particle-beam liquid chromatography mass spectrometry. Fresenius J Anal Chem 351:768–776, 1995.

304. PA Lehman, TJ Franz. A sensitive high-pressure liquid chromatography/particle beam/mass spectrometry assay for the determination of all-*trans*-retinoic acid and 13-*cis*-retinoic acid in human plasma. J Pharm Sci 85:287–290, 1996.

305. C Eckhoff, W Wittfoht, H Nau, W Slikker. Characterization of oxidized and glucuronidated metabolites of retinol in monkey plasma by thermospray liquid chromatography mass spectrometry. Biomed Environ Mass Spectrom 19:428–433, 1990.

306. JC Kraft, W Slikker, JR Bailey, LG Roberts, B Fischer, W Wittfoht, H Nau. Plasma pharmacokinetics and metabolism of 13-*cis* retinoic and all-*trans* retinoic acid in the cynomolgus monkey and the identification of 13-*cis* retinoyl beta-glucuronides—a comparison to one human case-study with isotretinoin. Drug Metab Dispos 19:317–324, 1991.

307. S Cotler, D Chang, L Henderson, W Garland, C Town. The metabolism of acitretin and isoacitretin in the in situ isolated perfused rat liver. Xenobiotica 22:1229–1237, 1992.

308. RB van Breemen, CR Huang. High-performance liquid chromatography electrospray mass spectrometry of retinoids. FASEB J 10:1098–1101, 1996.

309. MA Shirley, P Wheelan, SR Howell, RC Murphy. Oxidative metabolism of a rexinoid and rapid phase II metabolite identification by mass spectrometry. Drug Metab Dispos 25:1144–1149, 1997.

310. RB van Breemen, D Nikolic, X Xu, Y Xiong, M van Lieshout, CE West, AB Schilling. Development of a method for quantitation of retinol and retinyl palmitate in human serum using high performance liquid chromatography–atmospheric pressure chemical ionization mass spectrometry. J Chromatogr A 794:245–251, 1998.

311. R Andreoli, M Careri, P Manini, G Mori, M Musci. HPLC analysis of fat-soluble vitamins on standard and narrow-bore columns with UV, electrochemical and particle beam MS detection Chromatographia 44:605–612, 1997.

312. T Wingerath, D Kirsch, B Spengler, R Kaufmann, W Stahl. High-performance liquid chromatography and laser desorption/ionization mass spectrometry of retinyl esters. Anal Chem 69:3855–3860, 1997.

313. KA Humphries, RW Curley Jr. Triplet-sensitized photooxygenation of therapeutic retinoids. Pharm Res 8:826–831, 1991.

314. KL Salyers, ME Cullum, MH Zile. Glucuronidation of all-*trans*-retinoic acid in liposomal membranes. Biochim Biophys Acta 1152:328–334, 1993.

315. GE Eldred, MR Lasky. Retinal age pigments generated by self-assembling lysosomatropic detergents. Nature 361:724–726, 1993.

316. S Caccamese, D Garozzo. Odd-electron molecular ion and loss of toluene in fast atom bombardment mass spectra of some carotenoids. Org Mass Spectrom 25:137–140, 1990.

317. HH Schmitz, RB van Breemen, SJ Schwartz. Applications of fast atom bombardment mass spectrometry (FAB-MS) and continuous-flow FAB-MS to carotenoid analysis. Methods Enzymol 213:322–336, 1992.

318. RB van Breemen, HH Schmitz, SJ Schwartz. Fast atom bombardment tandem mass spectrometry of carotenoids. J Agric Food Chem 43:384–389, 1995.

319. R Kaufmann, T Wingerath, D Kirsch, W Stahl, H Sies. Analysis of carotenoids and carotenol fatty acid esters by matrix-assisted laser desorption ionization (MALDI) and MALDI-post-source-decay mass spectrometry. Anal Biochem 238:117–128, 1996.

320. T Wingerath, W Stahl, D Kirsch, R Kaufmann, H Sies. Fruit juice carotenol fatty acid esters and carotenoids as identified by matrix-assisted laser desorption ionization (MALDI) mass spectrometry. J Agric Food Chem 44:2006–2013, 1996.

321. RB van Breemen. Innovations in carotenoid analysis using LC/MS. Anal Chem 68:299A–304A, 1996.

322. TD McClure, DC Liebler. A rapid method for profiling the products of antioxidant reactions by negative ion chemical ionization mass spectrometry. Chem Res Toxicol 8:128–135, 1995.

323. TD McClure, DC Liebler. Electron capture negative chemical ionization mass spectrometry and tandem mass spectrometry analysis of beta-carotene, alpha-tocopherol and their oxidation products. J Mass Spectrom 30:1480–1488, 1995.

324. RB van Breemen. Liquid chromatography/mass spectrometry of carotenoids. Pure Appl Chem 69:2061–2066, 1997.

325. RF Taylor, PE Farrow, LM Yelle, JC Harris, IG Marenchic. Advances in HPLC and HPLC-MS of carotenoids and retinoids. In: NI Krinsky, MM Mathews-Roth, RF Taylor, eds. Carotenoids: Chemistry and Biology. New York: Plenum Press, 1990, pp 105–123.

326. F Khachik, G Englert, GR Beecher, JC Smith Jr. Isolation, structural elucidation,

and partial synthesis of lutein dehydration products in extracts from human plasma. J Chromatogr B Biomed Appl 670:219–233, 1995.

327. M Careri, A Mangia, P Manini, N Taboni. Determination of phylloquinone (vitamin K_1) by high performance liquid chromatography with UV detection and with particle beam-mass spectrometry. Fresenius J Anal Chem 355:48–56, 1996.

328. RB van Breemen, HH Schmitz, SJ Schwartz. Continuous-flow fast atom bombardment liquid chromatography/mass spectrometry of carotenoids. Anal Chem 65: 965–969, 1993.

329. RB van Breemen. Electrospray liquid chromatography-mass spectrometry of carotenoids. Anal Chem 67:2004–2009, 1995.

330. GJ van Berkel, F Zhou. Chemical electron-transfer reactions in electrospray mass spectrometry: effective oxidation potentials of electron-transfer reagents in methylene chloride. Anal Chem 66:3408–3415, 1994.

331. RB van Breemen, CR Huang, YC Tan, LC Sander, AB Schilling. Liquid chromatography/mass spectrometry of carotenoids using atmospheric pressure chemical ionization. J Mass Spectrom 31:975–981, 1996.

332. DC Liebler, TD McClure. Antioxidant reactions of beta-carotene: identification of carotenoid-radical adducts. Chem Res Toxicol 9:8–11, 1996.

333. R Buchecker, K Noack. Circular dichroism. In: G Britton, S Liaaen-Jensen, H Pfander, eds. Carotenoids. 1B: Spectroscopy. Basel: Birkhauser Verlag, 1995, pp 63–116.

334. K Bernhard, M Grosjean. Infrared spectroscopy. In: G Britton, S Liaaen-Jensen, H Pfander, eds. Carotenoids. 1B: Spectroscopy. Basel: Birkhauser Verlag, 1995, pp 117–134.

335. Y Koyama. Resonance Raman spectroscopy. In: G Britton, S Liaaen-Jensen, H Pfander, eds. Carotenoids. 1B: Spectroscopy. Basel: Birkhauser Verlag, 1995, pp 135–146.

336. G Englert. NMR Spectroscopy. In: G Britton, S Liaaen-Jensen, H Pfander, eds. Carotenoids. 1B: Spectroscopy. Basel: Birkhauser Verlag, 1995, pp 147–260.

337. F Mo. X-ray crystallographic studies. In: G Britton, S Liaaen-Jensen, H Pfander, eds. Carotenoids. 1B: Spectroscopy. Basel: Birkhauser Verlag, 1995, pp 321–342.

338. TC Bakker Schut, GJ Puppels, YM Kraan, J Greve, LL van der Maas, CG Figdor. Intracellular carotenoid levels measured by Raman microspectroscopy: comparison of lymphocytes from lung cancer patients and healthy individuals. Int J Cancer 74: 20–25, 1997.

339. RB Ramanauskaite, IGMJ Segers-Nolten, KJ de Grauw, NM Sijtsema, L van der Maas, J Greve, C Otto, CG Figdor. Carotenoid levels in human lymphocytes, measured by Raman microspectroscopy. Pure Appl Chem 69:2131–2134, 1997.

340. S Liaaen-Jensen. Combined approach: identification and structure elucidation of carotenoids. In: G Britton, S Liaaen-Jensen, H Pfander, eds. Carotenoids. 1B: Spectroscopy. Basel: Birkhauser Verlag, 1995, pp 343–354.

2
Vitamin Ds: Metabolites and Analogs

Glenville Jones
Queen's University, Kingston, Ontario, Canada

Hugh L. J. Makin
St. Bartholomew's and the Royal London School of Medicine and Dentistry, Whitechapel Campus, London, England

I. INTRODUCTION

The demonstration over three decades ago (1) of the conversion of vitamin D_3 into 25-hydroxyvitamin D_3 (25-OH-D_3) triggered a dramatic surge of interest in the basic biochemistry and physiology of vitamin D. The resultant elucidation of the metabolism and role of vitamin D in calcium homeostasis (1–5) has led to a change of emphasis from the study of the parent vitamin itself to the study of its metabolites. Older methods of bioassay, colorimetric, and ultraviolet (UV) absorption assay have been replaced by newer techniques based upon chromatography, radioligand procedures, and mass spectrometry (reviewed in 6,7). Present methods separate and specifically measure a wide variety of metabolites and chemical analogs instead of a heterogeneous "vitamin D" group. The increased sensitivity and specificity offered by modern assays have enabled the measurement of picogram quantities of vitamin D metabolites, thereby permitting these techniques to be applied to the analysis of vitamin D and its metabolites in physiological fluids. Chemical synthesis of vitamin D metabolites and analogs for clinical trials has led to improvements in the separation and analysis of the vitamin D family of compounds, which has also been beneficial to analytical chemists in the pharmaceutical and food industries.

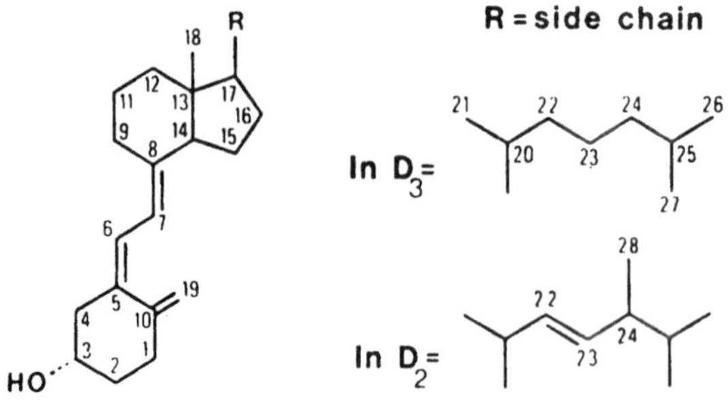

Figure 1 Basic formulae of vitamin D compounds.

A. Basic Formulae

Vitamin D and its metabolites are a group of 9,10-secosteroids with the basic structure shown in Figure 1. The A ring of the steroid nucleus is rotated about the C6 position. The conjugated *cis*-triene system gives rise to the characteristic UV absorption spectrum of vitamin D (molar extinction coefficient = 18,300; λ_{max} = 264 nm; λ_{min^-} = 228 nm). The D_3 series has a side chain derived from cholesterol; the D_2 series has a side chain derived from ergosterol containing an additional C22(23) double bond and a C24 methyl group.

B. Chemistry

Vitamin D_2 and vitamin D_3 are derived by photoirradiation from their respective 5,7-diene sterol precursors: 7-dehydrocholesterol or ergosterol. The intact sterol precursor, known as a provitamin, undergoes photolysis when exposed to UV light of wavelength 280 to 320 nm to yield a variety of photoirradiation products, the principal ones being previtamin D, tachysterol, and lumisterol, the structures of which are illustrated in Figure 2. The previtamin D undergoes spontaneous thermal rearrangements to vitamin D (8). Vitamin D has a number of other closely related isomers and derivatives that have little biological activity but can be formed during synthesis, derivatization, or handling.

   The isolation and identification of the metabolites of vitamin D have been followed by a keen interest in the chemical synthesis of the D vitamins. Clinical use of these new compounds has stimulated the development of more efficient syntheses (9). A popular approach to the synthesis of vitamin D metabolites involves the preparation of a suitably hydroxylated or substituted provitamin, conversion of this to the equivalent previtamin, followed by thermal isomerisation to the vitamin D, but other sophisticated procedures have been used where the molecule is synthesised in two halves (10–13).

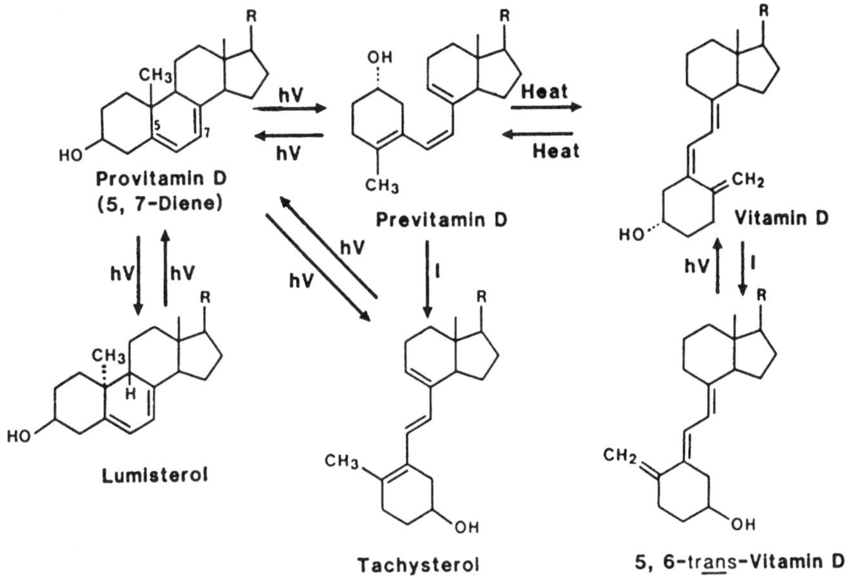

Figure 2 Irradiation products of pro-vitamin D.

C. Pro-drugs and Analogs

1. Pro-drugs

Table 1 lists some of the important pro-drugs of vitamin D. All of these compounds require a step (or more) of activation in vivo before they are biologically active. Included there is vitamin D_2, which is derived from the plant sterol, ergosterol, by irradiation. Since vitamin D_2 is found rarely in nature and is hard to detect in humans eating non-fortified foods, it can be considered to be an artificial form of vitamin D or pro-drug. Vitamin D_2 possesses two specific modifications of the side chain (see Table 1) but is still able to undergo the same series of activation steps as vitamin D_3, giving rise to 25-OH-D_2, $1\alpha,25$-$(OH)_2D_2$, and $24,25$-$(OH)_2D_2$. Two other pro-drugs, 1α-OH-D_3 and 1α-OH-D_2, were synthesized in the early 1970s (14,15) as alternative sources of $1\alpha,25$-$(OH)_2D_3$ and $1\alpha,25$-$(OH)_2D_2$, respectively, and in the process circumvent the renal 1α-hydroxylase enzyme, which was shown to be tightly regulated and prone to damage in renal disease.

The final compound in the list, dihydrotachysterol (DHT), has had a complex history as a pro-drug. Originally it was believed to be "active" when converted to 25-OH-DHT by virtue of an A-ring rotated through 180° such that 3β-hydroxyl function assumes a pseudo-1α-hydroxyl position (16). The mechanism of action of DHT has become less clear with the description of the extrarenal

Table 1   Vitamin D Pro-drugs

Vitamin D pro-drugs [ring structure][a]	Side chain structure (R)	Company	Possible target diseases	Mode of delivery	Reference
1α-OH-D₃ [3]		Leo	Osteoporosis	Systemic	14
1α-OH-D₂ [3] (Hectorol)		Bone Care Int.	Osteoporosis Hyperpara-thyroidism	Systemic	15
Dihydro-tachysterol₂ [2]		Duphar	Renal failure	Systemic	16, 17
Vitamin D₂ [1]		Various	Rickets Osteomalacia	Systemic Systemic	46

[a] Ring structures are indicated by the number in brackets and refer to the structures illustrated at the top of Table 2.
Source: Ref. 171.

metabolism of 25-OH-DHT to 1α,25-(OH)₂DHT and 1β,25-(OH)₂DHT, two further metabolites that have greater biological activity than either 25-OH-DHT or DHT itself (17). Both these 1-hydroxylated metabolites have been demonstrated to be present in human plasma after oral administration of DHT₂ (18).

2. Analogs

Table 2 lists some of the most promising analogs of 1α,25-(OH)₂D₃ already approved by governmental agencies or currently under development by various industrial and/or university research groups. Since the number of vitamin D analogs synthesized approaches the hundreds, the table is provided mainly to give a flavor of the structures experimented with, the thus far worldwide nature of the companies involved, and the broad spectrum of target diseases and uses.

The first generation of calcitriol analogs included molecules with fluorine atoms placed at metabolically vulnerable positions in the side chain and resulted in highly stable and potent "calcemic" agents such as 26,27-F6-1α,25-(OH)₂D₃. More recently, attention has focused on features that make the molecule more susceptible to clearance, such as in calcipotriol (MC903), where a C22=C23 double bond, a 24-hydroxyl function, and a cyclopropane ring have been introduced into the side chain or in 22-oxocalcitriol (OCT) where the 22-carbon has been replaced with an oxygen atom. Both modifications have given rise to highly promising analogs (19,20).

The C24 position is a favorite site for modification and numerous analogs

Table 2 Analogs of 1,25-(OH)$_2$D$_3$

Vitamin D analog (Ring structure)[a]	Side chain structure (R)	Company	Possible target diseases	Mode of delivery
1α,25(OH)$_2$D$_3$ [3]		Roche, Duphar	Hypocalcemia Psoriasis	Systemic Topical
26,27-F6- 1α,25(OH)$_2$D$_3$ [5]		Sumitomo- Taisho Discovery Penederm	Osteoporosis Hypoparathyroidism Osteoporosis Psoriasis	Systemic Systemic Systemic Topical
19-Nor- 1α,25(OH)$_2$D$_2$ [3]		Abbott	Hyperparathyroidism	Systemic
22-Oxacalcitriol (OCT) [3]		Chugai	Hyperparathyroidism Psoriasis	Systemic Topical
Calcipotriol (MC903) [3]		Leo	Psoriasis Cancer	Topical Topical
1α,25(OH)$_2$,16-ene- 23-yne-D$_3$ (Ro 23-7553) [6]		Roche	Leukemia	Systemic
EB1089 [3]		Leo	Breast cancer	Systemic
20-Epi- 1α,25(OH)$_2$D$_3$ [3]		Leo	Immune diseases	Systemic
KH1060 [3]		Leo	Immune diseases	Systemic
ED-71 [4]		Chugai	Osteoporosis	Systemic
1α,24(S)(OH)$_2$D$_2$ [3]		Bone Care International	Psoriasis	Topical
1α,24(R)(OH)$_2$D$_3$ (TV-02) [3]		Teijin	Psoriasis	Topical
24-Epi- 1α,25(OH)$_2$D$_2$ [3]		Elan	Osteoporosis	Systemic

[a] Ring structures are indicated by the number in brackets and refer to the structures given at the top of this table.

Source: Ref. 172.

contain 24-hydroxyl groups, e.g. $1\alpha,24(S)\text{-}(OH)_2D_2$ and $1\alpha,24(R)\text{-}(OH)_2D_3$ (21). Other analogs contain multiple changes in the side chain in combination, including unsaturation, 20-epimerization, 22-oxa replacement, homologation in the side chain, or terminal methyl groups. The resultant molecules such as EB1089 and KH1060 are attracting strong attention of researchers because of greatly increased potencies in vitro and are being pursued as possible anticancer and immunomodulatory compounds, respectively.

Few attempts have been made thus far to modify the nucleus of calcitriol. The Roche compound $1\alpha,25\text{-}(OH)_2\text{-}16\text{-ene-}23\text{-yne-}D_3$, reported to be an antitumor compound in vivo, possesses a D-ring double bond. Relatively recently, the A-ring substituted 2-hydroxypropoxy-derivative, ED71, has been tested as an antiosteoporosis drug. The Abbott compound, $19\text{-nor-}1\alpha,25\text{-}(OH)_2D_2$, lacks a 19-methylene group and is styled upon the in vivo active metabolite, $1\alpha,25\text{-}(OH)_2DHT$, formed from dihydrotachysterol (18), which retains biological activity though the C19 methylene is replaced by a C19 methyl. Many other compounds have been developed with rigid or altered *cis*-triene structures (22) or modifications of the 1α-, 3β-, or 25-hydroxyl functions not for the purpose of developing active molecules for use as drugs but in order to allow us to establish minimal requirements for biological activity in structure activity studies (23,24).

D. Biochemistry and Metabolism

Vitamin D_3 can be synthesized in the skin or derived from dietary sources. Skin synthesis of vitamin D_3 from 7-dehydrocholesterol involves a previtamin D intermediate (Fig. 2), is a nonenzymatic process involving light and heat, and therefore can be mimicked in the test tube. Once the previtamin D is formed in the upper layers of the skin as the result of the reaction of ultraviolet light, it filters down to the lower layers where it is slowly transformed thermally into vitamin D_3. Vitamin D_3 and its metabolites are primarily hydrophobic and thus require a transport protein to be carried in the aqueous environment of the bloodstream. A specific globulin known as vitamin D binding protein or DBP exists for the purpose of transporting vitamin D_3 from the skin to the liver for the first step of activation; to carry its initial product, 25-hydroxyvitamin D_3 ($25\text{-OH-}D_3$) to the kidney for the second step of activation, and also to carry the active form, $1\alpha,25\text{-}(OH)_2D_3$, to its sites of action.

Vitamin D from skin or dietary sources does not circulate for long in the bloodstream, but instead is immediately taken up by adipose tissue or liver for storage or activation. In humans, tissue storage of vitamin D can last for months or even years. Ultimately, the vitamin D_3 undergoes its first step of activation, namely 25-hydroxylation in the liver (Fig. 3). Over the years there has been some controversy over whether 25-hydroxylation is carried out by one enzyme or two and whether this cytochrome P450-based enzyme is found in the mitochondrial or microsomal fractions of liver (25). Currently, only one of these enzymes, the

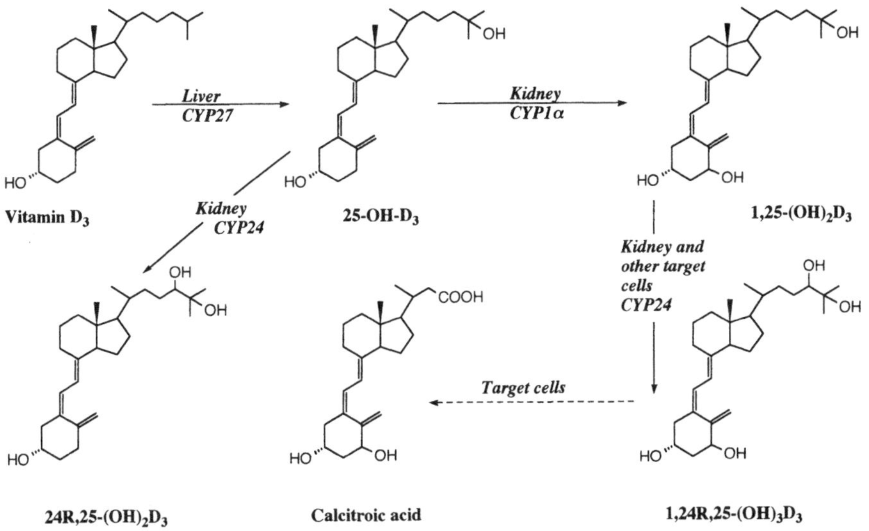

Figure 3 Metabolism of vitamin D_3.

mitochondrial form, has been purified to homogeneity, subsequently cloned from several species, and studied in any detail (26–28). The cytochrome P450 involved is known as CYP27 or P450c27 because it is a bifunctional cytochrome P450 which in addition to 25-hydroxylating vitamin D_3 also carries out the side-chain hydroxylation of intermediates involved in bile acid biosynthesis. Even though 25-hydroxylation of vitamin D_3 has been clearly demonstrated in cells transfected with CYP27, there is still some scepticism in the vitamin D field that a single cytochrome P450 can explain all the metabolic findings observed over the past two decades of research. These unexplained observations include:

1. Using the perfused rate liver, Fukushima et al. (29) demonstrated two 25-hydroxylase enzyme activities: a high-affinity, low-capacity form (presumably microsomal) and a low-affinity, high-capacity form (presumably mitochondrial; CYP27).

2. Regulation, albeit weak, of the liver 25-hydroxylase in animals given normal dietary intakes of vitamin D after a period of vitamin D deficiency (30) is not explained by a transcriptional mechanism since the gene promoter of CYP27 lacks a VDRE.

3. No obvious 25-OH-D_3 or 1α,25-$(OH)_2D_3$ deficiency in patients suffering from the genetically inherited disease cerebrotendinous xanthomatosis, where CYP27 is mutated. Although a subset of these patients can suffer from osteoporosis, this is more likely due to biliary defects leading to altered enterohepatic circulation of 25-OH-D_3 (31).

4. CYP27 does not appear to 25-hydroxylate vitamin D_2.

Very recently, a pig liver microsomal 25-hydroxylase was purified to homogeneity and subsequently cloned (32). However, it remains to be seen if this cytochrome P450, which belongs to the CYP2D family has a human counterpart, represents a physiologically important enzyme in vivo and therefore explains the list of anomalous findings listed above. One thing that the existence of CYP27 does explain is the occasional reports of extrahepatic 25-hydroxylation of vitamin D_3 (33). CYP27 mRNA has been detected in a number of extrahepatic tissues including kidney and bone (osteoblast) (34,35). The product of the 25-hydroxylation step, 25-OH-D_3, is the major circulating form of vitamin D_3 and in humans is present in plasma at concentrations in the range 10 to 40 ng/mL (25 to 125 nM). The main reason for the extended plasma half-life of 25-OH-D_3 is its strong affinity for DBP. Serum levels of 25-OH-D_3 therefore represent a measure of the vitamin D status of the animal in vivo and the analyst is obliged to provide efficient and sensitive assays for its measurement.

25-Hydroxyvitamin D_3 is converted to the active form of vitamin D known as calcitriol or $1\alpha,25$-dihydroxyvitamin D_3 ($1\alpha,25$-$(OH)_2D_3$). The second step of activation, 1α-hydroxylation, occurs in the kidney (36), and the synthesis of $1\alpha,25$-$(OH)_2D_3$ in the normal nonpregnant mammal appears to be the exclusive domain of that organ. The main evidence for this stems from clinical medicine where patients with chronic renal failure exhibit frank rickets or osteomalacia due to deficiency of $1\alpha,25$-$(OH)_2D_3$ caused by lack of 1α-hydroxylase, a situation that is reversible by $1\alpha,25$-$(OH)_2D_3$ administration. As the result of a tremendous amount of attention over the past two decades, the cytochrome P450, $CYP1\alpha$, representing the 1α-hydroxylase enzyme, was finally cloned from a rat renal cDNA library by St Arnaud's group in Montreal (37,38). This was rapidly followed by cloning of cDNAs representing mouse and human $CYP1\alpha$ (37,39,40) as well as the human and mouse genes (37,40–42). It had been known for some time that the kidney mitochondrial 1α-hydroxylase enzyme comprises three proteins: a cytochrome P450, ferredoxin, and ferredoxin reductase for activity and is strongly downregulated by $1\alpha,25$-$(OH)_2D_3$ and upregulated by PTH (43,44). The promoter for the $CYP1\alpha$ gene appears to contain the necessary regulatory elements (CREs) necessary to explain the observed physiological regulation at the transcriptional level. There had been claims that CYP27 can catalyse 1α-hydroxylation (34), but the physiological relevance of these observations must now be in question following the finding of the new cytochrome P450 by St Arnaud et al. (37). There is little doubt that $CYP1\alpha$ is important since its human gene colocalizes (37,42,45) to the chromosomal location of vitamin D dependency rickets type 1 (VDDR-I), a human disease state first proposed to be due to a mutation of the 1α-hydroxylase 25 years ago (46).

Over the past 20 years, it has been suggested that there are several physiological or pharmacological situations in which an extrarenal 1α-hydroxylase activity may exist. Placental 1α-hydroxylase was reported (47) but has proved to be difficult to purify and there have been suggestions that the activity may be artifactual. Adams and Gacad (48) have demonstrated the existence of a 25-OH-D_3-1α-hydroxylase in sarcoid tissue and this poorly regulated enzyme results in elevated plasma 1α,25-$(OH)_2D_3$ levels which in turn cause hypercalciuria and hypercalcemia in sarcoidosis patients. The induction of this 1α-hydroxylase in macrophages by cytokines and other growth factors has been postulated but its role, if any, in normal macrophages is unknown. The availability of molecular probes for the renal CYP1α has made it possible to detect this cytochrome and should now allow for exact characterisation of these various extrarenal 1α-hydroxylase activities. Indeed Jones et al. (210) have cloned the enzyme from human colon and lung sources and report the same sequence for extrarenal sources as the renal enzyme. The emphasis has now turned to the regulation of the CYP1α in renal and extrarenal sites in order to establish their physiological roles.

Irrespective of where 1α,25-$(OH)_2D_3$ is synthesized, as the sole hormonal form it represents the molecule which transduces the biological response inside vitamin D target cells. The physiological functions of 1α,25-$(OH)_2D_3$ are usually divided into (1) *classical roles*, which include the regulation of blood calcium and phosphate concentrations by actions at intestine, bone, and kidney; and (2) *nonclassical roles*, which include cell differentiation and antiproliferative actions on various cell lines, especially bone marrow (preosteoclast and lymphocyte), skin, and intestine. Though most of the classical physiological actions of vitamin D have been known from the early part of this century when dietary vitamin D deficiency was first demonstrated, the nonclassical roles have only emerged from more subtle studies involving experiments probing the mechanism of action of vitamin D at the molecular level.

All of these physiological functions of vitamin D are now known to be achieved largely through a steroid hormone-like mechanism involving a nuclear vitamin D receptor (VDR) which specifically regulates the transcription of vitamin D–dependent genes coding for proteins which in turn regulate cellular events such as intestinal calcium transport and cell division. In the steroid hormone model 1α,25-$(OH)_2D_3$ enters the cell by crossing the plasma membrane in a free form and binds strongly to the VDR inside the nucleus ($K_d = 2 \times 10^{-10}$ M). The liganded or occupied VDR specifically targets only vitamin D–dependent genes by interacting with a specific sequence found upstream of the vitamin D–depending gene. The sequence, known as a vitamin D responsive element (VDRE), is a tandem repeating oligonucleotide of six base pairs containing a 3-nucleotide spacer which is situated normally around 400 to 500 base pairs upstream of the 5' end of the vitamin D–responsive gene. A consensus VDRE, AGGTCAnnnAG-GTCA, is found in the rat and human osteocalcin genes, the rat calbindin-9K gene, and the mouse osteopontin gene, whereas more complex elements are found in the collagen type I gene and the pre-pro-PTH gene where they play a negative or suppressive role. Recent research has shown that VDR requires a heterodimeric partner called the retinoid X receptor (RXR) to transactivate genes. The current model suggests that occupied RXR-VDR dimer binds to the VDRE triggering a protein conformational change in the AF-2 domain of the C-terminus of the VDR which allows recruitment of positive transcription factors and/or shedding of transcriptional inhibitory factors which lead to increased formation of a transcription initiation complex and an increased rate of gene transcription. Figure 4 gives some idea of the complexity of this mechanism and the number of specific and general transcription factors involved.

Knowledge of the widespread distribution of the VDR protein in different tissues besides the classical targets of bone, intestine, and kidney encouraged researchers to look for other vitamin D–dependent processes. This was reinforced by the reports of disparate effects of vitamin D and its analogs and changes brought about in biological systems in vitro and nonclassical tissues in vivo. Though it is not always clear if these effects are physiological or pharmacological, these observations have certainly broadened our view of vitamin D. The molecular approach has also strengthened the case that vitamin D is more than a calcemic hormone. VDRE-containing genes now number around 50 or more, suggesting that vitamin D through $1\alpha,25$-$(OH)_2D_3$ and VDR may regulate many physiological processes besides intestinal calcium absorption, bone resorption, and bone matrix protein formation. Our view is that skin differentiation, immuno-

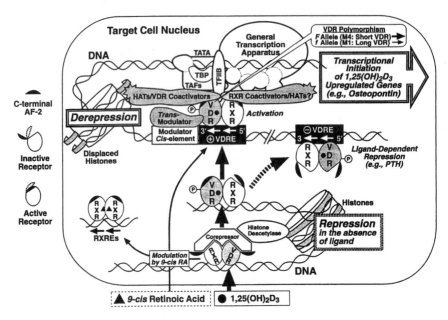

Figure 4   Working model for transcriptional control by $1,25$-$(OH)_2D_3$. (From Ref. 199, where further details can be obtained.)

suppression, and osteoclast recruitment are all processes significantly affected by vitamin D. Now that VDR knockout mice have been bioengineered, we can look forward to dissecting even more of the secrets of vitamin D at the molecular level (e.g., the role of vitamin D in reproduction since VDR knockouts have defective uterine development).

24-Hydroxylation of both 25-OH-D$_3$ and 1α,25-(OH)$_2$D$_3$ has been shown to occur in vivo to give rise to 24R,25-(OH)$_2$D$_3$ and 1α,24,25-(OH)$_3$D$_3$, respectively (49,50). The importance of this step has been immersed in controversy since it has been claimed that 24-hydroxylated metabolites might play a role in bone mineralization (51,52), and egg hatchability (53). Experimental evidence favors a different function for 24-hydroxylation—attenuation of calcitriol signal in target cells. This concept comes from four main lines of evidence:

1. The level of 24,25-(OH)$_2$D$_3$ does not appear to be regulated, reaching >100 ng/mL in hypervitaminotic animals (54).

2. There is no apparent 24,25-(OH)$_2$D$_3$ receptor similar to VDR within the steroid receptor superfamily.

3. Synthesis of vitamin D analogs blocked with fluorine atoms at the various carbons of the side chain (e.g., 24F$_2$-1α,25-(OH)$_2$D$_3$) results in molecules with the *full* biological activity of vitamin D in vivo.

4. 24-Hydroxylation appears to be the first step in a degradatory pathway demonstrable in vitro (55,56) (Fig. 3), which culminates in a biliary excretory form, calcitroic acid, observed in vivo (57).

The 25-OH-D$_3$-24-hydroxylase was originally characterized as a P450-based enzyme over 20 years ago (58) and more recently the cytochrome P450 species was purified and cloned by Okuda's group (59). The enzyme appears to 24-hydroxylate both 25-OH-D$_3$ and 1α,25-(OH)$_2$D$_3$, the latter with a 10-fold higher efficiency (60,61). However, since the circulating level of 25-OH-D$_3$ is ~1000 times higher than 1α,25-(OH)$_2$D$_3$, the role of the enzyme in vivo is not clear. The enzyme, particularly the renal form, which appears to be expressed at high constitutive levels in the normal animal, may be involved in the inactivation and clearance of excess 25-OH-D$_3$ in the circulation (54). On the other hand, the 24-hydroxylase may be involved in target cell destruction of 1α,25-(OH)$_2$D$_3$, especially since 1α,25-(OH)$_2$D$_3$ is a very good substrate for the 24-hydroxylase. Using a variety of cell lines representing specific vitamin D target organs (intestine: CaCo2 cells; osteosarcoma: UMR-106 cells; kidney: LLC-PK1 cells; keratinocyte: HPK1A and HPK1A-ras) a number of researchers have shown that 24-hydroxylation is the first step in the C24 oxidation pathway, a five-step, vitamin D–inducible, ketoconazole-sensitive pathway which changes the vitamin D molecule to water-soluble truncated products such as calcitroic acid (see Fig. 2) (56,57). In most biological assays, the intermediates and truncated products of this pathway possess lower or negligible activity. Furthermore, many of these compounds have little or no affinity for DBP, making their survival in plasma tenuous at best. The recent cloning of the cytochrome P450 component (CYP24) of the 24-hydroxylase enzyme has led to detection of CYP24 mRNA in a wide range of tissues, corroborating the earlier studies reporting widespread 24-hydroxylase enzyme activity in most if not all vitamin D target cells.

Additional studies have shown that mRNA transcripts for CYP24 are virtually undetectable in naïve target cells not exposed to $1\alpha,25$-$(OH)_2D_3$ but increase dramatically by a VDR-mediated mechanism within hours of exposure to $1\alpha,25(OH)_2D_3$ (62). In fact, the promoters of both human and rat CYP24 genes possess a double VDRE which has been shown to mediate the vitamin D inducibility of CYP24 enzyme in both species. It is therefore attractive to propose that not only is 24-hydroxylation an important step in inactivation of excess 25-OH-D_3 in the circulation but it is also involved in the inactivation of $1\alpha,25$-$(OH)_2D_3$ inside target cells. As such, one can hypothesize that C24 oxidation is a target cell attenuation or desensitization process which constitutes a molecular switch (see Fig. 5) to turn off vitamin D responses inside target cells (63). The recent development of CYP24 knockout animals (64,65) resulting in hypercalcemia, hypercalciuria, nephrocalcinosis, and premature death in 50% of null animals seems to support this hypothesis. On the other hand, surviving animals have unexplained changes in bone morphology which could suggest an alternative role for 24-hydroxylase in bone mineralization (64,65). These surviving CYP24 null animals might also be useful in demonstrating the importance of proposed backup systems such as 26,23-lactone formation to vitamin D catabolism. Calcitroic acid, the final product of $1\alpha,25$-$(OH)_2D_3$ catabolism, is probably not synthesized in

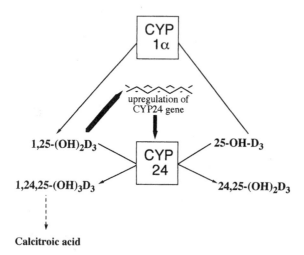

Figure 5 CYP24 as an attenuator of calcitriol signal in target cells.

liver because C24 oxidation does not occur in hepatocytes and therefore must presumably be transferred from target cells to liver via some plasma carrier. Though calcitroic acid has been found in various tissues in vivo (63), details of its transfer to bile have not been elucidated.

Many other vitamin D metabolites have been reported over the years but most are not much more than metabolic idiosyncrasies with little biological importance. The analyst should be familiar with a few of these because they can become quite abundant in certain biological samples, particularly those from animals given large doses of vitamin D, its metabolites, or analogs. Included in these is $25,26\text{-}(OH)_2D_3$, which was the first dihydroxylated metabolite to be identified in the late 1960s (24) and yet is still the most poorly understood. The metabolite is readily detectable in the plasma of animals given in large doses of vitamin D_3 and it retains strong affinity for DBP (66). However, its biological activity is inferior to other endogenous vitamin D compounds and it is presumed to be a minor catabolite. The knowledge that CYP27 is involved in vitamin D_3 inactivation, and that $26(27)\text{-}OH\text{-}D_3$ and $1\alpha,27(OH)_2D_3$ are formed from vitamin D_3 and $1\alpha\text{-}OH\text{-}D_3$ respectively in CYP27 transfection systems (27), when taken together suggests that 26(27)-hydroxylation may be a consequence of errant side chain hydroxylation.

The most abundant 26-hydroxylated analog appearing in vivo is the 26,23-lactone derivative of $25\text{-}OH\text{-}D_3$, $25\text{-}OH\text{-}D_3\text{-}26,23\text{-}lactone$, which accumulates in hypervitaminotic animals in vivo because of its extremely strong affinity for DBP (67). The route of synthesis of this metabolite is depicted in Figure 6 and research

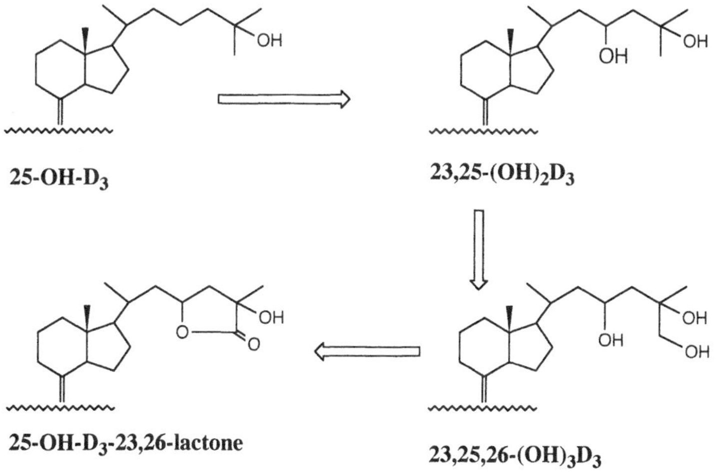

25-OH-D$_3$

23,25-(OH)$_2$D$_3$

25-OH-D$_3$-23,26-lactone

23,25,26-(OH)$_3$D$_3$

Figure 6 Biosynthesis of 25-OH-D$_3$-23,26-lactone (only C and D rings and side chain illustrated).

Table 3   Vitamin D and Its Natural Metabolites

Vitamin D metabolites [ring structure][a]	Side chain structure (R)	Site of synthesis	Relative VDR-binding affinity[b]	Relative DBP-binding affinity[c]	Reference
Vitamin D_3 [1]		Skin	≤0.001	3,180	
25-OH-D_3 [1]		Liver	0.1	66,800	1
1α,25-$(OH)_2D_3$ [3]		Kidney	100	100	36
24(R),25-$(OH)_2D_3$ [1]		Kidney	0.02	33,900	173
1α,24(R),25-$(OH)_3D_3$ [3]		Target tissues[d]	10	21	50
25(S),26-$(OH)_2D_3$ [1]		Liver?	0.02	26,800	66
25-OH-D_3-26,23-lactone [1]		Kidney	0.01	250,000	67

[a] Ring structures are indicated by the number in brackets and refer to those structures illustrated at the top of Table 2.
[b] Values reproduced from previously published data (201).
[c] Values reproduced from previously published data (202).
[d] Known target tissues include intestine, bone, kidney, skin, and the parathyroid gland.
Source: Ref. 171.

indicates that 26-hydroxylation follows 23-hydroxylation in this process (68). The site of this synthesis is probably extrahepatic, suggesting the existence of extrahepatic 26-hydroxylation. The recent evidence (69,70) that CYP24 carries out 23- as well as 24-hydroxylation of certain analogs raises the possibility that CYP24 may also be involved in 26,23-lactone formation. The role of 25(OH)D_3-26,23-lactone or its 1α,25-$(OH)_2D_3$ counterpart, which has also been reported (70a), are currently unknown. It is presumed that 26,23-lactone formation represents a backup degradatory pathway for 25-OH-D_3 and/or 1α,25-$(OH)_2D_3$. Studies with CYP24 knockout mice should test this hypothesis. Table 3 summarizes some of the pertinent biological properties of the common metabolites.

II. ANALYSIS

A. Introduction

The analysis of vitamin D and its metabolites is worthy of attention for many reasons. The measurement of the level of vitamin D metabolites provides important knowledge about the etiology, pathogenesis, and treatment of diseases involving disturbances of calcium and phosphorus metabolism. The analysis of vitamin D compounds in drug formulations, such as multivitamin tablets, injectable solutions, and new metabolite preparations, is required to ensure purity, stability, and potency of the compounds. There is at present tremendous interest in elucidating the full details of the metabolism and mode of action of vitamin D. The need for unequivocal identification of increasing numbers of metabolites in animals studies in vivo and in vitro is placing increased demands on the resolving capabilities of modern chromatographic techniques and the sensitivity of detection and measurement.

Part of the interest in the study of vitamin D stems from the availability of pure radioisotopically labeled derivatives of vitamin D and its metabolites and chemically synthesized metabolites. Present techniques of quality control can accurately determine purity and specific activity in radioactively labeled vitamin D, and chemical synthesis has been greatly aided by chromatographic techniques able to resolve stereoisomers of vitamin D, analogs, and metabolites (e.g., 71,72). Last but not least, the food and pharmaceutical industries have now introduced liquid chromatographic- and LC-MS, methods for vitamin D analysis of pharmaceutical preparations and vitamin D–supplemented human foods. Infant milk formulas and margarine provide the analyst with the task of measuring nanogram quantities of vitamin D in a matrix containing milligram or gram quantities of fat. Animal feeds also contain added vitamin D. It is likely that if vitamin D metabolites are to be used in the beef and dairy science industry on any major scale, there will be a need for accurate methods for the analysis of vitamin D and its metabolites in tissues and milk.

In this chapter we intend to concentrate on the use of LC and LC-MS methods for the analysis of vitamin D metabolites and analogs in biological fluids and in vitro cell cultures. The methodology used for these analyses does, however, have wider applications. Gas chromatography is no longer used for analysis unless attached to a mass spectrometer and thus GC-MS will be considered later on in this chapter as a means of method validation. The need for sample preparation, extraction, concentration, and partial purification prior to final analysis naturally depends upon the nature of the sample matrix and the methods of quantitation used. Minimal prepurification is required for concentrated samples of synthetic vitamin D compounds during chemical synthesis. At the other extreme, the analysis of vitamins D_2 and D_3 in foodstuffs involves saponification, extraction, and concentration prior to quantitation. The analysis of trace quantities of vitamin D metabolites, such as $1,25\text{-}(OH)_2D_3$, in human plasma requires the extraction of picogram quantities, purification by column chromatography, and finally, radioligand assay. Much attention has been paid to sample preparation, as a significant reduction in analysis time can be achieved by optimizing purification procedures.

Like all steroids, vitamin D and its metabolites are usually analysed in three stages, all of which are dependent on each other: [1] extraction of the metabolite of interest from the biological matrix; [2] separation of the metabolite of interest from other metabolites of vitamin D, other steroids, or nonspecific material which may interfere in stage 3; and [3] quantitation.

Clearly the specificity of stage 3, the method of quantitation chosen for a particular metabolite, will have a profound influence on the necessity for and extent of the prepurification required before quantitation. The development of immunoassays for steroid hormones has been extremely successful and many immunoassays are now available which can be carried out directly on plasma or on sample plasma extracts (73). Unfortunately the development of immunoassays for vitamin D and its metabolites has inexplicably not been so successful, and all such immunoassays require extensive prepurification before quantitation in order to achieve acceptable accuracy. Stage 3 does not end when an acceptable figure for the concentration of the analyte or analytes concerned is obtained. It is also necessary to show that the procedure used for quantitation, taken in conjunction with any preparatory purification, is precise (that is to say replicate measurements are close to the mean value—a coefficient of variation [CV] of <10%) and that the analytical result is accurate.

Accuracy implies that the measured concentration of the analyte is solely due to the analyte itself and not to closely related compounds or nonspecific interference. The assessment of accuracy is extremely difficult since there are seldom definitive methods available for vitamin D metabolites which enable this parameter to be measured. A number of devices have been used which purport to assess accuracy, such as addition of pure analyte and the demonstration of quantitative recovery and dilution of sample. In some instances such procedures do demonstrate deficiencies in the original analytical result, but even if recovery of pure standard is quantitative and/or dilution experiments are satisfactory, accuracy is not always guaranteed.

It seems to us that the only method available for the assessment of accuracy is by comparison to a definitive method, which is generally accepted to be isotope dilution mass spectrometry (74). Such methods are not widely available for vitamin D metabolites but will be discussed later on in this chapter. The use of such methodology is also subject to criticism in that the ''definitive'' GC-MS procedure may itself be flawed. It may be necessary in stage 3 to demonstrate the identity of the analyte, which can be done by a number of physicochemical means (such as UV and/or infrared spectra, nuclear magnetic resonance [NMR], mass spectrometry, etc.). These techniques have been reviewed (7) and will be dealt with later on in this chapter only insofar as they pertain to the solution of this problem in the field of vitamin D. There are a number of published reviews of vitamin D methodology which deal with both the early methodology and more recent developments (6,7,75–82).

B. Extraction

1. Solvent Extraction

Since vitamin D and its metabolites are fat-soluble sterols, partition into organic solvents provides significant purification by removal of water-soluble contaminants. Solvent systems for extraction fall into two basic categories (see Table 4): [1] total lipid extraction, for example, methanol-chloroform-water (2:1:0:8) (83) or ethanol-water (9:1) (84), and [2] selective lipid extraction, for example, ether, ethylacetate-cyclohexane (1:1), hexane, dichloromethane, or hexane-isopropanol (1:2).

Clearly, though solvents in the second class are desirable to minimize contamination of vitamin D extract, they pose problems in that they tend to provide efficient extraction for one particular metabolite or group of metabolites (e.g., the dihydroxylated metabolites) and poorly extract those of a different polarity. When simultaneous analysis of several vitamin D metabolites with a wide range of polarities is required, a total lipid extraction (83) may be necessary. The high efficiency of the Bligh and Dyer technique (83) for total lipid extraction (i.e., methanol:chloroform 2:1, v/v) is probably due to the formation of a monophasic dispersion of sample in extracting solvent, followed by return to the classic two-phase system by addition of extra chloroform and saturated KCl. All hydroxylated nonacidic vitamin D metabolites can be extracted quantitatively by this technique.

Substitution of methylene chloride for chloroform avoids the known carcinogenicity of chloroform and reduces evaporation times by taking advantage of the lower boiling point of methylene chloride. The use of ethanol is a rapid method for precipitation of protein and extraction of vitamin D metabolites but

Table 4 Extraction Procedures for Vitamin D and Its Metabolites

Solvent	Metabolite	Ref.
Methanol-chloroform (2:1)	General	83
Methanol-methylene chloride (2:1)	General	174
Ethyl acetate-cyclohexane (1:1)	General	85
Hexane-isopropanol (1:2)	General	86
Ethanol	D, 25-OH-D	175
Ethanol:water (9:1)	General	84
Hexane (after saponification)	D	176
Ether	25-OH-D	177
Methanol:water (Sep-Pak)	25-OH-D	93
Dichloromethane	$24,25\text{-}(OH)_2D_3$ $1,25\text{-}(OH)_2D_3$	89
Dichloromethane on Extrelut	$24,25\text{-}(OH)_2D_3$ $1,25\text{-}(OH)_2D_3$	90

becomes increasingly impractical as sample volumes exceed 1 mL. The Bligh
and Dyer method (83) is particular useful for total extraction from cell culture
medium, where the lipid content is so low and thus subsequent HPLC column
overloading is not a significant issue.

Two-phase liquid-liquid extraction of samples is the preferred choice of
sample preparation in the majority of applications. Vitamin D and its photoisomers are relatively nonpolar and can be extracted with hexane and other low-
polarity solvents, provided the matrix is broken down by saponification, etc. Acid
hydrolysis must be avoided due to the tendency to dehydrate the 1-hydroxyl and
to give rise to isomerization. In the biochemical areas of analysis, all types of
saponification are avoided, chiefly because of concern about the stability of vitamin D metabolites but also because of lack of convenience. More polar solvents,
such as modified hexane mixtures, cyclohexane-ethyl acetate (1:1) (85), hexane-
isopropanol (1:2) (86), or hexane-isopropanol-n-butanol (93:3:4) (87), are employed for the extraction of metabolites of vitamin D and to break protein-bound
vitamin D complexes. Horst et al. (88) use ether followed by methanol-methylene
chloride (1:3) in a lengthy but ingenious attempt to analyze a series of vitamin
D_2 and D_3 metabolites simultaneously. Care must be taken when using ether,
since this solvent may contain peroxides that can react with the *cis*-triene of
vitamin D. Methylene chloride is frequently used for the extraction of 24(R),25-
$(OH)_2D_3$ and 1,25-$(OH)_2D_3$ from plasma. Though the use of methylene chloride
was originally as a two-phase system with plasma (89), one modification is to
adsorb the plasma onto Kieselguhr packed in a column (commercial version,
Extrelut, Merck, Darmstadt, Germany) and elute by passing methylene chloride
through the column (90). A total of 80% to 90% extraction of both dihydroxylated
metabolites is obtained, and the amount of lipid extracted is so small as to make
additional cleanup steps unnecessary. An additional postextraction prechromatographic step that can be used is to backwash the extract with a low-ionic-strength,
weakly basic (pH 9 to 10) buffer, such as phosphate. This tends to remove any
acidic lipid components poorly ionized at neutral pH levels. It is often possible
to avoid this wash by extracting from alkaline medium.

2. Solid-Phase Extraction and Prepacked Cartridges

The use of an inexpensive, disposable, gravity flow minicolumn as a penultimate
step prior to LC serves several useful purposes:

1. It protects the expensive LC column from lipid, particulate matter, etc.
2. It eliminates remaining neutral and highly polar lipid contaminants,
further reducing the lipid load applied to the LC.
3. It can be used to fractionate the vitamin D metabolites into major
classes for subsequent analysis using different solvent systems.
4. It can serve as a support for solid-phase extractions.

Minicolumns have disadvantages, too:

1. Batch-to-batch and manufacturer-to-manufacturer variability, requiring frequent calibration checks.

2. The minicolumn sometimes provides unwanted resolution of D_2 and D_3 metabolites, particularly when using Sephadex-H20 and hydroxyalkoxypropyl Sephadex (HAPS), giving misleading results when using radioactive vitamin D metabolites as internal standards, unless larger fractions are collected to include both D_2 and D_3 metabolites.

3. Losses can occur without achieving any substantial purification of the sample.

4. Adsorbents used in the minicolumns can be dirty and introduce interference into subsequent analyses.

It is fair to say that many of these disadvantages of minicolumns have been overcome with the introduction of prepacked cartridges, particularly those based upon the syringe-type minicolumn. On balance then, the advantages of such cartridges outweigh the disadvantages and their introduction has revolutionized sample preparation, streamlining the procedures greatly and in some clinical assays eliminating the need for HPLC altogether (91). Cartridges have largely replaced the use of liquid-liquid extraction since they can act as a solid-phase extraction surface in the same way that Extrelut was used a decade ago. Cartridges can also be used as a separation tool to resolve vitamin D and its metabolites. They can be used for a combination of solid-phase extraction and separation procedures (82,97).

Cartridges are now available in a number of different chemistries ranging from straight-phase packings such as silica to exotic bonded phase packings (e.g., C18, or octadecasilane). An increasing number of different packing materials is becoming available from the manufacturers (AnalytChem and Waters), but the full use of these different packings has yet to be fully realized in the vitamin D field. Present-day procedures focus mainly on use of silica and C18 packings. For extraction of vitamin D, its analogs, and their metabolites C18 or ODS-bonded silica cartridges are the method of choice. Sep-Pak C18 is the reversed-phase analog of the silica cartridge and uses a bonded octadecasilane (C18) phase. It can also be used to fractionate plasma in the reverse manner to that described by Adams et al. (92). Dabek et al. (93), Turnbull et al (94), and Fraher et al. (95) have developed methods for 25-OH-D_2 and 25-OH-D_3 using Sep-Pak C18 for both extraction (94,95) and preliminary purification (93) of the sample. The method of Dabek et al. (93) utilizes the properties of methanol both as an extraction solvent and as a mobile phase. However, these procedures did not extract the parent compound, vitamin D. Hollis and Frank (96) described a technique that does remove vitamin D from serum or plasma using a modified solid-phase extraction technique.

Further investigation into the utility of ODS-bonded silica cartridges has led to a procedure that has been described as ''phase switching'' (97). This procedure uses a concept by which solid-phase extraction and subsequent normal-phase separation of vitamin D metabolites can be performed on a single cartridge. This study by Hollis (97) demonstrated that the type of ODS-microsilica packing used greatly affects the resolution one achieves with respect to metabolites of vitamin D. This concept is displayed in Figure 7. Using a fully end-capped ODS-silica cartridge results in little resolution between 24,25-(OH)$_2$D$_3$ and 1,25(OH)$_2$D$_3$ (Fig. 7A). However, using a partially end-capped ODS-silica material (ODS-OH), excellent resolution is achieved between these two important vitamin D$_3$ metabolites (Fig. 7B) (98), but it will be noted that using ODS-OH cartridges the resolution previously achieved between 25-OH-D$_3$ and 24,25-(OH)$_2$D$_3$ is lost. However, such a system has proved very useful in the purification of 1,25-(OH)$_2$D$_3$ prior to radioligand assay (91) and is now available as part of a commercial assay (Diasorin Ltd., Wokingham, Berks., U.K.; formerly Incstar Inc.). Other reversed-phase-type silica cartridges ($-$NO$_2$) have been used to separate 25-OH-D$_3$ and 1,25-(OH)$_2$D$_3$ (99). Straight-phase cartridges can thus be used to provide a convenient and simple method for preliminary separation of metabolites into different polarity fractions. Today further analysis is mainly accomplished by the use of HPLC.

Sep-Pak SIL cartridges (Waters Associates, Milford, MA) use high-efficiency silica particles and offer minimal variability. These cartridges have been used by Hollis and colleagues (100) to demonstrate the lack of vitamin D sulfate in human milk, and by Adams et al. (92) to fractionate classes of vitamin D metabolites in human plasma. Evaluations of the use of Sep-Pak SIL as a means of preliminary fractionating vitamin D and its metabolites have been published (92,101). Hollis et al. (100) demonstrated the use of these cartridges for the rapid separation of vitamin D and vitamin D sulfate. The plastic coating of the Sep-Pak is made of virgin polyethylene with minimal or zero amounts of plasticizers. Early versions of the Sep-Pak SIL were not completely devoid of UV-absorbing impurities, and as a result numerous interfering peaks were introduced into the sample during ''purification.'' Leaching may still occur from Sep-Paks with chlorinated hydrocarbons. Bond-Elut cartridges, on the other hand, are encased in surgical-grade polypropylene and leaching of plastics into the solvents has not been a problem. This issue becomes of great importance when using subsequent sensitive analytical techniques such as HPLC. In our hands, the separation of vitamin D metabolites on a Waters silica Sep-Pak cartridge is inferior to that achieved using the higher-quality microsilica supplied by Analytichem International as Bond-Elut cartridges (Fig. 8).

Immunoaffinity extraction has been used in the steroid field for some considerable time (102) but has only recently been introduced as part of methods for the analysis of vitamin D metabolites (103–105). Such systems involve the

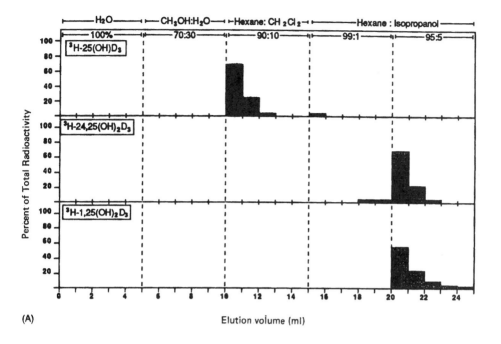

(A)

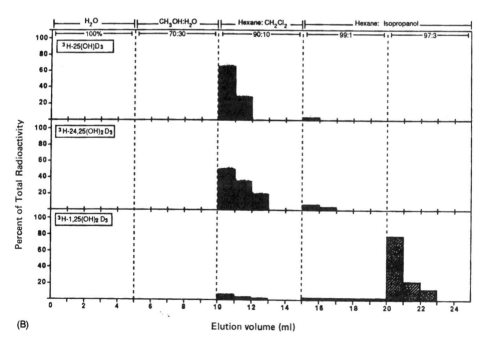

(B)

Figure 7   Elution of radiolabeled vitamin D_3 metabolites from (A) an ODS-BOND-ELUT cartridge and (B) an ODS-BOND-ELUT-OH cartridge. (From Refs. 96 and 98.)

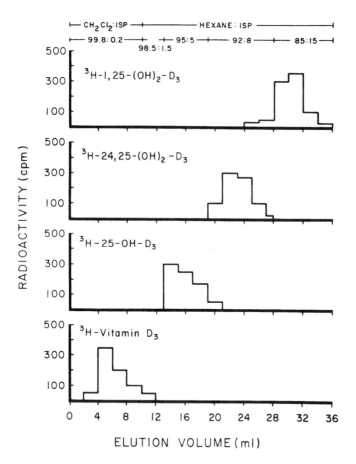

Figure 8   Elution profiles of radiolabeled vitamin D_3 and its metabolites on a silica BOND-ELUT cartridge. Each bar represents the amount of radioactivity in 2 mL eluent during batch elution with various mixtures of isopropanol in methylene chloride or hexane. (From Ref. 96.)

covalent linking of antibodies to the analyte of interest to column support material (e.g., Sepharose). As binding to the immunosorbent occurs in the aqueous phase, solvent extracts of the matrix must be redissolved in aqueous ethanol (10% ethanol) before application to the column containing the immunosorbent (105). A commercially available immunoextraction system for 1,25-$(OH)_2D_3$ in plasma delipidates with dextran sulfate and magnesium chloride prior to application to cartridges containing a solid-phase monoclonal antibody (103). The bound analyte is then eluted from the column. The specificity of extraction using such procedures is of course directly related to the specificity of the antibody used for this purpose which should ideally be a different antibody to that used for the quantitation.

C. Separation

1. General Considerations

One of the most common problems with LC is the presence of particulate matter in the samples for injection. Unchecked, this ultimately leads to clogging of the injector or, worse, the inlet frit of the column. Changing of the inlet frit of the column is a procedure to be avoided by the inexperienced chromatographer if possible, since, unless great care is taken, disturbance of the bed may occur, in which case column efficiency may be reduced. Thus, filtering of samples prior to injection is recommended. Various chromatographic companies now sell a variety of filtration kits for clarification of samples before HPLC. These feature 0.45 μm filters made of solvent-resistant materials (e.g., Teflon or PTFE) and suction or centrifugation to facilitate passage of solvent through the filter. Alternatively, samples for LC analysis can be transferred to a 5-mL conical screw-capped vial (Reactivial; Pierce Chemical Co., Rockford, IL) and centrifuged at low speed (around $500g$ for 5 min), although this is not always a completely satisfactory procedure.

LC analysis of vitamin D is now the method of choice, as over the last 15 years tremendous improvements in LC instrumentation have been made. Major advances have been made in pump technology which allow precise generation of isocratic and gradient solvent systems giving highly reproducible HPLC profiles. In addition, the improvement in automatic injection systems and the recent availability of automatic solid-phase cartridge injection systems make the introduction of the sample into the HPLC system more consistent, and the cartridge injection system negates the need for sample elution from the solid-phase extractant prior to LC, with all its attendent problems. In addition, improvements in detector sensitivity and column particle technology have made possible the detection of as little as 1 ng vitamin D under ideal conditions. Extra column peak broadening has been minimized by the design of eddy-free flow connectors and injectors and use of minimum dead-space tubing. While the modular approach to HPLC of the 1980s had its advantages, modern HPLC equipment is now increasingly marketed as an integrated computer-controlled system (e.g., the Alliance system marketed by Waters, Inc.), which can be very simply and directly connected to mass spectrometers.

2. Columns

The objective of all good LC analyses is to maximize resolution R, which is defined by the equation:

$$R = \frac{1}{4}\left(\frac{\alpha - 1}{\alpha}\right)\left(\sqrt{N}\right)\left(\frac{k'}{1 + k'}\right)$$
$$\text{Selectivity} \quad \text{Plates} \quad \text{Retention}$$

The value of N is determined by the size and nature of the particles used to make the column and by the quality of the packing procedures. Most vitamin D applications utilize particles in the 3- to 10-µm size range, and values of N in excess of 10,000 per column should be expected. The selectivity factor (α) and capacity factor (κ') can be altered by changing the chemical characteristics and the strength of the solvent, respectively. Examples of the effect of increasing R by changing either α or κ' are given below.

The nature of the type of analysis to be carried out and the design of the LC system may be important considerations in the selection of the type and size of the column to be used, since they may influence N. Some practical rules or considerations are given:

1. Use a column size appropriate to the volume of the injection, weight of solute, etc.
2. Use small volume injections since these are theoretically best for sharp peaks (i.e., large N). However, practical considerations of sample handling may dictate a compromise to a larger injection volume.
3. Use mobile phase as the sample solvent to avoid peak artifacts.
4. Standardize injection volumes for more reproducible results, since change in injection volume will influence the retention time and hence peak height area ratio. If peak areas are used for quantitation instead of peak heights, changes in injection volume have little effect.

Whereas a 25 × 0.21cm microbore LC column may be ideal for analysis of repeated 2- to 5-µL injections containing microgram quantities of a synthetic vitamin D metabolite using UV detection, it is not ideal for the analysis of 25-OH-D_3 in human plasma. Here the samples may still contain considerable amounts of lipid and a column with excellent loading capacity is the choice. Some of the newer 3µm particle columns (8cm × 0.62cm diameter, e.g., Zorbax) are ideal for the HPLC of biological samples since they are very tolerant of large injection volumes (100 to 200 µL) and large lipid load (i.e., whole extract from 2 to 5mL plasma), and yet give minimal sample losses and minimal effects of extracolumn band broadening. Several microparticulate (3 to 6µm) spherical silicas are now available: Lichrospher, Zorbax-SIL, and Spherisil. The advantage of these packings is that they pack more uniformly than irregularly shaped particles, and have less tendency to settle with time and pressure.

3. Solvents

Isopropanol-hexane mixtures provide a good solubility for the vitamins D and have been widely applied to the resolution of a variety of compounds ranging in polarity from photoirradiation mixtures of vitamin D to 1,24,25-$(OH)_3D_3$. Isocratic mixtures of 2.5% isopropanol in hexane separate vitamins D_2 and D_3 from

24-OH-D$_2$, 24-OH-D$_3$, 25-OH-D$_3$ (Fig. 9). Isopropanol-hexane (10:90) provides resolution of most of the known vitamin D metabolites (106) in a single chromatographic run (Fig. 10), and can also be useful for vitamin D$_2$ metabolites. These early separations can now be achieved on a single 25 × 0.46 or 25×0.6 cm column using similar solvent systems. The use of ternary solvent systems, incorporating methanol into the standard isopropanol-hexane mixture (107), greatly minimizes peak tailing, particularly of 24(R),25-(OH)$_2$D$_3$. Figure 11 shows the results of using a hexane-isopropanol-methanol (87:10:3) system solvent on the separation of 25-OH-D$_3$, 24(R),25-(OH)$_2$D$_3$, and 1,25-(OH)$_2$D$_3$. The polarity of the solvent can be reduced while keeping the proportion of isopropanol-methanol approximately the same.

Ikekawa and Koizumi (108) have described a gradient solvent system of

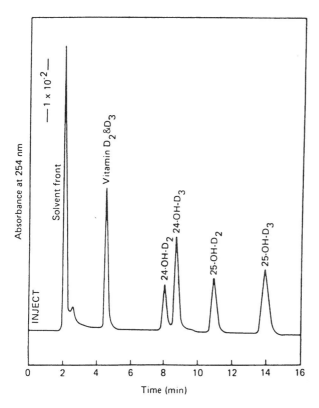

Figure 9 LC separation of vitamins D$_2$ and D$_3$, 24-OH-D$_2$, 24-OH-D$_3$, 25-OH-D$_2$, and 25-OH-D$_3$, using Zorbax SIL and eluting with 2.5% isopropanol in hexane. (From Ref. 106.)

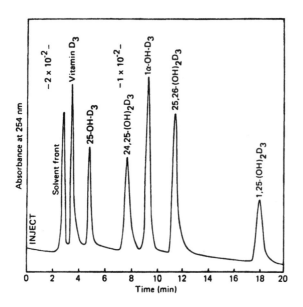

Figure 10 LC separation of vitamin D_3 and its metabolites on Zorbax-SIL using the solvent system 10% isopropanol in hexane. (From Ref. 106.)

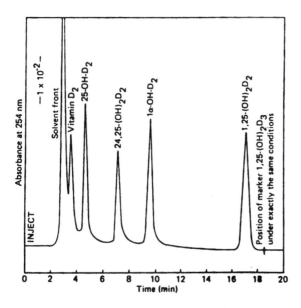

Figure 11 LC separation of 25-OH-D_3, 24(R),25-(OH)$_2$$D_3$, and 1,25(OH)$_2$$D_3$ on Zorbax-SIL (25 cm × 6.2 mm) using the solvent system hexane-isopropanol-methanol (87:10:3). (From Ref. 107.)

0.02% to 6% methanol in methylene chloride to resolve a mixture of dihydroxy metabolites on Zorbax-SIL (25×0.21 cm). This solvent system gives a higher theoretical plate count N than does the isopropanol-hexane-methanol mixture under otherwise identical conditions. However, the low viscosity of methylene chloride–methanol mixtures can lead to bubble problems with some solvent delivery systems. Replacing some of the methylene chloride with hexane and using a hexane–methylene chloride–methanol mixture (90:10:10) overcomes of the viscosity problems, but poor miscibility and poor sample solubility develop. The hexane–methylene chloride–methanol (90:10:10) mixture beautifully separates D_2 and D_3 metabolites (107) but not D_2 and D_3 themselves.

Figure 12 illustrates the influence of small changes in solvent composition (α) on the resolution of 1,25-$(OH)_2D_2$ and 1,25-$(OH)_2D_3$ using the same column (Zorbax-Sil, 25×0.62cm) while keeping the retention time approximately the same. One further solvent system that should be mentioned, which separates 25-OH-D_3-26,23-lactone and 24(R),25-$(OH)_2D_3$, is a methylene chloride–isopropanol (96.5:3.5) mixture (67). The lactone runs with a retention time almost identical to 24,25-$(OH)_2D_3$ with isopropanol-hexane mixtures but elutes with 25-OH-D_3 with a solvent mixture rich in methylene chloride.

Matthews et al. (109) were the first to report LC of vitamin D metabolites using the pellicular packing ODS-Permaphase and a gradient of methanol and water. With the advent of the microparticulate bonded phases, such as µBonda-

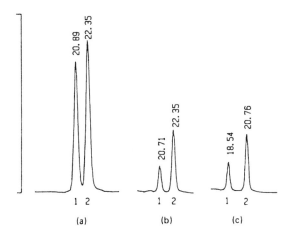

Figure 12 LC separation of 1,25-$(OH)_2D_2$ (peak 1) and 1,25-$(OH)_2D_3$ (peak 2) on Zorbax-SIL using (a) hexane-isopropanol (85:15); (b) hexane-isopropanol-methanol (87:10:3); and (c) hexane-ethanol-chloroform (80:10:10). Retention times in minutes are shown above peaks. Note the effect of changing solvent selectivity α on resolution. (From Ref. 107.)

pack C18 and Zorbax-ODS, analysts have successfully used reversed-phase LC for many compounds. However, in the vitamin D field, their use has been sparing for all but the separation of vitamins D_2 and D_3. Mixtures of methanol-water or acetonitrile-methanol-water (67) are preferred for vitamin D and its metabolites. Acetonitrile greatly reduces the viscosity of the solvent mixture and this results in lower back pressures. The limited use of reversed-phase packings stems from the aversion of vitamin D analysts to aqueous mobile phases, particularly when competitive binding assay of collected fractions is used for quantitation, since aqueous phases are difficult to evaporate, because they tend to dissolve support material, and, since they contain dissolved oxygen, may give rise to oxidation to the 5,7-diene system. Second, the resolution of the dihydroxylated compounds is not as good as that achieved using silica. Plate counts on bonded-phase columns are generally lower, pressures are higher, and column lifetimes are shorter than systems using straight-phase silica LC.

A recent application of ODS cartridges and HPLC columns is during the purification of calcitroic acid from tissue extracts. Here, the ionization properties of the weak acid group can be minimized and the molecule retarded by its interaction with the C18 hydrophobic surface. An example of such an interaction during sample preparation and quantitation on HPLC has been described (55). Mobile phases for calcitroic acid are rich in water and contain a small percentage of acetic acid or other acid to lower pH (acetonitrile-water-GAA, 40:60:1). Figure 13 shows an example of the use of reversed-phase LC using ODS for the resolution of metabolites of OCT (110). While such systems provide good resolution of more polar metabolites (P1–P5), this figure illustrates the poor resolution of the less polar metabolites (P6–P9), which can more readily be resolved by straight-phase LC (111). It should be noted here that aqueous solvent systems are required for LC-MS. These authors (110) have demonstrated the identity of P5 and P6 as side chain truncated metabolites (demonstrated by loss of radioactivity in Fig. 13) which are suggested to be conjugated subsequently with glucuronic acid in vivo to give peaks P1–P4.

There has been limited use of other packings. Straight-phase bonded packings offer excellent recoveries, but often the solvent strength has to be weakened significantly in comparison to that for silica in order to conserve adequate retention. μBondapak-CN and Zorbax-CN (112) have been used with some success for the separation of hydroxylated metabolites of vitamin D from those containing aldehyde or keto groups. One excellent example of this is the difficult separation of 25-OH-D_3-26,23-lactone and 24,25-$(OH)_2D_3$ referred to earlier. Zorbax-CN can be used to provide baseline resolution, the 26,23-lactone being strongly retarded by this packing and emerging much after the 24,25-$(OH)_2D_3$ (113). Contrast this with the other means of resolution: methylene chloride/isopropanol mixtures where the 26,23-lactone comigrates with 25-(OH)-D_3. The cyano packing has proven to be very useful in the recent identification of several oxo compounds

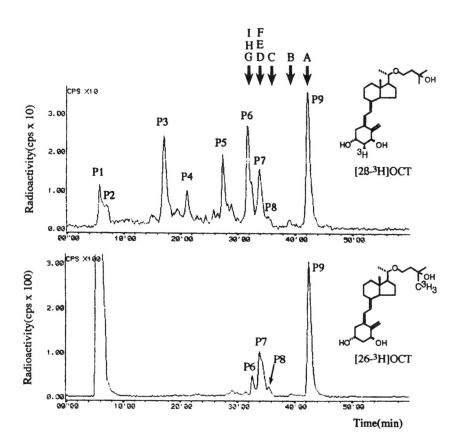

Figure 13 High-performance liquid radio chromatographic profile of plasma metabolites of 22-oxa-1α,25-(OH)$_2$D$_3$(OCT) 2h after administration of [^{3}H]OCT to rats. [^{3}H]OCT labeled at two different positions, [2β-^{3}H]OCT (upper panel) and [26-^{3}H]OCT (lower panel), were administered intravenously at the dose of 10 µg/kg to rats. Arrows (A–I) indicate the retention times of synthetic standards in the same system. (A) OCT; (B) pre-OCT; (C) 24-oxoOCT; (D) 20oxo-hexanor-OCT; (E) 24S(OH)OCT; (F) 24R(OH)OCT; (G) 20S(OH)-hexanor-OCT; (H) (25R)-26(OH)OCT; and (I) (25S)-26(OH)OCT. HPLC was carried out under the following conditions: column: YMC-Pack ODS-A A-313 (6.0 × 250 mm), mobile phase; (A): THF-H20 (1:9, v/v), (B): THF-CH$_3$OH-$_{H2O}$ (2:1:2, v/v/v), [(B): 0% (0 min); 75% (20 min); 75% (50 min); 100% (50.1 min)], flow rate: 1.0 mL/min, detection:radioactivity (CPS). (From Ref. 110.)

Table 5 LC Characterization of KH1060 Metabolites Generated in HPK1A-ras Cells

Metabolite	HPLC analysis, retention time (min)		Putative identity
	Z-SIL[a]	Z-CN[b]	
KH1060	9.35	7.80	KH1060
KH1060-Met 1	8.04	7.53	3-Epi-KH1060?
KH1060-Met 4	9.35	8.74	26-Oxo-KH1060
KH1060-Met 5	10.01	9.32	Octanor-20-oxo-KH1060
KH1060-Met 6	11.30	10.48	24a-Oxo-KH1060
KH1060-Met 7	12.05	9.03	Pentanor-24a-OH-KH1060?
KH1060-Met 8	12.85	10.04	Octanor-20-OH-KH1060
KH1060-Met 9	13.39	10.51	24a-OH-KH1060
KH1060-Met 13	17.66	11.88	24-OH-KH1060
KH1060-Met 14	17.66	12.69	24a-Oxo-26-OH-KH1060?
KH1060-Met 15	17.66	13.13	24-Oxo-26-OH-KH1060?
KH1060-Met 16	22.14	12.87	26-OH-KH1060
KH1060-Met 20	30.59	15.19	26a-OH-KH1060

[a] HPLC conditions: Zorbax-SIL (3 μm; 0.62 × 8 cm) column, 91/7/2 HIM (2.0 ml/min) solvent system.
[b] HPLC conditions: Zorbax-CN (6 μm; 0.46 × 25 cm) column, 91/7/2 HIM (1.5) ml/min) solvent system.
Source: Ref. 115.

found on the pathway from $1,25\text{-}(OH)_2D_3$ to calcitroic acid (114), and in the purification of metabolites of 25-hydroxydihydrotachysterol$_3$ (16). Table 5 illustrates the use of Zorbax-CN in the resolution of the metabolites of a calcitriol analog, KH1060 (115). The structure of this analog is illustrated in Table 2.

D. Detection and Quantitation (Not Using Mass Spectrometry)

1. Ultraviolet Absorbance (Fixed or Variable Wavelength)

Variable wavelength detectors are useful for distinguishing compounds in photo-irradiation mixtures, where they take advantage of the differences in the UV spectra of the various photoisomers. Normally, little is gained from the use of variable wavelength detectors over the fixed wavelength detectors in the analysis of vitamin D metabolites, since $\in 254$ is 90% of $\in 265$ and usually variable wavelength detectors are noisier at comparable sensitivity than fixed wavelength detectors due to an inherently weaker lamp intensity. Similarly, dihydrotachysterol $(\in 251 = 37,00)$ 37,000 metabolites with the 5,7-diene structure are best detected using a fixed wavelength detector because of the closeness of the 254-nm mercury line to the λ_{max}. However, optimizing UV absorbance is not the only reason to switch detection from 254 to 265 nm. Many sample matrices, whether biological or artificial, contain UV interferences of an aromatic nature absorbing in the 254-nm region, such as ketones and benzene derivatives. Effects of these can be minimized by switching to 265 nm or an even higher wavelength. Unfortunately, estrogens and several other compounds (e.g., peptides and phenolics) often absorb in the 270- to 290-nm range so that variable wavelength detectors offer no panacea.

But variable wavelength detectors can have a considerable value when, for example, isotachysterol derivatives are used for LC, since these isomers at 290 nm give approximately twice the absorbance of unisomerized vitamin D metabolites at 264 nm (116). Isotachysterols have been used to improve the sensitivity and specificity of an LC system for plasma 25-OH-D$_2$ and 25-OH-D$_3$. A simple method for 25-OH-D$_2$ and 25-OH-D$_3$ in plasma has been described using LC of isotachysterols and monitoring at 301 nm, where absorbance of vitamin D metabolites is minimal but where isotachysterols still absorb significantly (94). It should perhaps be noted that use of the phenyl-4-TAD adduct of vitamin D and its metabolites can also enhance UV absorbance by a factor of 5 and thus these adducts can be used to enhance sensitivity of detection by LC-UV systems (117). It is obviously possible to improve UV sensitivity further by a suitable selection of the moiety attached to the TAD, although attention must be paid to the separation of these adducts in the LC system chosen. If, however, such separation is compromised, it should always be possible to adopt the postcolumn derivatization procedure described by Vreeken et al. (118). Use of LC with UV detection for the measurement of vitamin D and its metabolites has been reviewed by Porteous et al. (6) and Makin et al. (7).

2. Diode-Array Scanning Spectrophotometric Detectors

UV detection has been hampered by its lack of specificity. In most cases the major criterion for peak identification has been the basis of comparison of retention time with known standards. Over the last 10 years increasing use has been made of a more sophisticated UV detector based upon the diode-array scanning spectrophotometer (119). Such an instrument is able to make repeated 200- to 400-nm scans of chromatographic effluent passing through the detector flow cell in as little as 100 msec or less. Using a microcomputer and associated data storage devices, vast amounts of chromatographic data can be acquired in a single 20-min chromatographic run. Jones et al. (120) applied this technique to the separation of vitamin D metabolites in extracts of kidney perfusates. As illustrated in Figure 14, the diode-array scanning spectrophotometer detector adds another dimension, wavelength, to the conventional absorbance versus time plot. Vitamin D metabo-

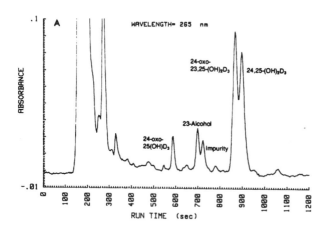

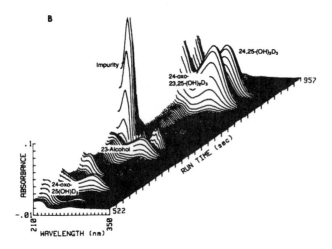

Figure 14   HPLC separation of metabolites of $24,25\text{-}(OH)_2D_3$ made in the perfused rat kidney. Extract of perfusate from kidneys perfused in the presence of 1250 nM $24,25\text{-}(OH)_2\text{-}D_3$. (A) Absorbance at 265 vs. run time. (B) Absorbance vs. wavelength vs. run time. (A) and (B) represent the same chromatographic run. Note in (A) that the peak marked as an impurity appears minor because the wavelength used for monitoring is 265 nm. The impurity peak absorbs strongly at 225 nm and in the 275- to 280-nm region. Note also that all metabolite peaks have the classical vitamin D absorption spectrum. Conditions: Zorbax-SIL (6.2 mm × 25 cm); hexane-isopropanol-methanol (94/5/1); 1.5 mL/min. (From Ref. 121.)

lite peaks can be recognized at retention times, 590, 700, 870, and 890 sec from their $\lambda_{max} = 265$ nm, $\lambda_{min} = 228$ nm properties without the use of known standards. In fact, the technique allowed Jones et al. (120) to recognize the existence of two new metabolites: 24-oxo-23,25-(OH)$_2$D$_3$ and 24,25,26,27-tetranor-23-(OH)-D$_3$, formed in the perfused kidney in vitro. It should be noted, however, that mass spectrometry in comparison to chemically synthesized material provided the means of identification of the metabolites in these studies (120,121). Another example can be seen in Jones et al. (16), where 3D plots have been used to identify metabolites of dihydrotachysterol$_3$ generated in vivo. DHT$_3$ has a characteristic tricuspid peak with maxima at 242, 251, and 26 nm, which stand out in the 3D plot.

Later generations of these detectors (e.g., Waters 996 diode array detector) offer excellent performance even at high sensitivity (0.01 AUFS) due to data bunching, smoothing, and noise reduction software. We have found diode-array detectors to be an extremely valuable analytical tool and much more than just a pretty picture. Irrespective of the type of spectral representation used, 3D plot, or isoabsorbance plot, the scanning spectrophotometry provides valuable information regarding the purity of vitamin D preparations. This is of tremendous value in predicting when, during a metabolite purification scheme, on-line LC-MS will be feasible and will not be hampered by impurities. Figures 15 and 16 illustrate that the 3D plot does not always indicate the presence of contaminating peaks and in the case illustrated the isoabsorbance plot was more informative, although scanning the peaks at the leading and trailing edges clearly revealed the same information.

3. Radioactivity Detector

Several manufacturers (Walac-Berthold, Beckman, Packard) have developed sensitive radioactivity monitors to measure radiolabeled compounds flowing through a flow cell placed between two photomultiplier tubes. Early models were bulky and used a spirally arranged Teflon tube between two plates as a flow cell. With this system a pulsatile, low-pressure pump and a post-LC column mixer were used to mix effluent with standard scintillation fluid. The mixture flowed through the flow cell, where counting efficiencies for tritium were high but residence time was short. The system was unreliable since it tended to get blocked when incompatible scintillation fluid-effluent mixtures gelled in the flow cell causing the system to leak cocktail and was sensitive to the pulses produced by the pump causing baseline instability. Later models replaced the pulsatile pump with a standard HPLC pump to pump an application-specific scintillation cocktail customized to the requirements of HPLC (clear, low viscosity, flow optimized), thereby reducing the source of the detector instability. The newer generation of detectors also possess improved sensitivity due to smoothing algorithms incorpo-

(a)

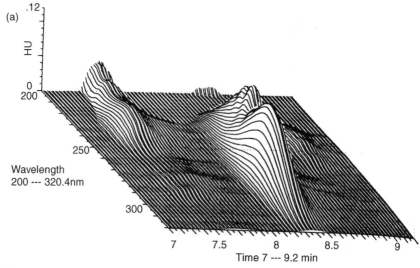

(b)

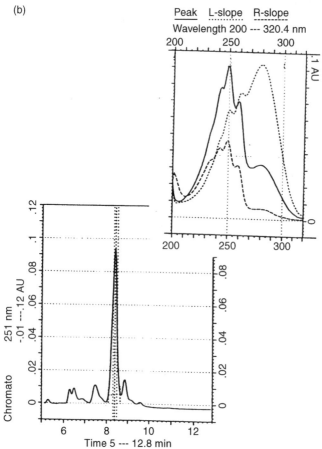

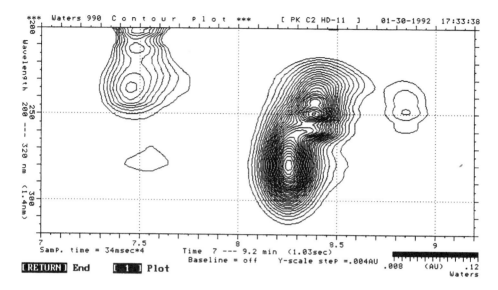

Figure 16   Contour plot of the UV spectrum of the same HPLC separation as that illustrated in Figure 15. It can clearly be seen from this plot that the major peak eluted between
8.1 and 8.6 min is composed of two components, one with a UV maximum at around
280 nm on the leading edge, and another with three UV maxima on the trailing edge. In
contrast, smaller, more symmetrical and uncontaminated peaks are seen at 7.47 min and
8.85 min. (From Ref. 7.)

rated as part of computer-based data-handling software. These modern scintillation counter–based detectors are thus very sensitive and reliable. An example
of data generated using this type of instrument is provided in Figure 17. One
disadvantage of this instrument is the peak-broadening caused by mixing effluent
and cocktail in the proportion 1:3 required for optimal efficiency.

   Although most instruments now use the scintillation fluid-based mixer approach, some manufacturers offer instruments with a 300-μL flow cell containing

---

Figure 15   Use of photodiode array detection of HPLC separation. (a) The photodiode
array detector can produce three-dimensional pictures such as that illustrated in this figure,
which represents part of an HPLC separation of the metabolites of 25-hydroxydihydrota-
chysterol$_3$ formed after in vitro incubation with HD-11 cells. The three-dimensional picture
suggests that what appears to be a single peak is in fact composed of two unresolved
compounds with very different spectra. This can perhaps be more easily appreciated in
(b), which shows the HPLC peak monitored at 251 nm and the UV spectra of the leading
edge of the peak, the peak apex, and the trailing edge. (From Ref. 7.)

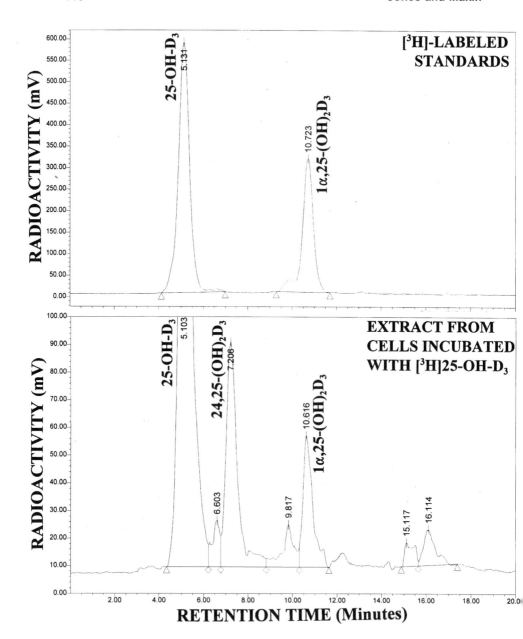

a solid glass yttrium salt–coated scintillator. This approach obviates the need for liquid scintillant and the result is that peaks are much sharper than when liquid scintillator is used. Adsorption of lipids to the scintillator is minimal, but eventually necessitates replacement of the flow cell and furthermore counting efficiency for ^{3}H is only in the 2% to 5% range. However, the detector provides excellent online counting efficiencies for [^{14}C]-labeled compounds of ~80%. In recent years only one (^{14}C)-labeled compound has been commercially available, this being [^{14}C]25-OH-D$_3$. Since most tracer studies use (^{3}H)-labeled vitamin D compounds, this type of flow cell is of limited value.

4. Electrochemical Detection

HPLC elutes can also be monitored by electrochemical (EC) means, and HPLC-EC systems have been described for the measurement of vitamin D in a number of situations. While this is an interesting and potentially valuable means of measuring vitamin D metabolites, and it does appear that at the moment these procedures are sufficiently sensitive for the measurement of plasma or serum concentrations, there is only one published method applied to plasma analysis. Table 6 gives a list of such methods which have been published in the last 10 years.

5. Fluorescence Detection

In order to enhance the sensitivity of detection of peaks emerging from an HPLC column other forms of detection have been investigated and one, using fluorescence detection, seems to provide significant improvements in sensitivity. Vitamin D and its metabolites are not themselves fluorescent and in order to utilize

---

Figure 17 Use of an in-line flow radioactive detector (Walac-Berthold) to detect [^{3}H]-labeled metabolites of [^{3}H]25-OH-D$_3$ produced by the human lung cancer cell line SW900. The metabolites were separated by HPLC using the usual Zorbax-SIL HIM 91-7-2 solvent system referred to in Table 5 and Ref. 115 at a flow rate of 1 mL/min. The LC eluent is first passed through a UV detector and then mixed with scintillation cocktail (Ready Flow, Beckman) in a 1:2 ratio prior to being passed through the radioactivity detector. Dead space between the two detector cells was determined to be 0.77 min. Metabolites include [^{3}H]24,25-(OH)$_2$D$_3$ and [^{3}H]1α,25-(OH)$_2$D$_3$. Putative products can be identified precisely by comparison of peak retention time to that of a chemically synthesized ^{3}H-labeled standard, as is illustrated for [^{3}H]1α,25-(OH)$_2$D$_3$ (10.616 vs. 10.723 min) or more approximately by comparison of retention time to that of a nonradioactive labelled standard using the UV detector. Retention times of nonradioactive standards in this system were 25-OH-D$_3$ = 4.38 min, 24,25-(OH)$_2$D$_3$ = 6.39 min, 1α,25-(OH)$_2$D3 = 9.88 min. Calibration of the instrument involves injection of pure ^{3}H-labeled standards containing known amounts of radioactivity; alternatively, radioactive peaks can be collected and counted off-line. Data courtesy of Dr. Angi Zhang.

Table 6 Recent Liquid Chromatography-Electrochemical Methods for the Measurement of Vitamin D and Its Metabolites

Metabolite	Ref.	Comments
7-Dehydrochloesterol	178	7-Dehydrocholesterol measured in skin samples as small as 5 mm diameter. Normal values 12-81 µg/g dry weight.
Vitamins D_2 and D_3	179	Studies oxidation at glassy carbon electrode and optimized LC-EC conditions. Linear response over range 10–200 ng injected.
Vitamins D_2 and D_3	180	Measured in liquid formula diet. Minimum detectable limit or S:N of 2 was 200 pg. Linear response 0.05–1 µg/ml. Good precision. Used oxidation followed by reduction at downstream electrode.
Vitamin D	181	Measured in milk (see 182).
1,25-$(OH)_2D_3$ and 25-OH-16-ene-23-yne-D_3	183	Post-column mixing with electrolyte. Claimed 1,000-fold increase in sensitivity over UV detection. Standards only.
Vitamin D_3 and 25-OH-D_3	182	Only using standards. Showed linear response 8–413 ng injected. Minimum detectable limit for D_3 8 ng and for 25-OH-D_3 25 ng. Reproducibility 3.2–5.8% CV.
25-OH-D_3, 24,25-$(OH)_2D_3$ and 25-OH-D_2 in plasma	184	Detection limits in pg range. Reproducibility <10% CV (between batch).

Source: Ref. 7.

this methodology some precolumn derivatization is required. A number of methods for 25-OH-D utilizing the formation of fluorescent triazoline adducts have been described. These adducts provide a degree of specificity since they recognize the conjugated *cis*-triene structure of vitamin D and form a bridge across the 6-19 carbons (see Fig. 18).

The value of substituted 1,2,4-triazoline-3,5-dione (TAD) as a reagent for the protection of the vitamin D *cis*-triene structure during chemical synthesis or modification has been known for many years (122). By linking TAD to fluores-

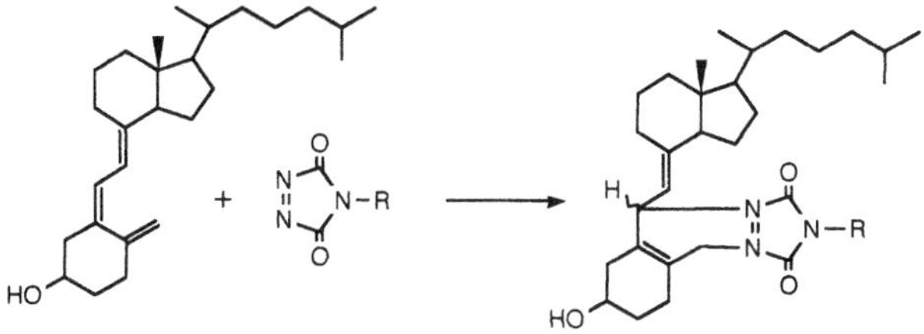

R group	Usage	Reference
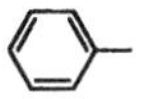 Phenyl	Static & dynamic FAB LC-MS using thermospray	117, 200
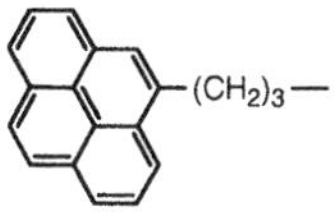 3-(1-pyrenyl)propyl	LC-fluorescent measurement of serum 25-OH-D$_3$	128
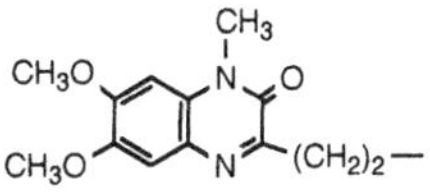 2-(6,7-dimethoxy-4-methyl-3,4-dihydroquinoxalinyl)ethyl [DMEQ]	LC-fluorescent measurement of 25-OH-D$_3$, 24,25-(OH)$_2$D$_3$ and 1,25-(OH)$_2$D$_3$	124, 126, 129 203,204
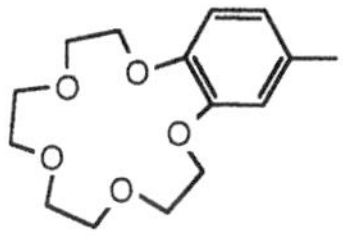 benzo-15-crown-5	LC-MS using electrospray	127
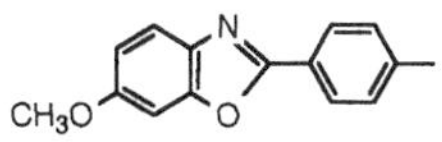 4-[4-(6-methoxy-2-benzoxazolyl)-phenyl]	LC using fluorescence detection of 7-dehydro cholesterol LC of vitamin D metabolites & conjugates	205 130
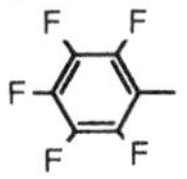 Hexafluoro-Phenyl	LC-MS using particle beam negative ionisation for analog Ro 24-2090	205

Figure 18 Examples of adducts of triazoline derivatives used for the measurement of vitamin D analogs and metabolites. (From Ref. 7.)

cent or UV absorbing moieties, adducts can be formed with vitamin D and its metabolites which can be used to enhance the sensitivity and specificity of detection after LC separation. TAD is not the only compound which can form such adducts with vitamin D, but a study of a number of such dienophilic compounds showed that phenyl-4-TAD was the reagent which most readily formed the appropriate adduct (118).

The development and synthesis of a number of derivatives of TAD and the nature of their reaction with vitamin D and its metabolites have been the subject of a number of investigations (118,123–127). Two methods for the measurement of 25-OH-D$_3$ using LC with fluorescence detection have been published, both procedures using precolumn formation of fluorescent adducts with derivatives of TAD. 25-OH-D$_3$ could be detected in serum at a concentration of 0.25 nmol/L (128) using as a derivatizing agent 3(1-pyrenyl)propyl-4-TAD. The second method (129), preliminary details of which were published in 1991 (124), used DMEQ-4-TAD (see Fig. 18) and assayed 25-OH-D$_3$ and 24,25-(OH)D$_3$ using 0.2 mL and 2 mL plasma, respectively. Both these methods require extraction, use of Sep-Pak SIL or C18 cartridges, formation of the appropriate adduct, and reversed-phase HPLC. Excess reagent was removed prior to HPLC using Baker NH$_2$ or Bond Elut PSA cartridges. The method of Shimizu et al. (129) interpolated an extra straight-phase LC separation after extraction and before derivatization, a step which was not found to be necessary by Jordan et al. (128). TAD adducts can be formed on either side of the vitamin D molecular and thus the hydrogen on C6 can be either α or β. The two isomers are separated in the reversed-phase LC system used by Shimizu et al. (129) but not in the system used by Jordan et al. (128). Details of the formation of TAD adducts with vitamin D and the formulae of some of the TAD derivatives used are given in Figure 18. LC systems for the separation of TAD adducts of pro-vitamin D, vitamin D, and 25-hydroxyvitamin D$_3$ and their glucuronide and sulfate conjugates have been described (130,131).

E. LC Detection and Quantitation by Offline Mass Spectrometry

1. Introduction

In previous editions we have dealt with GC-MS as a distinct separation and quantitation technique. Few novel GC-MS methods have been described over the last 10 years, although a simple method for 25-OH-D$_3$ involving on-column dehydration has been described (132). The increased sensitivity of high-resolution GC-MS has also been demonstrated by its application to the measurement of hexafluorocalcitriol in plasma with a sensitivity of 2 pg/mL (133). The major development has been its application as a means of structural determination or confirmation. We have therefore considered GC-MS as an offline means of LC detection and/or quantitation. Modern developments in the application of mass spectrometry have centered on the connection of the LC eluate directly to the mass spectrometer, and this application is considered as an online methodology. In this chapter we have not considered in detail direct probe MS analysis using El or Cl as no developments in this area have taken place. For convenience however a direct probe El(+) spectrum of 25-OH-D₃ is illustrated in Figure 19, which

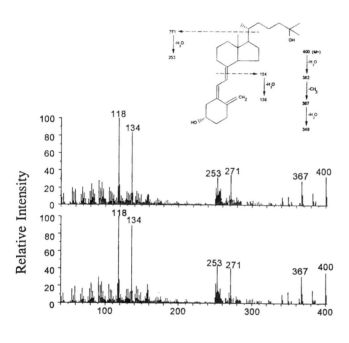

Mass / Charge

Figure 19 Mass spectrum of 25-OH-D₃ purified from Hep 3B cells (upper panel) compared to the mass spectrum of synthetic 25-OH-D₃ (lower panel). Hep 3B cells were incubated with vitamin D₃ (50 µM) for 48 h. Flasks were then extracted and the lipid extract dried under nitrogen and purified on HPLC (conditions: Zorbax SIL [6.2 mm × 25 cm], solvent HIM 96/3/3, flow rate 2 mL/min). A metabolite peak possessing the vitamin D chromophore and comigrating with synthetic 25-OH-D₃ was collected, purified further on a different HPLC system (conditions: Zorbax CN [4.6 mm × 25 cm], solvent HIM 94/5/1, flow rate 1 mL/min), and then dried under nitrogen and subjected to direct probe mass spectrometry using electron impact El(+) ionization. The putative 25-OH-D₃ gave the expected molecular ion with m/z 400; the other ions observed were consistent with the molecule being 25-hydroxylated (see inset fragmentation pattern). (From Ref. 207.)

shows a comparison between a standard and 25-OH-D$_3$ obtained from in vitro incubation studies. However, GC-MS is still the method of choice at the present time for structural determination, provided the compounds of interest are stable in the gas chromatograph. In the study of the metabolism of calcitriol analogs, this has become an increasing problem.

2. B-Ring Closure

A prerequisite for the use of gas-liquid chromatography is that the solute should be in the vapor phase during the process of partition between the mobile gas phase and the stationary liquid phase. For compounds of molecular mass in excess of around 200 Da, this requires analysis to be carried out at elevated temperatures, usually in excess of 100°C. At these temperatures, vitamin D and its hydroxylated metabolites undergo thermal rearrangement involving B-ring closure. In naturally occurring steroids (e.g., cholesterol), the 19-methyl attached to C10 is always β and is *trans* with respect to the 9α-hydrogen. When B-ring closure occurs in the vitamin D series, two isomers are formed, both of which are *cis* across the 9–10 carbon-carbon bond (i.e. C19β, 9β, isopyro; and C19α, 9α, pyro). The formation of these isomers was originally described in 1932 by Askew and his colleagues (134), who heated vitamin D in the test tube to around 190°C. Although isomerization took place rapidly at this temperature and was complete in about 4 h, ring closure has been shown to occur at all temperatures in excess of 125°C (135). It is irreversible and the ratio of pyro and isopyro remains at 2:1. This reaction occurs in the absence of oxygen, and it is now thought that the immediate precursor of these isomers is the previtamin (see Fig. 20). At the temperatures at which GLC is carried out, therefore, vitamin D and its metabolites are all converted irreversibly and quantitatively to a 2:1 mixture of pyro and isopyro isomers that separate in all the GLC systems so far examined (136–139). The formation of these two isomers during GLC was first demonstrated by Ziffer et al. (140).

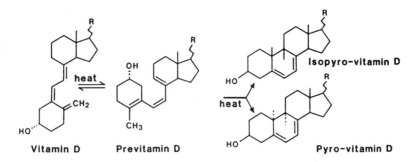

Figure 20   Thermal transformation of vitamin D to pyro and isopyro isomers.

3. Derivatives for GLC

As with all GLC systems for steroid analysis, the stability and volatility of vitamin D and its metabolites can be increased by the formation of derivatives on the polar hydroxyl groups. Free underivatized hydroxyl groups may give rise to adsorption during chromatography (141) and non-linearity of detector response. If vitamin D and particularly its polyhydroxylated metabolites are injected into a GC without derivatization, broad peaks are obtained, indicating some degree of adsorption. In addition, dehydration of the 25-hydroxyl can occur, giving rise to two extra peaks in addition to the pyro and isopyro isomers. This dehydration is variable and occurs with all hydroxylated steroids (142), but 100% conversation to the 25-dehydro derivative can sometimes be achieved by introducing powdered glass at the top of the GC column, thus allowing use of the dehydro peaks for quantitation.

The use of powdered glass to achieve dehydration is not always effective, however, and more consistent results can be obtained if required using aluminum powder at 400°C. Although it would be expected that pyro and isopyro isomers would still be formed during chromatography, only one major peak is observed for each metabolite during GC-MS, probably because the isopyro isomer has only low-intensity ions. The mass spectra of dehydrated 25-OH-D$_3$ is shown in Figure 21. Dehydration can occur in a number of ways; for example, loss of the

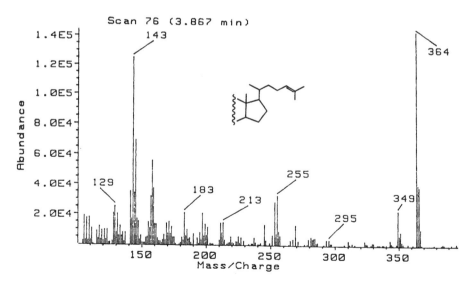

Figure 21 EI(+) mass spectrum of dehydration product of 25-OH-D$_3$. There are two possible dehydration products but only one is illustrated. Note the relatively high intensity of the molecular ion.

25-hydroxyl could lead to a double bond between C24 and C25 or between C25 and C26. Use of a high-resolution capillary column indicates that two isomers are formed during dehydration of 25-OH-D, and use of stable isotope-labeled compounds has confirmed that both the expected dehydration products are formed. Dehydration of dihydroxylated vitamin D metabolites also occurs, and the products from 24,25-$(OH)_2D_3$ and 25,26-$(OH)_2D_3$ are easily resolved during GC because in 25,26-$(OH)_2D_3$ both the 25- and 26-hydroxyls cannot be removed simultaneously and the remaining hydroxyl makes the product more polar than that from 24,25-$(OH)_2D_3$. Use of dehydration as a simple GC-MS method for plasma 25-OH-D_3 has been described (132). Modification or derivatization of vitamin D and its metabolites may offer mass spectrometric advantages, and, as an example of this, dehydration gives rise to more intense high mass ions which is a positive advantage in developing specific and sensitive MS assays (*vide infra*).

Formation of derivatives on the polar hydroxyl groups of vitamin D and metabolites, although it improves stability and volatility, does not overcome the problem of the B-ring closure and the formation of the two pyro and isopyro peaks. The formation of these isomers is quantitative, and either or both peaks can be used for measurement (143–147), but the separation of a number of metabolites in a single run can be complicated by the multiplicity of peaks. Because the 25-hydroxyl is sterically hindered, it is difficult to derivatize, and only two reagents have proved satisfactory for the low levels present in plasma—trimethylsilylimidazole (TMSI), and a mixture of bis-trimethylsilyltrifluoroacetamide (BSTFA):trimethylchlorosilane (TMCS) (3:1, v/v). Early work on the GC separation of vitamin D and its metabolites used conventional packed columns with relatively high carrier gas flows. Great improvement in separation is now achieved by using capillary columns, which can also be linked directly to the mass spectrometer as the carrier gas flow rate is low (1 to 2 mL/min). The end of the column can thus be inserted directly into the ion source of the spectrometer, eliminating the loss of analyte in separator systems, which were originally necessary to remove the carrier gas.

Trimethylsilyl ether derivatives have been widely used for GC-MS, but as can be seen from Figure 22, they are not always ideal derivatives for mass fragmentography (MF) in the vitamin D field, since intensities of ion fragments of high mass are low but they do offer some advantages in structural elucidation. The mass spectra of TMS ethers of 25-hydroxylated vitamin D_3 metabolites all contain a base peak at m/z 131 (the fragment obtained by side chain cleavage, between C24 and C25, containing C26, C27, and C25-O-TMS). 26-Hydroxylation increases the mass of this peak to m/z 219 because of the presence of the extra 26-trimethylsilanol group. Figure 22 also illustrates the EI(+) mass spectra obtained for a mixed t-BDMS/TMSi derivatives of 25-OH-D_3 in comparison to that obtained from 25-OH-D_3 per-TMSi. One further way of enhancing high-mass ions can be applied to quadrupole mass spectrometers, because they show

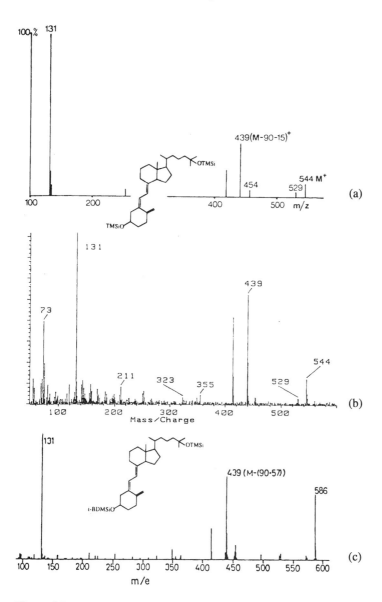

Figure 22 EI(+) mass spectra obtained from 25-OH-D$_3$ as (a) per-trimethylsilyl (TMS) ether on a single focusing magnetic sector GC-MS (LKB 2091); (b) per-TMS ether on a benchtop quadrupole system tuned to high mass; and (c) mixed TMS ether/t-butyldimethylsilyl ether on a magnetic sector GC-MS (LKB 2091). (From Ref. 208.)

mass discrimination and have electronic means of changing the intensity of high mass ions. Figure 22 also shows the enhancement of high-mass ions that can be achieved using a single benchtop quadrupole system for 25-OH-D$_3$-per-TMSi. Comparison of the mass spectra illustrated in Figure 19 with that in Figure 22 shows the value of derivatization on TMSi ether in that the number of hydroxyls (n) can easily be ascertained as the M$^+$ of the TMSi ether is $n \times 72$ daltons higher than the M$^+$ from the underivatized molecule. In addition, formation of TMSi ethers also enhances the distinction between oxo and hydroxy compounds.

An interesting derivative for 24,25-(OH)$_2$D$_3$ has been described by Lisboa and Halket (147a), who formed a cyclical boronate across the adjacent 24- and 25-hydroxyl groups, as previously described for steroids (148–150). Halket et al. (151) obtained mass-spectrometric and retention times for methyl- and n-butylboronate-3-trimethylsilyl ether derivatives of 24,25-(OH)$_2$D$_3$ and found that the use of the cyclic boronate derivative stabilized the molecule and gave rise to considerably enhanced intensity of ions of high-mass/charge ratio. Figure 23 compares the normalised mass spectra of 24,25-(OH)$_2$D$_3$-TMS with that obtained from the methylboronate TMS derivative, showing the greatly increased mass fragments at m/z 381, (M-131)$^+$, the fragment probably obtained by A-ring cleavage containing C2, C4, and C3-O-TMS, and m/z 407, $(M - 90 + 15)^+$. The use of n-butylboronate-trimethylsilyl derivatives also has the advantage that these derivatives of 24,25-(OH)$_2$D$_3$ and 25,26-(OH)$_2$D$_3$ separate during GC on a nonselective stationary phase (152), and such derivatives have been used in the development of mass-fragmentographic assays for both these vitamin D metabolites (153). Vicinal hydroxyls and 1,3-*cis*-diols can both form cyclic boronates, and n-butyl, n-phenyl, and n-methylboronates of 24,25-(OH)$_2$D$_3$ can easily be formed. Interestingly, only n-butyl- and n-phenylboronates can be formed with ease from 25,26-(OH)$_2$D$_3$; n-methylboronates appear not to be formed (152).

Use of other derivatives, such as cyclic boronates, can, as illustrated for 24,25-(OH)$_2$D$_3$, greatly enhance the intensities of higher-mass fragments more suitable for use in MF analysis. It will be noted that dehydration also provides intense high-mass ions, also suitable for improved sensitivity of measurement and enhanced specificity. Both mass spectra in Figure 23 were obtained using the pyro peak, since formation of these derivatives does not prevent the thermal cyclization of the B ring. Other derivatives have been used in attempts to improve the mass spectral characteristics, and studies have been carried out using t-butyldimethylsilyl (t-BDMS) ether derivatives (154), which have been shown to give greatly enhanced intensity of the (M-57)$^+$ ion fragment for steroids (155).

The mass spectrometer can be used as a detector for gas and liquid chromatography in two ways. The most sensitive method relies on the ability of the MS to focus on specific ion fragments. There are numerous ways of arranging this that depend upon the type of mass spectrometer being used. The spectrometer can be focused on a single ion fragment or on different ion fragments at different

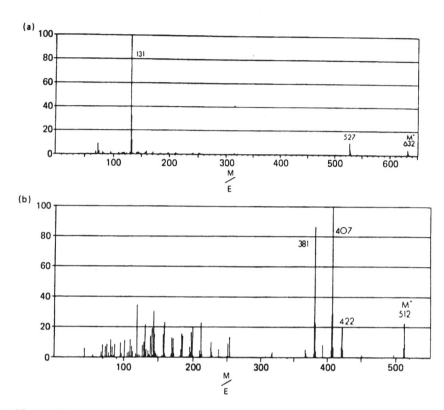

Figure 23 Normalized mass spectra of two derivatives of 24,25-(OH)$_2$D$_3$ during GC-MS on a 50-m OV-17 capillary column at 260°C. In each case the pyro peak was scanned. (a) 24(R),25-(OH)$_2$D$_3$-tris-trimethylsilyl ether; (b) 3-Trimethylsilyl ether-24,25-methylboronate ester derivative of 24(R),25-(OH)$_2$D$_3$. (From Ref. 151.)

times. The dwell time on each fragment is of course directly related to sensitivity. In a single-focusing magnetic sector machine, the focusing on different fragments can be achieved by altering the accelerating voltage so that different ions are sequentially focused on the electron multiplier. Modern laminated magnets now allow rapid alteration of the magnetic field without associated hysteresis effects, and focusing can thus be achieved by altering the current through the magnet. Quadrupole mass spectrometers operate on a different principle and focus ions by altering radiofrequency and direct current applied across four rods through the axis of which the ions travel. As mentioned previously, quadrupole mass spectrometers are assuming increasing importance in analytical laboratories. The process of mass fragmentography is the most sensitive mode of operation, but an alternative approach can be adopted. All modern mass spectrometers are now

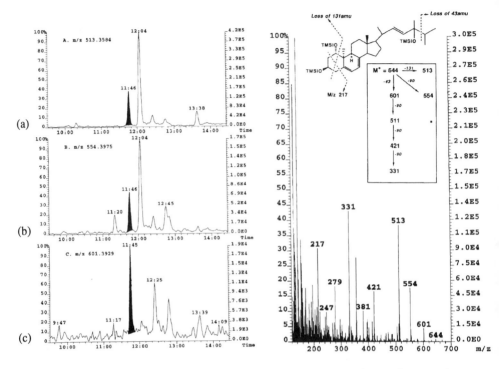

Figure 24 High-resolution mass fragmentography of an extract of serum from a patient taking vitamin D_2. When injected into the heated zone of the gas chromatograph, vitamin D molecules are cyclized to form pyro- and isopyroisomers. The mass spectral interpretation inserted in this figure has been applied to the pyroisomer. Ion chromatograms of per-trimethylsilylated (TMSi) $1\alpha,24-(OH)_2D_2$ monitoring three separate ions—m/z 513.3584 (a), m/z 554.3975 (b), and m/z 601.3929 (c)—showing the trace between 9 and 14 min. The peaks from the pyroisomer of $1\alpha,24-(OH)_2D_2$-TMSi are shaded. (From Ref. 209.)

equipped with sophisticated data acquisition and manipulation computing systems and thus are capable of scanning the eluent from the GC or LC at regular intervals (one to five scans per second) and storing the complete mass spectrum obtained. At the end of the GC/LC run data can be recalled and total-ion or single-ion chromatograms can be constructed. Such a system (mass chromatography) is not as sensitive as MF but provides a very valuable method of establishing the ion fragment to monitor for the best sensitivity and specificity. Sensitivity and specificity can be improved by the use of high-resolution GC-MS which is achieved by double-focusing instruments which give greatly improved signal: noise ratio by careful focusing on specified ions in MF mode. An example of the value of this technology is illustrated in Figure 24, which was used to demonstrate the presence of $1,24-(OH)_2D_2$ in human plasma.

F. LC Detection Using Online LC-MS

The introduction of liquid samples into the mass spectrometer has proved especially difficult. However, the advent of electrospray ionisation (ESI) and atmospheric pressure chemical ionisation (APCI) have overcome this problem and revolutionized the interfacing of HPLC with mass spectrometry. This success has pointed the way toward a major change in the mass spectrometric analysis of vitamin D, its analogs, and their metabolites. A disadvantage at the present time is that ionization using ESI and APCI is less efficient than the EI(+) used in GC-MS, but the immediate advantage is that it obviates the need for derivatization.

1. Direct Infusion

Metabolites separated by HPLC can be collected and directly infused into the mass spectrometer using standard microsyringe pumps. This technique provides valuable ESI or APCI mass spectra without the need to develop or modify existing HPLC systems which may not be immediately suitable for direct linkage to a mass spectrometer (i.e., straight-phase systems which may require postcolumn addition of aqueous solvent prior to the ion source of the MS). A second advantage is that direct infusion allows the performance of $(MS)^n$ which is possible only with ion trap technology. This technique allows the sample to be examined, mass spectra obtained, and specific ions to be selected and subjected to further ionization, giving rise to daughter ions (i.e., MS-MS or MS^2). This process can be repeated to look at granddaughter ions (MS^3) and so on. This enables a scheme of sequential fragmentation to be drawn up and indicates which ions generate the daughter and granddaughter ions. Such a technique should be very valuable in structural analysis.

A specific example of these methods is illustrated in Figure 25, where a metabolite of the CD-ring analog KS176 (156) was subjected to ESI(+) MS3 analysis using an ion trap MS system (Thermoquest, LCQ system, Hemel Hempstead, U.K.). This analog was, even when derivatized as a per-TMSi ether, was unstable in GC-MS systems. It can be seen in the MS1 that the MH^+ ion at m/z 391 readily dehydrates to a series of product ions at m/z 373, 355, and 337 $(MH-18-18-18)^+$, corresponding to the loss of the three hydroxyl groups. This successive fragmentation is confirmed by examination of MS2 and MS3 which represent successively the fragments of m/z 373 (MS2) and m/z 355 (MS3). Examination of MS1 indicates the presence of MNa^+ (m/z 413), MNH_4^+ (m/z 408) and MK^+ (m/z 429) adduct ions.

The HPLC can be directly linked to the mass spectrometer (HPLC-MS) and solute ions produced in a variety of different ways. In the vitamin D field, the techniques of thermospray, fast atom bombardment (FAB), ESI, and APCI have been used.

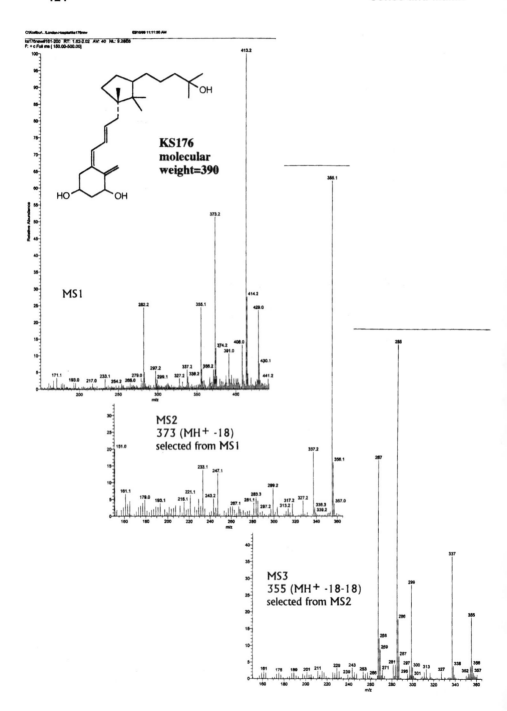

KS176
molecular
weight=390

MS1

MS2
373 (MH⁺ -18)
selected from MS1

MS3
355 (MH⁺ -18-18)
selected from MS2

2. Thermospray MS

Takamura et al. (157,158) have used this technique to identify the presence of vitamin D_2 in the shiitake mushroom. Watson et al. (159) used thermospray LC-MS to measure $1,25\text{-}(OH)_2D_3$ in mitochondrial incubations and also examined a plasma extract but concluded that while sensitivity was reasonable, the precision that would be attained using this technique was inadequate. LC-MS using thermospray of phenyl-4-TAD adducts of vitamin D_3, $25\text{-}OH\text{-}D_3$, $1,25\text{-}(OH)_2D_3$, $24,25\text{-}(OH)_2D_3$, and $1\alpha OHD_3$ has also been described by Vreeken et al. (118). Both positive and negative ion spectra were examined, and simple spectra with protonated and deprotonated molecular ions were seen. These workers reported better signal-to-noise ratios with negative ion spectra in distinction to the work of Watson et al. (159), who could find no signal when monitoring negative ions. Sensitivity of this procedure was good (1 to 7 nmoles/L) but not sufficient for use in the analysis of serum or plasma. Unlike other procedures, this method utilized postcolumn formation of the phenyl-4-TAD adduct which was effected by mixing the derivatizing reagent in a reaction coil after the LC separation but before MS.

Thermospray is said to be inherently unstable and Watson et al. (159) suggested that this technique might never be suitable for routine plasma vitamin D metabolite measurement. This view, however, was not endorsed by Vreeken et al. (118), who suggested that their system was sufficiently sensitive for the measurement of vitamin D and several of its metabolites but not for $1,25\text{-}(OH)_2D_3$, which would still require large volumes of plasma. Vicchio et al. (160) have used this technique to measure $25\text{-}OH\text{-}D_3$ and study its kinetics.

3. Continuous-Flow FAB

Formation of TAD adducts has also proved to be of value in improving the prospects of successful LC-MS and the use of continuous-flow FAB of phenyl-4-TAD adducts of vitamin D and its metabolites in a tandem mass spectrometer (117,200) has been described. The FAB mass spectra produced in M1 showed

---

Figure 25 MS-MS-MS of KS 176. Three successive fragmentations were carried out using an LCQ ion trap mass spectrometer (Thermoquest, Hemel Hempstead, U.K.) which allows for LC-MS^n, where n can equal 10. In MS1 the ions which arise from successive loss of from the MH^+ ion (m/z 391) are seen [$(MH^+\text{-}18) = 373$, $(MH^+\text{-}18\text{-}18) = 355$, and $(MH^+\text{-}18\text{-}18\text{-}18) = 337$]. These ions are then selected for subsequent fragmentation in MS2 and MS3. We suggest that the ion m/z 282 in MS1 arises by cleavage between C24 and C25 of m/z 337. Prominent ions in MS3, m/z 267, and 285 possibly arise from A-ring cleavages from m/z 337 and 355. The use of MS^n in this relatively simple and cheap ion trap LC-MS system provides much more structural information than that provided from a simple LC-MS.

an intense peak at m/z 298 and for 1-hydroxylated vitamin D metabolites this peak shifted to m/z 314. By monitoring these fragments in M2 after collision-induced dissociation of MH^+, selected by M1, sensitive and specific measurements of vitamin D and its metabolites could be effected detecting 100 to 300 pg of all metabolites examined. In this system the detection limit for 1,25-$(OH)_2D_3$ was 150 pg (with a signal:noise ratio of 2.3:1). This methodology has been used by Yamada's group to demonstrate the structure of the 23-glucuronide of 23S,25-$(OH)_2$-24-oxo-D_3 and show that it is a major biliary metabolite of 24,25-$(OH)_2D_3$ (161).

4. Electrospray (ESI) and Atmospheric Pressure Chemical Ionisation (APCI)

The introduction of ESI- and APCI-MS has improved the potential of LC-MS. The work of Wilson et al. (127) using crown ether derivatives of TAD has demonstrated that adducts with vitamin D using this reagent can be readily formed and that electrospray MS detection limits are close to physiological concentrations. Although at the present time these methods appear to have potentially sufficient sensitivity for use for the measurement of plasma concentrations, no fully evaluated method using LC-MS has been described, although capillary HPLC electrospray tandem mass spectrometry has been used to study in vitro metabolism of 1,25-dihydroxy-16-ene vitamin D_3 in rat kidney with detection limits in the 50 to 100-pg level (162). HPLC-MS is particularly useful in the field of vitamin D analysis since most metabolites of vitamin D can be resolved by a suitable combination of HPLC systems, and the added specificity of both derivatization and subsequent mass spectrometry would be a very powerful and specific means of quantitation.

It would be preferable if LC-MS with sufficient sensitivity could be carried out without the necessity for derivatization. Previous work has suggested that using LC-MS underivatized vitamin D compounds are not detected with satisfactory sensitivity. However, work in our laboratories has indicated that underivatized vitamin D analogs can be detected using LC-APCI-MS at the low nanogram level. APCI in our hands gives spectra which are in essence the same as those produced by ESI (see Fig. 26 and compare with MS1 spectrum in Fig. 25). LC-APCI-MS has recently been used to identify the structure of 25-OH-D_2 and 25-OH-D_3 glucuronides (163,164), and Ishigai et al. (165) have used LC-ESI-MS

Figure 26 LC-APCI-MS of KS 176. (a) Diode array detector output; (b) total ion current trace; (c) APCI (+) spectrum; (d) UV spectra obtained from an online photodiode array detector sampling at the peak apex. The formula of KS 176 is given in Figure 25. The interpretation of fragmentation seen in the mass spectrum shown is the same as that observed in MS1 of Figure 25. Data courtesy Dr. J. Lepore, Waters.

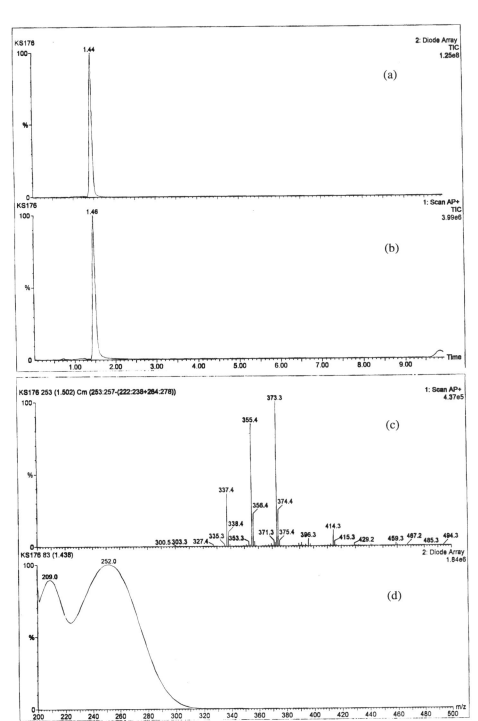

Table 7 Analytical Requirements for the Analysis of Vitamin D Metabolites and/or Analogs in Different Matrices

Analytical problem	Analytical requirements					Selected Solution
	Precision	Accuracy/ identity	Speed	Sensitivity	Purification	
1. Clinical (serum and other biological fluids). Known analyte in low concentration	+	+	+	+	Complex	SPE + saturation analysis (GC-MS for validation only)
2. In vitro (organs, cells or enzymes) identity		+			Simple	LC-MS-MS or GC-MS
3. Time course: Unknown analytes in sometimes complex but defined matrix.	+	+	+		Simple	LC-UV
3. Foodstuff: known analyte in complex matrix with large lipid content	+	+			Very complex	LC-UV
4. Pharmaceuticals (tablets and other preparations). Known analyte in defined matrix at high concentration.	+	+	+		Simple	LC-UV or LC-fluorescence (MS for validation)

Table 8 Gas Chromatography-Mass Spectrometry Methods for the Analysis of Vitamin D, Analogs, and Metabolites in Biological Fluids (1980–1999)

Analyte	Derivative used	Internal standard	Concentration	Ref.
D$_2$	TMSi	[6,19,19-^{2}H$_3$]-D$_2$	136 mg/mL in urine (268 mg/mL after enzyme hydrolysis)	185
D$_3$	TMSi	[3,4-^{2}H$_2$]-D$_3$*	120 mg/ml in urine (up to 500 ng/ml after enzyme hydrolysis) 4.5 ± 2.8 ng/ml in serum	185
25-OH-D$_3$	TMSi	[26(27)-^{2}H$_3$]-25-OH-D$_3$	21.4 ± 5.3 ng/ml in serum	185
25-OH-D$_3$	TMSi-tBDMS		15.3 ± 1.6 ng/mL	187
			19.2 ± 2.0 ng/mL	188
25-OH-D$_2$	TMSi	[26,27-^{2}H$_6$]-25-OHD$_3$	2.3–32.6 ng/mL in plasma	189
	TMSi	D$_2$	0.6–1.0 ng/mL	186
	TMSi	[26,27-^{2}H$_6$]-25-OHD$_3$	0.12–6.6 ng/mL in plasma	189
24,25-(OH)$_2$D$_3$	TMSi	Vitamin D$_2$	0.6–2.9 ng/mL	186
24,25-(OH)$_2$D$_3$	TMSi-nBBA	[26,27-^{2}H$_6$]-24,25-(OH)$_2$D$_3$	1.6 ± 0.6 ng/mL (UK summer) 3.1 ± 1.4 ng/mL (Australian summer)	152
	TMSi-nBBA	[6,19,19-^{2}H$_3$]-24,25-(OH)$_2$D$_3$	Rat serum (no values but linear response to added 24,25(OH)$_2$D$_3$)	190
25,26-(OH)$_2$D$_3$	TMSi-nBBA	[26,27-^{2}H$_6$]-25,26-(OH)$_2$D$_3$	Mean 0.39 ng/mL (UK plasma) mean 0.76 ng/mL range 0.3–1.3 ng/mL (Australian plasma)	153
24,25-(OH)$_2$D$_2$	TMSi-nBBA	[26,27-^{2}H$_6$]-25,26-(OH)$_2$D$_3$	ND in normal plasma. 4.1–29.5 ng/mL on D$_2$	189
25,26-(OH)$_2$D$_2$	TMSi-nBBA	[26,27-^{2}H$_5$]-25,26-(OH)$_2$D$_3$	ND in normal plasma 0.85–2.11 ng/mL on D$_2$	189
1,25-(OH)$_2$D$_3$	TMSi	[26-^{2}H$_3$]-25,26-(OH)$_2$D$_3$	Validation of calf thymus assay	191
1,25-(OH)$_2$D$_3$	TMSi	Vitamin D$_2$	Semiquantitative (three serum samples in range 22.0–38.1 pg/mL). Lower values than calf thymus assay.	192
[26,27-F6]-1,25-(OH)$_2$D$_3$	TMSi	[H$_2$][26,27-F6]-1,25(OH)$_2$D$_3$	28 ± 10 pg/mL in 12 volunteers 4 hr after 2 µg oral dose	133

* At a later stage [6,19,19-^{2}H$_3$]-D$_3$, [2,2,3,4,4,6-^{2}H$_6$]-D$_3$, and [2,2,3,4,4,6,19,19-^{2}H$_8$]-D$_3$ were used as internal standards.

for the measurement of 22-oxacalcitriol in mass serum. It is clear that the development of LC-MS systems is proceeding rapidly with improvements in sensitivity and ease of operation. We anticipate that LC-MS will become an increasingly valuable tool in the analysis of vitamin D and its chemical analogs. However, GC-MS can still provide valuable but complementary information (see 166) and should not be discarded simply on the grounds of convenience.

III. APPLICATIONS

We have considered four applications of the methodology reviewed in this chapter and the requirements of each are summarized in Table 7. It can be seen that each application utilizes a slightly different approach based upon the consideration of a number of analytical requirements which differ between each of the applications. Examples of each of these applications are considered in turn.

A. Clinical

In the main this application deals with the analysis of plasma or serum for 25-OH-D_3 and 1,25-$(OH)_2D_3$, two analytes which are present at low concentrations (25-OH-D_3 >> 1,25-$(OH)_2D_3$) in a relatively complex medium. It would appear at first sight that this presents a very difficult analytical challenge requiring the most extensive purification prior to quantitation. In the routine clinical laboratory, methods which are time-consuming and technically difficult are not as valuable as simpler and possibly, but not always, less specific methods. There is often a need in clinical situations for speed and simplicity and because of this, a considerable amount of development has taken place to produce simple and convenient assays in kit form, which depend upon the specificity of the saturation analysis quantitation procedure. The result is a method which couples solid-phase extraction with a radioreceptor assay which is both sensitive and specific. LC-MS meth-

---

Figure 27 The EI(+) mass spectra of metabolites of 1α-hydroxyvitamin D_3 (1α-OHD$_3$). GC-MS was carried out after derivatization to form the per-trimethylsilyl ethers as described in Ref. 115. Both pyro- and isopyroisomers of each metabolite were observed but the mass spectrum of the pyroisomer (the major peak) is shown in each case. The major ions (m/z 632 (M^+), m/z 542, 432, and 362 (not highlighted) (M^+ losing successive silanols) and m/z 501 (M^+ losing 131 by A-ring cleavage) are the same in all the spectra. M/z 217 is the characteristic ion always seen in these 1,2X-dihydroxylated steroids and m/z 251 (not highlighted) arises by side chain cleavage and subsequent loss of three silanols. It is possible, however, to distinguish each isomer from the characteristic fragmentation patterns illustrated for each above the appropriate spectrum. Only important ions have been highlighted. Full spectra for each isomer are available from the authors.

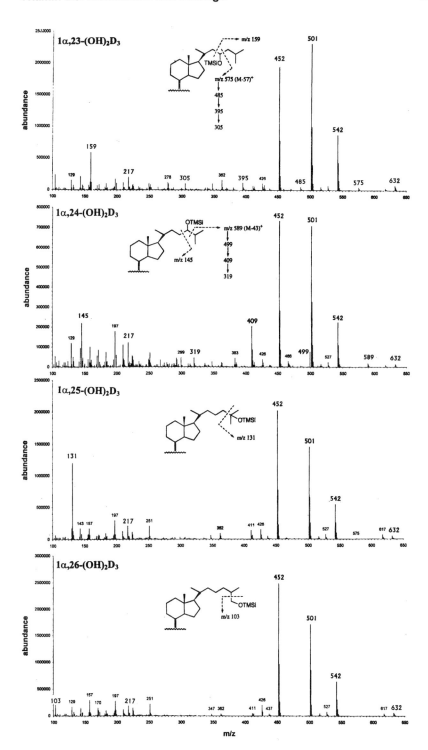

ods are, however, still used in some centres and this methodology has been discussed above (see section II.D.1). Some examples of the application of LC-UV to the measurement of vitamin D metabolites in human plasma can be seen in Shimada et al. (167,168), who describe a method for the measurement of 25-OH-D$_3$ and its sulfate in human plasma, and in Shimizu et al. (169), who describe a method for 25-OH-D$_3$ in plasma using 22-OH-tetranor-D3 as an internal standard.

Experience has shown that these assays can perform exceeding well but only if the user belongs to an external quality assurance scheme (EQAS) which monitors performance and uses target values determined by GC-MS (e.g., 170). Table 8 illustrates some GC-MS methods which have been published in the last 20 years, not all of which are suitable for use as reference methods.

B. In Vitro Studies

The metabolism of a large number of vitamin D analogs has been studied over the last 10 years and these studies have focused on two major aspects—firstly, identification of the principal metabolites formed in vitro using a variety of cell lines. This first application places particular emphasis on the ability of the detection methods to reveal details of metabolite structure which requires the use of GC-MS. Figure 27 illustrates the application of this technique to the identification of four monohydroxylated metabolites of 1α-OH-D$_3$ formed in a biological system. After extraction and derivatization, GC-MS was carried out and the spectra illustrated in Figure 27 were obtained. From these spectra it is possible to infer

---

Figure 28 Reproducibility of modern HPLC instrumentation. Multiple chromatographic profiles in this figure are overlaid to demonstrate the superior reproducibility of modern chromatographs, such as the Waters Alliance 2690 system. The experiment involves lipid extracts from human keratinocytes incubated with synthetic vitamin D analogs: 1α,25-(OH)$_2$D$_2$ and 1,24S-(OH)$_2$D$_2$ (10μM) for various times ranging from 0 to 48 h. Overlays show the remarkably consistent retention times that can be achieved for all components in the extract from the internal standard (labeled I.S., which is 1α-OH-D$_3$) with a retention time of 4.85 min through to the medium-derived component (retention time 27.2 min). When the LC method is combined with a consistent extraction procedure (giving remarkably similar I.S. peak heights), the biological time course can be seen to give valuable clues to metabolite identity and origin. In this relatively simple example, 1α,25-(OH)$_2$D$_2$ (RT = 11.42 min) gives a single product 1α,24,25-(OH)$_3$D$_2$ (RT = 18.05 min), whereas 1α,24S-(OH)$_2$D$_2$ (RT = 10.06 min) gives two products—1α,24,25-(OH)$_3$D$_2$ (RT = 17.99 min) and 1α,24,26-(OH)$_3$D$_2$ (RT = 19.18 min). In this case, peak identity was confirmed by offline GC-MS of derivatized (per-trimethylsilyl ethers) metabolites but currently available LC-MS instrumentation allows for online mass spectrometry of un-derivatized peaks, giving immediate information about peak identity.

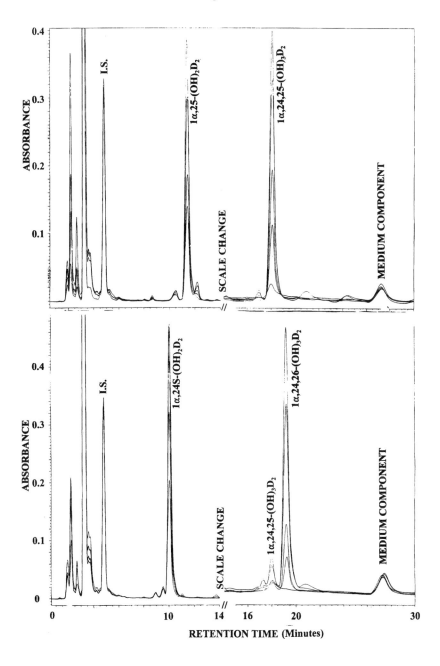

the structures of these metabolites. The approach can be extended to metabolites from other analogs of calcitriol.

The second aspect of this application is the need to provide reproducible time course data for metabolites of a number of analogs which can be compared with a necessary degree of confidence. This is most easily achieved using LC and requires that one set of data can be overlaid on another set which may have been produced at a later or earlier time. This can only be achieved if the HPLC used has a high degree of precision and retention times remain constant over long periods of time. Major advances have been made in pump, injector, and column technology, which makes this now possible to achieve. Such an example is provided in Figure 28 which shows the comparative metabolism of 1,24(S)-$(OH)_2D_2$ and 1,25-$(OH)_2D_2$ by HPK1A-ras cells analyzed using a Waters Alliance System. These two studies were performed on the same batch of keratinocytes at the same time but each analysis involves separate extraction replicates different time points and synthetic standards which extend the HPLC analysis time over a period of days. The reproducibility of retention times demonstrated in this figure is remarkable and could not have been achieved by previous generation of HPLC instruments.

C. Foodstuffs and Pharmaceuticals

Some applications in this area are provided in Jones et al. (81). Table 9 summarizes some HPLC methodology which has more recently been used for the mea-

Table 9   Some Liquid Chromatographic Methods for the Analysis of Foods and Pharmaceuticals for Vitamin D, Analogs, and Metabolites Published since 1990

Matrix	Analyte(s)	Methodology	Ref.
Food (milk powder and gruel, eggs, fish, milk and margarine)	Vitamins D_2 and D_3	Saponification, LC on silica gel, RP-HPLC-UV	193
Infant formula	Vitamin D	HPLC-UV	194
Milk	Vitamins D_2 and D_3	Saponification, Florisil LC and RP-HPLC-UV	195
Milk (liquid and powder)	Vitamin D_3	SPE on C18 cartridge and HPLC-UV	196
Infant formula and enteral products	Vitamin D	Liquid-liquid extraction, saponification, SPE, RP-HPLC-UV	197
Animal feeds	Vitamin D and pro-D_2	Sep-Pak cartridge extraction HPLC-UV	198

surement of vitamin D in a variety of matrices. Despite the complexity of some of these matrices the methodology used is based upon HPLC-UV alone and there appears to be no recent attempt to validate these procedures by mass spectrometry. There appear to be no publications using HPLC for the measurement of vitamin D, analogs, or metabolites in pharmaceutical preparations since 1990.

REFERENCES

1. JW Blunt, HF DeLuca, HK Schnoes. Biochemistry 7:3317, 1968.
2. AW Norman, J Roth, L Orci. Endocrinol Rev 3:331, 1982.
3. HL Henry, AW Norman. Annu Rev Nutr 4:493, 1984.
4. MR Haussler. Annu Rev Nutr 6:527, 1986.
5. HF DeLuca. FASEB J 2:224, 1988.
6. CE Porteous, RD Coldwell, DJH Trafford, HLJ Makin. J Steroid Biochem 28:785, 1987.
7. HLJ Makin, G Jones, MJ Calverley. In: HLJ Makin, DB Gower, DN Kirk, eds. Steroid Analysis. Glasgow: Blackie Academic & Professional, 1995, p 562.
8. E Havinga. Experientia 29:1181, 1973.
9. WG Salmond. In: AW Norman, K Schaefer, DV Herrath, H-G Grigoleit, JW Coburn, HF DeLuca, EB Mawer, T Suda, eds. Vitamin D: Basic Research and Its Clinical Application. Berlin: Walter de Gruyter, 1979, p 25.
10. S Yamada, M Ohmuri, H Takayama. Tetrahedron Lett 21:1859, 1979.
11. N Kobayashi, T Taguchi, N Kanuma, N Ikekawa, J-I Oschide. J Chem Soc Chem Commun 10:459, 1980.
12. MJ Calverley. Tetrahedron Lett 43:4609, 1987.
13. SJ Shiuey, JJ Partridge, MR Uskokovic. J Org Chem 53:1040, 1988.
14. DH Barton, RH Hesse, MM Pechet, E Rizzardo. J Am Chem Soc 95:2748, 1973.
15. HE Paaren, DE Hamer, HK Schnoes, HF DeLuca. Proc Natl Acad Sci USA 75: 2080, 1978.
16. G Jones, N Edwards, D Vriezen, C Porteous, DJH Trafford, J Cunningham, HLJ Makin. Biochemistry 27:7070, 1988.
17. F Qaw, MJ Calverley, NJ Schroeder, DJH Trafford, HLJ Makin, G, Jones. J Biol Chem 268:282, 1993.
18. NJ Schroeder, DJH Trafford, J Cunningham, G, Jones, HLJ Makin. J Clin Endocrinol Metab 78:1471, 1994.
19. J Abe-Hashimoto, T Kikuchi, T Matsumoto, Y Nishii, E Ogata, K Ikeda. Cancer Res 53:2534, 1993.
20. K Kragballe. J Cell Biochem 49:46, 1992.
21. S Strugnell, V Byford, HLJ Makin, RM Moriarty, R Gilardi, LW Levan, JC Knutson, CW Bishop, G Jones. Biochem J 310:233, 1995.
22. WH Okamura, MM Midland, AW Norman, MW Hammond, MC Dormanen, I Nemere. Ann NY Acad Sci 761:344, 1995
23. MJ Calverley, G Jones. In: RT Blickenstaff, ed. Antitumour Steroids. San Diego: Academic Press, 1992, p 193.

24. R Bouillon, WH Okamura, AW Norman. Endocrinol Rev 16:200, 1995.
25. MH Bhattacharyya, HF DeLuca. J Biol Chem 248:2969, 1973.
26. S Andersson, DL Davis, H Dahlback, H Jornvall, DW Russell. J Biol Chem 246: 8222, 1989.
27. Y-D Guo, S Strugnell, DW Back, G Jones. Proc Natl Acad Sci USA 90:8668, 1993.
28. KI Okuda, E Usui, Y Ohyama. J Lipid Res 36:1641, 1995.
29. M Fukushima, Y Suzuki, Y Tohira, Y Nishii, M Suzuki, S Sasaki, T Suda. FEBS Lett 65:211, 1976.
30. D Baran, ML Milne. J Clin Invest 77:1622, 1986.
31. VM Berginer, S Shany, D Alkalay, J Berginer, S Dekel, G Salen, GS Tint, D Gazit. Metabolism 42:69, 1993.
32. H Postlind, E Axen, T Bergman, K Wikvall. Biochem Biophys Res Commun 241: 491, 1997.
33. G Tucker, RE Gagnon, MR Haussler. Arch Biochem Biophys 155:47, 1973.
34. E Axen, H Postlind, J Wikvall. Biochem Biophys Res Commun 215:136, 1995.
35. F Ichikawa, K Sato, M Nanjo, Y Nishii, T Shinki, N Takahashi, T Suda. Bone 16: 129, 1995.
36. DR Fraser, E Kodicek. Nature 228:764, 1970.
37. R St-Arnaud, S Messerlian, JM Moir, JL Omdahl, FH Glorieux, J Bone Miner Res 12:1552, 1997.
38. FH Glorieux, JM Moir, S Messerlian, JL Omdahl, R St-Arnaud. In: AW Norman, R Bouillon, M Thomasset, eds. Vitamin D. Chemistry, Biology and Clinical Applications of the Steroid Hormone. Vitamin D Workshop, Riverside, California: University of California, 1997, p 127.
39. K-I Takeyama, S Kitanaka, T Sato, M Kobori, J Yanagisawa, S Kato. Science 277: 1827, 1997.
40. T Monkawa, T Yoshida, S Wakino, T Shinki, H Anazawa, HF DeLuca, T Suda, M Hayashi, T Saruta. Biochem. Biophys Res Commun 239:527, 1997.
41. GK Fu, AA Portale, WL Miller. DNA Cell Biol 16:1499, 1997.
42. GK Fu, D Lin, MY Zhang, DD Bikle, CH Shackleton, WL Miller, AA Portale. Mol Endocrinol, 11:1961, 1997.
43. RW Gray, JL Omdahl, JG Ghazarian, HF DeLuca. J Biol Chem 247:7528, 1972.
44. HL Henry. J Biol Chem 254:2722, 1979.
45. S Kitanaka, K Takeyama, A Murayama, T Sato, K Okumura, M Nogami, Y Hasegawa, H Niimi, J Yanigisawa, T Tanaka, K Sato. N Engl J Med 338:653, 1998.
46. D Fraser, SW Kooh, HP Kind, MF Holick, Y Tanaka, HF DeLuca. N Engl J Med 289:817, 1973.
47. GE Lester, TK Gray, ME Williams. In: DV Cohn, RV Talmage, JL Matthews, eds. Hormonal Control of Calcium Metabolism. Amsterdam: Excerpta Medica, 1981, p 376.
48. JS Adams, MA Gacad. J Exp Med 161:755, 1985.
49. MF Holick, HK Schnoes, HF DeLuca, RW Gray, IT Boyle, T Suda. Biochemistry 11:4251, 1972.
50. MF Holick, A Kleiner-Bossaller, HK Schnoes, PM Kasten, IT Boyle, HF DeLuca. J Biol Chem 249:6691, 1973.

51. A Ornoy, D Goodwin, D Noff, S Edelstein. Nature 276:517, 1978.
52. H Rasmussen, P Bordier. Metab Bone Dis Rel Res 1:7, 1978.
53. HL Henry, AW Norman. Science 201:835, 1978.
54. G Jones, D Vriezen, D Lohnes, V Palda, NS Edwards. Steroids 49:29, 1987.
55. G Makin, D Lohnes, V Byford, R Ray, G Jones. Biochem J 262:173, 1989.
56. GS Reddy, K-Y Tserng. Biochemistry 28:1763, 1989.
57. RP Esvelt, HK Schnoes, HF DeLuca. Biochemistry 18:3977, 1979.
58. JC Knutson, HF DeLuca. Biochemistry 13:1543, 1974.
59. Y Ohyama, M Noshiro, K Okuda. FEBS Lett 278:195, 1991.
60. Y Ohyama, K Okuda. J Biol Chem 266:8690, 1991.
61. M Tomon, HS Tenenhouse, G Jones. Endocrinology 126:2868, 1990.
62. T Shinki, CH Jin, A Nishimura, Y Nagai, Y Ohyama, M Noshiro, K Okuda, T Suda. J Biol Chem 267:13757, 1992.
63. D Lohnes, G Jones. J Nutr Sci Vitaminol Special issue. 75, 1992.
64. R St-Arnaud, A Arabian, R Travers, FH Glorieux. J Bone Miner Res 12:33, 1997.
65. R St-Arnaud, A Arabian, R Travers, FH Glorieux. In: AW Norman, R Bouillon, M Thomasset eds. Vitamin D. Chemistry, Biology and Clinical Applications of the Steroid Hormone. Vitamin D Workshop, Riverside, California: University of California, 1997, p 635.
66. T Suda, HF DeLuca, HK Schnoes, Y Tanaka, MF Holick. Biochemistry 9:4776, 1970.
67. RL Horst. Biochem Biophys Res Commun 89:286, 1979.
68. S Yamada, K Nakayama, H Takayama, T Shinki, Y Takasaki, T Suda. J Biol Chem 259:884, 1984.
69. M Akiyoshi-Shibata, T Sakaki, Y Ohyama, M Noshiro, K Okuda, Y Yabusaki. Eur J Biochem 224:335, 1994.
70. M Beckman, P Tadikonda, E Werner, JM Prahl, S Yamada, HF DeLuca. Biochemistry 35:8465, 1996.
70a. S Ishizuka, S Ishimoto, AW Norman. Biochemistry 23:1473, 1984.
71. G Jones, A Rosenthal, D Segev, Y Mazur, F Frolow, Y Halfon, D Rabinovich, Z Shakked. Biochemistry, 18:1094, 1979.
72. M Odrzywolska, M Chodynski, SJ Halkes, JP van de Velde, H Fitak, A Kutner. Chirality, 11:249, 1999.
73. MJ Wheeler. The Immunoassay Kit Directory, Vol 2, part 2. Steroid and Thyroid Hormones. Lancaster: Kluwer Academic, 1993.
74. RD Coldwell, DJH Trafford, MJ Varley, DN Kirk, HLJ Makin. Steroids, 55:418, 1990.
75. DA Seamark, DJH Trafford, HLJ Makin. J Steroid Biochem 14:111, 1981.
76. DD Bikle. Assay of Calcium Regulating Hormones. New York: Springer-Verlag, 1983.
77. RL Horst. In: R Kumar, ed. Vitamin D. Basic and Clinical Aspects. Boston: Martinus Nijhoff, 1985, p 423.
78. TL Clemens. Trends Endocrinol Metab 1:129, 1990.
79. MF Holick. J Nutr 120:1464, 1990.
80. RL Horst, TA Reinhardt, BW Hollis. Kidney Int 38:S28, 1990.
81. G Jones, DJH Trafford, HLJ Makin, BW Hollis. In: AP De Leenheer, WE Lambert,

HJ Nelis, eds. *Modern Chromatographic Analysis of Vitamins*. New York: Marcel Dekker, 1992, p 73.

82. BW Hollis. In: D Feldman, FH Glorieux, JW Pike, eds. Vitamin D. San Diego: Academic Press, 1997, p 587.
83. EG Bligh, WJ Dyer. Can J Biochem 37:911, 1957.
84. RE Belsey, HF DeLuca, JT Potts. J Clin Endocrinol Metab 38:1046, 1974.
85. R Bouillon, P DeMoor, EG Baggiolini, MR Uskokovic. Clin Chem 26:562, 1980.
86. MT Parviainen, KE Savolainen, PH Korhonen, EM Alhava, JK Visakorpi. Clin Chim Acta 114:233, 1981.
87. NJM Jongen, WJF Van der Vijgh, HJJ Willems, JC Netelenbos. Clin Chem 27: 444, 1981.
88. RL Horst, ET Littledike, JL Riley, JL Napoli. Anal Biochem 116:189, 1981.
89. JA Eisman, AJ Hamstra, BE Kream, HF DeLuca. Science 193:1021, 1976.
90. RS Mason, D Lissner, C Reek, S Posen. In: AW Norman, K Schaefer, DV Herrath, H-G Grigoleit, JW Coburn, HF DeLuca, EB Mawer, T Suda, eds. Vitamin D: Basic Research and Its Clinical Application. Berlin: Walter de Gruyter, 1979, p 243.
91. TA Reinhardt, RL Host, JW Orf, BW Hollis. J Clin Endocrinol Metab 58:91, 1984.
92. JS Adams, T Clemens, MF Holick. J Chromatogr 226:198, 1981.
93. JT Dabek, M Harkonen, O Wahlroos, H Adlercreutz. Clin Chem 27:1346, 1981.
94. H Turnbull, DJH Trafford, HLJ Makin. Clin Chim Acta 120:65, 1982.
95. LJ Fraher, G Jones, TL Clemens, S Adami, JLH O'Riordan. Acta Endocrinol 97(suppl 243), 1981 Abstract 8.
96. BW Hollis, NE Frank. J Chromatogr 343:43, 1985.
97. BW Hollis. Clin Chem 32:2060, 1986.
98. BW Hollis, T Kilbo. In: AW Norman, K Schaffer, eds. Vitamin D; Molecular, Cellular and Clinical Endocrinology. Berlin: Walter de Gruyter, 1988, p 710.
99. NJM Jongen, WJF Van der Vijgh, P Lips, SC Netelenbos. Nephron 36:230, 1984.
100. BW Hollis, BA Roos, HH Draper, PW Lambert. J Nutr 111:384, 1981.
101. AA Rhedwi, DC Anderson, GN Smith. Steroids 39:149, 1982.
102. SJ Gaskell, BG Brownsey. Clin Chem 29:677, 1983.
103. WD Fraser, BH Durham, JL Berry, EB Mawer. Ann Clin Biochem 34:632, 1997.
104. N Kobayashi, K Ueda, M Tsutsumi, Y Tabata, K Shimada. J Steroid Biochem Mol Biol 44:93, 1993.
105. N Kobayashi, H Mano, T Imazu, K Shimada. J Steroid Biochem Mol Biol 54:217, 1995.
106. G Jones, HF DeLuca. J Lipid Res 16:448, 1975.
107. G Jones. J Chromatogr Biomed Appl 221:27, 1980.
108. N Ikekawa, N Koizumi. J Chromatogr 119:227, 1976.
109. EW Matthews, PGH Byfield, KW Colston, IMA Evans, LS Galante, I MacIntyre. FEBS Lett 48:122, 1974.
110. M Ishigai, S Arai, Y Ishitani, K Kumaki. J Steroid Biochem Mol Biol 66:281, 1998.
111. S Masuda, V Byford, R Kremer, HLJ Makin, N Kubodera, Y Nishii, A Okazaki, T Okano, T Kobayashi, G Jones. J Biol Chem 271:8700, 1996.
112. AM Rosenthal, G Jones, SW Kooh, D Fraser. Am J Physiol 239:E12, 1980.

113. J Cunningham, RD Coldwell, G Jones, HS Tenenhouse, DJH Trafford, HLJ Makin. J Bone Miner Res 5:173, 1990.

114. D Lohnes, G Jones. J Biol Chem 262:14394, 1987.

115. FJ Dilworth, GR Williams, A-M Kissmeyer, JL Nielsen, E Binderup, MJ Calverley, HLJ Makin, G Jones. Endocrinology 138:5485, 1998.

116. DA Seamark, DJH Trafford, PG Hiscocks, HLJ Makin. J Chromatogr 197:271, 1980.

117. B Yeung, P Vouros, GS Reddy. J Chromatogr 645:115, 1993.

118. RJ Vreeken, M Honing, BLM van Baar, RT Ghijsen, GJ deJong, UAT Brinkman. Biol Mass Spectrom 22:621, 1993.

119. JC Miller, SA George, BG Willis. Science 218:241, 1982.

120. G Jones, M Kung, K Kano. J Biol Chem 258:12920, 1983.

121. G Jones, K Kano, S Yamada, T Furusawa, H Takayama, T Suda. Biochemistry 23:3749, 1984.

122. DJ Aberhart, AC-T Hsu. J Org Chem 41:2098, 1976.

123. K Shimada, T Oe, T Mizuguchi. Analyst 116:1393, 1991.

124. M Shimizu, S Yamada. In: AW Norman, R Bouillon, M Thomasset, eds. Vitamin D. Gene Regulation, Structure-Function Analysis and Clinical Application. Berlin: Walter de Gruyter, 1991, p 644.

125. M Shimizu, S Kamachi, Y Nishii, S Yamada. Anal Biochem 194:77, 1991.

126. M Shimizu, T Yamazaki, S Yamada. Bio-org Med Chem Lett 3:1809, 1993.

127. SR Wilson, ML Tulchinsky, YH Wu. Bio-org Med Chem Lett 3:1805, 1993.

128. PH Jordan, G Read, T Hargreaves. Analyst 116:1347, 1991.

129. M Shimizu, Y Gao, T Aso, K Nakatsu, S Yamada. Anal Biochem 204:258, 1992.

130. K Shimada, K Mitamura, H Kaji, M Morita. J Chromatogr Sci 32:107, 1994.

131. K Shimada, I Nakatani, K Saito, K Mitamura. Biol Pharm Bull 19:491, 1996.

132. RD Coldwell, DJH Trafford, HLJ Makin. J Mass Spectrom 30:348, 1995.

133. S Komuro, Nakatsuka, A Yoshitake, K Iba. Biol Mass Spectrom 23:33, 1994.

134. F Askew, RB Bourdillon, HM Bruce, RK Callow, J Philpot, TA Webster. Proc R Soc Ser B 107:76, 1932.

135. B Pelc, DH Marshall. Steroids, 31:23, 1978.

136. D Sklan, P Budowski, M Katz. Anal Biochem 56:606, 1978.

137. DA Seamark, DJH Trafford, HLJ Makin. J Steroid Biochem 13:1057, 1980.

138. PP Nair, C Bucana, S DeLeon, DA Yruner. Anal Chem 37:631, 1965.

139. PP Nair, S DeLeon. Arch, Biochem Biophys 128:663, 1968.

140. H Ziffer, WJA Vanden Heuvel, EOA Haahti, EC Horning. J Am Chem Soc 82:6411, 1960.

141. HLJ Makin, DJH Trafford. In: E Reid, ed. Methodological Surveys in Biochemistry, Vol 7. Chichester: Ellis Horwood, 1978, p 312.

142. DJH Trafford, RD Coldwell, HLJ Makin. J Pharm Biomed Anal 9:1095, 1992.

143. B Zagalak, HC Curtius, R Foschi, G Wipf, V Redweik, MJ Zagalak. Experientia, 34:1537, 1978.

144. I Bjorkhem, A Larsson. Clin Chim Acta 88:559, 1978.

145. I Bjorkhem, I Holmberg. Clin Chim Acta 68:215, 1976.

146. I Bjorkhem, I Holmberg, T Christiansen, JL Pedersen. Clin Chem 25:584, 1979.

147. JM Halket, BP Lisboa. Acta Endocrinol (Kbh) 87(suppl 251):120, 1978.

147a. BP Lisboa, JM Halket. In: A Frigerio, L Renoz, eds. Recent Developments in Chromatography and Electrophoresis. 1979, p 141.
148. CJW Brooks, BS Middleditch. Clin Chim Acta 34:145, 1971.
149. CJW Brooks, DJ Harvey. J Chromatogr, 54:193, 1971.
150. TA Baillie, CJW Brooks, BS Middleditch. Anal Chem 44:30, 1972.
151. JM Halket, I Ganschow, BP Lisboa. J Chromatogr 192:434, 1980.
152. RD Coldwell, DJH Trafford, HLJ Makin, MJ Varley, DN Kirk. Clin Chem 30: 1193, 1984.
153. RD Coldwell, DJH Trafford, MJ Varley, HLJ Makin, DN Kirk. J Chromatogr Biomed Appl 338:289, 1984.
154. B Lindback, T Berlin, I Bjorkhem. Clin Chem 33:1226, 1987.
155. G Phillipou, DA Bigham, RF Seamark. Steroids 26:516, 1975.
156. P De Clercq, C D'Halleweyn, S Gabriels, B Linclau, K Sabbe, B Sas, S Sebastian, H Van Dingenen, D Van Haver, Y Chen, S Ling, X Xu Y Wu, Y Wu, X Zhao, X Zhou, G-D Zhu, M Vandewalle, R Bouillon, A Verstuyft. In: AW Norman, R Bouillon, M Thomasset, eds. Vitamin D: Chemistry, Biology and Clinical Applications of the Steroid Hormone. Riverside: University of California, 1997, p 3.
157. K Takamura, H Hoshino, N Harima, T Sugahara, H Amano. J Chromatogr 543: 241, 1991.
158. K Takamura, H Hoshino, T Sugahara, H Amano. J Chromatogr 545:201, 1991.
159. DI Watson, KDR Setchell, R Ross. Biomed Chromatogr 5:153, 1991.
160. D Vicchio, A Yergey, K O'Brien, L Allen, R Ray, M Holick. Biol Mass Spectrom 22:53, 1993.
161. A Shimoyamada, S Tomiyama, M Shimizu, K Yamamoto, S Kunii, S Yamada. Biochim Biophys Acta 1346:147, 1997.
162. B Yeung, P Vouros, ML Siu-Caldera, GS Reddy. Biochem Pharmacol 49:1099, 1995.
163. K Shimada, Y Kamezawa, K Mitamura. Biol Pharm Bull 20:596, 1997.
164. K Shimada, K Mitamura, I Nakatani. J Chromatogr 690: 348, 1997.
165. M Ishigai, Y Asoh, K Kumaki. J Chromatogr 706:261, 1998.
166. M Ishigai, Y Ishitani, K Kumaki. J Chromatogr 704:11, 1997.
167. K Shimada, K Mitamura, N Kitama. Biomed Chromatogr 5:229, 1995.
168. K Shimada, K Mitamura, N Kitama, M Kawasaki. J Chromatogr 689:409, 1997.
169. M Shimizu, Y Iwasaki, H Ishida, S Yamada. J Chromatogr 672:63, 1995.
170. GD Carter, J Nolan, DJH Trafford, HLJ Makin. In: AW Norman, R Bouillon, M Thomasset, eds. Vitamin D: Chemistry, Biology and Clinical Applications of the Steroid Hormone. Riverside: University of California, 1997, p 737.
171. G Jones. In: JP Bilezikian, LG Raisz, GA Rodan, eds. Principles of Bone Biology. San Diego: Academic Press 1996, p 1069.
172. G Jones, SA Strugnell, HF DeLuca. Physiol Rev 78:1193, 1998.
173. MF Holick, HK Schnoes, HF DeLuca, T Suda, RJ Cousins. Biochemistry 10:2799, 1971.
174. PW Lambert, PB DeOreo, BW Hollis, IY Fu, DJ Ginsberg, BA Roos. J Lab Clin Med 98:536, 1981.
175. RE Belsey, HF DeLuca, JT Potts. J Clin Endocrinol Metab 33:554, 1971.
176. JN Thompson, G Hatina, WB Maxwell, S Duval. J Assoc Off Anal Chem 65:624, 1982.

177. EE Delvin, M Dussault, FH Glorieux. Clin Biochem 13:106, 1980.
178. JP Moody, CA Humphries, SM Allan, CR Paterson. J Chromatogr 530:19, 1990.
179. JP Hart, MD Norman, CJ Lacey. Analyst 117:1441, 1992.
180. H Hasegawa. J Chromatogr 605:215, 1992.
181. MMD Zamarreno, AS Perez, CG Perez, JH Mendez. J Chromatogr 623:69, 1992.
182. MMD Zamarreno, AS Perez, MCM Iglesias, JH Mendez. Anal Lett 26:2565, 1993.
183. L Lin, NM Metzler, IL Honigberg. J Liquid Chromatogr 16:3093, 1993.
184. S Masuda, T Okano, M Kamao, Y Kanedai, T Kobayashi. J Pharm Biomed Anal 15:1497, 1997.
185. B Zagalak, F Neuheiser, MJ Zagalak, T Kuster, HC Curtius, GU Exner, S Franconi, A Prader. In: A Frigerio, ed. Chromatography and Mass Spectrometry in Biomedical Sciences. Amsterdam: Elsevier, Vol 2, p 347, 1983.
186. DA Seamark, DJH Trafford, HLJ Makin. Clin Chim Acta 106:51, 1980.
187. T Berlin, L Emtestam, I Bjorkhem. Scand J Clin Lab Invest 46:723, 1986.
188. T Berlin, I Holmberg, I Bjorkhem. Scand J Clin Lab Invest 46:367, 1986.
189. RD Coldwell, DJH Trafford, MJ Varley, DN Kirk, HLJ Makin. Clin Chim Acta 180:157, 1989.
190. S Tomiyama, T Nitta, S Yamada. Steroids 59:559, 1994.
191. H Oftebrow, JA Falch, I Holmberg, E Hang. Clin Chim Acta 176:157, 1988.
192. PMK Poon, YT Mak, CP Pang Clin Biochem 26:461, 1993.
193. A Bognar. Z Lebensm Unters Forsch 194:469, 1992.
194. JT Tanner, SA Barnett, MK Mountford. J AOAC Int 76:1042, 1993.
195. AF Hagar, L Madsen, L Wales, HB Bradford. J AOAC Int 77:1047, 1994.
196. MM Delgado-Zammarreno, A Sanchez-Perez, MC Gomez-Perez, J Hernandez-Mendez. J Chromatogr 694:399, 1995.
197. MG Sliva, JK Sanders. J AOAC Int 79:73, 1996.
198. H Qian, M Sheng. J Chromatogr 825:127, 1998.
199. MR Haussler, GK Whitfield, CA Haussler, J-C Hsieh, PD Thompson, SS Selznick, CE Dominguez, PW Jurutka. J Bone Miner Res 13:325, 1998.
200. B Yeung, GS Reddy, P Vouros. Abstracts for the 40th American Society for Mass Spectrometry Conference on Mass Spectrometry and Allied Topics, Washington DC, 1992, p 1089.
201. P Stern. Calcif Tissue Int 33:1, 1981.
202. JE Bishop, ED Collins, WH Okamura, AW Norman. J Bone Miner Res 9:1277, 1994.
203. M Shimizu, X Wang, S Yamada. J Chromatogr 690:15, 1997.
204. XX Wang, M Shimizu, F Numano, H Asaoka, S Yamada. Anal Sci 13:255, 1997.
205. K Shimada, T Mizuguchi. J Chromatogr 606:133, 1992.
206. K Wang, PP Davis, T Crews, L Gabriel, RW Edom. Anal Biochem 243:28, 1996.
207. S Strugnell. PhD thesis, Queen's University, Kingston, Ontario, Canada, 1993.
208. I Bjorkhem, I Holmberg. Methods Enzymol 67:385, 1980.
209. EB Mawer, G Jones, M Davies, PE Still, V Byford, NJ Schroeder, HLJ Makin, CW Bishop, JC Knutson. J Clin Endocrinol Metab 83:2156, 1998.
210. G Jones, H Ramshaw, A Zhang, R Cook, V Byford, J White, M Petkovich. Endocrinology 140:3303, 1999.

3
Vitamin E

Hans J. Nelis, Eva D'Haese, and Karen Vermis
University of Ghent, Ghent, Belgium

I. INTRODUCTION

A. Historical Aspects (1–3)

In 1922, Evans and Bishop discovered a new nutritional factor, initially termed "factor X," later vitamin E and tocopherol, that was essential for a normal development of fetuses in pregnant rats. They found wheat germ oil to be a particularly rich source of this fat-soluble vitamin. Several compounds with vitamin E activity were subsequently isolated and named α-, β-, and γ-tocopherol. The structure of α-tocopherol was elucidated in 1937 and the next year its first synthesis was achieved. A variety of signs of vitamin E deficiency have been reported in birds and small mammals but rarely in humans. The exact biochemical function of vitamin E in the latter is subject to ongoing debate. In the early years of vitamin E research several bioassays were developed, e.g., the fetus resorption test in rats.

The once popular Emmerie-Engel analytical method involves the reduction of ferric to ferrous ions, reaction of the latter with 2,2'-dipyridyl and colorimetric measurement. For a long time this method remained the backbone of the majority of chemical assays for vitamin E.

B. Physiochemical Properties

1. Tocopherols, Tocotrienols, and Tocopheryl Esters (1,3,4)

Vitamin E is a collective term for tocopherols and tocotrienols. Their basic structural element is tocol or 2-methyl-2-(4',8',12'-trimethyltridecyl)-6-chromanol. The four naturally occurring tocopherols, α-, β-, γ-, and δ-tocopherol (abbreviated as α-T, β-T, γ-T, δ-T) as well as the synthetic 5,7-dimethyltocol (5,7-T) can be regarded as substituted tocols and differ in the number of methyl groups on the chroman ring, as illustrated in Figure 1. Their tocotrienol counterparts possess an unsaturated side chain containing three double bonds and are accordingly designated α-, β-, γ-, and δ-tocotrienol (abbreviated as α-T3, β-T3, γ-T3, δ-T3). β-, γ-Tocopherol and 5,7-dimethyltocol are positional isomers. Since tocopherols contain three asymmetric centers, i.e., at position 2 of the chroman ring and at carbons 4' and 8' of the terpenoid side chain, they can exist as eight diastereoisomers or four enantiomeric pairs. Natural α-tocopherol occurs in the 2R, 4'R, 8'R configuration. It was formerly called d-α-tocopherol and now (2R, 4'R, 8'R)-α-tocopherol or RRR-α-tocopherol. The (2S, 4'R, 8'R) diastereoisomer was designated as l-α-tocopherol and has been renamed 2-*epi*-α-tocopherol. Synthetic tocopherols are composed of different stereoisomers depending on the optical purity of the starting material. The synthesis starting from natural phytol yields (2RS, 4'R, 8'R)-tocopherol or 2-*ambo*-α-tocopherol, a mixture of the natural and the 2-*epi* forms. From synthetic phytol (isophytol) a total racemate, i.e., (2RS, 4'RS, 8'RS)-α-tocopherol or all-*rac*-α-tocopherol (formerly dl-α-tocopherol) is obtained. 4'-*ambo*-8'-*ambo*-α-tocopherol, prepared by hydrogenation of natural α-tocotrienol consists of four diastereoisomers with the R configuration at the 2 position.

All possible forms of α-tocopherol together with their chemical, designated, and trivial names are listed in Table 1. Tocotrienols naturally occur in the (2R), 3'-*trans*, 7'-*trans* configuration. Tocopherols are viscous, pale yellow oils that are soluble in fat, alcohols, and nonpolar organic solvents but not in water.

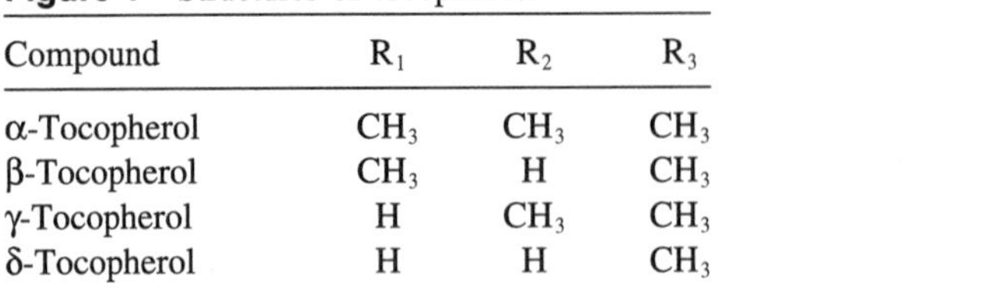

Figure 1 Structures of tocopherols.

Compound	R_1	R_2	R_3
α-Tocopherol	CH_3	CH_3	CH_3
β-Tocopherol	CH_3	H	CH_3
γ-Tocopherol	H	CH_3	CH_3
δ-Tocopherol	H	H	CH_3
5,7-Dimethyltocol[a]	CH_3	CH_3	H
Tocol[a]	H	H	H

[a] Synthetic compound.

Table 1 α-Tocopherol Stereoisomers and Racemates

Chemical name	Designated name	Trivial name
2 R 4′R 8′R	RRR-α-tocopherol	d-α-tocopherol
2 R 4′R 8′S		
2 R 4′S 8′S		
2 R 4′S 8′R		
2 S 4′R 8′R	2-*epi*-α-tocopherol	1-α-tocopherol
2 S 4′R 8′S		
2 S 4′S 8′S		
2 S 4′S 8′R		
2 RS, 4′R, 8′R	2-*ambo*-α-tocopherol	
2 R, 4′RS, 8′RS	4′-*ambo*-8′-*ambo*-α-tocopherol	
2 RS, 4′RS, 8′RS	all-*rac*-α-tocopherol	dl-α-tocopherol

In ethanol, their UV spectra display a maximum at 292 to 298 nm with $E_{1cm}^{1\%}$ values of 70 to 73.7 (292 nm), 86 to 87 (297 nm), 90 to 93 (298 nm), and 91.2 (298 nm) $mol^{-1}cm^{-1}$ for α-, β-, γ-, and δ-tocopherol, respectively. Acylation of the hydroxyl group shifts the maximum hypsochromically and hypochromically, as in α-tocopheryl acetate (α-TA) (λ_{max} = 284 nm, $E_{1cm}^{1\%}$ = 43.6 $mol^{-1}cm^{-1}$). The acetate and other esters have weak native fluorescence, unlike the free tocopherols, which emit light at 330 to 340 nm upon excitation at 295 nm or 210 nm. These tocopherol esters are more stable than the tocopherols but are biologically inactive before their in vivo conversion to the free forms. Particularly the acetate is frequently used in pharmaceuticals, infant formulas, and fish feed. α-Tocopheryl hydrogenosuccinate has also found some application. Owing to the presence of a hydroxyl group on the chroman ring, the free tocopherols possess strong reducing properties and, hence, can be readily oxidized, both chemically and electrochemically.

The order of in vitro antioxidant activity of E vitamers is generally considered α-T > β-T > γ-T > δ-T, although the exact reverse has also been reported.

2. Decomposition Products, Metabolites and Analogs of α-Tocopherol

The reaction of α-tocopherol with peroxyl radicals in vivo (see further D) leads to a number of oxidative decomposition products (5,6) (Fig. 2), of which α-tocopherolquinone (α-TQ) has been the most widely studied (7).

The latter can be further reduced to α-tocopherolhydroquinone, which is secreted in bile and feces as a glucuronide conjugate (5). Other oxidative urinary metabolites include α-tocopheronic acid and 2,5,7,8-tetramethyl-2-(2′-carboxy-

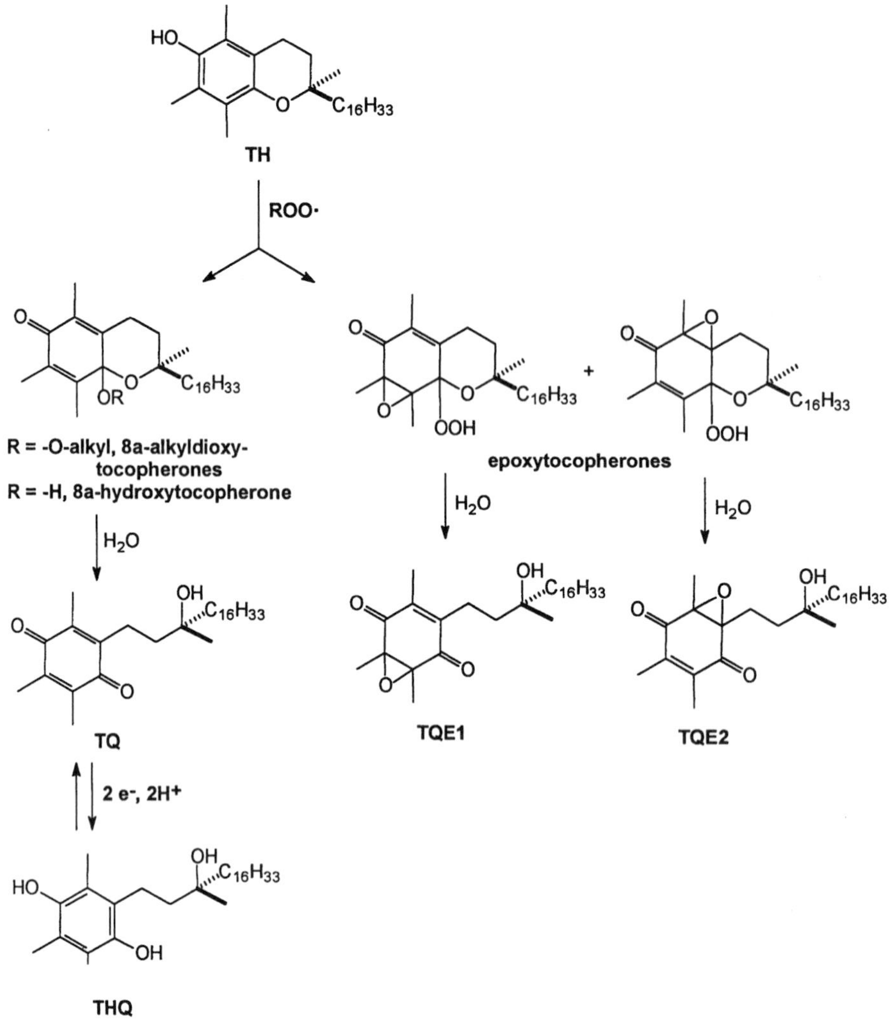

Figure 2 Structures of α-tocopherol and its oxidation products and pathways for product formation in peroxyl radical scavenging reactions. (From Ref. 6.)

ethyl)-6-hydroxychroman (α-CEHC) (8,9). However, less than 1% of the absorbed vitamin is excreted in the urine (5). α-Tocopherolquinone can also be formed by in vitro oxidation of α-tocopherol (4).

Studies on the oxidation chemistry of α-tocopherol are often performed using the model compound 2,2,5,7,8-pentamethyl-6-chromanol, an analog of α-tocopherol lacking the terpenoid side chain but having similar antioxidant properties (10). Advantages of this compound are its crystalline nature and its simplified spectral properties. Recently, synthetic water-soluble quaternary ammonium derivatives of α-tocopherol have been proposed as cardioselective antioxidants with therapeutic potential in connection with myocardial infarction (11,12). α-Tocopheryl nicotinate is used in cosmetic preparations to improve blood circulation and to stimulate hair growth (13).

C. Occurrence (1,3,5)

Tocopherols and tocotrienols are biosynthesized only by plants. In leaves, α-tocopherol is found mainly in the chloroplasts, whereas the γ- and δ-forms are extraplastidic. Seeds and grains also contain α-tocopherol but are often more abundant in γ-tocopherol. The richest sources of vitamin E are plant oils such as sunflower and wheat germ oil. Tocotrienols do not occur in green tissues but rather in the bran and germ of plants.

Since animals and humans are unable to biosynthesize vitamin E, they have to rely on an external supply of it through their diet. Consequently, animal tissues with the exception of adipose tissue and liver are poor in tocopherols. Tocotrienols are not resorbed to a significant extent. The recommended dietary allowance (RDA) of vitamin E is linked to the intake of polyunsaturated fatty acids (PUFAs) and is of the order of 10 mg α-tocopherol equivalents per day. α-Tocopherol is the predominant form in plasma (88%), whereas the β and γ forms contribute for only 2% and 10%, respectively, to the vitamin E levels. δ-Tocopherol is only detected after massive ingestion of, e.g., soybean fat emulsions. Tocopherols are nonspecifically transported by plasma β-lipoproteins and stored in membranes and adipose tissue.

D. Biological Function and Activity (3,5,14,15)

In spite of much speculation about yet to be elucidated cellular functions of vitamin E, the bulk of available evidence suggests that its biological effects are derived mostly or entirely from its antioxidant properties. As a scavenger of peroxyl radicals (R00°), vitamin E protects PUFAs from peroxidation and thus aids in the maintenance of the integrity of cellular membranes.

Lipid peroxidation proceeds through a chain reaction initiated by the attack of free radicals on PUFAs and propagated by reaction of the formed carbon-centered alkyl radicals (R°) with oxygen. The resulting peroxyl radicals can further abstract hydrogen atoms from PUFAs to produce finally hydroperoxides.

By trapping the peroxyl radicals vitamin E interrupts this process, and, hence, is rightly termed a chain-breaking antioxidant. In this process, α-tocopherol itself is converted to the α-tocopheroxyl radical, which reacts further with another peroxyl radical to produce tocopherones and finally α-tocopherolquinone (Fig. 2). The biological antioxidant action of vitamin E forms an integral part of a larger multicomponent cellular defense system including such elements as superoxide dismutase, catalase, glutathione peroxidase, and selenium. Apart from its protective effects on membrane lipids, vitamin E is also thought to play a role in counteracting the biological effects of oxyradicals. The latter have been linked to aging, cancer, Parkinson's disease, cataract, atherosclerosis, and cardiovascular diseases. Finally, vitamin E appears to be essential for the maintenance of a normal neurological structure and function.

In principle, vitamin E deficiency can result from insufficient dietary intake, malabsorption or excessive consumption in case of oxidative stress in biomembranes. While the former is virtually nonexistent in Western society, the impaired ability to utilize dietary fats may create hypovitaminosis E in certain risk populations, e.g., patients with pancreatic dysfunction or defects in lipoprotein metabolism and particularly in premature infants. Deficiency disorders in the latter include bronchopulmonary dysplasia, retrolental fibroplasia (retinopathy), intraventricular hemorrhage, hemolytic anemia, and neuromuscular anomalies.

As a measure of their biological activity, all E vitamers can be assigned a potency, determined in a bioassay. One international unit (IU) is defined as the activity of 1 mg of *all-rac*-α-tocopheryl acetate. Accordingly, the biopotencies of RRR-α-tocopheryl acetate, RRR-α-tocopherol and *all-rac*-α-tocopherol have been calculated to be 1.36, 1.49, and 1.10 IU, respectively. The chirality of the C-2 atom is crucial to the activity, the 2-epimers having only one-third of the potency of the 2R forms. Biopotencies of tocopherols and tocotrienols can be expressed relative to that of α-tocopherol. Reported values (in %) are 100 (α-tocopherol), 40 (β-tocopherol), 8 (γ-tocopherol), 1 (δ-tocopherol), 21 (α-tocotrienol), and 4 (β-tocotrienol). This order parallels that of the relative antioxidant activity of the various vitamers (see B).

E. Analytical Aspects

1. The Need for Assaying Vitamin E

Vitamin E Status of Humans.   As hypovitaminosis E is a rare event (see D), a routine assessment of the vitamin E status in humans is not required but only justified for risk groups (e.g., prematures) and as part of studies on the chemoprevention of diseases presumably involving oxidative stress as an etiological factor, e.g., cancer, atherosclerosis, and cataract. Knowledge about the in vivo fate of vitamin E (see C) will guide in the choice of samples to be analyzed. The most common, easily obtainable biological matrix for the determination of vitamin E and of α-tocopherol in particular is plasma/serum. Plasma levels of γ-tocopherol, which is more abundant in foods, are of little importance except in trials involving supplementation with α-tocopherol, where the γ-/α-tocopherol ratio may serve as an index for assessing compliance (16). However, the reliability of plasma/serum α-tocopherol as an indicator of the vitamin E status has been challenged (17). In view of the strong correlation that exists between vitamin E levels and those of total lipids and cholesterol, only a tocopherol/lipid ratio appears to be meaningful (17). Alternatively, the tocopherol contents of erythrocytes and platelets may even be more relevant because of their independence of the plasma lipid level and their direct relation to the site of action of vitamin E—i.e., biomembranes (17,18). The analysis of adipose tissue and liver may provide valuable information about vitamin E storage and body reserves and lipid metabolism disorders, e.g., abetalipoproteinemia (19). An assessment of the antioxidant status at the cellular level in relation to a risk for cancer and cataract would require samples of target tissues such as lung biopsies (20) and human lenses (21). Buccal mucosa cells have been proposed as a replacement for tissue samples that are difficult to collect (22). The antioxidant status of a cell may also be reflected by the ratio of α-tocopherol/α-tocopherolquinone because the latter is supposedly linked to the extent of oxidative stress (7). Finally, studies on the chemopreventive action of various antioxidants on cancer call for a simultaneous determination of vitamin E and other isoprenoid compounds, including vitamin A, carotenoids, and ubiquinones (23).

Vitamin E in Foods, Feeds, and Infant Formulas (3). The determination of tocopherols and tocotrienols in foods and feeds serves a dual purpose, namely nutritional assessment and quality control. To establish the vitamin E nutritional value of foods such as cereals, oils, margarine, and vegetables as well as infant formulas, both the naturally present tocopherols and the esters used for fortification should be quantitated.

In the past this was often done by hydrolyzing the latter and measuring the ''total'' tocopherol. The quality aspect relates to identity and antioxidant stability. Knowledge of the relative abundance of the various tocopherols and tocotrienols may aid in the characterization of vegetable oils and be useful for purity assessment or to detect possible adulteration. The contents of various E vitamers reflect the antioxidant potential in the foods and, hence, their shelf life. In addition, they also provide information about the possible deleterious effects (oxidation) of food processing and storage.

2. The Need for Chromatography

The need to determine tocopherols and tocotrienols in plasma/serum, red blood cells, platelets, tissues, foods, and feeds has been rationalized above. However, in these complex materials vitamin E is mostly only a minor constituent of the lipid fraction. Consequently, nonchromatographic assays such as the notoriously nonspecific Emmerie-Engel method require extensive sample pretreatment to remove interfering coextractants. In the 1950s and 1960s, colorimetric assays were coupled with chromatographic procedures, particularly open column (CC) and thin-layer chromatography (TLC), used mainly for cleanup purposes.

In 1962, gas liquid chromatography (GC) was welcomed as a breakthrough for the separation of tocopherols. However, when applied to biological materials, other lipid constituents, particularly cholesterol, which is often present in large excess, interfered with the quantitation of tocopherols. In addition, β- and γ-tocopherol proved very difficult to separate. Both problems could be solved by combining GC with yet another chromatographic technique such as CC and/or TLC. The resulting complex sample pretreatment schemes illustrate once again the need for a powerful chromatographic technique to quantitate vitamin E in biological materials. This requirement was fulfilled with the advent of high resolution chromatographic techniques, such as capillary GC, and particularly high-performance liquid chromatography (HPLC). The latter approach obviates the need for an exhaustive removal of lipids from sample extracts because, unlike in GC, they are no longer detected in the chromatographic run. Apart from providing specificity, chromatography has a second important function in vitamin E assays—the separation of the individual tocopherols and tocotrienols, their esters, and sometimes the stereoisomers. They all differ in biopotency (see D), so a determination of "total vitamin E" would not be meaningful in connection with nutritional assessment. The practice of hydrolyzing added α-tocopheryl acetate to α-tocopherol and quantitating the total α-tocopherol (hydrolyzed + natively present) likewise neglects these differences. In human samples, α-tocopherol is the predominant form of vitamin E and a systematic determination of γ-tocopherol has little relevance. However, its exclusive quantitation in foods is insufficient to predict their vitamin E activity (24). Furthermore, as mentioned in E.1, the relative abundance of all tocopherols and tocotrienols may serve as a fingerprint for the characterization of vegetable oils. Even more important, the various forms contribute to a different extent to the antioxidant stability of foods (see B.1). The discrimination of the stereoisomers of α-tocopherol is of interest to study their relative abundance and possible preferential accumulation in plasma and tissues.

Finally, modern chromatographic techniques will also provide the sensitivity (= low detection levels), lacking in nonchromatographic methods but necessary to quantitate vitamin E in very small samples such as blood from neonates and tissue biopsies.

3. Analytical Challenges and Aim of This Review

When the first edition of this book appeared in 1985, the field of vitamin E analysis was still in full expansion (25). Many GC and HPLC methods available at the time relied on complex sample pretreatment schemes, including saponification, CC, and TLC to remove interfering lipids, often at the expense of recovery and analyte stability.

Positional isomers had only been resolved in a few GC and TLC systems and by adsorption HPLC, with its known disadvantages relative to reversed-phase HPLC. All tocopherols and tocotrienols had rarely been separated in a single chromatographic run. Now, nearly 15 years later, these challenges have apparently ceased to exist. The majority of modern HPLC methods are based on simple extraction procedures, have a high degree of specificity for the target compounds, and effortlessly separate tocopherols and tocotrienols, including β- and γ-tocopherol. So the question can rightly be raised as to which innovative analytical concepts and/or new applications have warranted the influx of such a vast amount of recent analytical papers on vitamin E.

In this chapter, an attempt will be made to highlight the new trends in this area. The approach that was chosen is to briefly summarize the state of the art before 1989, based on the chapters on vitamin E in the two previous editions of this book and an earlier review paper (25–27) and to confront this with the new developments since then. Sections II through V are only concerned with vitamin E *sensu strictu*—i.e., tocopherols and tocotrienols. The chromatography of stereoisomers, oxidation products, metabolites, and tocopherol analogs is treated separately in sections VI.A through VI.C.

II. THIN-LAYER CHROMATOGRAPHY

A. General (28,29)

1. Introduction

Although TLC is an old technique, new developments including HP-TLC and densitometric quantitation have given it a new impetus and a more timely image. It would appear that the great variety of available stationary phases and possible developing solvent combinations should be able to meet all analytical challenges for vitamin E. However, the susceptibility of tocopherols and tocotrienols toward oxidation appears to constitute an argument against using a technique where the analytes are exposed to air on a large surface area and for prolonged periods of time.

2. Systems

Tocopherols and tocotrienols can be separated on silica gel but also, sometimes with greater efficacy, on special phases such as secondary magnesium phosphate or mixtures of silica gel/alumina with zinc carbonate. The tocopherols migrate in the order α-T > β-T > γ-T > δ-T (descending Rf values) and run ahead of the more polar tocotrienols. Reversed-phase systems based on silica gel/kieselguhr impregnated with paraffin or on C18 bonded phases also afford good resolution between the homologs. In contrast, the TLC separation of β- and γ-tocopherol is difficult. It has only been achieved in a two-dimensional mode or, occasionally in one-dimensional systems using complex, multicomponent mobile phases.

3. Detection

On TLC layers containing an inorganic fluorescent indicator the E vitamers appear as dark spots under UV light. Their detectability is considerably enhanced in the presence of sodium fluorescein. Alternatively, they can be visualized by spraying with chromogenic reagents such as α,α'-dipyridyl/Fe^{3+} (Emmerie-Engel reagent), bathophenanthroline/Fe^{3+}, phosphomolybdic acid, or a variety of other, specially designed mixtures.

4. Quantitation

Off-line quantitation involves the nondestructive localization of the spots of interest, elution of the intact tocopherols/tocotrienols with an organic solvent and finally colorimetry or gas chromatography. For a colorimetric determination the eluate is mixed with, e.g., the Emmerie-Engel reagent, whereas in case of gas chromatography the eluted fractions are evaporated to dryness and the residue is reconstituted and derivatized. For densitometry, the native UV absorbance or fluorescence of the E vitamers can be exploited. Alternatively, tocopherols can be converted to the more strongly absorbing tocopherolquinones or to colored spots by spraying with the above-mentioned reagents. Densitometric scanning has been carried out both in the transmittance and the reflectance mode.

5. Sample Pretreatment

Details on the different approaches for sample pretreatment in TLC, HPLC, and GC of vitamin E are given in III.A.5. In TLC of vitamin E, schemes are often complex, involving saponification, solvent extraction, and open column chromatography.

B. Status Before 1989 (25,28)

In the 1985 edition of this book, TLC received the credit of "occupying an important position among analytical methods in the vitamin E field." Before the advent of modern liquid chromatography and given the inability of the older GC systems to separate β- and γ-tocopherol, assays aiming at the differentiation of positional isomers nearly exclusively relied on TLC. Furthermore, TLC was frequently used as a lipid prefractionation step for complex biological samples prior to the quantitation of the E vitamers by colorimetry or GC. However, due to the poor resolving power of conventional TLC and the nonselective detection procedures, extensive sample purification was usually required. The situation changed with the introduction of HP-TLC and densitometry in the vitamin E area, which permitted a considerable simplification of the sample pretreatment. It became possible to subject unsaponifiable lipid fractions of serum, foods, and oils to TLC without cleanup. Finally, TLC found application for the separation of tocopherols from their oxidized decomposition products and from other fat soluble vitamins.

C. Status After 1989

A 1996 review concluded that "few recent" (published after 1980) papers are dealing with TLC of vitamin E (29). The once promising approach of HP-TLC densitometry has not been further followed up for vitamin E.

Two TLC methods for vitamin E still relying on the old approach of off-line quantitation were described in the early 1990s. Koswig and Mörsel separated α-, γ-, and δ-tocopherols in diluted, unsaponified vegetable oils on Kieselgel G plates, eluted the intact tocopherols with ethanol, and subjected each fraction to spectrophotometry at 292 to 295 nm (30). Hachula and Buhl determined α-tocopherol in capsules and soybean oil by TLC on silica gel after dilution of samples in acetone (31). The eluted tocopherol was quantitated colorimetrically using a bathophenanthroline/Fe^{3+}/bromophenol blue chromogenic reagent. A separation of all four tocopherols on Kieselguhr G impregnated with paraffin oil with visualization on the basis of the Emmerie-Engel reaction has been reported by Sliwiok et al. (32). These investigators also resolved the 2-epimers of α-tocopherol on a Chiralplate developed with a mixture of propanol-water-methanol (8.5:1.0:0.5, v/v/v). In general, the scarcity of recent applications suggests that, in spite of the availability of HP-TLC and densitometry and of its historical significance, TLC of vitamin E has finally lost its momentum.

III. LIQUID CHROMATOGRAPHY (LC)

A. General Concepts (25–27)

1. Introduction

Tocopherols and tocotrienols appear to unite all necessary physicochemical properties to make them the ideal analytes for a liquid chromatographic separation and quantitation. They are nonpolar, nonvolatile, unstable, and easily detectable owing to their favorable UV, fluorescent, and electrochemical characteristics. Their nonpolar nature together with the absence of silanol sensitive functional groups minimizes unwanted chromatographic phenomena such as peak tailing and low efficiency. Unlike GC, LC of vitamin E proceeds at room temperature and does not require derivatization to improve its chromatographic properties or to enhance its detectability. The combination of the powerful resolving capacity of microparticulate LC columns with selective fluorescence, electrochemical, and to a lesser extent absorption detection obviates the need for the complex sample pretreatment schemes that plagued the old GC and TLC methods for vitamin E. In LC, sample cleanup can be kept minimal because coextracted lipids will remain undetected and, hence, will not interfere with the chromatography of the target compounds.

2. Systems

Normal and reversed-phase chromatography have their own specific areas of application, with regard to both the separation of E vitamers and their quantitation in biological samples. It should be noted that in our terminology the term normal phase covers both adsorption chromatography on plain silica and chromatography on polar bonded phases. In keeping with its well-known stereochemical selectivity, silica is capable of resolving positional isomers. The least polar α-tocopherol has the lowest retention of all E vitamers, and each tocotrienol elutes after its tocopherol equivalent, yielding an elution order of α-T < α-T3 < β-T < β-T3 < γ-T < γ-T3 < δ-T < δ-T3. Optimal resolution between β- and γ-tocopherol and between all eight vitamers is obtained using binary mobile phases containing a hydrocarbon as a base solvent and an alcohol or an ether as a polar modifier. As it is rarely necessary to distinguish between β- and γ-tocopherol and never necessary between tocotrienols in human biological samples, normal-phase chromatography is primarily indicated for the analysis of foods and vegetable oils, where this need does exist (see 1.E.1). An additional advantage in this regard is that dilutions of oils or extracts of foods in hexane are compatible with the nonpolar eluent and, hence, can be directly injected. Among the polar bonded phases amino- and aminocyano, unlike cyano columns, are also capable of separating positional isomers.

Tocopherol and tocotrienol homologs can be separated on reversed-phase columns but the positional isomers coelute. The tocotrienols as a group elute ahead of the tocopherols, the latter being retained in the order δ-T < β + γ-T < α-T. Elution is most commonly carried out with mixtures of a polar organic solvent with water. Sample extracts are preferably dissolved in the chromatographic solvent or a solvent with weaker eluotropic strength to avoid a reduction of chromatographic efficiency. However, solubility problems might occur in samples containing a high lipid load. In such case there would be a rationale for using nonaqueous reversed-phase (NARP) chromatography on highly retentive packing materials with high carbon load and/or other retention-enhancing properties. On such materials stronger nonaqueous eluents yield a retention of nonpolar compounds comparable to that on less retentive phases eluted with semiaqueous solvents.

A second feature of these columns is that within a series of structural analogs they enhance the retention of the more polar (substituted) derivatives relatively more than that of their unsubstituted counterparts, resulting in their eluting more closely to each other. Thus, NARP has proven useful to chromatograph lipophilic compounds of widely divergent polarity isocratically, as for example retinol, α-tocopherol, and β-carotene. In other reversed-phase systems gradient elution would be required to produce the same effect. Gradient elution is also indicated to chromatograph simultaneously tocopherols, their more polar oxidation products, and their less polar dimeric forms. In general, reversed-phase columns are more stable, are faster to equilibrate, and yield more reproducible retention times than silica. Unlike normal-phase columns, they lend themselves to the concurrent chromatographic separation of tocopherols, retinoids, and carotenoids, which is meaningful in connection with studies on the chemopreventive effects of isoprenoid compounds with antioxidant activity on various diseases, notably cancer. Consequently, reversed-phase chromatography is the method of choice in routine analysis, except, as stated above, for vegetable oils and certain foods, where β- and γ-tocopherol and possibly β- and γ-tocotrienol should be differentiated. A relative disadvantage of reversed-phase systems is their incompatibility with the direct injection of biological extracts in hexane, the most common extraction solvent. However, the loss of time due to the need to evaporate the solvent is outweighed by the gain in detectability resulting from the extra concentration factor. Finally, reversed phase chromatography is uniquely suited to yield the ultimate detection sensitivity because electrolytes can be incorporated in the semi-aqeous eluents, so that electrochemical detection becomes feasible.

3. Detection

To detect tocopherols and tocotrienols advantage can be taken of their UV characteristics, their native fluorescence or their electrochemical activity (see 1.B.1). Although the UV absorbance of the E vitamers is relatively weak, detection at the absorption maximum (292 to 298 nm) using a variable wavelength detector affords sufficient sensitivity for most applications. Many investigators prefer detection at 280 nm using a cheaper, fixed-wavelength detector. A better signal-to-noise ratio and stability may partly compensate for the loss in absolute detectability caused by working at a nonoptimal wavelength. Rarely, 210 nm is chosen to avoid the interference of carotenoids in biological extracts. For the simultaneous detection of vitamin E, other fat-soluble vitamins (A, D, K), carotenoids, and ubiquinones, no common wavelength can be found to accommodate all compounds, so monitoring at multiple wavelengths is required. This can be achieved in a single run using multichannel detectors, wavelength programming, or diode array detection. When only vitamin E (mostly α-tocopherol) and vitamin A (retinol) are to be detected, a single wavelength of 280 or 292 nm is an acceptable compromise that leads to maximum sensitivity for the most weakly absorbing species (tocopherol) while not compromising too much the detectability of retinol (λ_{max} 325 nm), having a much higher absorptivity. Another but rarely adopted option is to connect a UV and a fluorescence detector in series. Over the years, fluorescence detection of vitamin E has nearly equaled absorbance detection in popularity. In most assays, the long wavelength maximum (292 nm) is used for excitation, but exciting at 205 nm yields a substantial gain in sensitivity, obviously at the expense of selectivity. A programmable fluorescence detector can be considered for the simultaneous determination of retinol and α-tocopherol at their respective wavelengths of excitation and emission. The gain in sensitivity and selectivity brought about by fluorescence detection has permitted a reduction in the size and extent of pretreatment of biological samples. Thus, saponification and double-phase extraction have been replaced by simpler single-phase extractions, without cleanup or concentration step (see III.E). Unprecedented sensitivity can be achieved with electrochemical detection, yielding detection limits in the picogram or, according to some reports, even subpicogram range. In addition, as tocopherols are oxidizable at low potential, excellent selectivity is also obtained. This has enabled an even further downscaling of the sample size and simplification of the isolation procedure.

Although a single example exists of a normal-phase method with postcolumn addition of an electrolyte, electrochemical detection of vitamin E is in principle restricted to reversed-phase chromatography, because only polar eluents can contain a sufficient concentration of a supporting electrolyte. Both amperometric and coulometric detectors have found application. Coulometry has an added value in that it can concurrently detect tocopherols, tocopherolquinones, and other quinones. To this end, a two-cell configuration placed in series is required. The first one acts as a reactor cell in which the quinones are reduced; in the second one the tocopherols and prereduced quinones are oxidized to produce the actual signal. In practice, electrochemical detection is the only approach that is capable of demonstrating low levels of α-tocopherolquinone in biological samples, because this compound is nonfluorescent and has poor UV absorptivity.

4.   Quantitation

To quantitate the E vitamers in LC, both external and internal standardization can be considered. The external standard approach involves the identical processing of unknown and standard samples supplemented with known concentrations of the analytes but in the absence of an internal standard. Calibration curves are constructed by plotting absolute peak heights or areas versus the corresponding concentrations. The use of modern sampling valves has largely eliminated imprecision caused by variations in injection, and the stability and reproducibility of LC systems for vitamin E also guarantee minimal chromatographic variability. Hence, external standardization may give good precision, provided the fluctuations and losses associated with the sample pretreatment can be kept under control. It is indeed more common to compensate for these variations by using a suitable internal standard. Accordingly, chromatographic responses will be expressed as peak height or peak area ratios. A good internal standard should be structurally similar to the analyte(s), have comparable physicochemical properties such as polarity, extractability, and detectability, be absent in the sample, and be stable during the sample preparation.

Few vitamin E–like compounds satisfy all these criteria but 5,7-dimethyltocol (Fig. 1) appears to be the best candidate. It possesses the basic tocol structure, differs from α-tocopherol by only one methyl group and from β- and γ-tocopherol only in the position of methyl substituents. Hence, as its polarity approaches that of the compounds of interest, particularly in terms of the number of carbon atoms, it will elute closely to them. Furthermore, its UV, fluorescence, and electrochemical properties are comparable to those of the natural tocopherols, and, being synthetic, it is absent in all biological samples. Unfortunately, up to recently, it has been very hard to acquire. In practice, tocol enjoys the highest popularity as an internal standard for tocopherols. However, although being structurally similar to them, it has a number of shortcomings. Its structure differs by 3 methyl groups from that of α-tocopherol so that the two compounds will not elute closely to each other in reversed-phase systems. In addition, it is reportedly unstable during saponification and is poorly extracted in the presence of an excess of lipids.

The second most frequently used internal standard for α-tocopherol is α-tocopheryl acetate. The two compounds possess the same carbon backbone but differ in that the 6-chromanol group is either free or esterified, respectively. Consequently, the difference in elution position will not be excessively high in reversed-phase systems. However, α-tocopheryl acetate has virtually no retention in normal-phase systems and is neither strongly fluorescent nor electrochemically active, so its use is virtually restricted to reversed-phase chromatography with absorption detection. It cannot be used in conjunction with isolation methods involving saponification as it is hydrolyzed to free α-tocopherol.

E vitamers that do not commonly occur in biological samples of animal and human origin, particularly δ-tocopherol, have also been suggested as internal standards for α-tocopherol. Retinyl acetate is often used in connection with the simultaneous determination of retinol and α-tocopherol, but sometimes also for the latter alone. It is combined with α-tocopheryl acetate and possibly a third internal standard if β-carotene is also included as an analyte. Finally, the synthetic tocopherol analog 2,2,5,7,8-pentamethyl-6-chromanol, which resembles a tocol lacking the isoprenoid chain has some practical advantages (see I.B.2.) but differs substantially from α-tocopherol in the number of carbon atoms, thus eluting too remote from the latter in reversed-phase systems. The question whether the above internal standards compensate not only for losses and uneven recovery during extraction but also for analyte instability is subject to debate, particularly in view of the differences that exist between tocopherol homologs in their susceptibility toward oxidative decomposition (see I.B.1.). The need for more suitable vitamin E analogs as candidate internal standards remains.

5. Sample Preparation

In principle, any lipid extraction procedure is useful for the isolation of tocopherols and tocotrienols from biological materials. Homogenous samples including plasma and vegetable oils do not require any pretreatment prior to the extraction itself. Solid samples are homogenized in an aqueous or semiaqueous medium or directly in the organic extracting solvent. Aqueous suspensions of erythrocytes and platelets are extracted as liquids. Five categories of sample preparation methods can be distinguished.

1.   Single-phase or monophasic extraction. In single-phase extraction, a liquid sample is either dissolved (e.g., vegetable oils in hexane) or diluted (e.g., plasma) using a miscible solvent. From aqueous biological samples, e.g., plasma, the proteins precipitate and are removed by centrifugation. After clarification, an aliquot of the unconcentrated semiorganic extract is injected.

2.   Double-phase or two-phase extraction. Double-phase extraction involves the partitioning of the analytes between an aqueous or semiaqueous phase and a water-immiscible organic solvent. The actual extraction is nearly always coupled with protein precipitation using a water-miscible organic solvent, mostly alcohol. Sometimes a treatment with SDS precedes the addition of the alcohol, which reportedly results in cleaner extracts, particularly from erythrocytes. The organic layer is isolated, evaporated to dryness in a stream of nitrogen or under vacuum, and the residue is redissolved in a solvent that is compatible with the chromatographic eluent. Alternatively, an aliquot of the organic layer can be directly injected without concentration, but this is obviously at the expense of sensitivity.

3.   Saponification. Lipid-rich samples such as oils, tissues, foods, and feeds may be saponified prior to double-phase extraction. In the course of this process not only the triglycerides and the phospholipids but also the possibly present tocopheryl esters are hydrolyzed. The procedure involves heating the sample with alcoholic potassium hydroxide, mostly in the presence of an antioxidant, followed by double-phase extraction, whereby the potassium salts of the liberated fatty acids and the glycerol remain in the aqueous phase. This decreases the overall organic load in the extract, which improves the selectivity and avoids contamination of the chromatographic column.

4.   Solid-phase extraction (SPE). In solid-phase extraction, the analytes are retained on short, disposable cartridges packed with microparticulate adsorbents or bonded phases. After sample application, the minicolumns are washed with a solvent to remove poorly retained matrix components, followed by displacement of the analyte(s) with a strong solvent. SPE may be used as a cleanup

in connection with solvent extraction and, as such, replace the once popular but now obsolete open column chromatographic procedures for vitamin E on macroparticulate materials.

5. Supercritical fluid extraction (SFE). Supercritical fluids are formed when a gas is held above its critical temperature while at the same time being compressed to a pressure exceeding a critical value. Their unique physicochemical properties make them particularly suitable for the extraction of solid samples such as foods. Owing to their low viscosity and the absence of surface tension, they readily penetrate into a matrix. Their high solvating power results in a rapid dissolution of the solutes, and the high diffusion coefficients of the latter in supercritical fluids permit a rapid mass transfer out of the matrix. Supercritical CO_2 is the most widely used extractant, having a critical temperature and pressure of $31°C$ and 73 atm, respectively. It is an excellent extractant for nonpolar compounds such as tocopherols and tocotrienols and in the presence of a low percentage of an alcohol for more polar analytes as well. Special instrumentation is required including a high-pressure pump to deliver the supercritical fluid at a constant, controllable pressure to the extraction vessel. The latter is placed in an oven or heating block to maintain it at a temperature above the critical temperature of the supercritical fluid. Both static and dynamic approaches, the latter involving constant renewal of the extractant, exist. SFE can be employed off-line or on-line by coupling to capillary GC, supercritical fluid chromatography (SFC), or HPLC.

As E vitamers are easily oxidizable, the use of an antioxidant during sample pretreatment would seem to be warranted. However, the effect of this practice on analyte recovery varies. There appears to be none for human plasma, unlike, e.g., for rat plasma and buccal mucosal cells. In contrast, the addition of an antioxidant is essential in connection with the analysis of erythrocytes to prevent dramatic losses of tocopherols due to their interaction with coextracted, iron-containing pigments. Likewise, the harsh conditions of saponification (alkaline pH, high temperature) do require this protective measure. As antioxidants, pyrogallol, ascorbic acid, and butylated hydroxytoluene are most commonly used.

B. Status up to 1989 (25–27)

1. Systems

Estimations in 1988–1989 about the portion of reversed-phase based LC assays for vitamin E ranged from $> 70\%$ to 90%. In approximately 70% of reversed-phase systems methanol-water mixtures were used as mobile phases. Occasionally, ethanol, acetonitrile, or isopropanol was substituted for methanol. Nonaqueous reversed-phase chromatography occupied a marginal position and found application particularly for the simultaneous chromatography of retinol, α-tocopherol, and β-carotene. Mobile phases in this case included mixtures of acetonitrile-methanol with chloroform or dichloromethane. The separation of β- and γ-tocopherol proved impossible by reversed phase, unlike by adsorption chromatography on silica or, alternatively but less commonly, on amino- or aminocyano-alkyl bonded phases. Normal-phase eluents were typically composed of an alkane (mostly hexane, rarely heptane or iso-octane) with a small concentration of a polar modifier, particularly isopropanol or di-isopropyl ether but sometimes another alcohol (methanol, ethanol, butanol), another ether (THF, methyl-t-butyl ether), or chloroform/dichloromethane. The first separation of all eight tocopherols and tocotrienols was achieved on a pellicular silica column in 1974 and took 80 min. For comparison, on a microparticulate silica column all forms except δ-tocotrienol were separated 3 years later in 14 min, but the resolution between β- and γ-tocotrienol was incomplete.

2. Detection

Absorbance and fluorescence detection have always formed the backbone of most vitamin E assays, but electrochemical detection has been steadily advancing since the mid-1980s. A 1985 review still stated that "electrochemical detectors have not yet experienced a real breakthrough," a conclusion that was supported by the two papers that had appeared at the time. However, 3 years later, some 16 outstanding applications of electrochemical detection in connection with the determination of tocopherols in plasma/serum, erythrocytes, and tissues had been reported. A 20-fold gain in sensitivity with reference to fluorescence detection was achieved (detection limits of 0.1 and 2 ng, respectively). With some procedures the incredibly low amounts of 0.4 and 0.65 pg of α-tocopherol had allegedly been detected, which represented an impressive 3 orders of magnitude gain in sensitivity compared to absorbance detection.

3. Sample Preparation and Applications

The determination of α- and to a lesser extent γ-tocopherol in human samples including plasma/serum, erythrocytes, platelets, and tissues was mainly carried out by reversed-phase chromatography in conjunction with any of the three detection modes. Some examples illustrate the progress in sample pretreatment procedures that was made by replacing absorbance by fluorescence and electrochemical detection. While a 1979 pioneering method for the determination of tocopherols in serum based on absorbance detection still required a 200-µL sample and double-phase extraction with a concentration step, fluorescence detection permitted a single-phase extraction of as little as 50 µL serum with 3 mL of methanol, in the absence of any concentration. In a similar approach with electrochemical detection, only 10 µL of plasma proved sufficient. Although many assays of tocopherols in biological fluids prior to 1990 continued to rely on double-phase extractions, single-phase approaches using such solvents as methanol, ethanol, isopropanol, or acetone were gaining importance owing to the extra sensitivity and selectivity provided by fluorescence and electrochemical detection. However, the injection of unpurified extracts may result in poor peak shape, lower efficiency, and possibly long-term column deterioration. One investigator used double-phase extraction in butanol-ethyl acetate or acetonitrile-butanol-ethyl acetate and injected the extracts directly on a reversed-phase column. Tissue samples were mostly saponified prior to extraction, although double-phase extraction after SDS pretreatment and single-phase extraction followed by solid-phase cleanup were also reported.

The analysis of oils, foods, and feeds was more commonly carried out using normal-phase chromatography, at least insofar as positional isomers and/or tocotrienols had to be separated. Samples were either saponified or, as in the case of oils, mixed with hexane or the chromatographic solvent and directly injected. Saponification also proved useful for infant formulas, in which α-tocopherol is determined mostly by reversed-phase chromatography.

4. Quantitation

The advantages and disadvantages of candidate internal standards for tocopherols have been discussed in III.A.4. In practice, tocol and α-tocopheryl acetate were used in the majority of cases, followed by retinyl acetate in assays including also retinoids and carotenoids.

C. Status After 1989

1. Introduction

In the 10 years that elapsed since the second edition of this book, a vast amount of literature on the LC separation and quantitation of vitamin E has continued to appear. The majority of these recent LC methods still rely on the same basic concepts of separation, detection, and sample preparation that got firmly established between 1980 and 1989.

2. Systems

Reversed-Phase. Reversed-phase chromatography continues to form the backbone of most assays of tocopherols and, rarely, tocotrienols in biological materials. Its popular status in the vitamin E area has been rationalized in III.A.2. When the methods for the simultaneous determination of tocopherols and retinoids/carotenoids are also taken into account, reversed-phase systems outnumber their normal-phase counterparts by a factor 2. A survey of reversed-phase systems for the separation and quantitation of tocopherols, tocopheryl esters, tocotrienols, and α-tocopherolquinone is presented in Table 2. Methods specifically

Table 2 Reversed-Phase Liquid Chromatographic Systems for Tocopherols and Tocotrienols

Type of mobile phase	Column	Analytes	Application	Ref.
1. Methanol-water or other alcohol-water	C18 Resolve 5 μm 150 × 3.9 mm	α-T, γ-T, α-TA, vitamin K1	serum	33 (1989)
	Partisil PXS 10 ODS-3 10 μm 250 × 4.6 mm or μ Bondapak C18 10 μm 300 × 3.9 mm	α-T, γ-T	lipoproteins lipoproteins, plasma	34 (1989) 35 (1991)
	Spherisorb ODS-2 — 150 × 4.6 mm	α-T, γ-T, δ-T	model compounds	36 (1991)
	Partisphere 5 C18 5 μm 110 × 4.7 mm	α-T, γ-T, δ-T, α-TA, retinol, carotenoids	plasma	38 (1992)
	YMC-PACK A-302 S-5 120A ODS 5 μm 150 × 4.6 mm	α-T, β-T, γ-T, δ-T, retinol	rat liver	39 (1992)
	μ Bondapak C18 —	α-T	milk products	40 (1993)
G	Supelcosil LC-18 5 μm 250 × 4.6 mm	α-T, α-TA, retinol, vitamin D, retinyl acetate	serum	41 (1994)
	Taxsil PFP 5 μm 250 × 4.6 mm	α-T, β-T, γ-T, δ-T	oils, pharmaceutical preparations	42 (1994)

Column	Compounds	Application	Ref. (Year)
Lichrospher RP 18 5 μm 125 × 4 mm	α-T, α-TA, tocol, retinol, retinyl acetate	serum, plasma	43 (1995)
LiChrosorb RP-8 10 μm 250 × 4.6 mm	α-T, α-TA, retinol, retinyl acetate, retinyl palmitate, vitamin D_3	multi-vitamin preparations, milk, foods	44 (1995)
Hypersil ODS 5 μm 150 × 4.6 mm	α-T, γ-T, δ-T, tocol	aquatic organisms	45 (1996)
Spherisorb C18 ODS 2 3 μm 100 × 4.6 mm	α-T, β-T, γ-T, δ-T	liver tissue	46 (1996)
Hypersil ODS 5 μm 150 × 4.6 mm	α-T, γ-T, δ-T, tocol	microalgae	47 (1997)
μ Bondapak C18 10 μm 250 × 4 mm	α-T, γ-T, δ-T, retinol, β-carotene	margarine	48 (1997)
LiChrosorb RP18 5 μm 120 × 4.6 mm	α-T, γ-T, δ-T, α-TA	platelets, endothelial cells	49 (1997)
Nucleosil 120-5 C8 5 μm 250 × 2 mm or 4 mm	α-T, α-TA, retinol, retinyl acetate, retinyl palmitate, vitamin D_3	infant formulas	50 (1997)
Asahipak ODP (ODP VA) or Curosil PFPS 5 μm 250 × 4.6 mm	α-T, β-T, γ-T, δ-T, 5,7-T	model compounds	51 (1997)

Table 2 Continued

Type of mobile phase	Column	Analytes	Application	Ref.
*1	Suplex pKb-100 5 µm 250 × 4.6 mm	α-T, γ-T, α-TA, retinol, retinyl acetate, carotenoids	plasma	52 (1997)
	Spherisorb ODS II 5 µm 250 × 4.6 mm	α-T, α-tocopherol oxybutyric acid, δ-T	liver cells, HC-60 human leukemia cells	53 (1998)
	µ-Bondapak C18 10 µm 300 × 3.9 mm	α-T, α-TA	serum	54 (1998)
2. Methanol	Ultrasphere C$_8$ 5 µm 250 × 4.6 mm	α-T, α-TA	liver	55 (1989)
	LiChrosorb RP18 5 µm 250 × 4 mm	α-T, α-TA	plasma, erythrocytes, platelets	56 (1990)
	C18 Rad-PAK 5 µm 250 × 4.6 mm	α-T, γ-T, δ-T	foods	57 (1990)
	C18 5 µm 250 × 4.6 mm	α-T, γ-T	CSF	58 (1991)
	Nucleosil ODS 5 µm 150 × 4.6 mm	α-T, α-TA	erythrocytes	18 (1992)
	Ultrapack C18 5 µm 250 × 4.6 mm	α-T, γ-T, α-TA	lipoproteins, serum	59 (1992)

Mobile phase	Column	Compounds	Sample	Ref. (Year)
3. Acetonitrile-water or acetonitrile-alcohol-water	Spherisorb ODS-2 3 μm 100 × 2.1 mm or Lichrospher 100 RP-18 5 μm 125 × 4 mm	α-T, vitamin D_3, vitamin K_1, retinyl palmitate	plasma	60 (1994)
	Spherisorb ODS-2 5 μm 250 × 4.6 mm	α-T, α-TA	erythrocytes, plasma	61 (1994)
	Spherisorb ODS-2 3 μm 100 × 2.1 mm	α-T, retinol, vitamin D_2, vitamin K_1	parenteral nutritions	62 (1994)
	LiChrosorb RP-18 10 μm 200 × 4.6 mm	α-T, α-TA	plasma	63 (1995)
	Novapack C18 4 μm 150 × 3.9 mm	α-T, γ-T, δ-T	lipoproteins, plasma	64 (1996)
	Pecosphere 3 × 3C C18	α-T, α-TA	red blood cells	65 (1997)
	LiChrospher WP 200 (C30) 3 μm 250 × 4.6 mm	α-T, β-T, γ-T, δ-T, α-TA	model compounds	66 (1998)
	Superspher RP-18 4 μm 250 × 4 mm	α-T, α-TA, tocol	blood, plasma, breast milk	67 (1989)
*2	C18 Resolve 5 μm 150 × 3.9 mm	α-T, β-T, γ-T, δ-T	vegetable oils	68 (1990)

Table 2 Continued

Type of mobile phase	Column	Analytes	Application	Ref.
*3	Biophase ODS 5 μm —	α-T, γ-T, α-TA, carotenoids, retinoids	plasma, serum	69 (1990)
	Kromasil C₁ 5 μm 100 × 4.6 mm	α-T	plasma	70 (1997)
	Nucleosil 100-5-C18 5 μm 250 × 4 mm	α-T, α-TA	serum	71 (1997)
4. Alcohol-water or alcohol-acetonitrile or alcohol mixture + electrolyte	OD-224 RP-18 5 μm 220 × 4.6 mm	α-T, vitamin A, vitamin D₃	milk	72 (1992)
	Superspher 100 RP-18 4 μm 250 × 4 mm	α-T, β-T, γ-T, δ-T	serum	73 (1993)
	Ultrasphere ODS 5 μm 250 × 4.6 mm	α-T, γ-T, δ-T, α-TQ, tocol	erythrocytes, platelets	74 (1994)
	Burdick & Jackson OD5 5 μm 250 × 4.6 mm	α-T, PMC, γ-T, δ-T	ocular lenses	75 (1994)
	Spherisorb ODS 5 μm 250 × 4.6 mm	α-T, β-T, γ-T, δ-T, α-T3, β-T3, γ-T3, δ-T3	olive oil	76 (1995)
	SuperPac Pep-S RP_{C2/C18} 5 μm 250 × 4 mm	α-T, γ-T, γ-T3, carotenoids, ubiquinones	plasma	77 (1995)

	Column	Mobile phase	Analytes	Sample	Ref. (Year)
	MC Medical C18, 3 μm, 80 × 4.6 mm		α-T, β-T, γ-T, δ-T, α-TQ	plasma, erythrocyte membranes	78 (1996)
G	Ultrasphere ODS C-18, 5 μm, 250 × 4.6 mm		α-T, γ-T, α-T3, γ-T3, ubiquinones, ubiquinols	tissues	79 (1996)
G	Vydac 201 TP 54, 5 μm, 250 × 4.6 mm	5. Alcohol (mixture) or acetonitrile-water plus ammonium acetate (+sometimes THF)	α-T, γ-T, tocol, retinol, carotenoids	serum	80 (1990)
	Ultrasphere ODS, 5 μm, 250 × 4.6 mm		α-T, γ-T, δ-T, retinol, carotenoids	plasma	81 (1991)
	Ultracarb ODS, 5 μm, 250 × 4.6 mm		α-T, α-TN, retinol, carotenoids	plasma, tissues	82 (1992)
	Novapak C18, 4 μm, 300 × 3.9 mm		α-T, γ-T, retinol, retinyl palmitate, carotenoids	buccal mucosal cells, plasma	114 (1992)
G	Spherisorb ODS1, 5 μm, 250 × 4.6 mm		α-T, retinol, retinyl acetate, carotenoids	plasma	83 (1993)
G	Bakerbond C18, 5 μm, 250 × 4.6 mm		α-T, γ-T, δ-T, tocol, retinoids, carotenoids	serum, foods	84 (1993)
	Ultrasphere ODS, 5 μm, 250 × 4.6 mm		α-T, β-T, γ-T, δ-T, α-T3, β-T3, γ-T3, δ-T3	plasma, tissues	85 (1994)
	Nucleosil 100-5 C18, 5 μm, 250 × 4.6 mm		α-T, γ-T, δ-T, retinol, retinyl acetate, carotenoids	plasma	86 (1994)

Table 2 Continued

Type of mobile phase		Column	Analytes	Application	Ref.
	G	Pecosphere-3 C18 83 × 4.6 mm 5 μm	α-T, γ-T, tocol, retinol, retinyl esters, carotenoids	human lenses	87 (1995)
		Zorbax C8 150 × 4.6 mm 5 μm	α-T, α-TA, retinoids	silicone oil	88 (1996)
		LiChrospher 100 RP-18 5 μm	tocol, α-T, γ-T, δ-T, retinol	plasma	89 (1997)
6. NARP binary mixtures: e.g., alcohols, alcohol-acetonitrile; ternary mixtures: e.g., acetonitrile-methanol-dichloromethane, acetonitrile-methanol-tetrahydrofuran		Zorbax ODS — 250 × 4.6 mm	α-T, β-T, γ-T, δ-T, α-T3, γ-T3, δ-T3	model compounds, oils	91 (1989)
		Zorbax ODS 5 μm 250 × 4.6 mm	α-T, β-T, γ-T, δ-T, α-TA	foods	24 (1989)
		LiChrosorb RP-18 5 μm 250 mm	α-TA, retinyl acetate, retinyl palmitate, vitamin D_3, vitamin K_1	vitamin mix	92 (1989)
		Vydac C18 — 250 × 4.6 mm	α-T, β-T, γ-T, δ-T, α-T3, γ-T3, δ-T3	model compounds, oils	91 (1989)
	G	Zorbax C18 5 μm 250 × 4.6 mm	α-T, retinyl acetate, retinol, retinyl palmitate	serum, plasma	93 (1989)
		Spheri-5-RP-18 5 μm 220 × 4.6 mm	α-TA, retinol, retinyl acetate, retinyl palmitate, carotenoids	serum, vegetables	94 (1990)
		Hitachi Gel 3057 ODS 3 μm 150 × 4.0 mm	α-T	lipoproteins	95 (1990)

Column	Compounds	Matrix	Ref. (Year)
Ultrasphere ODS 5 μm 150 × 4.6 mm	α-T, retinol, α-carotene, β-carotene	serum	96 (1991)
*4　Hypersil ODS 3 μm 125 × 4.6 mm or Nucleosil 120-3 C18 120 × 4.6 mm	α-T, α-TA, carotenoids	plasma, tissues	97 (1991)
Resolve C-18 5 μm 300 × 3.9 mm	α-T, γ-T, retinol, retinyl esters, carotenoids	serum	98 (1993)
Resolve C18 5 μm 300 × 3.9 mm	α-T, γ-T, α-TA, retinol, carotenoids	serum	99 (1993)
Nova-Pak C-18 4 μm 250 × 3.9 mm	α-T, γ-T, α-TA, retinol, carotenoids	serum	100 (1994)
Ultramex ODS C18 5 μm 150 × 4.6 mm	α-T, retinol, retinyl esters, carotenoids	serum	101 (1994)
RP-18 5 μm 125 × 4 mm	α-T, β-T, γ-T, δ-T	oils	102 (1994)
LiChrospher CH-8 5 μm 250 × 4.0 mm	α-TA, retinyl palmitate	pharmaceutical preparations	103 (1995)
Spherisorb ODS-2 5 μm 300 × 0.32 mm	α-T, vitamin D_3, vitamin K_1, retinyl palmitate	plasma	104 (1995)

Table 2 Continued

Type of mobile phase	Column	Analytes	Application	Ref.
	Nucleosil C18 5 μm 150 × 4.0 mm	α-T, γ-T, α-TA, retinol, retinyl palmitate, carotenoids	serum	105 (1995)
	BST Rutin C18 10 μm 250 × 4 mm	α-T, β-T, γ-T, α-TA, retinol, retinyl acetate	plasma	106 (1996)
	Ultrasphere C18 5 μm 250 × 4.6 mm	α-T, α-TA, retinol, retinyl acetate, carotenoids	lung tissue, pulmonary macrophages	107 (1996)
	Spherisorb RP-18 5 μm 220 × 4.6 mm	α-T, α-TA, retinol	serum, urine, pharmaceutical preparations	108 (1997)
	C18 5 μm 250 × 4.6 mm	α-T, retinoids, vitamin D_2, D_3, vitamin K_1, K_2, K_3	milk	109 (1997)
	Vydac 201TP54 5 μm 250 × 4.6 mm	α-T, γ-T, δ-T, α-T3, β-T3, γ-T3, δ-T3, carotenoids, retinol	palm oil	110 (1997)
	Bakerbond C18 5 μm 250 × 4.6 mm	α-T, δ-T, α-TA	garlic	111 (1997)
	Nucleosil ODS 1 5 μm 250 × 3.2 mm	α-T, α-TA, retinol, retinyl acetate, carotenoids	plasma	112 (1998)
	ODS Supelcosil LC 18 5 μm 250 × 4.6 mm	α-T, α-TA, retinol, retinyl acetate, carotenoids	serum	113 (1998)

7. Gradient elution with nonaqueous eluent mixture (in final conditions)	Spherisorb ODS-2 5 μm 125 × 4.6 mm	α-T, β-T, γ-T, δ-T, α-TA	paprika	37 (1992)
	Superspher RP18 4 μm 25 × 4 mm	α-T, tocol, retinol, retinyl palmitate	plasma	115 (1993)
	Lichrospher 100 RP-18 5 μm 125 × 4 mm	α-T, vitamin D_3, vitamin K_1, retinyl palmitate	plasma	60 (1994)
	Spherisorb ODS-2 3 μm 100 × 2.1 mm	α-T, vitamin D_3, vitamin K, retinyl palmitate	plasma	60 (1994)
	Adsorbosphere HS C18 3 μm 150 × 4.6 mm or 100 × 4.6 mm	α-T, tocol, retinol, carotenoids	serum	116 (1997)
	Vydac 218TP54 5 μm 250 × 4.6 mm	α-T, γ-T, tocol, carotenoids	plasma	90 (1997)
	Microsorb MV 3 μm 100 × 4.6 mm	α-T, γ-T, retinoids, carotenoids	serum, liver tissue, fruits, vegetables	117 (1998)

Abbreviations: α-T, β-T, γ-T, δ-T: α-, β-, γ-, δ-tocopherol; α-T3, β-T3, γ-T3, δ-T3: α-, β-, γ-, δ-tocotrienol; α-TA: α-tocopheryl acetate; α-TQ: α-tocopherolquinone; α-TN: α-tocopheryl nicotinate; 5,7-T: 5,7-dimethyltocol; PMC: 2,2,5,7,8-pentamethyl-6-hydroxychroman; G: gradient elution.
Special eluent compositions: *1: methanol-t-butyl methyl ether-water; *2: acetonitrile-water-tetrahydrofuran; *3: acetonitrile-chloroform-isopropanol-water; *4: acetonitrile-methanol-dichloromethane-water.

designed for tocopherol metabolites, decomposition products (including α-tocopherolquinone), or pharmacological analogs are not listed (see VI.B and VI.C). A significant part of the methods also cover retinoids (retinol, retinyl acetate and retinyl palmitate) and carotenoids, particularly β-carotene. A great variety of column materials has been used, mostly of 5 μm particle size. However, the use of smaller particles, e.g., 4 μm (64,67,73,100,114,115) and 3 μm (46,60,62,66,78,95,97,116,117), is becoming more widespread.

Considering only tocopherols and α-tocopheryl acetate as analytes, 70% of the systems inventoried in Table 2 employ an alcohol or an alcohol-water mixture as the eluent. Less than 20% of the systems are based on NARP. The method of Tan and Brzuskiewicz is outstanding in this regard (91). It separates six tocopherols and tocotrienols on Zorbax ODS eluted with a mixture of acetonitrile-methanol-dichloromethane (60:35:5, v/v/v) in approximately 9 min, but β- and γ-tocopherol coelute, as expected (Fig. 3b). It is interesting to compare this separation to that obtained by Schüep, who used a semiaqueous quaternary mobile phase consisting of acetonitrile–THF–methanol–1% ammonium acetate (85) (Fig. 4). Tan et al. (91) rightly state that NARP is useful primarily for the differentiation of tocopherols, retinoids, carotenoids, and ubiquinones, as evidenced from section 6 in Table 2. However, they also recommend this approach for tocopherols and tocotrienols as a way to avoid interferences in biological samples.

The superior compatibility of NARP with lipid-rich extracts has been emphasized above (see III.A.2). To maximally exploit this advantage, one investigator even used a reversed-phase column in conjunction with a nonpolar eluent adopted from normal-phase systems for vitamin E, i.e., hexane-isopropanol (57). However, the elution order (α-T < γ-T < δ-T) suggested indeed normal-phase interaction mechanisms. Elution of the very nonpolar carotenoids in reversed phase also requires nonaqueous eluents (NARP), either binary (e.g., acetonitrile-methanol) or ternary (e.g., acetonitrile-methanol in combination with dichloromethane or THF) (Table 2, section 6). However, on some reversed-phase columns these eluents afford insufficient retention of the more polar compounds such as retinol and α-tocopherol, thus necessitating the use of gradient elution to bridge the polarity difference with the carotenoids. To this end, many systems listed in Table 2 (section 7) use semiaqueous starting and nonaqueous final conditions, e.g., a gradient from methanol-water to ethyl acetate-propanol or THF-methanol. Furthermore, many eluents contain ammonium acetate (Table 2, section 5), which as a column deactivating agent is beneficial for the chromatography of retinoids and carotenoids but, as far as known, does not affect that of tocopherols. The presence of other electrolytes, e.g., perchlorates in the eluents (Table 2, section 4) serves the purpose of increasing their conductivity to allow electrochemical detection. Modern reversed-phase systems readily separate tocol, α-T, γ-T, δ-T, and α-TA, but fail to distinguish between β-T and γ-T. Wahyuni

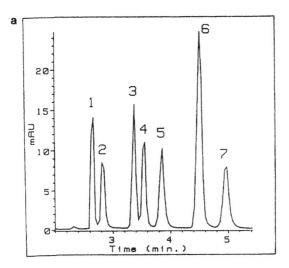

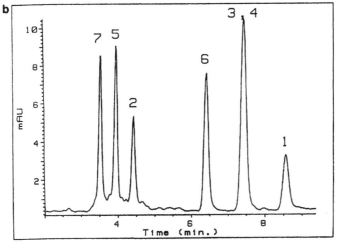

Figure 3 Separation of tocopherols and tocotrienols by (a) normal-phase and (b) re-versed-phase chromatography. Chromatographic conditions: (a) Column, Zorbax SIL, 250 × 4.6 mm; mobile phase, hexane:isopropanol (99:1, v/v), (b) Column, Zorbax ODS, 250 × 4.6 mm; mobile phase, acetonitrile:methanol:dichloromethane (60:35:5, v/v); detection, UV, 295 nm. Peak identification: 1, α-tocopherol; 2, α-tocotrienol; 3, β-tocopherol; 4, γ-tocopherol; 5, γ-tocotrienol; 6, δ-tocopherol; 7, δ-tocotrienol. (From Ref. 91.)

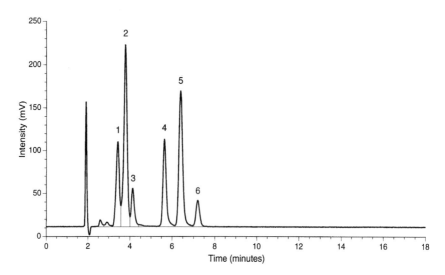

Figure 4 Separation of a standard mixture of tocopherols by reversed-phase HPLC. Chromatographic conditions: Column, Ultrasphere ODS, 5 μm, 250 × 4.6 mm; mobile phase, acetonitrile:tetrahydrofuran:methanol:1% ammonium acetate (684:220:68:28, v/v/v/v); detection, fluorescence, λ_{exc}: 298 nm, λ_{em}: 328 nm. Peak identification: 1, δ-tocotrienol; 2, β- and γ-tocotrienol; 3, α-tocotrienol; 4, δ-tocopherol; 5, β- and γ-tocopherol; 6, α-tocopherol. (From Ref. 85.)

was the first to resolve this pair by reversed-phase chromatography using a 75-cm capillary column packed with a polymeric C18 phase and elution with acetonitrile-hexane (91.5:8.5, v/v) (118). A partial, poorly efficient separation in a 30-min run was achieved by Warner and Mounts on a C18 Resolve column using a ternary mobile phase consisting of acetonitrile-THF-water (68). These investigators attributed the improved separation to their injecting extracts in hexane, assuming that the analytes were temporarily deposited, apparently at a different rate, on top of the column, due to the immiscibility between hexane and the eluent.

    Another partial separation was achieved on an uncommon column material of Japanese origin (YMC-PACK A-302 S-5 ODS) eluted with propanol-water (35:65, v/v) (39). The most complete reversed-phase separation so far of β- and γ-tocopherol is based on a pentafluorophenyl (PFPS) bonded phase and methanol-water as the eluent, as described by Richheimer (42) (Fig. 5). In their fundamental study on the separation of tocopherols on reversed phases, Abidi and Mounts corroborated the results of Richheimer with respect to the superiority of the PFPS phase for the separation of positional isomers, including 5,7-dimethyl-

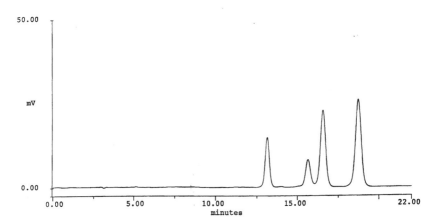

Figure 5 Separation of a standard mixture of tocopherols on a pentafluorophenyl reversed-phase material. Chromatographic conditions: Column, MetaChem PFP, 5 μm, 250 × 4.6 mm; mobile phase, methanol:water (92:8, v/v); detection, UV, 290 nm. Peak identification (from left to right): δ-tocopherol, β-tocopherol, γ-tocopherol, α-tocopherol. (From Ref. 42.)

tocol (51). On all "conventional" ODS phases tested by these investigators, β- and γ-tocopherol coeluted, except when the analytes were converted to acetate, butyrate, benzoate, or pentafluoropropionate esters. To separate the five homologs, including 5-7-dimethyltocol on a PFPS phase, a 40-min run was required. They also obtained a partial resolution on an octadecyl polyvinyl alcohol (OD-PVA) phase, but at the expense of very long analysis times. Unlike on the YMC-PACK A-302, β-T exhibits less retention than γ-T on a PFPS column, which reportedly has a unique selectivity for aromatic compounds. Strohschein et al. reported a partial separation of β- and γ-tocopherol on a C30 reversed-phase (LiChrospher WP 200, 3 μm) using methanol as the eluent (Fig. 6) (66). This column material had been previously developed for the differentiation of carotenoid isomers.

Normal Phase. As rationalized in III.A.2, the strength of normal-phase chromatography for vitamin E lies in its ability to separate all tocopherols and tocotrienols, including the positional isomers, particularly in connection with the analysis of vegetable oils and foods. Silica continues to be the most popular column material for this purpose, but polar bonded phases have increasingly gained a foothold (Table 3). Diol phases in particular can be readily substituted for silica and can be eluted with similar binary mobile phases containing a hydrocarbon as a base solvent and an alcohol, an aliphatic ether, or a cyclic ether as

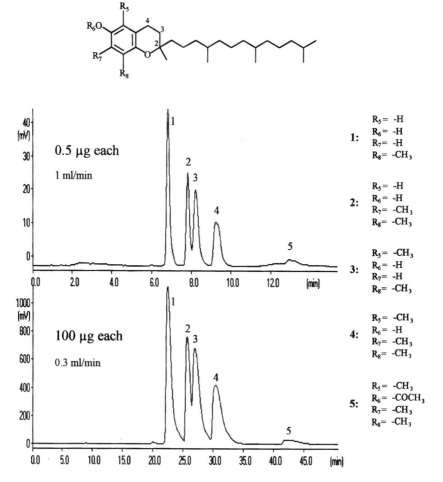

Figure 6 Separation of tocopherol isomers on a C30 reversed-phase column. Chromatographic conditions: Column, LiChrospher WP 200, 3 μm, 250 × 4.6 mm; mobile phase; methanol; detection, UV, 295 nm. Peak identification: 1, δ-tocopherol; 2, γ-tocopherol; 3, β-tocopherol; 4, α-tocopherol; 5, α-tocopheryl acetate. (From Ref. 66.)

a polar modifier (123,125,137,138,140). Reported analysis times required to resolve isocratically all eight tocopherols and tocotrienols on silica vary from 17 min (85) to 25 min (76,127) (Fig. 7). Tan and Brzuskiewicz accomplished a separation of all tocopherols and tocotrienols except β-tocotrienol in 6 min on Zorbax Sil eluted with hexane-isopropanol (91) (Fig. 3a).

Comparable separations on diol columns are equally efficacious but may take longer, e.g., 45 min (137) to 55 min (123). The method of Abidi stands out because it is the only one that resolves the three isomeric dimethyltocols on polar bonded phases (138), as illustrated in Figure 8. Previously, the same investigators had achieved a similar separation on a cyclodextrin-bonded silica column, eluted with mixtures of cyclohexane and an aliphatic or cyclic ether (130). α-Tocopheryl acetate, retinyl palmitate, and β-carotene have little retention in normal-phase chromatography because of their nonpolar nature (139). Nevertheless, gradient elution and flow programming have permitted their simultaneous chromatography with tocopherols and tocotrienols on silica (119,131,133), diol (125), and nitro (134) columns. To ensure a sufficient degree of retention for α-tocopheryl acetate and β-carotene, the initial eluotropic strength should be kept minimal. Accordingly, the gradient proceeds from pure hexane to mixtures of hexane with a polar modifier. Although reversed phase is obviously better suited for the simultaneous separation of tocopherols, retinoids, and carotenoids, normal phase retains the advantage of its compatibility with extracts in organic solvents, thus precluding the need for an evaporation/concentration step.

3. Detection

As shown in Tables 4 through 9, the well established three detection modes based on absorption, fluorescence and electrochemistry continue to be used, though to a different extent. The choice of detector will depend on the selectivity and the sensitivity requirements of the assay. Both requirements are in turn related to the choice of the approach for sample preparation, i.e., whether a cleanup and concentration step are included, as well as to the nature and properties of the analytes. Thus, the following tendencies are recognizable in Tables 4 through 9. Overall, among the methods inventoried, UV detection has the major share, i.e., 55% versus 36% for fluorescence and 9% for electrochemical detection. UV detection is primarily useful for those samples that do not put stringent demands on selectivity and sensitivity—in other words, that are not too complex and contain relatively high levels of E vitamers. For example, 72% of the assays in plasma rely on UV detection. When retinoids and carotenoids are included as analytes, detection at multiple wavelengths is indicated. All listed methods for vitamin E in pharmaceuticals (Table 9), where few matrix or detectability problems occur, also employ UV detection. In contrast, for the analysis of vegetable oils (Table 7), fluorescence detection predominates (65% of cases versus 22% for UV detection), one of the reasons being that the commonly applied single-phase extractions afford less selectivity and sensitivity. Fluorescence detection is nearly always conducted using the long wavelength of excitation (295 nm) of the tocopherols.

Table 3 Normal-Phase Liquid Chromatographic Systems for Tocopherols and Tocotrienols

Mobile phase	Column	Analytes	Application	Ref.
1. *Binary mixtures*				
Hydrocarbon + alcohol (e.g., isopropanol)	Zorbax Sil 7 μm 250 × 4.6 mm	α-T, β-T, γ-T, δ-T, α-T3, γ-T3, δ-T3	model compounds	91 (1989)
Hydrocarbon + aliphatic ether (e.g., *tert*-butyl-methyl-ether)	HS-Silica 3 μm 100 × 4.0 mm	α-T, β-T, γ-T, δ-T, α-T3, β-T3, γ-T3, retinoids, β-carotene	butter, margarine	119 (1990)
Hydrocarbon + cyclic ether (e.g., dioxane, tetrahydrofuran)				
Hydrocarbon + ester (e.g., ethyl acetate)	MCH-5 silica gel — 250 × 4.6 mm	α-T, β-T, γ-T, δ-T	serum	120 (1991)
	LiChrosorb Si 60 5 μm 125 × 4.0 mm	α-T, β-T, γ-T, δ-T, α-T3, retinol	liver	121 (1992)
	LiChrosorb Si 60 5 μm 250 × 4 mm	α-T, β-T, γ-T, δ-T	oils	122 (1992)
	LiChrospher 100 Diol 5 μm 250 × 4 mm	α-T, β-T, γ-T, δ-T, α-T3, β-T3, γ-T3, δ-T3	oils, cereals	123 (1992)
	LiChrosorb Si 60 5 μm 250 × 4.0 mm	α-T, retinol	plasma	124 (1993)
G	LiChrospher 100 Diol 5 μm 250 × 4 mm	α-T, β-T, γ-T, δ-T, α-TA, α-T3, β-T3, γ-T3, δ-T3	cereals, milk, infant formulas	125 (1993)

	Column	Analytes	Matrix	Ref. (Year)
	LiChrosorb Si 60, 5 µm, 250 × 4.6 mm	α-T, β-T, γ-T, δ-T	food	126 (1993)
	Supelcosil LC-Si, 5 µm, 250 × 4.6 mm	α-T, β-T, γ-T, δ-T, α-T3, β-T3, γ-T3, δ-T3	rice bran	127 (1993)
	Hypersil Silica, 5 µm, 100 × 2.1 mm	α-T, β-T, γ-T, δ-T	margarine	128 (1993)
	LiChrosorb Si 60, 5 µm, 125 × 4 mm	α-T, β-T, γ-T, δ-T, α-T3, β-T3, γ-T3, δ-T3	tissue, plasma, serum	85 (1994)
	Ultrasphere Si, 5 µm, 250 × 4.6 mm	α-T, β-T, γ-T, δ-T, 5,7-T	oils	129 (1994)
	Cyclobond I or II, 5 µm, 200 × 4.6 mm	α-T, β-T, γ-T, δ-T, 5,7-T	oil	130 (1994)
G	Ultrasphere Si, 5 µm, 250 × 4.6 mm	α-T, β-T, γ-T, δ-T, retinol, carotenes	dairy products	131 (1994)
	LiChrosorb Si 60, 5 µm, 250 × 4 mm	α-T	tissue	132 (1995)
	LiChrosorb Si 60, 5 µm, 250 × 4 mm	α-T, β-T, γ-T, δ-T, α-T3, β-T3, γ-T3, δ-T3	vegetable oils	76 (1995)
G	Nova-Pak silica, 4 µm, 150 × 3.9 mm	α-T, retinol, β-carotene	feeds, tissues, serum	133 (1995)

Table 3 Continued

Mobile phase	Column	Analytes	Application	Ref.
G	Nucleosil 100-5 NO$_2$ 5 μm	α-T, β-T, γ-T, δ-T, α-TA, α-T3, β-T3, γ-T3, δ-T3, retinoids, β-carotene	foods	134 (1995)
	250 × 4 mm Merck Si60	α-T, β-T, γ-T, δ-T	essential oils	135 (1996)
	250 × 4 mm Resolve silica 5 μm	α-T	beef muscle	136 (1996)
	150 × 3.9 mm LiChrosorb Diol 5 μm	α-T, β-T, γ-T, δ-T, α-T3, β-T3, γ-T3, δ-T3	margarine, vegetables, infant foods	137 (1996)
	250 × 4.6 mm LiChrosorb Diol 10 μm	α-T, β-T, γ-T, δ-T, 5,7-T	model compounds	138 (1996)
	250 × 4.6 mm Chromega Diol 5 μm	α-T, β-T, γ-T, δ-T, 5,7-T	model compounds	138 (1996)
	250 × 4.6 mm μ Bondapak NH$_2$ 10 μm	α-T, β-T, γ-T, δ-T, 5,7-T	model compounds	138 (1996)
	300 × 3.9 mm LiChrosorb Si 60 5 μm	α-T, γ-T, δ-T, α-TA, retinyl palmitate	infant formula	139 (1997)
	250 × 4.6 mm			

Column	Dimensions	Compounds	Sample	Ref. (year)
Supelcosil LC-Diol 5 μm	250 × 4.6 mm	α-T, β-T, γ-T, δ-T, 5,7-T; α-T3, β-T3, γ-T3, δ-T3	tissue, oil	140 (1997)
LiChrosorb Si 60 5 μm	250 × 4.6 mm	α-T, γ-T, δ-T	margarine	141 (1998)

2. Ternary mixtures
Hydrocarbon + alcohol + aliphatic or cyclic ether

Column	Dimensions	Compounds	Sample	Ref. (year)
LiChrospher Si 60 —	120 × 4.6 mm	α-T, β-T, γ-T, δ-T	infant formulas	142 (1989)
Develosil Si 60-5 5 μm	250 × 4.6 mm	α-T, β-T, γ-T, δ-T	food, seeds, oils	143 (1993)
Chromatorex-SI 5 μm	250 × 4.6 mm	α-T, β-T, γ-T, δ-T, tocol	vegetable oils	144 (1994)
Nucleosil-NH₂ 3 μm	250 × 4.5 mm	α-T, PMC	tissues, plasma	145 (1994)

Hydrocarbon + alcohol or ether + aliphatic acid

Column	Dimensions	Compounds	Sample	Ref. (year)
Supelco LC-CN 5 μm	250 × 4.6 mm	α-T, retinol	feed	146 (1991)
Econosphere silica 3 μm	150 × 4.6 mm	α-T, α-TA, retinoids, β-carotene	dairy foods	147 (1996)

3. Quaternary mixtures

Column	Dimensions	Compounds	Sample	Ref. (year)
Supelcosil LC-Si 5 μm	250 × 4.6 mm	α-T, β-T, γ-T, δ-T, α-T3, β-T3, γ-T3, δ-T3	rice bran	127 (1993)

Abbreviations: see Table 2.

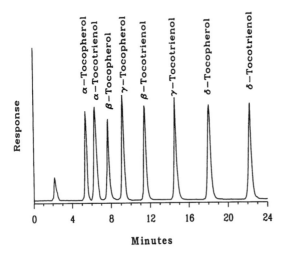

Figure 7 Separation of tocopherols and tocotrienols by normal-phase chromatography on silica. Chromatographic conditions: Column, Supelcosil LC-Si, 5 μm, 250 × 4.6 mm; mobile phase, iso-octane:ethyl acetate (97.5:2.5, v/v); detection, fluorescence, λ_{exc} 290 nm, λ_{em} 330 nm. Peak identification: indicated in the figure. (From Ref. 127.)

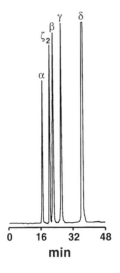

Figure 8 Separation of tocopherols by normal-phase chromatography. Chromatographic conditions: Column, Chromega Diol, 5 μm, 250 × 4.6 mm; mobile phase, hexane:di-isopropyl ether (90:10, v/v); detection, fluorescence, λ_{exc} 298 nm, λ_{em}, 345 nm. Peak identification: α, β, γ, δ-tocopherols; ζ_2, 5,7-dimethyltocol. (From Ref. 138.)

However, next to the work of Hatam and Kayden (cited in 26), Seta et al. have employed 215 nm for excitation in their assay of α-tocopherol in lipoproteins to improve the sensitivity (95).

In the past, electrochemical detection has found extensive application for the simultaneous determination of α-tocopherol, α-tocopherolquinone, and ubiquinones in plasma, erythrocytes, and platelets (26,27). The use of coulometry with a dual cell configuration to this end has been explained above (see III.A.3). One recent example of this type of assay is the paper by Finckh et al. (77). Electrochemical detection appears to have been ignored lately as a tool to push the detectability to its ultimately low limit. This approach permits a reduction of the sample size, which is of interest, e.g., for monitoring of premature infants. Such miniaturized assays were discussed in a 1988 review paper (26). Finally, evaporative light-scattering detection has been used in connection with the analysis of oils (68,129,155).

4. Quantitation

The choice of internal standards in recent assays for tocopherols appears to be governed more by considerations of availability than by the requirements of structural and physicochemical similarity. Based on the latter criteria, 5,7-dimethyltocol would still be the most appropriate choice, but its commercial nonavailability has always been prohibitive. However, several papers have recently listed a U.S.-based company as a supplier of this valuable compound (51,129,130,138,140) and, accordingly, have used it as an internal standard (129,140). α-Tocopheryl acetate ranks as the No. 1 internal standard, used in approximately 43% of the methods surveyed in Tables 4 through 9. It is followed in importance by tocol (22% of methods) and retinyl acetate (12% of methods), the latter particularly in connection with the simultaneous determination of tocopherols, retinoids, and carotenoids (83,86,93,117). Other proposed internal standards include δ-tocopherol (53,64,73,75,140), γ-tocotrienol (77), α-tocopheryl nicotinate (82), and 2,2,5,7,8-pentamethyl-6-chromanol (148,149). The unfavorable polarity difference between the latter and α-tocopherol has been mentioned above (see III.A.4). Although the advantages of an internal standard are generally recognized, several methods still rely on external standardization (see, e.g., Table 4 for assays in plasma and serum).

5. Sample Preparation

Tables 4 through 9 list the applications of both LC and GC (see further IV) methods for the determination of tocopherols and tocotrienols in a variety of matrices. The five basic approaches for sample preparation outlined in III.A.5 are represented throughout. An evaluation of the different procedures reveals the following general trends:

Table 4 Chromatographic Determination of Tocopherols and Tocotrienols in Blood Plasma and Serum

Sample preparation	Analytes	Analytical technique (ref)	Detection[a]	Internal standard	Ref.
1. Single-phase extraction					
Addition of water miscible organic solvent, clarification, injection	α-T, δ-T	RP-HPLC	fluorescence $\lambda_{exc} = 292$, $\lambda_{em} = 335$ nm	—	32 (1989)
	α-T, α-TA	RP-HPLC	UV, 292 nm	—	63 (1995)
2. Double-phase extraction					
Addition of water miscible organic solvent, extraction with water immiscible organic solvent, analysis of organic phase (either or not after evaporation and reconstitution of the residue)	α-T, α-TA	RP-HPLC	UV, 280 nm	tocol	67 (1989)
	α-T, retinol, retinyl palmitate	RP-HPLC	UV, 300 nm	retinyl acetate	93 (1989)
	α-T, γ-T, vitamin K₁	RP-HPLC	UV, 292 nm	α-TA	33 (1989)
	α-T	RP-HPLC	UV, 280 nm	α-TA	56 (1990)
	α-T, β-T, γ-T, δ-T, retinol, carotenoids	RP-HPLC	P: fluorescence $\lambda_{exc} = 298$, $\lambda_{em} = 328$ nm	—	81 (1991)
	α-T, retinol, α-, β-carotene	RP-HPLC	P: UV, 292 nm	α-TA	96 (1991)
	α-T, α-TA, retinoids, carotenoids	RP-HPLC	P:UV, 292 nm	—	97 (1991)
	α-T, β-T, γ-T, δ-T	NP-HPLC	UV, 292 nm	—	120 (1991)
	α-T, γ-T, δ-T, retinol, carotenoids	RP-HPLC	P: UV, 290 nm	α-TA	38 (1992)
	α-T, γ-T, retinoids, carotenoids	RP-HPLC	UV, 300 nm	—	114 (1992)

Analytes	Method	Detection	Internal standard	Ref. (year)
α-T, retinol, carotenoids	RP-HPLC	P:UV, 292 nm	α-TN	82 (1992)
α-T, γ-T, retinol, carotenoids	RP-HPLC	P: UV, 290 nm	α-TA	99 (1993)
α-T, β-T, γ-T	RP-HPLC	coulometry, +0.35V	δ-T	73 (1993)
α-T, retinol, carotenoids	RP-HPLC	P: UV, 292 nm	retinyl acetate	83 (1993)
α-T, γ-T, retinoids, carotenoids	RP-HPLC	UV, 300 nm	retinyl hexanoate	98 (1993)
α-T, retinoids	RP-HPLC	UV, 305 nm	tocol	115 (1993)
α-T, γ-T, δ-T, retinoids, carotenoids	RP-HPLC	fluorescence $\lambda_{exc} = 295$, $\lambda_{em} = 335$ nm	tocol	84 (1993)
α-T, retinol	NP-HPLC	fluorescence $\lambda_{exc} = 296$, $\lambda_{em} = 328$ nm	—	124 (1993)
α-T, α-TA stereoisomers	Chiral phase HPLC	UV, 284 nm	PMC	148 (1993)
α-T, retinol, vitamin D	RP-HPLC	UV, 265 nm	α-TA	41 (1994)
α-T, retinoids, carotenoids	RP-HPLC	UV, 300 nm	retinyl butyrate	101 (1994)
α-T	RP-HPLC	UV, 292 nm	α-TA	61 (1994)
α-T, γ-T, retinol, carotenoids	RP-HPLC	P: UV, 280 nm	α-TA	100 (1994)
α-T, γ-T, δ-T, retinol, carotenoids	RP-HPLC	P: UV, 290 nm	retinyl acetate	86 (1994)
α-T, retinyl palmitate, vitamin D$_3$, vitamin K$_1$	RP-HPLC	UV, 284 nm	—	60 (1994)

Table 4 Continued

Sample preparation	Analytes	Analytical technique (ref)	Detection[a]	Internal standard	Ref.
	α-T, β-T, γ-T, δ-T, α-T3, β-T3, γ-T3, δ-T3	RP-HPLC NP-HPLC	fluorescence $\lambda_{exc} = 298$, $\lambda_{em} = 328$ nm → RP fluorescence $\lambda_{exc} = 295$, $\lambda_{em} = 330$ nm → NP	—	85 (1994)
	α-T, α-TA	RP-HPLC	UV, 292 nm	tocol	43 (1995)
	α-T, γ-T, retinol, carotenoids	RP-HPLC	P: fluorescence $\lambda_{exc} = 295$, $\lambda_{em} = 330$ nm	α-TA	105 (1995)
	α-T, γ-T, ubiquinones, carotenoids	RP-HPLC	coulometry, −0.60V amperometry +0.60V	γ-T3	77 (1995)
	α-T, retinol, carotenoids	NP-HPLC	P: UV, 294 nm	—	133 (1995)
	α-T, β-T, γ-T, δ-T, α-TQ	RP-HPLC	coulometry −0.45V +0.40V	—	78 (1996)
	α-T, retinol, carotenoids	RP-HPLC	P: UV, 290 nm	α-TA	107 (1996)
	α-T	NP-HPLC	UV, 290 nm or fluorescence $\lambda_{exc} = 290$, $\lambda_{em} = 330$ nm	PMC	149 (1996)

Analyte	Method	Detection	Internal standard[a]	Ref. (Year)
α-T	RP-HPLC	fluorescence λ_{exc} = 292, λ_{em} = 325 nm; UV, 284/292 nm	tocol; α-TA	71 (1997)
α-T, γ-T, δ-T, retinol	RP-HPLC	P: UV, 292 nm or fluorescence λ_{exc} = 298, λ_{em} = 330 nm	tocol	89 (1997)
α-T, retinol, carotenoids	RP-HPLC	P: UV, 292 nm	tocol	116 (1997)
α-T, β-T, γ-T, retinol, carotenes	RP-HPLC	UV, 292 nm	α-TA	106 (1997)
α-T, retinol, carotenoids	RP-HPLC	P: UV, 290 nm	α-TA	112 (1998)
α-T, γ-T, δ-T	RP-HPLC	UV, 280 nm fluorescence λ_{exc} = 290, λ_{em} = 325 nm	α-TA	54 (1998)
α-T, retinoids, carotenes	RP-HPLC	UV, 292 nm	α-TA	113 (1998)

3. Miscellaneous

Analyte	Method	Detection	Internal standard[a]	Ref. (Year)
Double-phase extraction with CH$_2$Cl$_2$–isopropanol–acetic acid — α-T, γ-T, retinoids, carotenoids	RP-HPLC	P: UV, 290 nm	retinyl acetate	117 (1998)

4. SPE

Analyte	Method	Detection	Internal standard[a]	Ref. (Year)
α-T, α-TA	RP-HPLC	UV, 290 nm	anthraquinone	108 (1997)

[a] Detection of tocopherols only.

Abbreviations: RP, reversed-phase; NP, normal-phase; P, wavelength programming or PDA detector; SPE, solid-phase extraction.
Other abbreviations: see Table 2.

Table 5 Liquid Chromatographic Determination of Tocopherols and Tocotrienols in Blood Elements and Fractions

Sample preparation	Analytes	Analytical technique (ref)	Detection[a]	Internal standard	Ref.
1. *Erythrocytes*					
Double-phase extraction: addition of water miscible organic solvent, extraction with water immiscible organic solvent, analysis of organic phase (either or not after evaporation and reconstitution of the residue)	α-T, α-TA	RP-HPLC	UV, 280 nm	tocol	67 (1989)
	α-T	RP-HPLC	UV, 280 nm	α-TA	56 (1990)
	α-T	RP-HPLC	UV, 280 nm	α-TA	18 (1992)
	α-T	RP-HPLC	UV, 292 nm	α-TA	61 (1994)
	α-T, γ-T	RP-HPLC	fluorescence $\lambda_{exc} = 295$, $\lambda_{em} = 330$ nm	δ-T	64 (1996)
	α-T	RP-HPLC	UV, 285 nm	α-TA	65 (1997)
	α-T stereoisomers (as acetates)	Chiral phase HPLC	UV, 284 nm	PMC	148 (1993)
Saponification, double-phase extraction: addition of water miscible organic solvent, saponification, extraction with water immiscible solvent	α-T, γ-T, δ-T, α-TQ	RP-HPLC	coulometry − 0.25V + 0.55V	tocol	7 (1993)

2. Platelets

Double-phase extraction: addition of water miscible organic solvent, extraction with water immiscible organic solvent, analysis of organic phase (either or not after evaporation and reconstitution of the residue)	α-T	RP-HPLC	UV, 292 nm	α-TA	61 (1994)
Saponification, double-phase extraction: addition of water miscible organic solvent, saponification, extraction with water immiscible solvent	α-T, γ-T, δ-T, α-TQ	RP-HPLC	coulometry −0.25V, +0.55V	tocol	74 (1994)

3. Lipoproteins

Single-phase extraction: addition of water miscible organic solvent, clarification, injection	α-T, γ-T	RP-HPLC	fluorescence λ_{exc} = 295, λ_{em} = 330 nm	—	34 (1989) 35 (1991)
Double-phase extraction: addition of water miscible organic solvent, extraction with water immiscible organic solvent, analysis of organic phase (either or not after evaporation and reconstitution of the residue)	α-T	RP-HPLC	fluorescence λ_{exc} = 215, λ_{em} = 320 nm	—	95 (1990)
	α-T, γ-T	RP-HPLC	UV	α-TA	59 (1992)
	α-T, γ-T, carotenoids	RP-HPLC	P: UV, 300 nm	retinyl hexanoate	150 (1996)
	α-T, γ-T	RP-HPLC	fluorescence λ_{exc} = 295, λ_{em} = 330 nm	δ-T	64 (1996)

Abbreviations: see Tables 2 and 4.

Table 6 Liquid Chromatographic Determination of Tocopherols and Tocotrienols in Human and Animal Tissues

Sample preparation	Matrix	Analytes	Analytical technique	Detection[a]	Internal standard	Ref.
1. Addition of water miscible solvent, clarification, injection	liver	α-T, β-T, γ-T, δ-T	RP-HPLC	fluorescence $\lambda_{exc} = 290$, $\lambda_{em} = 330$ nm	—	46 (1996)
2. Addition of water miscible solvent, column purification	rat liver	α-T	RP-HPLC	UV, 280 nm	α-TA	55 (1989)
3. Extraction with water immiscible organic solvent	eye lens	α-T, γ-T	RP-HPLC	amperometry +0.8V or +0.6V	δ-T	75 (1994)
4. Addition of water miscible organic or immiscible organic solvent, analysis of organic phase	rat liver, lung, mamma	α-T, α-TA, retinoids, carotenoids	RP-HPLC	P: UV, 292 nm	—	97 (1991)
	eye lens	α-T, γ-T, retinoids, carotenoids	RP-HPLC	fluorescence $\lambda_{exc} = 292$, $\lambda_{em} = 330$ nm	tocol	87 (1995)
	lung	α-T, retinoids, carotenoids	RP-HPLC	P: UV, 290 nm	α-TA	107 (1996)
	pig liver, heart	α-T, β-T, γ-T, δ-T, α-T3	NP-HPLC	fluorescence $\lambda_{exc} = 296$, $\lambda_{em} = 330$ nm	5,7-T	140 (1997)
	liver, rat tissues, liver	α-T, γ-T, retinoids, carotenoids	RP-HPLC	P: UV, 290 nm	retinyl acetate	117 (1998)
	liver	α-T, α-tocopherol-oxybutyric acid	RP-HPLC	fluorescence $\lambda_{exc} = 210$, $\lambda_{em} = 300$ nm	δ-T	53 (1998)

5. Addition of water miscible organic solvent, saponification, extraction with water immiscible organic solvent	liver	α-T, β-T, γ-T, δ-T, retinol	NP-HPLC	fluorescence λ_{exc} = 295, λ_{em} = 330 nm	—	121 (1992)
	adipose tissue, brain, liver, adrenal gland	α-T stereoisomers (as acetates)	chiral phase HPLC	UV, 284 nm	PMC	148 (1993)
	liver	α-T, β-T, γ-T, δ-T, α-T3, β-T3, γ-T3, δ-T3	NP-HPLC RP-HPLC	fluorescence λ_{exc} = 295, λ_{em} = 330 nm fluorescence λ_{exc} = 298, λ_{em} = 328 nm	—	85 (1994)
	muscle, adipose tissue	α-T	NP-HPLC	fluorescence λ_{exc} = 295, λ_{em} = 330 nm	—	132 (1995)
	mouse brain, heart, kidney, liver, skin	α-T, γ-T, α-T3, γ-T3, ubiquinones, ubiquinols	RP-HPLC	UV, 275 nm amperometry +0.5V	—	79 (1996)
6. Enzymatic digestion, addition of water miscible solvent, extraction with water immiscible solvent	skin, breast, colon, lung	α-T, retinol, carotenoids	RP-HPLC	P: UV, 292 nm	α-TN	82 (1992)

Abbreviations: see Tables 2 and 4.

Table 7 Determination of Tocopherols and Tocotrienols in Oils and Fats

Sample preparation	Matrix	Analytes	Analytical technique	Detection	Internal standard	Ref.
1. Dilution in organic solvent and injection of aliquot	vegetable oils	α-T, β-T, γ-T, δ-T	RP-HPLC	ELSD	—	68 (1990)
	plant seed oils	α-T, β-T, γ-T, δ-T, α-T3, β-T3, γ-T3, δ-T3	NP-HPLC	fluorescence $\lambda_{exc} = 295$, $\lambda_{em} = 330$ nm	—	123 (1992)
	oils and butter	α-T, β-T, γ-T, δ-T	NP-HPLC	fluorescence $\lambda_{exc} = 293$, $\lambda_{em} = 326$ nm	cholestane	122 (1992)
	margarine	α-T, β-T, γ-T, δ-T	NP-HPLC	fluorescence $\lambda_{exc} = 290$, $\lambda_{em} = 330$ nm	—	128 (1993)
	fish oils	α-T	GC	FID, 360°C	—	151 (1993)
		α-T	SFC	FID, 300°C		
	vegetable oils	α-T, β-T, γ-T, δ-T	NP-HPLC	fluorescence $\lambda_{exc} = 290$, $\lambda_{em} = 330$ nm ELSD	5,7-T	129 (1994)
	vegetable oils	α-T, β-T, γ-T, δ-T, α-T3, β-T3, γ-T3, δ-T3	RP-HPLC NP-HPLC	Amperometry, +0.6V fluorescence: $\lambda_{exc} = 290$, $\lambda_{em} = 330$ nm	—	76 (1995)
	margarine	α-T, β-T, γ-T, δ-T, α-T3, β-T3, γ-T3, δ-T3, retinoids	NP-HPLC	P: fluorescence: $\lambda_{exc} = 295$, $\lambda_{em} = 330$ nm	—	134 (1995)
	edible oils and fats	α-T	SFC	FID, 340°C	tridecanoin	153 (1995)

Matrix	Analytes	Method	Detection	Internal standard	Reference
essential oils	α-T, β-T, γ-T, δ-T	GC, NP-HPLC	FID, 310°C; UV, 295 nm; fluorescence: $\lambda_{exc} = 296$, $\lambda_{em} = 330$ nm	hexacosanol; 5,7-T, (δ-T)	135 (1996); 140 (1997)
palm oil, dietary oils	α-T, β-T, γ-T, δ-T, α-T3, β-T3, γ-T3, δ-T3	NP-HPLC			
margarine, vegetable oil products, oils, margarine	α-T, γ-T, δ-T	NP-HPLC	fluorescence: $\lambda_{exc} = 290$, $\lambda_{em} = 330$ nm	—	141 (1998)
	α-T, α-TA	GC	FID	2-t-butyl-4-methylphenol 5-α-cholestane	154 (1998)
2. Saponification and extraction with organic solvent					
oils	α-T, γ-T, δ-T	RP-HPLC	fluorescence: $\lambda_{exc} = 290$, $\lambda_{em} = 330$ nm	α-TA	24 (1989)
butter and margarine	α-T, β-T, γ-T, δ-T, α-T3, β-T3, γ-T3, retinoids, β-carotene	NP-HPLC	P: fluorescence: $\lambda_{exc} = 290$, $\lambda_{em} = 330$ nm	—	119 (1990)
oils	α-T, γ-T, δ-T	RP-HPLC	fluorescence: $\lambda_{exc} = 295$, $\lambda_{em} = 330$ nm	—	57 (1990)
oils, butter	α-T, β-T, γ-T, δ-T	GC	FID, 300°C	cholestane	122 (1992)
vegetable oils	α-T, β-T, γ-T, δ-T	SFC	UV, 290 nm	tocol	144 (1994)
soybean oil	α-T, β-T, γ-T, δ-T, α-T3, β-T3, γ-T3, δ-T3	NP-HPLC	fluorescence: $\lambda_{exc} = 290$, $\lambda_{em} = 330$ nm	—	152 (1994)

Table 7 Continued

Sample preparation	Matrix	Analytes	Analytical technique	Detection	Internal standard	Ref.
	vegetable oils	α-T, β-T, γ-T, δ-T	NP-HPLC	fluorescence: λ_{exc} = 290, λ_{em} = 325 nm	tocol	144 (1994)
	vegetable oils	α-T, β-T, γ-T, δ-T, α-T3, β-T3, γ-T3, δ-T3	NP-HPLC	fluorescence: λ_{exc} = 290, λ_{em} = 330 nm	—	76 (1995)
	margarine	α-T, β-T, γ-T, δ-T, α-T3, β-T3, γ-T3, δ-T3	NP-HPLC	fluorescence: λ_{exc} = 295, λ_{em} = 330 nm	—	137 (1996)
	margarine, vegetable oils	α-T, γ-T, δ-T	RP-HPLC	UV, 280 nm	—	48 (1997)
3. SPE	oils	α-T	HPSEC	ELSD	—	155 (1992)
4. Miscellaneous direct silylation of oils	edible oils and fats	α-T, β-T, γ-T, δ-T	LC-GC	FID	cholesterol	156 (1993)
transesterification of triglycerides	edible oils and fats	α-T, α-TA	GC	FID, 250°C	5-α-cholestane	157 (1996)
enzymatic hydrolysis of oil	palm oil	α-T, γ-T, δ-T	RP-HPLC	fluorescence: λ_{exc} = 298, λ_{em} = 328 nm	—	110 (1997)

ELSD: evaporative light scattering detection.
Other abbreviations: see Tables 2 and 4.

Table 8 Determination of Tocopherols and Tocotrienols in Foods and Feeds

Sample preparation	Matrix	Analytes	Analytical technique	Detection[a]	Internal standard	Ref.
1. Extraction with organic solvent, injection of an aliquot or evaporation and reconstitution of the residue	infant formulas, milk	α-T, α-TA	RP-HPLC	UV, 280 nm	tocol	67 (1989)
	cereals	α-T, β-T, γ-T, δ-T, α-T3, β-T3, γ-T3, δ-T3	NP-HPLC	fluorescence:: $\lambda_{exc} = 295$, $\lambda_{em} = 330$ nm	—	123 (1992)
	paprika	α-T, β-T, γ-T, δ-T, α-TA	RP-HPLC	UV, 280 nm	—	37 (1992)
	infant formulas, cereals, beverages	α-T, β-T, γ-T, δ-T, α-T3, β-T3, γ-T3, δ-T3, α-TA	NP-HPLC	fluorescence: $\lambda_{exc} = 295$, $\lambda_{em} = 330$ nm	—	125 (1993)
	fruits, vegetables, grains	α-T, γ-T, δ-T, retinoids, carotenoids	RP-HPLC	fluorescence: $\lambda_{exc} = 295$, $\lambda_{em} = 335$ nm	tocol	84 (1993)
	eggs	α-T, β-T, γ-T, δ-T	NP-HPLC	fluorescence: $\lambda_{exc} = 295$, $\lambda_{em} = 325$ nm	—	126 (1993)
	feed premix	α-T, α-TA	GC	FID	squalane	158 (1993)
	dried apricots	α-T, α-TA, retinoids	RP-HPLC	UV, 280 nm, MS	—	44 (1995)
	dairy foods	α-T (α-TA), retinoids	NP-HPLC	P: fluorescence $\lambda_{exc} = 295$, $\lambda_{em} = 330$ nm	(α-TA) BHA	147 (1996)
	infant formulas	α-T, γ-T, δ-T, α-TA, retinyl palmitate	NP-HPLC	P: fluorescence $\lambda_{exc} = 285$, $\lambda_{em} = 310$ nm	—	139 (1997)
	garlic	α-T, δ-T, α-TA	RP-HPLC GC	UV, 292 nm, MS	—	111 (1997)
	soups	α-T, α-TA	GC	FID	2-t-butyl-4-methylphenol, 5-α-cholestane	154 (1998)

Table 8 Continued

Sample preparation	Matrix	Analytes	Analytical technique	Detection[a]	Internal standard	Ref.
2. Saponification, extraction with water immiscible organic solvent	infant formulas	α-T, β-T, γ-T, δ-T	NP-HPLC	fluorescence: λ_{exc} = 292, λ_{em} = 320 nm	—	142 (1989)
	vegetables	α-T	NP-HPLC GC	UV, 292 nm FID	—	159 (1989)
	fruits, fruit juices, chocolate, nuts, salad dressings, vegetables, teas, cheese	α-T, γ-T, δ-T	RP-HPLC	fluorescence: λ_{exc} = 290, λ_{em} = 330 nm	α-TA	24 (1989)
	milk, infant formulas, egg yolk	α-T, γ-T, δ-T	RP-HPLC	fluorescence: λ_{exc} = 295, λ_{em} = 330 nm UV, 212 nm	—	57 (1990)
	rodent feed	α-T, retinol	NP-HPLC	P: UV, 296 nm	—	146 (1991)
	milk, milk powder	α-T, vitamin A, vitamin D₃	RP-HPLC	UV, 280 nm	—	72 (1992)
	seeds, grains, nuts	α-T, β-T, γ-T, δ-T	NP-HPLC	UV, 298 nm	—	143 (1993)
	rice bran	α-T, β-T, γ-T, δ-T, α-T3, γ-T3, δ-T3	NP-HPLC	fluorescence: λ_{exc} = 290, λ_{em} = 330 nm	—	127 (1993)

					Ref.
milk, milk powder	α-T	RP-HPLC	UV, 295 nm	—	40 (1993)
cheese, milk	α-T, β-T, γ-T, δ-T, retinol, carotenoids	NP-HPLC	P: fluorescence: λ_{exc} = 280, λ_{em} = 325 nm	—	131 (1994)
cereal flours, vegetables	α-T, β-T, γ-T, δ-T, α-T3, β-T3, γ-T3	GC	MS	squalane	160 (1995)
milk	α-T, α-TA, retinoids	RP-HPLC	P: UV, 280 nm	—	44 (1995)
beef muscle	α-T	NP-HPLC	fluorescence: λ_{exc} = 296, λ_{em} = 325 nm	—	136 (1996)
dairy products	α-T, β-T, γ-T, δ-T, retinol, carotenoids	NP-HPLC RP-HPLC	P: fluorescence: λ_{exc} = 280, λ_{em} = 325 nm	—	161 (1996)
vegetables, infant formulas	α-T, β-T, γ-T, δ-T, α-T3, β-T3, γ-T3, δ-T3	NP-HPLC	fluorescence: λ_{exc} = 295, λ_{em} = 330 nm	—	137 (1996)
milk	α-T, vitamin A, vitamin D, vitamin K	RP-HPLC	UV, 250 nm	—	109 (1997)
3. SFE					
wheat germ powder	α-T, β-T	SFC NP-HPLC	UV, 295 nm or 290 nm		162 (1989)
seeds	α-T	SFC	FID		163 (1991)
seeds	α-T, β-T, γ-T, δ-T	RP-HPLC	UV, 290 nm	—	102 (1994)
seeds	α-T	SFC	FID	—	153 (1995)

BHA: butylhydroxyanisole.
Other abbreviations: see Tables 2 and 4.

Table 9 Determination of Tocopherols and Tocotrienols in Pharmaceuticals and Cosmetics

Sample preparation	Matrix	Analytes	Analytical technique	Detection	Internal standard	Ref.
1. SFE	ointment	α-TA, retinyl palmitate	SFC	UV, 284 nm	—	164 (1993)
	tablet formulations	α-TA, retinyl palmitate	RP-HPLC	P: UV, 280 nm	—	103 (1995)
	pharmaceutical preparations	α-TA	SFC	FID	—	165 (1998)
2. SPE	pharmaceutical preparations	α-T, α-TA, retinol	RP-HPLC	UV, 290 nm	anthraquinone	108 (1997)
3. Extraction with organic solvent, injection of aliquot or evaporation and reconstitution of the residue	vitamin mix	α-TA, retinyl esters, vitamin D_3, vitamin K_1	RP-HPLC	UV, 248 nm	—	92 (1989)
	parenteral nutrition	α-TA, retinol, vitamin D_2, vitamin K_1	RP-HPLC	P: UV, 284 nm	—	62 (1994)
	multivitamin preparations	α-T, α-TA, retinyl esters, vitamin D_3	RP-HPLC	P: UV, 280 nm MS	—	44 (1995)
4. Dilution	cosmetics	α-TN	RP-HPLC	UV, 210 nm	α-TA	13 (1992)
	pharmaceutical preparations	α-T, β-T, γ-T, δ-T	RP-HPLC	UV, 290 nm	—	42 (1994)

Abbreviations: see Tables 2 and 4.

Single-Phase Extraction. For the analysis of plasma/serum (Table 4) the convenient approach of single-phase extraction, first described in the pioneering paper of Lehmann in 1982 (cited in 26), has not been followed up except for one application (63). It has still been used by the same author and coworkers as part of assays of tocopherols in lipoproteins (34,35). In contrast, single-phase extraction is the method of choice when high concentrations of E vitamers have to be determined in oils and fats, which can simply be dissolved and directly injected in a normal-phase system (Table 7, section 1). To avoid emulsion formation due to the presence of water in the sample, Ye et al. added anhydrous $MgSO_4$ as a desiccating agent (141).

Double-Phase Extraction. Double-phase extraction with hexane or another organic solvent following precipitation of proteins with ethanol remains the preferred procedure in the majority of GC and LC assays of vitamin E in any biological matrix (Tables 4,5,6,8), except oils and fats (Table 7). For example, among the sample pretreatment schemes listed in Table 4 for plasma/serum, double-phase extraction has a share of $> 85\%$. Cooper et al. claim a superior recovery of vitamin E from human plasma by adding magnesium chloride and sodium tungstate prior to protein precipitation with methanol (70). A pretreatment with SDS has been claimed to result in cleaner extracts from plasma and erythrocytes. This approach was adopted in several papers published in the mid-1980s (cited in 26). Recently, it has been applied to the determination of tocopherols in human buccal mucosa cells (114). After double-phase extraction, the organic extract is usually evaporated to dryness and the residue is reconstituted. However, a few methods still rely on the approach of double-phase extraction with relatively polar solvent mixtures containing butanol, ethyl acetate, and acetonitrile, followed by the direct injection of the organic layer in a reversed-phase system (38,73,113).

Saponification Followed by Double-Phase Extraction. Saponification is almost never used for the analysis of plasma/serum, unlike erythrocytes and platelets (7,74,148). It also forms an integral part of a substantial number of the vitamin E assays in tissues, oils, fats, foods and feeds (Tables 6 through 8). Alternatively, oils can be enzymatically hydrolyzed, which would minimize losses of thermolabile oxidizable compounds such as tocopherols (110).

Solid-Phase Extraction (SPE). SPE-based chromatographic methods for vitamin E are virtually nonexisting. In view of the past popularity of open column chromatography this may be surprising. However, the latter was used for cleanup purposes, whereas SPE is supposed to replace two-phase extraction. In one method for the simultaneous determination of water and fat-soluble vitamins in plasma, protein-free extracts were subjected to SPE on a C18 cartridge (108). The water-soluble compounds passed through the cartridge, whereas the fat-soluble ones were retained and subsequently eluted with methanol. However, SPE has no obvious advantage over two-phase extraction of vitamin E.

Supercritical Fluid Extraction (SFE). SFE has been used for the extraction of oils (102,153), pharmaceutical preparations (103,164,165), seeds (163), and wheat germ (162). It has been coupled to supercritical fluid chromatography (SFC) (153,162–165) and HPLC (102,103).

The use of antioxidants during extraction remains widespread. Antioxidants are mostly added for the analysis of blood elements (Table 5), nearly always during saponification (see, e.g., Table 8, section 2) but increasingly also in connection with the analysis of plasma/serum (56,67,77,80,84,86,89,90,97). The three compounds listed in III.A.5 but particularly BHT (77,80,84,86,89,90,97) have all found application. Tirmenstein et al. pretreated tissue samples with a solution of SDS supplemented with ascorbic acid (53). The relevance of adding an antioxidant to an aqueous sample before extraction to protect a liposoluble compound may be questioned.

6. Applications

The applications surveyed in Tables 4 through 9 illustrate the general principles of vitamin E assays outlined in I.E.1. Specifically, each matrix puts a different emphasis on the E vitamers to be determined. Thus, in serum/plasma α-tocopherol is clearly the main compound of interest. Accordingly, reversed-phase chromatography with UV detection is the indicated technique for this purpose. In addition, as part of the assessment of the antioxidant status of humans, tocopherols are determined concurrently with retinoids (retinol, retinyl palmitate) and carotenoids (particularly β-carotene). α-Tocopherol is also the principal target compound in erythrocytes and platelets but here, predictably, the quantitation of α-tocopherolquinone may also be meaningful as an indicator of oxidative stress (7). The need to assay this minor constituent in turn justifies coulometric detection. The analysis of tissues is complementary to that of plasma and red blood cells and mainly concerns the determination of α-tocopherol as well as retinoids, carotenoids, and ubiquinones (Table 6).

Tables 7 and 8, pertaining to the analysis of oils, fats, foods, and feeds, show a different picture in that the assessment of the nutritional value of these materials also requires the determination of tocopherols other than the α-form, as well as the tocotrienols. Accordingly, normal-phase HPLC, GC, and SFC occupy a more important position in this context than reversed-phase HPLC. In foods, notably infant formulas (67,125,139), in feeds (158), and in pharmaceutical preparations (44,92,103,108,164), α-tocopheryl acetate is also frequently included as an analyte, whereas in dairy products vitamin E may be determined concurrently with other fat-soluble vitamins, particularly A and D (72,109,161), sometimes β-carotene (161), and vitamin K (109).

IV. GAS LIQUID CHROMATOGRAPHY (GC)

A. General Concepts

1. Introduction

Before the advent of modern liquid chromatography, gas liquid chromatography (GC) was one of the most widely used analytical techniques for vitamin E, ranking in importance second after colorimetry (25). Yet, it always had to struggle with unfavorable phenomena, related to the physicochemical properties of the analytes but also to the composition of the biological matrices to be analyzed. The low volatility of tocopherols necessitates high temperatures and long analysis times, with a concomitant risk of thermodegradation. Most older GC methods were plagued by the interference of unsaponifiable lipid constituents in extracts, owing to the use of universal rather than selective detectors. A major problem in this regard has always been the coelution of α-tocopherol and cholesterol, which is present in large excess in samples of human and animal origin. As a result, sample pretreatment schemes were mostly complex and time-consuming.

2. Systems

Tocopherols can be separated using a variety of stationary phases ranging from nonpolar (e.g., SE-30), to moderately polar (e.g., OV-17) and polar ones (e.g., Silar-10C). In GC, the elution order of tocopherols is normally according to their degree of methyl substitution, i.e., δ-T < β-T + γ-T < α-T. Tocotrienols elute after their tocopherol counterparts. Capillary, unlike packed columns are able to resolve positional isomers. Tocopherols can be chromatographed natively or, more commonly, after derivatization (see IV.A.5).

3. Detection

Both flame ionization (FID) and argon detectors, the latter now obsolete, provide a sufficiently low detectability for routine applications. Detection limits (amounts injected) using FID have been estimated as 20 and 1 ng of α-tocopherol for packed and capillary columns, respectively (26). If additional sensitivity and selectivity are required, GC-MS is the obvious choice (see V).

4. Quantitation

Both external and internal standardization can be used in GC assays of vitamin E. Often, non–vitamin E–like compounds are used as internal standards, solely on the basis of their retention characteristics. Examples of bad choices include higher alkanes, fatty acid esters, and sterols. Of the structurally related compounds α-tocopheryl propionate (25), tocol, and 5,7-dimethyltocol (25) appear to satisfy most criteria of a good internal standard.

Table 10 Gas Chromatographic Systems for Determination of Tocopherols and Tocotrienols

Stationary phase	Carrier gas, column temperature	Detection	Analytes	Derivatization	Application	Ref.
Packed columns						
SE-30	Ar, He, Air, 270°C	FID	α-T	TMS	vegetables	159 (1989)
Capillary columns						
1. Nonpolar stationary phases						
DB-5	He, gradient	FID	α-T, β-T	—	oil	166 (1990)
DB-5	H₂, gradient	FID	α-T, β-T, γ-T, δ-T	HFB	feeds	167 (1991)
DB-5	H₂, gradient	FID	α-T, β-T, γ-T, δ-T	—	oils	122 (1992)
SE-54:OV-61 (5:1)	H₂, gradient	FID	α-T, β-T, γ-T, δ-T	TMS	oils and fats	156 (1993)
CP Sil 5 CB	H₂, gradient	FID	α-T, α-TA	—	feed premixes	158 (1993)

Column	Conditions	Detection	Compounds	Derivatization	Sample	Ref.
Ultra-1	He, gradient	MS SIM	α-T, β-T, γ-T, δ-T, α-T3, β-T3, γ-T3	TMS	foods of vegetable origin	160 (1995)
DB-5ms	gradient	MS SIM	α-T + oxidation products	TMS	rat liver microsomes	6 (1996)
Alltech heliflex AT1	H₂, gradient	FID	α-T, β-T, γ-T, δ-T	—	essential oils	135 (1996)
HP-1	N₂, gradient	FID	α-T, α-TA	—	oils and fats	157 (1996)
Permabond OV-1	He, 290°C	MS	α-T	TMS	serum	71 (1997)
HP-1	N₂, gradient	FID	α-T, α-TA, other antioxidants	—	oils, foods	154 (1998)
2. Medium polar stationary phases						
Alltech RSL-300	H₂, gradient	FID	α-T	—	margarine	128 (1993)
3. Polar stationary phases						
Silar 10C	165°C	FID	2 stereoisomers of α-T	acetate, methyl ether	plasma, tissues	145, 149, 168 (1990)

Abbreviations: TMS, trimethylsilyl; HFB, heptafluorobutyryl; FID, flame ionization detection; MS, mass spectrometry; SIM, selected ion monitoring. Other abbreviations: see Tables 2 and 4.

5. Sample Preparation

In principle, all approaches for sample pretreatment outlined for LC (see III.A.5 and III.C.5) are applicable to GC. However, there is a tendency toward more elaborate cleanup in GC, the reasons being poor detector selectivity (see IV.A.3) and low resolving power of packed, as opposed to capillary, columns. Accordingly, to decrease the lipid load of extracts, saponification as well as one or more steps to prefractionate the unsaponifiable fraction are often included. Derivatization is optional but usually beneficial in terms of analyte stability, volatility, and peak shape. Silylation (TMS ethers) is most common but, alternatively, acetate, propionate, butyrate, trifluoroacetate, and pentafluoropropionate esters can also be prepared. However, esterification requires heating and longer reaction times, whereas silylation proceeds rapidly at room temperature.

B. Status Before 1989 (25,27)

The decreasing importance of GC as an analytical technique for vitamin E becomes evident from the number of systems/applications inventoried in the first and second editions of this book. In the 1985 edition, 64 systems, 61 of which used packed columns, were listed. However, the number of papers that appeared between 1985 and 1989, included in the second edition was only five. A milestone in the history of GC of vitamin E was the first and, to the best of our knowledge so far only separation, in 1967, of β- and γ-tocopherols as their quinone derivatives on a packed column, using a special binary liquid phase. The separation of TMS ethers of both isomers was achieved in 1975 on a 32-m open tubular column coated with a polar PZ-176 liquid phase and, later on a 30-m DB-5 capillary column. All tocopherols and tocotrienols were successfully resolved in as little as 15 min by capillary GC on a 20 m OV-17 column. GC was also the first method to distinguish (partially) between stereoisomers (see VI.A).

GC assays of vitamin E have always suffered from the interference of cholesterol. A separation from α-tocopherol proved feasible on capillary columns and rarely on packed columns but at the expense of long analysis times. More commonly, cholesterol was exhaustively removed from saponified extracts using digitonin precipitation, open column chromatography on florisil, alumina or celite-digitonin and TLC. Only capillary GC allowed to determine tocopherols (as their TMS ethers) in plasma using the simple solvent extraction approaches typical of LC assays.

C. Status After 1989

1. Systems

The decreasing trend in the application of GC to vitamin E that became apparent in the mid-1980s continues. A survery of systems published after 1989 is given in Table 10. With only one current application (159), packed column GC has fallen into total oblivion. Among the stationary phases used in capillary GC the nonpolar ones predominate. A polar Silar 10C phase has been used for the partial separation of stereoisomers (145,149,168) (see VI.A). Among the tocopherol derivatives, TMS ethers are the most common ones.

2. Detection

Apart from GC-MS methods, which will be discussed separately (see V), detection is carried out by FID.

3. Quantitation

Many procedures are based on the external standard approach. The few applications that do involve internal standardization use vitamin E-unlike compounds such as hexaconazole (135) and cholestane (122,157).

4. Sample Preparation and Applications

Recent GC-FID methods have found application for the determination of tocopherols in foods (128,154,159), feeds (167), oils (122,154,156,157), and feed premixes (158), but no longer for that in biological materials from human origin, except in combination with LC for the differentiation of stereoisomers (145,149).

V. MASS SPECTROMETRY (MS)

A. General Concepts

1. Introduction

In principle, the added value of MS detection in GC and LC of vitamin E could be threefold. This powerful technique provides superior selectivity and sensitivity as well as additional information on the identity of chromatographic peaks. As demonstrated earlier, owing to the use of fluorescence and electrochemical detection, selectivity, and sensitivity are no longer a real concern in LC of vitamin E, unlike in GC with FID. Consequently, there appears to be rarely a need to use MS for quantitative purposes, except in those cases where the analysis of extremely small complex samples, e.g., tissue biopsies, puts stringent demands on sensitivity and selectivity. However, MS is also useful for the off-line confirmation of presumed tocopherol peaks in LC.

2. Systems

Electron impact (EI) as well as chemical ionization (CI), field desorption (FD), and fast atom bombardment (FAB), have all found application. The advantage of the latter three soft ionization techniques is that they result in less fragmentation of tocopherols and, hence, allow monitoring of the characteristic molecular ions. MS has been coupled to both GC and LC as well as to MS (tandem MS). Tocopherols have been analyzed with (as TMS ethers) and without derivatization.

3. Quantitation

In selected ion monitoring (SIM), the choice of deuterated analogs of tocopherols, e.g., α-tocopherol-d_6 or α-tocopherol-d_{13}, as internal standards is straightforward. Sometimes no internal standard is used.

B.   Status up to 1989 (25,27)

1. GC-MS

Three outstanding papers on the quantitation of α-tocopherol in minute samples by GC-MS still deserve to be mentioned because recent studies of this scope and quality are lacking. SIM with α-tocopherol-d_{13} as internal standard was applied to the analysis of lung (5 to 50 µg) (20) and ocular tissue (5 to 200 µg) (21), using direct silylation on the sample, without prior extraction of the analyte. Detection limits were in the subpicogram range. In contrast, a determination of underivatized α-tocopherol in rat and tuna liver involved the complex sample pretreatment typical of the older GC assays for vitamin E, including saponification, double-phase extraction and column chromatography on digitonin-celite (169).

2. LC-MS

In the previous edition of this book, LC-MS was still included in the section ''new developments and future trends.'' Its application to the determination of tocopherols and tocotrienols in a vegetable oil using moving belt LC-MS and EI or CI was reported.

3. Tandem MS

Tandem MS was used for the identification of rare monomethyltocols and of γ-tocotrienols and δ-tocopherol, the latter two in cyanobacteria.

C.   Status After 1989

1. GC-MS

Few papers on the quantitative GC-MS of vitamin E have appeared since 1989. This approach has only once been recommended as a reference method for vitamin E. Kock et al. described an isotope dilution capillary GC-MS method using α-tocopherol-d_6 as an internal standard for the determination of α-tocopherol in serum (71). Two LC methods were compared with this ID-GC-MS procedure. De Luca et al. used capillary GC-MS for the quantitation of TMS ethers of tocopherols and tocotrienols together with phenolics in foods of vegetable origin (160). SIM was applied by Liebler to the identification and quantitation of a range of α-tocopherol oxidation products in rat liver microsomes including α-tocopherolquinone and various epoxy-α-tocopherolquinones, which are of interest in connection with peroxyl radical scavenging reactions (see Fig. 2) (6). Silyl ethers were prepared and deuterated analogs of the various compounds of interest were used as internal standards. The potential of GC-MS for off-line structure elucidation is illustrated in the work of Shin, who characterized tocopherols and tocotrienols isolated from natural sources by semipreparative HPLC (152). Malik et al. used GC-MS to confirm the identity of a presumed α-tocopherol peak in the HPLC of garlic extracts (111).

GC-MS has also found application for the elucidation of the structures of urinary metabolites of α-tocopherol (8,9) and to distinguish between deuterated tocopherols as part of a study on the bioavailability of α-tocopheryl acetate in humans (170).

2. LC-MS

Contrary to the predictions in 1989 (27), LC-MS has not experienced a breakthrough as an analytical technique in the vitamin E area. One research group reported two studies on the determination of liposoluble vitamins in foods and infant formulas by LC-particle beam MS (44,50). Caimi and Brenna coupled HPLC with combustion isotope ratio mass spectrometry for the analysis of mixtures of liposoluble vitamins, including α- and γ-tocopherol (171).

3. SFC-MS

MS can also be coupled with supercritical fluid chromatography (SFC, see VI.F). The four tocopherols have been separated on capillary columns and identified by MS in commercial antioxidant samples and deodorizer distillates (172).

VI. SPECIAL APPLICATIONS AND RECENT TECHNIQUES

A. Separation of Isomers

The separation of positional isomers, i.e., β- and γ-tocopherol and to a lesser extent 5,7-dimethyltocol, has been addressed above in different paragraphs. In summary, a complete resolution of β- and γ-tocopherol can be achieved by normal-phase chromatography on silica, as well as on diol, amino, and aminocyano bonded phases (Table 3). The three positional isomers have been separated on silica (129), diol (138), amino (138), and cyclodextrin (130) columns. In reversed phase, a partial resolution between the β and γ forms has been obtained by capillary LC on a polymeric C18 phase (118), by conventional LC on an uncommon column material (39), by injecting the samples in a solvent that is immiscible with a semiaqueous mobile phase (68) and on a C30 column (66). Baseline separation has been reported on a pentafluorophenyl (PFPS) bonded phase (42). On this material, β- and γ-tocopherol could likewise be resolved from 5,7-dimethyltocol (51). Abidi and Mounts also achieved a partial separation of the positional isomers without derivatization on an octadecyl polyvinyl column and on ODS bonded phases after their conversion to various esters (51). In GC, positional isomers can be separated as TMS ethers on capillary columns or, exceptionally as quinones on packed ones, containing binary stationary phases (25,27). Finally, β- and γ-tocopherol have been resolved by supercritical fluid chromatography (144) (see VI.F) and by micellar electrokinetic chromatography (173) (see VI.H).

One paper deals with the separation of *cis-trans* isomers of α-tocotrienol on a permethylated β-cyclodextrin phase using water-acetonitrile as the eluent (174). The separation of the eight stereoisomers of α-tocopherol continues to represent a challenge in that so far a single chromatographic technique has failed to accomplish this. To develop and validate chromatographic methods in this regard various mixtures can be analyzed, i.e., *all-rac*-α-tocopherol (eight stereoisomers), 2-*ambo*-α-tocopherol (2R + 2S isomers) and 4′-*ambo*-8′-*ambo*-tocopherol (four stereoisomers) (see 1.B.1 and Table 1 for the terminologies). The aim of applying the methods to biological materials such as plasma and tissues is to study the possible preferential accumulation of certain stereoisomers. Neither GC or LC is capable of distinguishing between all eight α-tocopherol stereoisomers but the two techniques are complementary and, hence, can be applied consecutively. Gas chromatography affords the separation of 2R, 4′R, 8′R-α-tocopherol (natural) from its 2-*epi* analog (2S, 4′R, 8′R-α-tocopherol) in 2-*ambo*-α-tocopherol and of the four diastereoisomers with the R configuration at C-2 (RRR, RSR, RRS, and RSS), present in 4′-*ambo*-8′-*ambo*-α-tocopherol. All-*rac*-α-tocopherol is resolved in 4 peaks, each consisting of two stereoisomers.

Although dating back to 1981 the pioneering work of Slover and Thompson deserves to be mentioned in this context because it still represents one of the most elegant GC separations of α-tocopherol stereoisomers (175). These investigators used a 115-m SP 2340 capillary column and prepared TMS ethers of the tocopherols. Recent GC systems for stereoisomers use α-tocopheryl methyl ethers and a Silar 10C stationary phase (145,149,168) according to Cohen (176). In HPLC on chiral phases *all-rac*-α-tocopheryl acetate is differentiated in four peaks with compositions depending on the nature of the chiral column. On polytriphenylmethacrylate (PTMA) coated silica and Chiralcel OD, these four peaks consist of the RSR + RSS, RRR + RRS, SSS + SSR, and SRS + SRR pairs, respectively (145,149,168). In contrast, on Chiralpak OD the four fractions have a different composition, i.e., RRR, RRS, RSR, RSS (peak 1), SSS, SSR (peak 2), SRR (peak 3), and SRS (peak 4) (148). Chromatography of α-tocopherol stereoisomers as methyl ethers instead of acetates on Chiralcel OD yields 5 fractions: peak 1, all stereoisomers with the S-configuration at C-2 (SSR, SSS, SRS, and SRR); peak 2, RSS; peak 3, RRS; peak 4, RRR; peak 5, RSR (Fig. 9) (145).

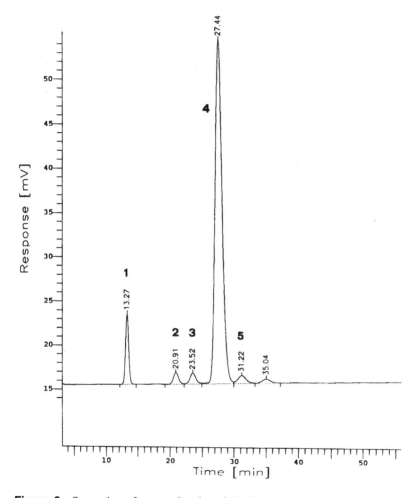

Figure 9 Separation of α-tocopheryl methyl ether stereoisomers from a purified rat liver extract by chiral-phase HPLC. Chromatographic conditions: Column, Chiralcel OD, 250 × 4.6 mm; mobile phase, n-hexane; detection, UV, 220 nm. Peak identification: 1, all four 2S stereoisomers (SSR + SSS + SRS + SRR); 2, RSS−; 3, RRS−; 4, RRR−; 5, RSR-α-tocopheryl methyl ether. (From Ref. 145.)

To resolve all eight stereoisomers, the systems based on PTMA silica or Chiralcel OD for the separation of α-tocopheryl acetates have been combined with GC (145,149,168). First, the LC separation of the acetates is carried out to yield four fractions. Collected peaks are pooled in an LC-A fraction, containing all 2R stereoisomers (peaks 1 + 2) and an LC-B fraction composed of the 2S stereoisomers (peaks 3 + 4). Both LC-A and LC-B are then further resolved by GC, after conversion of the acetate to methyl ether derivatives, into the eight individual components. Application of this method to biological samples of rats revealed that, after administration of *all-rac*-α-tocopheryl acetate the 2R stereoisomers accumulated preferentially (70% to 80%) in plasma and tissues, as opposed to their 2S counterparts, which represented a minor (14% to 30% of total) fraction (149). The four 2R stereoisomers, including the natural RRR form, were equally enriched.

The separation of all stereoisomers has been simplified by conducting the preliminary LC fractionation using α-tocopheryl methyl ethers instead of acetates. Thus, subsequent reconversion to another derivative for GC is no longer required. In addition, all individual R isomers are already resolved by LC (Fig. 9), so that only the S isomers have to be further differentiated by GC (Fig. 10).

In another study, on the distribution of α-tocopherol stereoisomers in rats,

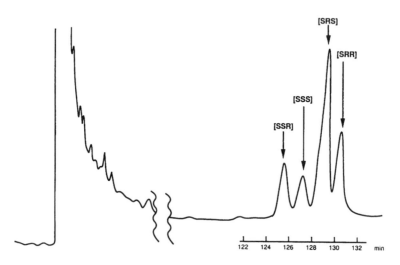

Figure 10 Gas chromatographic separation of the four 2S-α-tocopheryl-methyl ether stereoisomers after separation from a rat liver extract with HPLC on Chiralcel OD. Chromatographic conditions: Column, Silar 10C, 100 m × 0.3 mm; temperature, 165°C; injector temperature 260°C; detection, FID, 220°C. Peak identification: indicated in the figure. (From Ref. 145.)

a similar conclusion with regard to the preferential accumulation of the R forms was reached (148). However, here the four individual 2R stereoisomers were not separated because only LC fractionation on Chiralpak OD was used.

B. Separation of Oxidation Products

Oxidation products of α-tocopherol are formed in vitro photochemically or chemically by reaction with free radical generating compounds such as certain azo derivatives. In vivo they result from the reaction with peroxyl radicals (see 1.D). However, the presence of small quantities of oxidized "metabolites" in biological extracts may be due to artefactual formation during sample pretreatment rather than to genuine biotransformation. The pioneering work on the chromatographic separation of oxidation products of α-tocopherol carried out before 1979 has been reviewed by Baca et al. (10) and Molnar and Koswig (177). One of the most useful HPLC methods in this regard is that of Ha and Csallany, who separated α-tocopherol from five of its oxidation products in a 50 min isocratic run (178). To study the oxidation of α-tocopherol, its more convenient analog 2,2,5,7,8-pentamethyl-6-chromanol has been used as a model compound. Nine oxidation products, including dimers, trimers, quinone, and dione derivatives were separated from the parent compound by normal-phase gradient elution with detection at 290 nm (10). Yamauchi and coworkers studied the products resulting from the reaction of α-tocopherol with 2,2'-azo-*bis* (2,4-dimethylvaleronitrile) and with methyl linoleate-peroxyl radicals using reversed-phase gradient and normal-phase isocratic liquid chromatography, respectively (179–181). Molnar and Koswig exposed α-tocopherol to γ-irradiation and identified a number of oxidative breakdown products, e.g., α-tocopherolquinone and α-tocopherone, resulting from an initially formed cationic radical (177). They applied reversed-phase chromatography in conjunction with a rapid-scanning spectrophotometric detector to aid in the identification. HPLC with chemiluminescent detection has been used to demonstrate 8a-hydroperoxy-tocopherone in photo-oxidized α-tocopherol (182). Other, less selective and less sensitive methods have failed to detect this compound, which in vivo could allegedly be reconverted to α-tocopherol by ascorbic acid.

The main oxidative metabolite of α-tocopherol, α-tocopherolquinone, has been quantitated in plasma, erythrocytes and platelets by reversed-phase chromatography and coulometric detection in the redox mode (7,74,78), as outlined in III.A.3. Levels of α-tocopherolquinone never exceeded 1% of those of α-tocopherol. The most thorough study of the reaction of α-tocopherol with peroxyl radicals, generated by azo-*bis*(amidinopropane) is that by Liebler et al., who used GC-MS to quantitate α-tocopherolquinone, 5,6-epoxy-α-tocopherolquinone, and 2,3-epoxy-α-tocopherolquinone, as well as α-tocopherolhydroquinone in rat liver microsomes (6) (Fig. 2).

In urine, α-tocopherol is excreted after transformation to α-tocopheronic acid, α-tocopheronolactone, and two other acids. Recently, a new metabolite, i.e., α-CEHC (2,5,7,8-tetramethyl-2-(2′-carboxyethyl)-6-hydroxychroman), has been isolated and identified by GC-MS (8,9).

C. Assay of Pharmacologically Active Analogs of α-Tocopherol

α-Tocopheryl nicotinate has been determined in cosmetic preparations by reversed-phase chromatography using a mixture of water, methanol, and propanol as the mobile phase (13). Samples were dissolved in propanol, and α-tocopheryl acetate was used as an internal standard, which eluted before the nicotinate. Two papers report the HPLC of quaternary ammonium analogs of α-tocopherol, tested as cardioselective antioxidants (11,12). For the purity control, UV detection at 280 nm was appropriate, whereas electrochemical detection afforded superior sensitivity and selectivity for a determination in plasma and heart tissue. Tirmenstein et al. prepared α-tocopheroloxybutyric acid, an ether derivative of α-tocopherol which, like its ester counterpart α-tocopheryl hemisuccinate, is credited with antitumor activity. This compound has been determined in liver and human leukemia cells (53).

D. Narrow-Bore and Capillary LC

Narrow-bore (2 mm ID) columns have occasionally been recommended to obtain enhanced sensitivity in assays of mixtures of fat-soluble vitamins, including retinol, retinyl acetate, retinyl palmitate, vitamin D, α-tocopheryl acetate, and vitamin K (50,60,62). Not all of these compounds can be detected fluorimetrically or electrochemically, so that programmed UV detection was the obvious choice. Absorption detection, being concentration sensitive will yield higher signals in connection with narrow bore columns because of reduced peak dispersion. This feature also explains their easier coupling with GC (see VI.E) and MS. Thus, narrow-bore LC in combination with GC and with particle beam MS has been applied to the determination of tocopherols in oils and fats (128,156) and of fat-soluble vitamins in infant formulas (50). Capillary LC is particularly advantageous when only very small biological samples are available, e.g., blood from babies. A method for the determination of α-tocopherol, cholecalciferol, vitamin K, and retinyl palmitate in bovine plasma using a 300×0.32 mm ID fused silica column packed with a 5-μm reversed-phase material has been reported (104). Although the absolute detection limits (29, 1, 2.6, and 2.9 pg for the above four compounds, respectively) substantially superseded those obtained on conventional columns, the factor limiting the detectability in terms of analyte concentration in the sample is the maximum injection volume, which is proportional to the column length and the square ID of the column and inversely proportional to the square root of the column plate number.

In the above study, the injection volume ranged from 60 to 300 nL. Consequently, for biological samples the eventual gain in sensitivity will depend on the degree of miniaturization that is practically feasible. A sample volume of 60 nL has been claimed in the cited paper, but no details on the downscaling of the extraction procedure, described for 1-mL samples, are given (104).

E.   On-line Coupled LC-GC

Although many examples exist of simple HPLC methods for the determination of vitamin E in fats and oils (see Table 7), some of these analyses may present problems of selectivity. Saponification with its known disadvantages of being time-consuming and causing possible degradation of the analytes often provides an answer to this problem. As an alternative, on-line coupled LC-GC may be considered. Here, a preseparation is conducted on a narrow-bore LC column and fractions of interest are transferred via an interface to the GC, which resolves the tocopherols from the initially co-eluting substances. Micali reported such on-line LC-GC method to determine tocopherols in margarine (128). Samples were dissolved in hexane and subjected to LC on a 10 × 0.21 cm Hypersil Sil column. β-, γ-, and δ-Tocopherol could be directly quantitated using fluorimetric detection, but the impure α-tocopherol peak had to be further differentiated by GC. An advantage of the nonselective FID is that it allows the detection of other lipids as well. Thus, on-line LC-GC methods have been developed for the determination in oils and fats of minor components such as alcohols, esters, sterols, and tocopherols (156). The samples were directly acylated or silylated and prefractionated by LC on a 10 × 0.2 cm silica column. A window of interest was cut from the chromatogram and transferred to the GC for further separation and quantitation. LC has also been coupled off-line to GC and GC-MS, e.g., for the identification of tocopherols and tocotrienols after their isolation from vegetable oils and cereals by semipreparative HPLC (152) and for the resolution of α-tocopherol stereoisomers (see VI.A).

F.   Supercritical Fluid Chromatography (SFC)

On a semipreparative scale, SFC has shown its value for the isolation of tocopherols and tocotrienols from wheat germ oil, a process which may be of industrial interest (162,166). It consists of four steps, i.e., supercritical fluid extraction (SFE) of the oil, preconcentration, chromatography on a silica column with UV detection at 290 nm, and fractionation. Tocopherols were considerably enriched (up to 70% to 85% purity) in the collected fractions relative to the crude samples, in which they represented only minor constituents among an excess of triglycerides. Efficiency and economy (solvent consumption) were improved by operating the SFC in the recycle mode.

Advantages of SFC include the on-line nature of the process of extraction, chromatography, and fractionation, the mild conditions of both SFE and SFC, and the low cost and nonpolluting properties of CO_2. Semipreparative SFC coupled with SFE has been used to fractionate the lipids of palm oil, including triglycerides, diglycerides, fatty acids, carotenoids, tocopherols, and tocotrienols (183). Analytical SFC has found application for the assay of vitamin E in oils (151,153,183,184), seeds (163), and deodorizer distillates (172), mostly in combination with that of other fat soluble compounds such as retinyl esters, triglycerides, sterols, and fatty acids. This technique has also been used for the determination of α-tocopheryl acetate in pharmaceutical preparations (164,165). SFC can be conducted either on capillary GC columns with FID detection (151,153,163, 172, 184,185) or, alternatively, on LC columns with FID (165) or UV detection (144,164,183).

The combination of SFC with MS has been briefly addressed in V.C.3. Mobile phases in SFC consist of CO_2 and a polar modifier such as ethanol. Oils and fats are either extracted by SFE or dissolved in a nonpolar solvent prior to SFC. A thorough study dedicated to SFC of tocopherols is that of Yarita et al. (144). These investigators examined the retention behavior of α-, β-, γ-, and δ-tocopherol on a silica ODS column eluted with mixtures of CO_2 and methanol. At low modifier concentrations (0.5% of methanol), β- and γ-tocopherol were baseline separated. Increasing the modifier content of the eluent improved the resolution of the other tocopherols but at the expense of that of the β and γ pair. The four homologs elute according to the number of methyl groups, but the elution order of β- and γ-tocopherol depends on the concentration of the modifier.

This elegant system has been applied to the analysis of vegetable oils, after saponification and extraction with petroleum ether. Tocol was used as an internal standard. Examples of tocopherol profiles of vegetable oils are depicted in Fig. 11. There was a good agreement between the SFC results and those obtained by HPLC on the same column using fluorescence detection. Other investigators also accomplished a separation by SFC of the different E vitamers, e.g. in marine oil (184) and deodorizer distillates (172).

G.   Column Switching

Column-switching techniques for the determination of drugs in biological fluids usually involve on-line solid-phase extraction on reversed-phase columns coated with albumin or modified with peptides and enzymes. To determine α-tocopherol and retinol in serum, Japanese investigators developed a unique sample preparation involving treatment of the serum with SDS and ethanol, deproteinization and trapping of the analytes on a BSA-80TS precolumn, followed by stepwise

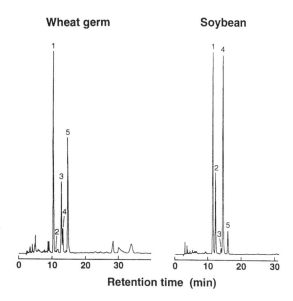

Figure 11 Separation of tocopherols in vegetable oils by SFC. Chromatographic conditions: Column, L-column, ODS, 5 μm, 250 × 4.6 mm; mobile phase, CO_2:methanol, (99.5:0.5, w/w); detection, UV, 290 nm. Peak identification: 1, tocol (internal standard); 2, δ-tocopherol; 3, β-tocopherol; 4, γ-tocopherol; 5, α-tocopherol. (From Ref. 144.)

elution and transfer through column switching of the fractions of interest to the analytical column (186,187). The recovery of α-tocopherol in this procedure was 86%.

H. Capillary Zone Electrophoresis

Capillary zone electrophoresis (CZE) of vitamin E is hampered by its hydrophobic, nonionic character. However, this technique becomes applicable in its modified variants micellar electrokinetic chromatography (MEKC), microemulsion electrokinetic chromatography (MEEKC), and hydrophobic electrokinetic chromatography (HIEKC). In these approaches, pseudostationary phases are generated in the forms of micelles (e.g., from anionic detergents such as SDS), microemulsions (e.g., by emulsifying heptane in a water-butanol mixture containing a surfactant as a stabilizer), or an adsorbed layer of quaternary ammonium ions, respectively. Separation is based on the different partition of the analytes between the aqueous and the pseudo-stationary phases.

The three approaches, MEKC (173,188), MEEKC (189), and HIEKC (190), have been used mainly for the separation of the fat soluble vitamins A, D, and

E. Only Spencer and Purdy applied MEKC specifically to the various E vitamers (173). These investigators added β-cyclodextrins to a buffer containing SDS micelles. The four natural vitamin E homologs, including β- and γ-tocopherol, were resolved in 30 min. In contrast, positional isomers could not be separated by reversed-phase HPLC using cyclodextrins as additives to the mobile phase.

VII. FUTURE TRENDS

The present review has shown that the traditional concepts of LC assays of vitamin E that were established in the 1980s have stood firm. Although various analytical developments predicted in the 1989 edition of this book have materialized in the 10 years elapsed since then, this has occurred on a limited scale. Examples include the use of 3-μm particles in LC, narrow-bore and capillary LC, coupled GC-LC separations, supercritical fluid chromatography, and capillary zone electrophoresis. Contrary to the expectations, LC-MS has not gained real importance. In keeping with the trend already observed in 1989, GC with FID and TLC are becoming marginal as analytical techniques for vitamin E. The advantages of CZE over HPLC are not obvious because, with one exception, the former technique has only been used for the separation of fat-soluble vitamins in general. The applicability of CZE for the determination of vitamin E in a complex matrix has not been investigated.

So far it looks as if the well-established LC methods, possibly complemented by GC-MS for demanding applications, will continue to form the backbone of vitamin E assays in the years to come. The ongoing appearance of dozens of LC methods with often only slight procedural variations points in this direction.

REFERENCES

1. W Friedrich. Vitamins. Berlin, New York: Walter de Gruyter, 1988, pp 217–283.
2. KE Mason. The first two decades of vitamin E story. In: LJ Machlin, ed. Vitamin E. A Comprehensive Treatise. New York: Marcel Dekker, 1980, pp 1–6.
3. GFM Ball. Bioavailability and Analysis of Vitamins in Foods. London: Chapman & Hall, 1998, pp 194–239.
4. S Kasparek. Chemistry of tocopherols and tocotrienols. In: LJ Machlin, ed. Vitamin E. A Comprehensive Treatise. New York: Marcel Dekker, 1980, pp 7–65.
5. GF Combs Jr. The Vitamins. Fundamental Aspects in Nutrition and Health. San Diego: Academic Press, 1992, pp 179–203.
6. DC Liebler, JA Burr, L Philips, AJL Ham. Gas chromatography–mass spectrometry analysis of vitamin E and its oxidation products. Anal Biochem 236:27–34, 1996.

7. GT Vatassery, WE Smith, HT Quach. A liquid chromatographic method for the simultaneous determination of α-tocopherol and tocopherolquinone in human red blood cells and other biological samples where tocopherol is easily oxidized during sample treatment. Anal Biochem 214:426–430, 1993.
8. A Schönfeld, M Schultz, M Petrzika, B Gassmann. A novel metabolite of RRR-α-tocopherol in human urine. Die Nahrung 37:498–500, 1993.
9. M Schultz, M Leist, M Petrzika, B Gassmann, R Brigelius-Flohé. Novel urinary metabolite of α-tocopherol, 2,5,7,8-tetramethyl-2(2′-carboxyethyl)-6-hydroxychroman, as an indicator of an adequate vitamin E supply. Am J Clin Nutr 62(suppl): 1527S–1534S, 1995.
10. M Baca, C Suarna, PT Southwell-Keely. Separation of the α-tocopherol model compound 2,2,5,7,8-pentamethyl-6-chromanol from its oxidation products by high performance liquid chromatography. J Liq Chromatogr 14:1957–1966, 1991.
11. J Verne-Mismer, M Lamard, J Wagner. Evaluation of deactivated reversed phases for the analysis of an N,N,N-trimethylethanaminium analogue of α-tocopherol. Application to its purity control and determination in biological samples. J Chromatogr 645:251–258, 1993.
12. KY Chan, DA Dusterhoft, T-M Chen. Determination of MDL 74,405, a synthetic analogue of α-tocopherol, in dog plasma and heart tissue by high-performance liquid chromatography with electrochemical detection. J Chromatogr 656:359–365, 1994.
13. A Baruffini, E De Lorenzi, C Gandini, M Kitsos, G Massolini. High-performance liquid chromatographic determination of α-tocopheryl nicotinate in cosmetic preparations. J Chromatogr 593:95–97, 1992.
14. AT Diplock, LJ Machlin, L Packer, WA Pryor (eds). Vitamin E. Biochemistry and Health Implications. Ann NY Acad Sci 570, 1989.
15. L Packer, J Fuchs, eds. Vitamin E in Health and Disease. New York: Marcel Dekker, 1993.
16. GJ Handelman, LJ Machlin, K Fitch, JJ Weiter, EA Dratz. Oral α-tocopherol supplements decrease plasma γ-tocopherol levels in humans. J Nutr 115:807–813, 1985.
17. M Mino, M Kitagawa, S Nakagawa. Red blood cell tocopherol concentrations in a normal population of Japanese children and premature infants in relation to the assessment of vitamin E status. Am J Clin Nutr 41:631–638, 1985.
18. C Sierra, MC Pastor, M de Ramón. Liquid chromatography determination of α-tocopherol in erythrocytes. Clin Chim Acta 208:119–126, 1992.
19. HJ Kayden, LJ Hatam, MG Traber. The measurement of nanograms of tocopherol from needle aspiration biopsies of adipose tissue: normal and abetalipoproteinemic subjects. J Lipid Res 24:652–656, 1983.
20. DW Thomas, RM Parkhurst, DS Negi, KD Lunan, AC Wen, AE Brandt, RJ Stephens. Improved assay for α-tocopherol in the picogram range, using gas chromatography–mass spectrometry. J Chromatogr 225:433–439, 1981.
21. RJ Stephens, DS Negi, SM Short, FJGM Van Kuijk, EA Dratz, DW Thomas. Vitamin E distribution in ocular tissues following long-term dietary depletion and supplementation as determined by microdissection and gas chromatography–mass spectrometry. Exp Eye Res 47:237–245, 1988.

22. NR Badcock, DA O'Reilly, CB Pinnock. Liquid chromatographic determination of retinol and α-tocopherol in human buccal mucosal cells. J Chromatogr 382:290–296, 1986.

23. WC Willett, BF Polk, BA Underwood, MJ Stampfer, S Pressel, B Rosner, JO Taylor, K Schneider, CG Hames. Relation of serum vitamins A and E and carotenoids to the risk of cancer. N Engl J Med 310:430–434, 1984.

24. CJ Hogarty, C Ang, RR Eitenmiller. Tocopherol content of selected foods by HPLC/fluorescence quantitation. J Food Comp Anal 2:200–209, 1989.

25. HJ Nelis, VORC De Bevere, AP De Leenheer. Vitamin E: tocopherols and tocotrienols. In: AP De Leenheer, WE Lambert, MGM De Ruyter, eds. Modern Chromatographic Analysis of the Vitamins. New York: Marcel Dekker, 1985, pp 129–200.

26. AP De Leenheer, HJ Nelis, WE Lambert, RM Bauwens. Chromatography of fat-soluble vitamins in clinical chemistry. J Chromatogr 429:3–58, 1988.

27. JK Lang, M Schillaci, B Irvin. Vitamin E. In: AP De Leenheer, WE Lambert, HJ Nelis, eds. Modern Chromatographic Analysis of the Vitamins. 2nd ed. New York: Marcel Dekker, 1989, pp 153–195.

28. AP De Leenheer, WE Lambert, HJ Nelis. Lipophilic vitamins. In: J Sherma, B Fried, eds. Handbook of Thin-Layer Chromatography. New York: Marcel Dekker, 1991, pp 993–1019.

29. AP De Leenheer, WE Lambert. Lipophilic vitamins. In: J Sherma, B Fried, eds. Handbook of Thin-Layer Chromatography. 2nd ed. New York: Marcel Dekker, 1996, pp 1055–1077.

30. S Koswig, J-Th Mörsel, Vergleichende Untersuchungen zur quantitativen Bestimmung von Tocopherolen. Die Nahrung 34:89–91, 1990.

31. U Hachula, F Buhl. Determination of α-tocopherol in capsules and soybean oil after chromatographic separation. J Plan Chromatogr 4:416, 1991.

32. J Sliwiok, B Kocjan, B Labe, A Kozera, J Zalejska. Chromatographic studies of tocopherols. J Plan Chromatogr 6:492–494, 1993.

33. BE Cham, HP Roeser, TW Kamst. Simultaneous liquid-chromatographic determination of vitamin K_1 and vitamin E in serum. Clin Chem 35:2285–2289, 1989.

34. BA Clevidence, J Lehmann. Alpha- and gamma-tocopherol levels in lipoproteins fractionated by affinity chromatography. Lipids 24:137–140, 1989.

35. BA Clevidence, R Ballard-Barbash. Tocopherol contents of lipoproteins from frozen plasma separated by affinity chromatography. Lipids 26:723–728, 1991.

36. DL Luscombe, AM Bond. Determination of tocopherols by reverse-phase liquid chromatography and electrochemical detection at a surface-oxide modified platinum microelectrode, without added electrolyte. Talanta 38:65–72, 1991.

37. P Viñas, N Campillo, MH Córdoba. Direct determination of tocopherols in paprika and paprika oleoresin by liquid chromatography. Mikrochim Acta 106:293–302, 1992.

38. BL Lee, SC Chua, HY Ong, CN Ong. High-performance liquid chromatographic method for routine determination of vitamins A and E and β-carotene in plasma. J Chromatogr 581:41–47, 1992.

39. Y Satomura, M Kimura, Y Itokawa. Simultaneous determination of retinol and tocopherols by high-performance liquid chromatography. J Chromatogr 625:372–376, 1992.

40. C Vidal-Valverde, R Ruiz, A Medrano. Effects of frozen and other storage conditions on α-tocopherol content of cow milk. J Dairy Sci 76:1520–1525, 1993.
41. L Aksnes. Simultaneous determination of retinol, α-tocopherol, and 25-hydroxyvitamin D in human serum by high-performance liquid chromatography. J Pediatr Gastroenterol Nutr 18:339–343, 1994.
42. SL Richheimer, MC Kent, MW Bernart. Reversed-phase high-performance liquid chromatographic method using a pentafluorophenyl bonded phase for analysis of tocopherols. J Chromatogr 677:75–80, 1994.
43. M Jezequel-Cuer, G Le Moël, J Mounié, J Peynet, C Le Bizec, MH Vernet, Y Artur, A Laschi-Loquerie, S Troupel. Dosage de l'α-tocophérol sérique ou plasmatique par chromatographie liquide haute performance: optimisation du mode opératoire. Ann Biol Clin 53:343–352, 1995.
44. M Careri, MT Lugari, A Mangia, P Manini, S Spagnoli. Identification of vitamins A, D and E by particle beam liquid chromatography-mass spectrometry. Fresenius J Anal Chem 351:768–776, 1995.
45. J-Z Huo, HJ Nelis, P Lavens, P Sorgeloos, AP De Leenheer. Determination of vitamin E in aquatic organisms by high-performance liquid chromatography with fluorescence detection. Anal Biochem 242:123–128, 1996.
46. JF Koprivnjak, KR Lum, MM Sisak, R Saborowski. Determination of α-, γ(+β)-, and δ-tocopherols in a variety of liver tissues by reverse-phase high pressure liquid chromatography. Comp Biochem Physiol 113B:143–148, 1996.
47. J-Z Huo, HJ Nelis, P Lavens, P Sorgeloos, AP De Leenheer. Determination of E vitamers in microalgae using high-performance liquid chromatography with fluorescence detection. J Chromatogr 782:63–68, 1997.
48. JI Rader, CM Weaver, L Patrascu, LH Ali, G Angyal. α-Tocopherol, total vitamin A and total fat in margarines and margarine-like products. Food Chem 58:373–379, 1997.
49. C Leray, M Andriamampandry, G Gutbier, J Cavadenti, C Klein-Soyer, C Gachet, J-P Cazenave. Quantitative analysis of vitamin E, cholesterol and phospholipid fatty acids in a single aliquot of human platelets and cultured endothelial cells. J Chromatogr 696:33–42, 1997.
50. R Andreoli, M Careri, P Manini, G Mori, M Musci. Narrow bore columns with UV, electrochemical and particle beam MS detection. Chromatographia 44:605–612, 1997.
51. SL Abidi, TL Mounts. Reversed-phase high-performance liquid chromatographic separations of tocopherols. J Chromatogr 782:25–32, 1997.
52. JR Lane, LW Webb, RV Acuff. Concurrent liquid chromatographic separation and photodiode array detection of retinol, tocopherols, all-*trans*-α-carotene, all-*trans*-β-carotene and the mono-*cis* isomers of β-carotene in extracts of human plasma. J Chromatogr 787:111–118, 1997.
53. MA Tirmenstein, BW Watson, NC Haar, MW Fariss. Sensitive method for measuring tissue α-tocopherol and α-tocopheryloxybutyric acid by high-performance liquid chromatography with fluorometric detection. J Chromatogr 707:308–311, 1998.
54. D Hoehler, AA Frohlich, RR Marquardt, H Stelsovsky. Extraction of α-tocopherol from serum prior to reversed-phase liquid chromatography. J Agric Food Chem 46:973–978, 1998.

55. HR Patel, SP Ashmore, L Barrow, MS Tanner. Improved high-performance liquid chromatographic technique for the determination of hepatic α-tocopherol. J Chromatogr 495:269–274, 1989.
56. C Caye-Vaugien, M Krempf, P Lamarche, B Charbonnel, J Pieri. Determination of α-tocopherol in plasma, platelets and erythrocytes of type I and type II diabetic patients by high-performance liquid chromatography. Int J Vitam Nutr Res 60:324–330, 1990.
57. HE Indyk. Simultaneous liquid chromatographic determination of cholesterol, phytosterols and tocopherols in foods. Analyst 115:1525–1530, 1990.
58. GT Vatassery, MJ Nelson, GJ Maletta, MA Kuskowski. Vitamin E (tocopherols) in human cerebrospinal fluid. Am J Clin Nutr 53:95–99, 1991.
59. G Carcelain, F David, S Lepage, D Bonnefont-Rousselot, J Delattre, A Legrand, J Peynet, S Troupel. Simple method for quantifying α-tocopherol in low-density + very-low-density lipoproteins and in high-density lipoproteins. Clin Chem 38:1792–1795, 1992.
60. D Blanco Gomis, V Escotet Arias, LE Fidalgo Alvarez, MD Gutiérrez Alvarez. Simultaneous determination of vitamins D_3, E and K_1 and retinyl palmitate in cattle plasma by liquid chromatography with a narrow-bore column. J Chromatogr 660:49–55, 1994.
61. MJ Gonzalez-Corbella, N Lloberas-Blanch, AI Castellote-Bargallo, MC Lopez-Sabater, M Rivero-Urgell. Determination of α-tocopherol in plasma and erythrocytes by high-performance liquid chromatography. J Chromatogr 660:395–400, 1994.
62. D Blanco, M Pajares, VJ Escotet, MD Gutiérrez. Determination of fat-soluble vitamins by liquid chromatography in pediatric parenteral nutritions. J Liq Chromatogr 17:4513–4530, 1994.
63. S Torrado, EJ Caballero, R Cadorniga, J Torrado. A selective liquid chromatography assay for the determination of d1-α-tocopherol acetate in plasma samples. J Liq Chromatogr 18:1251–1264, 1995.
64. E Teissier, E Walters-Laporte, C Duhem, G Luc, J-C Fruchart, P Duriez. Rapid quantification of α-tocopherol in plasma and low- and high-density lipoproteins. Clin Chem 42:430–435, 1996.
65. D Moyano, MA Vilaseca, M Pineda, J Campistol, A Vernet, P Póo, R Artuch, C Sierra. Tocopherol in inborn errors of intermediary metabolism. Clin Chim Acta 263:147–155, 1997.
66. S Strohschein, M Pursch, D Lubda, K Albert. Shape selectivity of C30 phases for RP-HPLC separation of tocopherol isomers and correlation with MAS NMR data from suspended stationary phases. Anal Chem 70:13–18, 1998.
67. A Celardo, A Bortolotti, E Benfenati, M Bonati. Measurement of vitamin E in premature infants by reversed-phase high-performance liquid chromatography. J Chromatogr 490:432–438, 1989.
68. K Warner, TL Mounts. Analysis of tocopherols and phytosterols in vegetable oils by HPLC with evaporative light-scattering detection. J Am Oil Chem Soc 67:827–831, 1990.
69. LA Kaplan, JA Miller, EA Stein, MJ Stampfer. Simultaneous, high-performance liquid chromatographic analysis of retinol, tocopherols, lycopene and α- and β-carotene in serum and plasma. Methods Enzymol 189:155–167, 1990.

70. JDH Cooper, R Thadwal, MJ Cooper. Determination of vitamin E in human plasma by high-performance liquid chromatography. J Chromatogr 690:355–358, 1997.

71. R Kock, S Seitz, B Delvoux, H Greiling. Two high performance liquid chromatographic methods for the determination of α-tocopherol in serum compared to isotope dilution-gas chromatography-mass spectrometry. Eur J Clin Chem Clin Biochem 35:371–378, 1997.

72. MM Delgado Zamarreño, A Sánchez Pérez, C Gómez Pérez, J Hernández Méndez. High-performance liquid chromatography with electrochemical detection for the simultaneous determination of vitamin A, D_3 and E in milk. J Chromatogr 623: 69–74, 1992.

73. C Sarzanini, E Mentasti, M Vincenti, M Nerva, F Gaido. Determination of plasma tocopherols by high-performance liquid chromatography with coulometric detection. J Chromatogr 620:268–272, 1993.

74. GT Vatassery. Determination of tocopherols and tocopherolquinone in human red blood cell and platelet samples. Methods Enzymol 234:327–331, 1994.

75. KP Mitton, JR Trevithick. High-performance liquid chromatography-electrochemical detection of antioxidants in vertebrate lens: glutathione, tocopherol, and ascorbate. Methods Enzymol 233:523–539, 1994.

76. F Dionisi, J Prodolliet, E Tagliaferri. Assessment of olive oil adulteration by reversed-phase high-performance liquid chromatography/amperometric detection of tocopherols and tocotrienols. J Am Oil Chem Soc 72:1505–1511, 1995.

77. B Finckh, A Kontush, J Commentz, C Hübner, M Burdelski, A Kohlschütter. Monitoring of ubiquinol-10, ubiquinone-10, carotenoids, and tocopherols in neonatal plasma microsamples using high-performance liquid chromatography with coulometric electrochemical detection. Anal Biochem 232:210–216, 1995.

78. H Takeda, T Shibuya, K Yanagawa, H Kanoh, M Takasaki. Simultaneous determination of α-tocopherol and α-tocopherolquinone by high-performance liquid chromatography and coulometric detection in the redox mode. J Chromatogr 722:287–294, 1996.

79. M Podda, C Weber, MG Traber, L Packer. Simultaneous determination of tissue tocopherols, tocotrienols, ubiquinols and ubiquinones. J Lipid Res 37:893–901, 1996.

80. WA MacCrehan. Determination of retinol, α-tocopherol and β-carotene in serum by liquid chromatography. Methods Enzymol 189:172–181, 1990.

81. D Hess, HE Keller, B Oberlin, R Bonfanti, W Schüep. Simultaneous determination of retinol, tocopherols, carotenes and lycopene in plasma by means of high-performance liquid chromatography on reversed phase. Int J Vitam Nutr Res 61:232–238, 1991.

82. DW Nierenberg, SL Nann. A method for determining concentrations of retinol, tocopherol, and five carotenoids in human plasma and tissue samples. Am J Clin Nutr 56:417–426, 1992.

83. Z Zaman, P Fielden, PG Frost. Simultaneous determination of vitamins A and E and carotenoids in plasma by reversed-phase HPLC in elderly and younger subjects. Clin Chem 39:2229–2234, 1993.

84. KS Epler, RG Ziegler, NE Craft. Liquid chromatographic method for the determination of carotenoids, retinoids and tocopherols in human serum and in food. J Chromatogr 619:37–48, 1993.

85. W Schüep, R Rettenmaier. Analysis of vitamin E homologs in plasma and tissue: high-performance liquid chromatography. Methods Enzymol 234:294–302, 1994.

86. MH Bui. Simple determination of retinol, α-tocopherol and carotenoids (lutein, all-*trans*-lycopene, α- and β-carotenes) in human plasma by isocratic liquid chromatography. J Chromatogr 654:129–133, 1994.

87. K-J Yeum, A Taylor, G Tang, RM Russell. Measurement of carotenoids, retinoids and tocopherols in human lenses. Invest Ophthalmol Vis Sci 36:2756–2761, 1995.

88. MJ Del Nozal, JL Bernal, P Marinero. Simultaneous HPLC determination of cholesterol, α-tocopherol, retinol, retinal and retinoic acid in silicone oils used as vitreous substitutes in eye surgery. J Liq Chromatogr Rel Technol 19:1151–1167, 1996.

89. Y Göbel, C Schaffer, B Koletzko. Simultaneous determination of low plasma concentrations of retinol and tocopherols in preterm infants by a high-performance liquid chromatographic micromethod. J Chromatogr 688:57–62, 1997.

90. O Sommerburg, L-Y Zang, FJGM Van Kuijk. Simultaneous detection of carotenoids and vitamin E in human plasma. J Chromatogr 695:209–215, 1997.

91. B Tan, L Brzuskiewicz. Separation of tocopherol and tocotrienol isomers using normal- and reverse-phase liquid chromatography. Anal Biochem 180:368–373, 1989.

92. E Stary, AMC Cruz, CA Donomai, JL Monfardini, JTF Vargas. Determination of vitamins A-palmitate, A-acetate, E-acetate, D_3 and K_1 in vitamin mix by isocratic reverse phase HPLC. J High Resol Chromatogr 12:421–423, 1989.

93. JL Rudy, F Ibarra, M Zeigler, J Howard, C Argyle. Simultaneous determination of retinol, retinyl palmitate, and α-tocopherol in serum or plasma by reversed-phase high-performance liquid chromatography. LC-GC 7:969–971, 1989.

94. B Olmedilla, F Granado, E Rojas-Hidalgo, I Blanco. A rapid separation of ten carotenoids, three retinoids, alpha-tocopherol and d-alpha-tocopherol acetate by high performance liquid chromatography and its application to serum and vegetable samples. J Liq Chromatogr 13:1455–1483, 1990.

95. K Seta, H Nakamura, T Okuyama. Determination of α-tocopherol, free cholesterol, esterified tocopherols and triacylglycerols in human lipoproteins by high-performance liquid chromatography. J Chromatogr 515:585–595, 1990.

96. J Arnaud, I Fortis, S Blachier, D Kia, A Favier. Simultaneous determination of retinol, α-tocopherol and β-carotene in serum by isocratic high-performance liquid chromatography. J Chromatogr 572:103–116, 1991.

97. T Van Vliet, F Van Schaik, J Van Schoonhoven, J Schrijver. Determination of several retinoids, carotenoids and E vitamers by high performance liquid chromatography. Application to plasma and tissues of rats fed a diet rich in either β-carotene or canthaxanthin. J Chromatogr 553:179–186, 1991.

98. AB Barua, HC Furr, D Janick-Buckner, JA Olson. Simultaneous analysis of individual carotenoids, retinol, retinyl esters, and tocopherols in serum by isocratic nonaqueous reversed phase HPLC. Food Chem 46:419–424, 1993.

99. AB Barua, D Kostic, JA Olson. New simplified procedures for the extraction and simultaneous high-performance liquid chromatographic analysis of retinol, tocopherols and carotenoids in human serum. J Chromatogr 617:257–264, 1993.

100. C-Z Chuang, G Trosclair, A Lopez-S. Adaptation of a carotenoid procedure to

analyze carotenoids, retinol, and alpha-tocopherol simultaneously. J Liq Chromatogr 17:3613–3622, 1994.

101. AL Sowell, DL Huff, PR Yeager, SP Caudill, EW Gunter. Retinol, α-tocopherol, lutein/zeaxanthin, β-cryptoxanthin, lycopene, α-carotene, *trans*-β-carotene, and four retinyl esters in serum determined simultaneously by reversed-phase HPLC with multiwavelength detection. Clin Chem 40:411–416, 1994.

102. FM Lanças, MEC Queiroz, ICE da Silva. Seed oil extraction with supercritical carbon dioxide modified with pentane. Chromatographia 39:687–692, 1994.

103. S Scalia, G Ruberto, F Bonina. Determination of vitamin A, vitamin E, and their esters in tablet preparations using supercritical fluid extraction and HPLC. J Pharm Sci 84:433–436, 1995.

104. D Blanco Gomis, V Escotet Arias, LE Fidalgo Alvarez, MD Gutiérrez Alvarez. Determination of fat-soluble vitamins by capillary liquid chromatography in bovine blood plasma. Anal Chim Acta 315:177–181, 1995.

105. L Yakushina, A Taranova. Rapid HPLC simultaneous determination of fat-soluble vitamins, including carotenoids, in human serum. J Pharm Biomed Anal 13:715–718, 1995.

106. A Somogyi, M Herold, A Blázovics, E Szaleczky, P Pusztai, A Rosta. Vitamin A and E determination in blood plasma by isocratic high pressure liquid chromatography. Mod Chem 133:545–551, 1996.

107. CA Redlich, JN Grauer, AM Van Bennekum, SL Clever, RB Ponn, WS Blaner. Characterization of carotenoid, vitamin A, and α-tocopherol levels in human lung tissue and pulmonary macrophages. Am J Respir Crit Care Med 154:1436–1443, 1996.

108. IN Papadoyannis, GK Tsioni, VF Samanidou. Simultaneous determination of nine water and fat soluble vitamins after SPE separation and RP-HPLC analysis in pharmaceutical preparations and biological fluids. J Liq Chromatogr Rel Technol 20:3203–3231, 1997.

109. BY Gong, JW Ho. Simultaneous separation and detection of ten common fat-soluble vitamins in milk. J Liq Chromatogr Rel Technol 20:2389–2397, 1997.

110. G Lietz, CJK Henry. A modified method to minimise losses of carotenoids and tocopherols during HPLC analysis of red palm oil. Food Chem 60:109–117, 1997.

111. MN Malik, MD Fenko, AM Shiekh, HM Wisniewski. Isolation of α-tocopherol (vitamin E) from garlic. J Agric Food Chem 45:817–819, 1997.

112. D Talwar, TKK Ha, J Cooney, C Brownlee, DSJ O'Reilly. A routine method for the simultaneous measurement of retinol, α-tocopherol and five carotenoids in human plasma by reverse phase HPLC. Clin Chim Acta 270:85–100, 1998.

113. MA Abahusain, J Wright, JWT Dickerson, MA El-Hazmi, HY Aboul Enein. Determination of retinol, α-tocopherol, α- and β-carotene by direct extraction of human serum using high performance liquid chromatography. Biomed Chromatogr 12:89–93, 1998.

114. Y-S Peng, Y-M Peng. Simultaneous liquid chromatographic determination of carotenoids, retinoids and tocopherols in human buccal mucosal cells. Cancer Epidemiol Biomark Prev 1:375–382, 1992.

115. A Bortolotti, G Lucchini, MM Barzago, F Stellari, M Bonati. Simultaneous determination of retinol, α-tocopherol and retinyl palmitate in plasma of premature new-

borns by reversed phase high-performance liquid chromatography. J Chromatogr 617:313–317, 1993.

116. J-P Steghens, AL van Kappel, E Riboli, C Collombel. Simultaneous measurement of seven carotenoids, retinol and α-tocopherol in serum by high-performance liquid chromatography. J Chromatogr 694:71–81, 1997.

117. AB Barua, JA Olson. Reversed-phase gradient high-performance liquid chromatographic procedure for simultaneous analysis of very polar to nonpolar retinoids, carotenoids and tocopherols in animal and plant samples. J Chromatogr 707:69–79, 1998.

118. WT Wahyuni, K Jinno. Separation of tocopherols on various chemically bonded phases in microcolumn liquid chromatography. J Chromatogr 448:398–403, 1988.

119. G Micali, F Lanuzza, P Currò. Simultaneous determination of β-carotene, retinol, retinyl esters and tocopherols by high performance liquid chromatography in butter and margarine. Riv Ital Sost Grasse 67:409–412, 1990.

120. AA Al-Meshari, MA Aleem. Separation of β- and γ-tocopherols in serum by high performance liquid chromatography. Biomed Chromatogr 5:83–85, 1991.

121. R Rettenmaier, W Schüep. Determination of vitamins A and E in liver tissue. Int J Vitam Nutr Res 62:312–317, 1992.

122. F Ulberth, H Reich, W Kneifel. Zur Analytik von Tocopherolen—Ein Methodenvergleich zwischen HPLC und GC. Fat Sci Technol 94:51–54, 1992.

123. M Balz, E Schulte, H-P Thier. Trennung von Tocopherolen und Trocotrienolen durch HPLC. Fat Sci Technol. 94:209–213, 1992.

124. MA Belisario, G Azar, G Oriani, GP Pizzuti, L Sacchetti. Simultaneous evaluation of vitamins A and E in human plasma by normal phase HPLC. Boll Soc Ital Biol Sper 69:641–647, 1993.

125. MK Balz, E Schulte, H-P Thier. Simultaneous determination of α-tocopheryl acetate, tocopherols and tocotrienols by HPLC with fluorescence detection in foods. Fat Sci Technol 95:215–220, 1993.

126. A Abdollahi, NS Rosenholtz, JL Garwin. Tocopherol micro-extraction method with application to quantitative analysis of lipophilic nutrients. J Food Sci 58:663–666, 1993.

127. T-S Shin, JS Godber. Improved high-performance liquid chromatography of vitamin E vitamers on normal-phase columns. J Am Oil Chem Soc 70:1289–1291, 1993.

128. G Micali, F Lanuzza, P Currò. Analysis of tocopherols in margarine by on-line HPLC-HRGC coupling. J High Resol Chromatogr 16:536–538, 1993.

129. GW Chase Jr, CC Akoh, RR Eitenmiller. Analysis of tocopherols in vegetable oils by high-performance liquid chromatography: comparison of fluorescence and evaporative light-scattering detection. J Am Oil Chem Soc 71:877–880, 1994.

130. SL Abidi, TL Mounts. Separations of tocopherols and methylated tocols on cyclodextrin-bonded silica. J Chromatogr 670:67–75, 1994.

131. G Panfili, P Manzi, L Pizzoferrato. High-performance liquid chromatographic method for the simultaneous determination of tocopherols, carotenes, and retinol and its geometric isomers in Italian cheeses. Analyst 119:1161–1165, 1994.

132. A Pfalzgraf, H Steinhart, M Frigg. Rapid determination of α-tocopherol in muscle and adipose tissues of pork. Z Lebensm Unters Forsch 200:190–193, 1995.

133. RB McGeachin, CA Bailey. Determination of carotenoid pigments, retinol, and α-tocopherol in feeds, tissues, and blood serum by normal phase high performance liquid chromatography. Poultry Sci 74:407–411, 1995.

134. M Balz, E Schulte, H-P Thier. Simultaneous determination of retinol esters and tocochromanols in foods using nitro-column HPLC. Fat Sci Technol 97:445–448, 1995.

135. B Fayet, L Fino, C Tisse, M Guérère. Caractérisation et dosage des tocophérols et des gallates dans les huiles essentielles. Sci Aliment 16:83–93, 1996.

136. Q Liu, KK Scheller, DM Schaefer. Technical note: a simplified procedure for vitamin E determination in beef muscle. J Anim Sci 74:2406–2410, 1996.

137. EJM Konings, HHS Roomans, PR Beljaars. Liquid chromatographic determination of tocopherols and tocotrienols in margarine, infant foods and vegetables. J Assoc Off Anal Chem 79:902–906, 1996.

138. SL Abidi, TL Mounts. Normal phase high-performance liquid chromatography of tocopherols on polar phases. J Liq Chromatogr Rel Technol 19:509–520, 1996.

139. GW Chase Jr, RR Eitenmiller, AR Long. Liquid chromatographic analysis of all-rac-α-tocopheryl acetate, tocopherols, and retinyl palmitate in SRM 1846. J Liq Chromatogr Rel Technol 20:3317–3327, 1997.

140. JKG Kramer, L Blais, RC Fouchard, RA Melnyk, KMR Kallury. A rapid method for the determination of vitamin E forms in tissues and diet by high-performance liquid chromatography using a normal-phase diol column. Lipids 32:323–330, 1997.

141. L Ye, WO Landen Jr, J Lee, RR Eitenmiller. Vitamin E content of margarine and reduced fat products using a simplified extraction procedure and HPLC determination. J Liq Chromatogr Rel Technol 21:1227–1238, 1998.

142. S Tuan, TF Lee, CC Chou, QK Wei. Determination of vitamin E homologues in infant formulas by HPLC using fluorometric detection. J Micronutr Anal 6:35–45, 1989.

143. J-D Su. Studies of rapid determination of vitamin E and its application on natural foodstuffs. J Food Drug Anal 1:61–70, 1993.

144. T Yarita, A Nomura, K Abe, Y Takeshita. Supercritical fluid chromatographic determination of tocopherols on an ODS-silica gel column. J Chromatogr 679:329–334, 1994.

145. G Riss, AW Kormann, E Glinz, W Walther, UB Ranalder. Separation of the eight stereoisomers of all-rac-α-tocopherol from tissues and plasma: chiral phase high-performance liquid chromatography and capillary gas chromatography. Methods Enzymol 234:302–310, 1994.

146. LG Rushing, WM Cooper, HC Thompson Jr. Simultaneous analysis of vitamins A and E in rodent feed by high-pressure liquid chromatography. J Agric Food Chem 39:296–299, 1991.

147. AK Hewavitharana, AS van Brakel, M Harnett. Simultaneous liquid chromatographic determination of vitamins A, E and β-carotene in common dairy foods. Int Dairy J 6:613–624, 1996.

148. T Ueda, H Ichikawa, O Igarashi. Determination of α-tocopherol stereoisomers in biological specimens using chiral phase high-performance liquid chromatography. J Nutr Sci Vitaminol 39:207–219, 1993.

149. H Weiser, G Riss, AW Kormann. Biodiscrimination of the eight α-tocopherol stereoisomers results in preferential accumulation of the four 2R forms in tissues and plasma of rats. J Nutr 126:2539–2549, 1996.
150. S Vogel, JH Contois, SC Couch, CJ Lammi-Keefe. A rapid method for separation of plasma low and high density lipoproteins for tocopherol and carotenoid analyses. Lipids 31:421–426, 1996.
151. C Borch-Jensen, A Staby, J Mollerup. Supercritical fluid chromatographic analysis of a fish oil of the sand eel (*Ammodytes* sp.). J High Resol Chromatogr 16:621–623, 1993.
152. T-S Shin, JS Godber. Isolation of four tocopherols and four tocotrienols from a variety of natural sources by semi-preparative high-performance liquid chromatography. J Chromatogr 678:49–58, 1994.
153. P Manninen, P Laakso, H Kallio. Method for characterization of triacylglycerols and fat-soluble vitamins in edible oils and fats by supercritical fluid chromatography. J Am Oil Chem Soc 72:1001–1008, 1995.
154. M González, E Ballesteros, M Gallego, M Valcárcel. Continuous-flow determination of natural and synthetic antioxidants in foods by gas chromatography. Anal Chim Acta 359:47–55, 1998.
155. AI Hopia, VI Piironen, PE Koivistoinen, LET Hyvönen. Analysis of lipid classes by solid-phase extraction and high-performance size-exclusion chromatography. J Am Oil Chem Soc 69:772–776, 1992.
156. A Artho, K Grob, C Mariani. On-line LC-GC for the analysis of minor components in edible oils and fats—the direct method involving silylation. Fat Sci Technol 95: 176–180, 1993.
157. E Ballesteros, M Gallego, M Valcárcel. Gas chromatographic determination of cholesterol and tocopherols in edible oils and fats with automatic removal of interfering triglycerides. J Chromatogr 719:221–227, 1996.
158. S Kmostak, DA Kurtz. Rapid determination of supplemental vitamin E acetate in feed premixes by capillary gas chromatography. J Assoc Off Anal Chem 76:735–741, 1993.
159. M Horbowicz. High-performance liquid chromatography for determination of α-tocopherol in vegetables. Acta Agrobot 42:197–205, 1989.
160. C De Luca, E Quattrucci, S Passi. A gas chromatography-mass spectrometry method for the analysis of tocopherol and low-molecular weight phenol contents in foods of vegetable origin. Current status and future trends. Proceedings of Euro Food Chem VIII, Vienna, 1995, pp 74–78.
161. P Manzi, G Panfili, L Pizzoferrato. Normal and reversed-phase HPLC for more complete evaluation of tocopherols, retinols, carotenes and sterols in dairy products. Chromatographia 43:89–93, 1996.
162. M Saito, Y Yamauchi, K Inomata, W Kottkamp. Enrichment of tocopherols in wheat germ by directly coupled supercritical fluid extraction with semipreparative supercritical fluid chromatography. J Chromatrogr Sci 27:79–85, 1989.
163. RM Hannan, HH Hill Jr. Analysis of lipids in aging seed using capillary supercritical fluid chromatography. J Chromatogr 547:393–401, 1991.
164. M Masuda, S Koike, M Handa, K Sagara, T Mizutani. Application of supercritical fluid extraction and chromatography to assay fat-soluble vitamins in hydrophobic ointment. Anal Sci 9:29–32, 1993.

165. A Salvador, MA Jaime, M de la Guardia, C Becerra. Supercritical fluid extraction and supercritical fluid chromatography of vitamin E in pharmaceutical preparations. Anal Commun 35:53–55, 1998.
166. M Saito, Y Yamauchi. Isolation of tocopherols from wheat germ oil by recycle semi-preparative supercritical fluid chromatography. J Chromatogr 505:257–271, 1990.
167. F Ulberth. Simultaneous determination of vitamin E isomers and cholesterol by GLC. J High Resol Chromatogr 14:343–344, 1991.
168. M Vecchi, W Walther, E Glinz, T Netscher, R Schmid, M Lalonde, W Vetter. Chromatographische Trennung und quantitative Bestimmung aller acht Stereoisomeren von α-Tocopherol. Helv Chim Acta 73:782–789, 1990.
169. CR Smidt, AD Jones, AJ Clifford. Gas chromatography of retinol and α-tocopherol without derivatization. J Chromatogr 434:21–29, 1988.
170. RV Acuff, SS Thedford, NN Hidiroglou, AM Papas, TA Odom Jr. Relative bioavailability of RRR- and all-*rac*-α-tocopheryl acetate in humans: studies using deuterated compounds. Am J Clin Nutr 60, 397–402, 1994.
171. RJ Caimi, JT Brenna. High-sensitivity liquid chromatography-combustion isotope ratio mass spectrometry of fat-soluble vitamins. J Mass Spectrom 30:466–472, 1995.
172. JM Snyder, SL Taylor, JW King. Analysis of tocopherols by capillary supercritical fluid chromatography and mass spectrometry. J Am Oil Chem Soc 70:349–354, 1993.
173. BJ Spencer, WC Purdy. Comparison of the separation of fat-soluble vitamins using β-cyclodextrins in high-performance liquid chromatography and micellar electrokinetic chromatography. J Chromatogr 782:227–235, 1997.
174. AM Drotleff, W Ternes. Separation and characterization of *cis-trans* isomers of α-tocotrienol by HPLC using a permethylated β-cyclodextrin phase. Z Lebensm Unters Forsch A 206:9–13, 1998.
175. HT Slover, RH Thompson Jr. Chromatographic separation of the stereoisomers of α-tocopherol. Lipids 16:268–275, 1981.
176. N Cohen, CG Scott, C Neukom, RJ Lopresti, G Weber, G Saucy. Total synthesis of all eight stereoisomers of α-tocopheryl acetate. Determination of their diastereoisomeric and enantiomeric purity by gas chromatography. Helv Chim Acta 64:1158–1173, 1981.
177. I Molnar, S Koswig. Investigation of γ-irradiation of α-tocopherol and its related derivatives by high-performance liquid chromatography using a rapid scanning spectrophotometer. J Chromatogr 605:49–62, 1992.
178. YL Ha, AS Csallany. Separation of α-tocopherol and its oxidation products by high performance liquid chromatography. Lipids 23:359–361, 1988.
179. R Yamauchi, T Matsui, Y Satake, K Kato, Y Ueno. Reaction products of α-tocopherol with a free radical initiator, 2,2'-azo-*bis*(2,4-dimethylvaleronitrile). Lipids 24:204–209, 1989.
180. R Yamauchi, T Matsui, K Kato, Y Ueno. Reaction of α-tocopherol with 2,2'-azo-*bis*(2,4-dimethylvaleronitrile) in benzene. Agric Biol Chem 53:3257–3262, 1989.
181. R Yamauchi, T Matsui, K Kato, Y Ueno. Reaction products of α-tocopherol with methyl linoleate-peroxyl radicals. Lipids 25:152–158, 1990.

182. T Miyazawa, T Yamashita, K Fujimoto. Chemiluminescence detection of 8a-hydro-peroxy-tocopherone in photooxidized α-tocopherol. Lipids 27:289–294, 1992.

183. YM Choo, AN Ma, H Yahaya, Y Yamauchi, M Bounoshita, M Saito. Separation of crude palm oil components by semipreparative supercritical fluid chromatography. J Am Oil Chem Soc 73:523–525, 1996.

184. A Staby, C Borch-Jensen, S Balchen, J Mollerup. Quantitative analysis of marine oils by capillary supercritical fluid chromatography. Chromatographia 39:697–705, 1994.

185. A Staby, C Borch-Jensen, S Balchen, J Mollerup. Supercritical fluid chromatographic analysis of fish oils. J Am Oil Chem Soc 71:355–359, 1994.

186. K Adachi, N Katsura, Y Nomura, A Arikawa, M Hidaka, T Onimaru. Serum vitamin A and vitamin E in Japanese black fattening cattle in Miyazaki prefecture as determined by automatic column-switching high performance liquid chromatography. J Vet Med Sci 58:461–464, 1996.

187. H Moriyama, H Yamasaki, S Masumoto, K Adachi, N Katsura, T Onimaru. Rapid determination of vitamins A and E in serum with surfactant as a diluent by column-switching high-performance liquid chromatography. J Chromatogr 798:125–130, 1998.

188. CP Ong, CL Ng, HK Lee, SFY Li. Separation of water- and fat-soluble vitamins by micellar electrokinetic chromatography. J Chromatogr 547:419–428, 1991.

189. RL Boso, MS Bellini, I Mikšik, Z Deyl. Microemulsion electrokinetic chromatography with different organic modifiers: separation of water- and lipid-soluble vitamins. J Chromatogr 709:11–19, 1995.

190. S Pedersen-Bjergaard, KE Rasmussen, T Tilander. Separation of fat-soluble vitamins by hydrophobic interaction electrokinetic chromatography with tetradecylammonium ions as pseudostationary phase. J Chromatogr 807:285–295, 1998.

4
Vitamin K

Günter Fauler, Wolfgang Muntean, and Hans Jörg Leis
University of Graz, Graz, Austria

I. INTRODUCTION

Since the early 1930s, when Henrik Dam (1) described a compound named Koagulation Vitamin, or vitamin K (VK), which prevented hemorrhagic disease in chickens, many of its biochemical functions have been clarified, but up to now some of them are still not completely understood. The role that VK plays in blood coagulation seems more or less elucidated, but many questions remain unanswered, as, for example, the actual role of VK in bone metabolism. Chromatography in all its variations played an important role from first purification and detection of the compounds in the 1930s up to recent studies of quantitation in plasma, food, or other biological samples. The complexity of diverse matrices, combined with relative low concentrations of the vitamins in biological samples, besides the sensitivity of the molecule, makes modern chromatography essential for any analytical assay. That is the reason why sample preparation and cleanup, followed by specific chromatographic and detection methods, are common to all investigations in this field. To give an overview of running practices in pertinent literature, we will focus mainly on papers published since 1989.

A. Structures; Physical and Chemical Properties

Vitamin K is a collective name for a group of molecules with the same 2-methyl-1,4-naphthoquinone structure, but with differences in the side chain on position 3 of the naphthoquinone. The number of carbon atoms in this side chain is used to characterize the different molecules. In this nomenclature $VK_{1(20)}$ means 20 carbon atoms in the side chain, whereas MK_n is a synonym for the number of n

(mainly 4-13) isoprene units, each consisting of five carbon atoms, in the side chain (Structure 4 in Fig. 1). Three molecules, each a representative of a group of K vitamins, are described in more detail in the following section.

1.   Vitamin $K_{1(20)}$ (VK1), or phylloquinone, is the generic name of 2-methyl-3-eicosa-2′-ene-1,4-naphthoquinone. The (2′-*trans*-) phytyl side chain in position 3 of the naphthoquinone nucleus contains 20 carbon atoms, structured in four isoprene units, the last three of them totally reduced (see Structure 1 in Fig. 1). If a synthetic VK1 is analyzed, always a mixture of *cis/trans*-isomers can be detected, deriving from synthetic phytol, which consists of a mixture of these isomers. VK1 in pure form an orange oil and a fat-soluble vitamin is insoluble in water, slightly soluble in ethanol, and can be easily solved in ether, hexane,

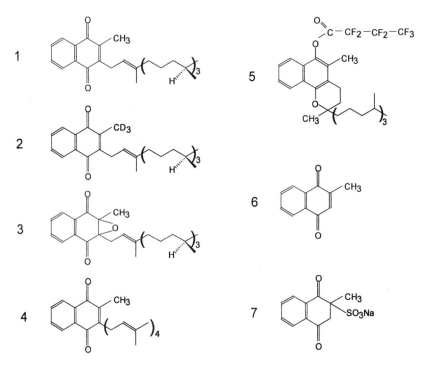

Figure 1   Structural formulas of some K vitamers: (1) vitamin $K_{1(20)}$, (VK1), phylloquinone or 2-methyl-3-eicosa-2′-ene-1,4-naphthoquinone; (2) trideuterium labeled VK1, VK1-d_3, 2-methyl-d_3-3-eicosa-2′-ene-1,4-naphthoquinone as internal standard for GC-MS analysis; (3) VK1-epoxide; (4) menaquinone-4, MK4, 2-methyl-3-tetraprenyl-1,4-naphthoquinone, as an example of the MKn series; (5) VK1-heptafluorobutyryl ester after reductive acylation of VK1; (6) menadione, vitamin K_3 (VK3), or 2-methyl-1,4-naphthoquinone; (7) menadione sodium bisulfite (MSB), water-soluble form of VK3.

or chloroform. The molecule is sensitive to daylight (forming isomers [2]) and will be destroyed in a strong alkaline medium, but is stable in a slightly acidic medium and under oxidizing conditions. Under reductive conditions (e.g., hydrogen gas and platinum as catalyst, etc.) the naphthoquinone is converted to the corresponding hydroquinone, a highly fluorescent molecule. Natural VK1, formed mainly by green plants, is the main form of K vitamins in human food and acts in coagulation of blood (3). Epoxidation of the double bond between position 2 and 3 of the naphtoquinone nucleus leads to $VK_{1(20)}$-epoxide (Structure 3 in Fig. 1), a compound also involved in blood coagulation (4).

2. Menaquinone-n (MK_n), or vitamin $K_{2(n)}$, stands for 2-methyl-3-multiprenyl-1,4-naphthoquinone, where n indicates the amount of isoprene units in the side chain, a number naturally between 4 and 13. Sometimes one (or more) of the double bonds in the side chain is saturated. Physical and chemical properties of menaquinones are similar to those of VK1, with the exception of the lipohilic character, due to different design of the side chain. MK_ns are formed by bacteria (e.g., *Escherichia coli, Staphylococcus aureus, Eubacterium lentum* etc.) of the intestine, the most common configuration is all-*trans*. Other stores of MK_ns were found to be animal liver (mainly MK7–13) and bone (5).

3. Menadione, or vitamin K_3 (VK3), is the synonym for 2-methyl-1,4-naphthoquinone, and does not occur in nature. It has to be synthesized, and acts as an intermediate in synthesis of K vitamins. VK3, as a provitamin, can be transformed to MK4 by micro-organisms. Menadione has been shown to exhibit antitumor activity in rodent and human tumor cells (6–9). A water-soluble form of the provitamin VK3 (menadione sodium bisulfite, MSB; see Structures 6 and 7 in Fig. 1) is often added to animal feed, and after treatment with aqueous Na_2CO_3 (destroying the bisulfite salt) and/or irradiation with UV-radiation (10), it can be chromatographed and analyzed in usual ways.

B. Biochemistry, Physiological Behavior, and Diagnostic Importance

Vitamin K was detected about six decades ago by Henrik Dam when he investigated the role of cholesterol in chicken feed. He reported hemorrhagic symptoms in chickens fed an ether-extracted diet. The antihemorrhagic and fat-soluble compound was called vitamin K, or phylloquinone. Since those days the role of VK1 in coagulation of blood has been investigated, leading to a deeper understanding of the nutritional and clinical role of vitamin K. It was shown that $VK_{1(20)}$ in mammals and birds is an essential cofactor in the posttranslational carboxylation reaction of glutamic acid residues to γ-carboxyglutamic acid residues (GLA) in a number of blood clotting factors, namely factors II (prothrombin), VII, IX, and X (11). The introduction of γ-carboxygroups provides the procoagulant factors with characteristic calcium-binding properties that are essential for activation and function of these clotting factors. Later discoveries of three other VK1-dependent plasma proteins (proteins C, S, and Z) have been made by searching for further homologous proteins in plasma with similar domain structures and γ-carboxyglutamic acid residues. The four procoagulant factors and the coagulation inhibitors C and S undergo, after synthesis by hepatic ribosomes, further posttranslational modifications, including γ-glutamyl carboxylation and glycosilation, which enables them taking part in coagulation cascade. More details of this complex reactions are shown in a review by Roberts et al. (3). The cellular function of VK1 in γ-glutamyl carboxylation has been shown to be closely linked to a multienzyme metabolic cycle, where VK1 is converted to VK-hydroquinone followed by an oxidation to VK-epoxide and final reconversion to VK1. This "vitamin K–epoxide cycle" and the associated enzyme reactions are too complex to be given in detail here, but they are reported in more detail in a number of review articles— e.g., Shearer (4,12), Suttie (13), Gardill (14), whereas Kuliopulos et al. (15), using supercritical fluid chromatography and mass spectrometry, show the role of molecular oxygen in these reactions. Vervoort et al. (16) investigated the antioxidant activity of the VK-cycle in microsomal lipid peroxidation. Several authors found VK1-concentrations in normal human plasma from 20 pg/mL up to 3 ng/mL, although concentrations between 200 ng/mL and 1000 ng/mL are most common when selective HPLC detection procedures (4,12,17) are used.

Another group of VK1-dependent GLA proteins that do not participate in coagulation, but which seem to be implicated in calcium homeostasis, have been detected the same way (i.e., searching for proteins with similar domain structures and γ-carboxyglutamic acid residues) in a wide variety of animal tissues such as bone, kidney, lungs, etc. Until today only one of these proteins, called osteocalcin, is well characterized. It is one of the most abundant noncollagenous proteins in the extracellular matrix of bone. Osteocalcin binds strongly to hydroxyapatite crystals and is a potent inhibitor of hydroxyapatite formation. The role of vitamin K in bone metabolism is described in all known details by Vermeer et al. (18), and a connection of VK to arteriosclerosis is drawn by Jie et al. (19). Although the precise function of osteocalcin, as well as those of many other GLA-proteins, is under investigation (e.g., Hara et al. [20]), osteocalcin-measurement has become a diagnostic marker of osteoblast activity in bone.

The main nutritional sources of VK1 are green leafy vegetables, certain legumes, and some vegetable oils. Since the development of routine HPLC detection methods with sufficient specificity and sensitivity, some research groups are involved in determination of different K vitamins (VK1, MK_n, and VK3) in food matrices, including infant formulas and milk. Tables of VK concentrations of diverse nutritional sources are shown, for example, by Booth et al. (21,22) or by other authors (23–25), and numerous reviews (26–32) have been published dealing with dietary and nutritional aspects of K vitamins.

In contrast, menaquinones seem to have a lower importance in human nutrition, as they occur in considerable concentration only in animal liver or in some cheeses. However, whether human beings can utilize the pool of menaquinones synthesized by intestinal microflora remains an open question. It is difficult to explain how these highly lipohilic MK_ns (mainly MK9-11 [33]), tightly bound to the bacterial cytoplasmic membrane, could be absorbed, but it has been shown that the MK_n content in human liver (90% of human liver vitamin K content) has similar relative proportions and interindividual variations to those in the intestine (12,34,35). Recently two assays have been published dealing with K vitamins in human liver (36) and MK_ns in bacterial cultures, stools, and intestinal contents (37).

Since the observations that a methyl group in position 3 of the naphtoquinone and an unsubstituted benzene ring (17) are essential for its activity, and that the first isoprene unit in the side chain plays an important role in the geometrical anchoring of the ubiquinones to the proteins (38), the interest of analysis of all of the VK compounds is increasing. However, the liver is supposed to be the largest store of K vitamins in human. But the content of K vitamins in plasma, intestinal microflora, and bone (MK6-8 values as high as in liver[!] [39]) are not to be forgotten, although many questions are still unanswered in these fields (40,41).

A deficiency in VK leads to the production of des-γ-carboxylated clotting factors or PIVKAs (proteins induced by vitamin K absence), such as des-γ-carboxyprothrombin, lacking the ability to bind calcium. Calcium binding is essential for the biologic activity of these clotting factors. Therefore, VK deficiency leads to a prolongation of clotting tests, such as the prothrombin time. PIVKAs can be determined more specifically by clotting and immunologic methods and reflect the grade of vitamin K-deficiency in the clotting system.

Decrease of activity of vitamin K dependent clotting factors can induce severe bleeding in humans. While VK deficiency is rare in adults (with the exception of intestinal malabsorption syndromes and the therapeutic use of agents interfering with the VK epoxide cycle such as warfarin (42) to achieve anticoagulation in patients suffering from thrombembolic diseases), VK deficiency poses a big problem in neonates and young infants. Because VK transfer via the placenta is small and its content in breast milk low, neonates would regularly develop VK deficiency that might lead to severe, and in many cases, intracranial bleeding (hemorrhagic disease of the newborn [43–46]). In addition, some infants develop bleeding due to VK deficiency several weeks after birth (late-onset hemorrhagic disease). This manifestation of VK deficiency is poorly understood, since it cannot be due to poor intake alone.

Because of the possible devastating consequences of VK deficiency in the newborn period, all neonates receive oral or parenteral VK supplementation for prophylaxis against VK deficiency bleeding (47–49). Publications by Golding et al. (50,51) connecting this general VK prophylaxis to an increase in cancer in children has further boosted interest in the field of VK studies. This increase finally was enabled through the development of modern liquid chromatographic and gas chromatographic methods with high specificity and sensitivity to detect low concentrations of VK within a complex matrix.

C. Analytical Problems Resulting from the Physical and Chemical Properties

As mentioned above, major problems in vitamin K analysis are caused by its sensitivity to daylight on one side and the destroying capacity of strong alkalines on the other side. Due to this sensitivity all sample preparation steps should be carried out only under subdued light, or samples should at least be protected from strong light by wrapping them tightly or using amber glassware. To remove traces of detergents from washed glasses one should rinse them extensively, or heat glasses above 500°C in the oven.

Another problem is caused by low concentration of K vitamins within complex matrices. In case of plasma sample preparation, for example, often only a limited volume of plasma is available, especially when plasma of newborn babies is investigated. On the other hand, if one uses larger sample volumes, coextracted compounds may interfere with detection of the vitamins, or at least lead to extensive contamination of the chromatographic materials, which results in time-consuming cleaning processes and/or a shorter lifetime of chromatographic columns.

In the case of milk samples the problem occurs that the K vitamins, together with triglycerides, are encapsulated by membranes. By breaking up these membranes, one should avoid strong alkaline media. In the chromatographic determination, following lipase hydrolysis, one has to prepurify samples of an excess of coextracted lipid material.

II. THIN-LAYER CHROMATOGRAPHY

During the past decades thin-layer chromatography (TLC), as a chromatographic method for determination of vitamin K, was replaced mainly by high-performance liquid chromatography (HPLC) and, to a lesser degree, by gas chromatography (GC). The reason for this decline is found in the drawbacks of classical TLC, such as low speed, poor sensitivity, and incompatibility with on-line detection and quantitation methods. This, and the fact that vitamin K normally occurs in low concentrations within a complex matrix, is a reason why TLC is drawn back as the method of choice for chromatographic assays.

The relevant literature up to 1993–1994 has been reviewed in the first two editions of this volume (2,17) and by Fried et al. (52). In the last few years, some groups used TLC for special applications (53,54), or as a cleanup step for HPLC

(55–57) or for GC (58) methods. Das et al. (54), who found MK7 in lipid extracts from clostridium bacteria, used two TLC steps for cleanup and detection. The first was done with preparative silica TLC and with benzene/hexane-1/1 as eluent, followed by a second silica TLC, developed with petroleum ether/diethyl ether-85/15. They used MK4 and ubiquinone-50 as standards to localize MK7. Sakamoto et al. (57) and Hirauchi et al. (55) used the same second TLC system as a cleanup step for K vitamins in animal tissue, the latter group with the difference of working with fluorescence-coated silica; other groups (53,56,58) decided to use a reversed-phase thin-layer assay (RPTLC). Thin-layer plates with chemically bonded phases (mainly C_{18}) which are similar to those used in HPLC, have been developed with methylene chloride/methanol-70/30 as eluent. As an example, Madden et al. (53) found VK1 in bovine liver samples by using this method for chromatography but with time-consuming cleanup procedures, such as Florisil columns and Sep-Pak cartridges. As a matter of fact, extensive cleanup steps are a common necessity when working with TLC as a chromatographic method. Furthermore, in nearly every TLC assay final conformation of vitamin K structure with different, mainly mass selective (53,54,56,58), methods had to be done.

To localize K vitamins on TLC plates researchers normally use two different quantitation systems—destructive (with regard to the molecule structure), and nondestructive. Destructive detection of quinones is based on the use of spray reagents to visualize the vitamins. A number of different reagents have been frequently used in earlier work and are reviewed by Lefevere et al. (2), but they are limited in terms of sensitivity, precision, on-line methods, etc.

Nondestructive detection of K vitamins can be performed either by elution of separated bonds (localized by a standard spot) with chloroform and following determination of K vitamin concentration by UV detection (53) or mass selective detection (54,58). Densitometric scanning of TLC plates, based on either absorption, fluorescence, or fluorescence quenching of coated plates (55), is another method for nondestructive detection and can be used with external as well as internal standardization. The introduction of this quantitation method resulted in an improvement in terms of speed, sensitivity, and precision of TLC but could not prevent the advance of HPLC applications in past decades.

III. HIGH-PERFORMANCE LIQUID CHROMATOGRAPHY

Comparing the frequency of all chromatographic assays done in analytical laboratories, modern liquid chromatography has claimed a leading position in common analysis, and in vitamin K analysis in particular. Since the development of high-quality chromatographic columns, inclusive narrow-bore columns for small sample volumes, together with automatization of injection and quantitation, high-performance liquid chromatography (HPLC) has gained major advances over all other vitamin K assays. A benefit of this method is the elimination of the risk of thermodegradation of the underivatized compound, or protection from light during the chromatographic process. The chemical properties of vitamin K, such as its lipophilic and neutral character, tolerate a great variety of different HPLC systems with regard to stationary phases as well as detection systems. However, the main problems of usual vitamin K analysis, low concentrations of the vitamin in a complex biological matrix, and presence of a great number of lipid compounds, are still present. How these problems have been managed with HPLC assays and, less frequently, with other methods of liquid chromatography (e.g., supercritical fluid chromatography [SFC]) will be evaluated in the next section.

A. Internal Standardization, Isolation, and Cleanup

Any chromatographic assay for vitamin K detection and quantitation requires an individual procedure for compound isolation and cleanup, depending on the matrix of interest and on the analytical system of choice. In this context it is of great importance to suggest the use of an appropriate internal standard (IS), performing a complicated and multiple-step analysis. An effective IS, owing to general requirements of, for example, structural analogy, absence in the matrix, and equal stability and extraction recoveries, compensates for losses during workup procedures and works as a control for preparation of unknown samples. An ideal IS, added to the sample in the first step of preparation (i.e., even before freezing a plasma sample, if necessary), should coelute with the compound of interest during isolation and chromatographic purification but must be separated from it in the final analytical detection system. If any further reaction is hyphenated to chromatography (i.e., postcolumn reduction of the vitamin), the IS of choice should have the same properties as the compound(s) of interest. Under these circumstances, in our opinion, it is not acceptable to ignore the use of an IS in multistep chromatographic analysis, in particular under the favorable fact of saving numbers of recovery controls for reliability of analytical data.

Final design and choice of IS depend on biological matrix and on the specific analytical system used in the assay. A selection of ISs, used in recent HPLC analysis, is given in Table 1. A limited number of mainly three different compounds is used for IS, and are distinguishable from each other in the side chain in position 3 of the naphthoquinone. These compounds, with structural analogies to vitamin K, namely vitamin $K_{1(25)}$ (the same as VK1, but one reduced isoprene unit more), $K_{1(H2)}$ (VK1 with hydration of the double bond in position 2 of the phytyl chain), and MK_n (menaquinones with three to seven isoprene units in the side chain), show the same chemical properties but slightly different chromatographic properties as compared to the compounds of interest. When mass selective detection systems are used, a special opportunity of internal standardization is feasible, called stable isotope dilution, which will be discussed later. It should be noted that many researchers dealing with sample matrices like tissue, foods, and so on, often do not use ISs, because chromatograms are not pure enough to use an appropriate IS.

In Table 1, sample preparation and cleanup procedures for vitamin K analysis using HPLC detection are shown, collected in groups of various sample matrices. A remarkable uniformity exists in the preparation of plasma samples. Almost every research group used identical solvents (ethanol in a volume ratio of 1:4) for denaturation of VK transport proteins (lipoproteins of the VLDL fraction) as well as for extraction of the vitamins from plasma (hexane in a volume ratio up to 20, depending on sample volume). MacCrehan et al. (77) and Sakon et al. (90) used isopropanol for denaturation, which is said to provide better extraction recoveries, but coextracted polar compounds may interfere with the vitamins in the final chromatography. The uniformity of this isolation process may be a result of former experiments, using strong acids, alkalines, or different extraction solvents and methods, which are summarized and discussed by Lambert et al. (17) in the second edition of this volume.

After extraction into hexane, a solid-phase extraction, using silica and nonpolar eluents, or sometimes reversed-phase extraction with polar solvents, is common in many assays. Some research groups used a semipreparative HPLC method (silica column and acetonitrile [62,67,68] or di-isopropylether [89] in hexane) with UV detection for cleanup. Blanco-Gomis et al. (60,61), introducing narrowbore HPLC or capillary liquid chromatography for small-sample analysis of vitamins, were able to get appropriate results without solid-phase cleanup by reextracting the hexane layer with methanol:water 9:1 to remove interfering lipids, as described elsewhere (24,88,108).

A number of groups (21,22,78–87,113) used a specific procedure for cleaning samples from interfering lipids, which was called liquid-phase reductive extraction and was introduced by Haroon et al. (78). After Sep-Pak extraction, a solution, containing zinc chloride, acetic acid, and acetonitrile was added to the eluted solution. After reduction to the corresponding hydroquinones by addition of zinc metal, the more polar hydroquinones were extracted into acetonitrile. The hexane layer (containing interfering lipids) was discarded and the acetonitrile layer was reextracted with water and hexane. Through this extraction process the hydroquinones were reoxidized to the hexane-soluble quinones, the hexane layer was evaporated, the compounds were redissolved in hexane, and the vitamin sample was subjected to HPLC analysis.

A considerable degree of conformity can be seen also in preparation of milk samples. Due to the large amount of triglycerides present in milk and some infant formulas, enzymatic hydrolysis with pancreas lipase is common to many assays. The triglycerides, together with the K vitamins, are encapsulated by membranes consisting of proteins and phospholipids. These fat globule membranes, which inhibit the hydrolytic action of lipase, should be broken before enzymatic

Table 1 Sample Preparation for Vitamin K Analysis

Sample volume (mL)	Internal standard (IS)	Denaturation (volume ratio)	Liquid/liquid extraction (volume ratio)	Cleanup Solid phase	Eluent	Reference
1. Serum, plasma						
0.05–0.5	$K_{1(25)}$	Ethanol	Hexane	Sep-Pak C_8	Acetonitrile/isopropanol/dichloromethane 70/10/20	59
1.0	—	Ethanol (2)	Hexane (3)		Extraction with methanol/water 9/1	60
1.0	—	Ethanol (2)	Hexane (3)		Extraction with methanol/water 9/1	61
2.0–5.0	MK6	Ethanol (1)	Hexane (1)	1. Sep-Pak silica 2. Semiprep. HPLC: silica UV detection	Hexane/diethylether 97/3 0.34% Acetonitrile in hexane	62
1.0–3.0	MK7	Ethanol	Hexane	Silica		63
0.5–1.0	$K_{1(H2)}$	Ethanol (1)	Petroleum/diethylether (1/1) (4)	Silica	Heptane/ethyl acetate 99/1	64
1.0	—	Ethanol (4)	Hexane (6)	Sep-Pak silica	Hexane/diethylether 96/4	65,66
0.2–0.5	$K_{1(25)}$	Ethanol (2)	n-Hexane (6)	Semiprep. HPLC: silica UV detection (248 nm)	0.15% Acetonitrile in hexane	67,68
1.0	$K_{1(H2)}$	Ethanol (3)	Hexane (7)	Sep-Pak silica	Hexane/diethylether 95/5	69
2.0	$K_{1(25)}$	Ethanol (2)	Hexane (6)	1. Silica 2. C_{18}	Hexane/diethylether 93/3 Methanol/methylene chloride 80/20	70
0.5	MK6	Ethanol (2)	Hexane (6)	—	—	71
1.0	MK3	Ethanol	Hexane/diethylether (1/1) (6)	1. Silica 2. Alumina	Hexane/benzene 1/2, 15 mL Hexane/benzene 3/1, 5 mL	72–74

2.0	—	Ethanol (1.5)	Hexane (2.5)	Sep-Pak silica	Hexane/diethylether 96/4	75
0.050	—	Ethanol (8)	Hexane (20)	—	—	76
1.0	MK4	2-Propanol	Hexane (2)	Silica	Hexane/diethylether 97/3	77
0.5–1.0	$K_{1(H2)}$	Ethanol (2)	Hexane (6)	1. Sep-Pak silica 2. Liquid phase reductive extraction	Hexane/diethylether 97/3	78–87
0.5–1.0	$K_{1(H2)}$	Ethanol (2)	Hexane (5)		Methanol/water 9/1 extraction	88
2.0	$K_{1(25)}$	Ethanol (2)	Hexane (3.75)	Semiprep. HPLC: silica UV detection (254 nm)	n-Hexane/di-isopropyl-ether 98.5/1.5	89
1.0	—	2-Propanol (3)	Hexane (10)	Sep-Pak silica	Hexane/diethylether 93/7	90
0.2	MK6	2-Propanol (2)	Hexane (18)	—	—	91
2. Milk, infant formulas						
3.0 g milk powder or 15.0 g ready formula	Cholesterylphenyl-acetate	Lipase, ethanol/methanol 95/5 (2/3)	Hexane (1)	Semiprep. HPLC: silica UV detection (269 nm)	Hexane/isopropanol 99.9/0.1	92
10.0 g milk, (1 g powder)	$K_{1(H2)}$	Lipase, ethanol (1)	Hexane (3)	—	—	93
20.0 mL milk	—	Lipase, ethanol (5), NaOH	n-Pentane (3*100 mL)	Semiprep. HPLC: C_{18} UV detection (248 nm)	Methanol/acetonitrile 1/1	94
5.0–10.0 mL human milk	MK7	Isopropanol/hexane 3/2 (3)		1. Silicar CC-4 2. Bio-sil-silica	Iso-octane/methylene chloride/isopropanol 75/25/0.02 Hexane/chloroform 75/25	95,96
25 mL milk (1 g powder)	—	Lipase, ethanolic NaOH	Hexane (2*50 mL)	—	—	97
1.0 g powdered formula	—	—	SFC		CO_2	98
0.5 mL human milk	$K_{1(25)}$	Lipase, ethanol (8)	n-Hexane	Semiprep. HPLC: silica UV detection (254 nm)	n-Hexane/ di-isopropyl-ether 98.5/1.5	99

Table 1 Continued

Sample volume (mL)	Internal standard (IS)	Denaturation (volume ratio)	Liquid/liquid extraction (volume ratio)	Cleanup		Reference
				Solid phase	Eluent	
3. Tissue, foods, oils						
1 g animal tissue	—	66% 2-Propanol (5)	n-Hexane (6)	1. Sep-Pak silica 2. TLC: Silica Gel 60 F_{254}	n-Hexane/diethylether 96/4 (5) Light petroleum/diethyl-ether 85/15	55
Bacterial suspension	—	2-Propanol (1)	Hexane (2)	—	—	100
Fibroblast incubation	—	Methanol	Chloroform, 2* extr.	—	—	101
0.1 g rat liver	—	Isopropanol (4)	Hexane (10)	TLC: silica	Petroleum ether/diethyl-ether 85/15	57
1 g liver	—	66% isopropanol (5)	Hexane (6)	1. Sep-Pak silica 2. TLC: silica gel 60 F_{254}	n-Hexane/diethylether 96/4 (5) Petroleum ether/diethyl-ether 85/15	102,103
0.5 mL liver homogenate	$K_{1(H2)}$	Ethanol (4)	Hexane (7)	Silica gel 60	Hexane/ethyl acetate 98/2	104,105
4 g human placenta	—	66% isopropanol (2.5)	Hexane (6)	1. Sep-Pak silica 2. TLC: silica gel 60 F_{254}	n-Hexane/diethylether 97/3 Petroleum ether/diethyl-ether 85/15	106
0.5 bone powder	—	Demineralization (EDTA), lipase	Chloroform/methanol 3/1, 6*40 mL	Sep-Pak silica	n-Hexane/diethylether 97/3 (10 mL)	39

Sample	Analyte	Saponification / solvent	Extraction	Cleanup	Mobile phase	Ref.
1 g feces	—	Water/chloroform/methanol 10/12.5/25	Hexane (3*4 mL)	Sep-Pak silica	n-Hexane/diethylether 97/3	33
1–5 g vegetable	—	Methanol (10 mL), sodium carbonate solution 80°C		Filtration		107
2 g vegetable	K_{1(H2)}		Hexane	Reextraction Filtration	Methanol/water 9/1	24,108
5.7 g plants	—	Soxhlet extraction	Dichloromethane			109
0.25–0.5 g food	K_{1(25)}	Isopropanol, 9 mL	Hexane, 6 mL	1. Silica 2. C_{18}	Hexane/diethylether 97/3 Methanol/methylen chloride 80/20	110
0.25–0.5 g food	K_{1(H2)}	2-Propanol, 9 mL	Hexane, 6 mL	1. Silica 2. Liquid phase reductive extraction	Hexane/diethylether 97/3	21
0.5–1.0 g oil	MK4			Semiprep. HPLC: silica UV detection (248 nm)	n-Hexane/diethylether 99/1	111
100 mg soybean oil	K_{1(25)}	Ethanol (10), NaCl solution	Hexane (30)			112
50–100 µL lipid emulsion	K_{1(H2)}	Isopropanol, 9 mL	Hexane, 6 mL	1. Silica 2. Liquid phase reductive extraction		113
0.5 g vitamin premix	—	Dimethylsulfoxide, 20 mL	Hexane, 25 mL	—	—	114
0.5 g rat chow	—		SFC		CO_2	115,116
1 g animal feed	—	Water/methanol 6/4 (10)	n-Pentane, 3*10 ml	—	—	117

hydrolysis can start. For this purpose some groups sonicate (92,99) or vortex (93) samples prior to or during incubation with lipase. Of course all other reaction conditions for hydrolysis have to be optimized, together with the avoidance of strong light and alkali, although Careri et al. (107) demonstrated that the use of Na_2CO_3 had no influence on the recovery of vitamin K from a vegetable matrix.

After hydrolysis of triglycerides, denaturation (ethanol) and extraction (hexane) of milk samples are similar to those procedures used in preparation of plasma samples. Due to high concentrations of coextracted lipophilic compounds, a semipreparative HPLC cleanup step often is used in these preparations. Indyk et al. (92) and Lambert et al. (99) used adsorption HPLC and hexane/isopropanol mixtures for cleanup and a reversed-phase HPLC for detection, whereas Isshiki et al. (94) used a C_{18} reversed-phase system with a methanol/acetonitrile mixture for cleanup and a C_2 or C_3 reversed phase for detection. Canfield et al. (95,96) worked with two open-column chromatography systems (silica) to isolate vitamin K compounds prior to HPLC detection, whereas Schneiderman et al. (98) introduced a completely different method for isolation of these compounds. They used supercritical fluid extraction (SFE) with CO_2 as solvent to determine VK1 in powdered infant formulas. Details of this particular method will be discussed later in this chapter.

Due to the great diversity of matrices described in the tissue, foods, oils section of Table 1, isolation of K vitamins is less uniform than those described in other sections, although denaturation and extraction steps are often similar. For denaturation of tissue samples mainly isopropanol mixtures have been used, whereas hexane is again the solvent of choice for extraction procedures. Sometimes individual sample preparations have been used for specific matrices. For example, Hodges et al. (39), looking for VK content in bones, demineralized bone powders with EDTA and hydrolyzed chloroform/methanol extracts with lipase before processing common denaturation and extraction. Poulsen et al. (109) analyzed quinones from extracts of shepherd's purse, obtained by Soxhlet extraction of the herb with dichloromethane. Finally, Indyk (114) used a dimethylsulfoxide/hexane mixture for extraction of VK1 from a vitamin premix, whereas Schneidermann et al. (115) again used SFE (CO_2) for detection of K vitamins in rat chow. The great diversity of biological matrices and methods used also may be a reason why internal standardization is often neglected in sample preparation.

Following denaturation and extraction, extensive cleanup procedures, using silica chromatography in complex and different chromatographic systems (TLC, Sep-Pak, HPLC, etc.), have been done in the great majority of research in this type of matrix. For further details of these specific cleanup steps see Table 1.

B. Analytical Chromatography

Following isolation and various cleanup procedures, final detection and quantitation of vitamin K compounds can be performed in numerous ways with respect to stationary phases and, above all, with respect to detection modes. Due to chemical and physical properties of the quinone molecules, many opportunities are offered for different liquid chromatographic systems. Regarding stationary phases, clear preference is shown to reversed-phase chromatography because of a large variety of possible packing materials and solvent mixtures. On the other hand, due to the lipophilic character of the vitamin, adsorption chromatography is limited to hexane (with slight modifications) as solvent, which normally does not provide good separation from interfering lipids. Regarding detection modes, there are also a certain number of possibilities, but the majority of analysis is done with fluorescence detection, after different methods for reduction of the quinone compounds to highly fluorescing hydroquinones.

The running methods for liquid chromatography, as well as all current detection modes, are described in more detail in the following section and are summarized in Table 2, with the exception of mass selective methods, which are discussed later in this chapter.

1. HPLC and Adsorption Chromatography

Although sample cleanup on open silica columns or Sep-Pak silica cartridges (and so on) removes parts of the lipids that show different polarity in comparison to vitamin K, many equipolar lipids are collected together with the vitamin fraction. Using silica as stationary phase, only little variability is possible in the choice of solvent, due to the lipophilic character of the vitamin K molecules. Hexane with a content of 3% to 5% diethylether or minimal amounts of acetonitrile, ethyl acetate, or diisopropylether (up to 1%) can be used in these systems, with the consequence of a relative poor separation from interfering lipids. That is why in the last few years adsorption chromatography (inclusive semipreparative silica HPLC) is used only as a cleanup step in sample preparation for reversed-phase HPLC (see Table 1).

2. HPLC and Reversed-Phase Chromatography

In recent years final HPLC analysis of vitamin K has been done using reversed-phase columns. A great variety of packing materials is offered, allowing the use of adapted solvent mixtures for individual problems and detection techniques. Normally commercial columns, packed with bonded silica material of 5μm, are used. Column length varies between 10 and 30 cm, but most frequently 15- to 25-cm columns are used. The usual diameter of the columns is 4.6 mm, but also narrow-bore columns (2 to 3 mm diameter [59,60]) or even capillary columns (0.32 mm diameter [61]) are used in special applications. Indyk et al. (92) analyzed VK1 in milk and infant formulas by the use of a 8 × 10 cm radial compression module. Highly retentive octadecyl-silyl (ODS) phases are used in most cases as packing materials for columns; only a few applications used octyl bonded

Table 2 Final HPLC Step, Detection Modes, and Detection Limits for VK Analysis

Analytical column (cm*cm; μm)	HPLC eluent (flow rate: mL/min)	Derivatization procedure	Detection (nm)	Compound(s) of interest (IS)	Detection limit	Matrix	Reference
UV detection							
Spherisorb ODS-102 (10*0.21; 5)	Acetonitrile/isopropanol/ dichloromethane 70/10/20	—	248	VK1 (VK_{1(25)})	—	Plasma	59
Ultrasphere ODS (25*0.46; 5)	Methanol/water 70/30 (0.8)	—	265	VK3 (carbazole)	10 ng/mL	Plasma	9
Spherisorb ODS-2 (30*0.032; 5)	Methanol/tetrahydrofuran 80/20 (5μL/min)	—	250	VK1	2.6 pg	Plasma	61
Spherisorb ODS-2 (10*0.21; 3)	Methanol/water methanol/THF gradient	—	250	VK1	0.42 ng	Plasma	60
Resolve C_{18} Radial compression module (8*10; 5)	Methanol/isopropanol/ ethyl acetate/water 450/350/145/135 (2.0)	—	269,277	VK1 (cholesteryl phenyl-acetate)	1 ng	Milk	92
Chrompack RP18	Acetonitrile/isopropanol/ water 100/8/1–5	—	254	VK1, MK4, VK3 (tocopherolacetate)	—	Liver microsomes	126
Zorbax ODS (25*0.46)	Methanol/di-isopropylether 7/2 (1.0)	—	248–270	MK6–10, demethyl-menaquinones	—	Enterobacteria	127
LiChrosorb RP-8 (25*0.46; 10)	Methanol (0.6)	—	247	VK1	0.1 ng	Vegetable	107
Electrochemical detection							
Hypersil C8 (25*0.45)	Methanol/Na acetate buffer (0.05 M) 97/3	—	−1.3V/ +0.15V	VK1 (MK6)	0.5 ng	Plasma	62
Partisil ODS-2 (3) (25*0.46; 5)	Methanol/ethanol/60% perchloric acid, 600/400/1.2; 0.05 M NaClO₄ (1.0)	—	0.45V/ +0.35V	VK1, MK6-9 (MK4)	0.2–1 ng	Milk	94

Column	Mobile phase	Electrode/Detection	Potential/λ	Compounds	Detection limit	Sample	Ref.
Radial Pak C18	Ethanol/hexane/water 90/6.5/3.5, 25 mM TBAP	—	−0.6V/+0.2V	VK1 (MK7, ^{3}H-VK1)	0.63 ng	Milk	95,96
OD-224 RP 18 (22*0.46; 5)	Methanol/water 99/1, 2.5 mM acetic acid/Na acetate (1.25)	—	−1.1V/+0.7V	VK1	3.1 ng	Milk	97
Hypersil C8 (25*0.46; 5)	Methanol/Na acetate buffer (0.05 M) 97/3 0.1 mM EDTA	—	−1.3V/+0.05V	VK1, MK6–8	—	Bone	39
Vydac 201 TP54 (25*0.46; 5)	Methanol/Na acetate buffer (0.05 M) 95/5, (1.0)	—	−1.1V/+0.0V	VK1 (MK4)	50 pg	Oil	111
Spherisorb ODS (20*0.4; 5)	Methanol/0.2 M acetic acid/Na acetate 40/60 (1.0)	—	−0.3V/+0.2V	VK3	15 ng	Water	128
CLC-ODS (15*0.6; 5)	Methanol, 85%, 0.05 M NaClO$_4$ (1.0)	5% Pt on alumina	+0.7V	VK3 as IS for idebenone	—	Serum, brain	129

Fluorescence detection: electrochemical reduction

Column	Mobile phase	Electrode/Detection	Potential/λ	Compounds	Detection limit	Sample	Ref.
Nucleosil C$_{18}$ (15*0.46; 5)	Ethanol/water 92.5/7.5, 0.25% NaClO$_4$ (1.0)	Coulometric reduction +0.25V/−0.55V	320/430	VK1, VK1-epo, MK$_n$	5–8 pg	Plasma, tissue	55,65,66
MOS Hypersil C8 (15*0.39; 5)	Methanol/water 95/5, 0.03 M NaClO$_4$	Coulometric reduction, +1.0V (?)	240/418	VK1	—	Plasma	130
MOS Hypersil C8 (15*0.39; 5)	Methanol/water 93/7, 0.03 M NaClO$_4$, (1.0)	Coulometric reduction, −1.0V	240/418	VK1 (K$_{1(H2)}$)	40 pg/mL	Serum	64
Novapak-C18 (15*0.4; 4)	Acetonitrile/ethanol 95/5, 5mM NaClO$_4$	Coulometric reduction, −0.8V	320/430	VK1, VK1-epo (K$_{1(25)}$)	15 ng/L	Serum	67,112

Table 2 Continued

Analytical column (cm*cm; μm)	HPLC eluent (flow rate: mL/min)	Derivatization procedure	Detection (nm)	Compound(s) of interest (IS)	Detection limit	Matrix	Reference
Nucleosil C18 (15*0.46; 5)	92.5–97.5% ethanol, 0.25% NaClO$_4$	Coulometric reduction, –1.0V	320/430	VK1, MK$_n$	—	Plasma, liver	102
Bondabak-C18 (30*0.4)	Methanol/methylene chloride 95/5, 0.3M NaClO$_4$ (1.0)	Coulometric reduction, –0.65V	320/418	VK1 (K$_{1(H2)}$)	—	Serum	69
Fluorescence detection: chemical reduction							
RoSIL C18 (15*0.32; 5)	Methanol/ethyl acetate 96/4, (0.85)	(CH$_3$)$_4$NB$_3$H$_8$	325/430	VK1 (K$_{1(25)}$)	150 PG	Serum	68,89,99,104, 131–135
Nucleosil 5C-18 (15*0.46)	Ethanol 92.5%	0.025% NaBH$_4$, in ethanol	320/430	VK1, MK$_n$	50 pg	Plasma, placenta	106
Nucleosil C18 (20*0.46)	Ethanol/water 87.5/12.5	0.1% NaBH$_4$, in ethanol	320/430	VK1, MK4, MK7	—	Plasma	90
Novapak C18 (15*0.39)	Ethanol/water 95/5 (0.7)	0.1% NaBH$_4$, in ethanol	—	VK1, MK$_n$	0.02–0.05 μ/g	Feces	33–35
Hypersil ODS (25*0.46; 5)	Methanol/dichloromethane 80/20, 10 mM ZnCl, 5mM NaAc, 5 mM acetic acid (1.0)	Zn particles	248–320/ 418–430	VK1, VK1-epo, MK$_n$ (K$_{1(H2)}$, K$_{1(15)}$)	25 pg	Plasma, food, etc.	21,22,78–87, 113,136,137

Column	Mobile phase	Reduction	Detection	Compounds	Detection limit	Sample	Refs
Hypersil-ODS (25*0.46; 5)	Methanol/dichloromethane 90/10, 10 mM ZnCl, 5mM NaAc, 5 mM acetic acid (1.0)	Zn powder	243–320/430	VK1, MK$_n$, VK3 (K$_{1(H2)}$)	0.04 ng/mL	Plasma, food	24,88,93,117
Nucleosil 5C18 (25*0.46)	Methanol/ethanol/water 1/2/0.06 (1.2)	PtO$_2$/H$_2$	254/430	VK1, MK$_n$	25–150 pg	Plasma	72,73
Nucleosil 5C18 (15*0.46)	Ethanol/methanol 1/4, 0.25% NaClO$_4$ (1.0)	RC-10 Pt reduction column	320/430	VK1, MK$_n$	5–10 pg	Plasma	57,75,76
Wakosil-II 5C18 (25*0.46; 5)	Ethanol/methanol/water 50/47/3	Pt-reducing column	254/430	MK4, (MK6)	1 ng/mL	Plasma	91
201TP54-C18 (25*0.46; 5)	Ethanol/methanol 4/6 (1.0)	10% Pt on Alumina	242/430	VK1, (MK4)	7 pg	Serum	77
Nucleosil 5C18 (25*0.46)	Gradient from 100% methanol to 80% 2-propanol/ethanol 4/1	Platinum-black catalyst	320/430	VK1, MK$_n$	5–40 pg	Liver	103
Fluorescence detection: photoinduced reduction							
Zorbax ODS (25*0.46; 5)	Methanol/isopropanol 60/40 (0.78)	Photochem. reactor or 150W Xe lamp	244/>370	Quinones	0.055 pmol	Plant extracts, plasma	109,114

phases (e.g., RP-8 [107] or Hypersil C8 [64,130]) and methanol/water mixtures as eluent, resulting in shorter elution times.

The choice of eluent, to a high degree, depends on the method of detection. When UV detection is used, nonaqueous eluents should be used, whereas water should be added to dissolve electrolytes (e.g., $NaClO_4$ or acetate buffers) that are necessary for chemical and electrochemical reduction or detection. A high percentage of ethanol, on the other hand, is necessary for the use of fluorescence detection after chemical reduction because of stability problems of the borohydride complex in methanol. To avoid these problems, Lambert et al. (17,89) introduced tetramethylammonium octahydridotriborate $[(CH_3)_4NB_3H_8]$ as reducing reagent, which was found to be more stable and effective in methanol mixtures. Nearly all research groups used normal isocratic flow rates of about 1 mL/min at ambient temperatures, except those working with special narrow-bore columns. Two groups (60,103) eluted the vitamin compounds with different solvent gradients.

Currently used columns, eluents, and derivatization procedures and conventional detection methods are summarized in Table 2. Mass selective detection methods for liquid chromatography are shown in an extra paragraph, due to the increasing importance and specificity of these methods.

3. Other Separation Methods (SFC and SFE)

Only a few alternatives to HPLC are given to liquid chromatography. One way of analysis is shown by Gil Torro et al. (118), where they described a spectrofluorimetric determination of VK3 by a solid-phase zinc reactor, immobilized in a flow injection assembly. They reached a limit of detection of 5 ng/mL by a remarkable throughput of 70 (!) samples of pharmaceutical formulations per hour. Another trial (119) is shown in the determination of VK1 by adsorptive stripping square-wave voltmetry at a hanging mercury drop electrode, but most prospective alternatives, in our opinion, seem to be supercritical fluid extraction (SFE) and supercritical fluid chromatography (SFC). The basic principles of these systems are as follows.

The matrix is brought into an extraction chamber which, in this case, is filled with CO_2 gas and pressurized to about 8000 psi with simultaneous heating at 65°C for a few minutes (98,115,116). At these conditions CO_2 is working as an apolar solvent which is able to extract highly lipophilic compounds. After this SFE process the exit valve is opened to bleed out the extract-laden CO_2, and the extract is collected on silica gel by simultaneous evaporation of the CO_2 gas. Without any further purification the vitamin compounds can be quantitated, as usual, on HPLC. If samples are introduced into a heated and density-programmed analytical silica column (coated with, e.g., cyanopropyl-methyl-polysiloxane) of 10 m length (15), instead of an extraction chamber, the compounds can be extracted and chromatographed simultaneously (SFC). The advantages of this method are that one can perform extraction and chromatography in one step in the dark, in inert gas atmosphere and without the problem of solvent evaporation and sample concentration during cleanup procedures. Especially in combination with mass selective detection, these methods seem to be a useful alternative to HPLC, although up to this point instrumentation does not seem to be ideal.

C. Detection

As described above, the possibilities of variations in the procedures for extraction and chromatography of K vitamins are limited. Due to the shown properties of the quinone compounds, however, many more alternatives are given in detection procedures of final liquid chromatography. Chromatographers have the choice among UV detection, fluorescence detection (after treatment with different reduction systems to get the corresponding hydroquinones), electrochemical detection, and last but not least, mass selective detection. Also, the moment of addition of derivatization reagents, if necessary, can be varied (direct into eluent, or postcolumn, by the use of a second pumping system, etc.). Which of these systems is to be chosen will be directed mainly by the individual tasks and the available technical equipment.

1. UV Detection

The molecules of the K vitamins show relatively poor UV properties ($\epsilon = 19,900$ L/mol $\times$ cm at 248 nm [17]). As a result, the absorbance of the quinone nucleus is rather low and, furthermore, the maximum of absorbance (248 nm) is situated at a nonselective wavelength, where numerous interferences are difficult to eliminate, especially when nonaqueous eluents are used. Due to these properties, detection limits, as shown in Table 2 (detection limits are given per unit of matrix [e.g., ng/mL plasma] or amount injected into the system), are relatively high with UV detection. Therefore, this instrument setup, although relatively simple in construction and handling, is chosen only when special sensitive chromatography columns (narrow-bore [59,60] or capillary liquid chromatography [61] columns) are used, or when K vitamins are determined in complex investigations (i.e., for simultaneous analysis of eight fat-soluble vitamins [120], or comparison of the vitamins K_1, K_2 and K_3 as cofactors for VK-dependent carboxylase [126]) and in special applications (determination of anticancer drug VK3 in plasma [9]), where higher concentrations of the compounds are expected.

2. Electrochemical Detection

In amperometric detection, initially the naphthoquinone nucleus of the K vitamins has been reduced to the corresponding hydroquinone by applying a sufficiently negative potential to an electrode, resulting in a measurable current, proportional to the amount of reduced vitamin K. However, even traces of oxygen cause interferences of high background currents and baseline drift. This and a possible passivity of the electrode led to the introduction of a dual electrochemical detector system in redox mode, as described in Canfield at al. (95). By this system the vitamin molecules are reduced to the hydroquinones on the first (upstream) electrode and are then detected on the second (downstream) electrode by reoxidation to the vitamins. This system of serially arranged electrodes eliminates interferences of oxygen because the reduction of oxygen is irreversible under these conditions. For reduction of VK a minimum potential of −0.4V is necessary (VK epoxide needs potentials of lower than −1.0V), whereas on the downstream electrode a potential of 0.0V up to +0.7V has been used. To provide the system with sufficient electrolytes, aqueous mobile phases have been used (regularly 1% to 5% water in methanol). McCarthy et al. (121) published recently a review of VK determination in plasma using electrochemical detection.

3. Fluorescence Detection

A great number of reports dealing with vitamin K analysis in the last 10 years was performed using reversed-phase HPLC in combination with fluorescence detection. Since vitamin K does not show native fluorescence, the quinones must be converted to the corresponding fluorescing hydroquinones (excitation 320 nm, emission 430 nm). This reduction can be performed either electrochemically, wet chemically, using different reducing reagents, or, with minor importance, by photochemical decomposition.

Electrochemical Reduction   Since Langenberg et al. in 1984 introduced the method of determination of VK1 (122) and VK1 epoxide (123) by postcolumn electrochemical reduction and fluorimetric detection, some groups have used this system with slight modifications and improvements. Reduction of the compounds is similar to the formerly discussed detection method, nearly all groups used NaClO4 as electrolyte in water-miscible eluents with small additions of water. On-line fluorimetric detection was performed by excitation of 240 nm to 320 nm and emission of 418 nm to 430 nm. Applications to the system are done by Hirauchi et al. (55,65,66); the design of their analytical cell is shown in (66), providing a limit of detection of 5 to 8 pg per injection. Some other groups (64,67,69,71,102,112,130) used similar systems for quantitation of VK1 in serum and plasma.

Chemical Reduction   Chemical reduction of quinones can be performed either by adding a reducing reagent to the HPLC eluent, or on a solid phase reactor. Wet-chemical reduction of quinones initially was reported in Japanese by Abe et al. (124), by postcolumn addition of an ethanolic NaBH4 solution to the eluent, and performing reduction in a stainless-steel reaction coil at room temperature. Other authors (33–35,90,100,106,108) also used this system for different matrices, with slight modifications mainly regarding reaction temperatures, and using ethanol-water mixtures as eluents.

To eliminate main drawbacks of this system (e.g., long elution times and decomposition of reagent in aqueous mobile phases) and to simplify this technique (a second pump, air bubbling, and debubbler), Lambert et al. (89,131) introduced an improved method. They used a new reagent, tetramethylammonium octahydridotriborate [$(CH_3)_4NB_3H_8$], which decomposes much more slowly but needs higher reaction temperatures. These findings show the main advantage that the reducing reagent can be added directly into the mobile phase, analytical chromatography is performed at room temperature (no reduction), and postcolumn reduction (without a second pumping system) is performed at 70 to 75°C in a "knitted coil" reactor. The special design of this reactor minimizes peak broadening and eliminates air handling, resulting in high reaction yields and low limits of detection, but with the disadvantage that VK1 epoxide is not reduced with sufficient yield to be quantified at physiological levels. All details and comparisons to other methods were described by Lambert et al. (68,89,99,131), leading to a more frequent use of this method by different research groups (104,105,132–135). An alternative to wet-chemical reduction is given in postcolumn reduction of quinones to hydroquinones on a solid-phase reactor.

Research groups at the Tufts University of Boston (21,22,78–81) developed a method for fluorimetric detection of K vitamins after postcolumn reduction with solid zinc: a solid-phase postcolumn reactor dry-packed with zinc particles is placed between chromatography column and fluorimeter. To the nonaqueous mobile phase (methanol:dichloromethane 8:2) are added zinc chloride (10 mM), sodium acetate, and acetic acid (5 mM each, to generate hydrogen gas for reduction of quinones). This system was found to reduce K vitamins (95% conversion) to hydroquinones in a faster and more sensitive way, without passivation of reactor and without the need of a second pumping system (78–80). Haroon et al., were able to determine VK1 (down to 50 pg/mL), MKn (100 pg), and VK1 epoxide from different matrices (21,22,78–81), and a number of researchers used the same HPLC system thereafter (82–87,113,136,137) for widespread vitamin K investigations, although some used the system with slight modifications (24,88,93,117).

Alternative solid-phase reactors are shown by Shino (72), Hiraike et al. (75), and Usui et al. (103) using platinum as catalyst for reduction of quinones but in different reduction systems. The first group and followers (72–74,91) used hydrogen-saturated mobile phase; Hiraike et al. (75–77) used alcoholic $NaClO_4$ and the last group a catalytic alcohol reduction on powdered platinum black columns (103), with more complexity and operational difficulties of the HPLC system, by mobile-phase sparging for hydrogen saturation or oxygen removal, but with relatively low detection limits. Recently, MacCrehan (77) used 10% platinum on alumina as catalyst in a simpler alcohol-reduction system.

A review describing determination of VK in plasma using HPLC with chemical reduction and fluorimetric detection was published by Davison et al. (125).

Photochemical Decomposition   When vitamin K is irradiated by strong light (e.g., 150 W xenon lamp [114]), the molecule is photodegraded to different compounds, one of which is the fluorescent hydroquinone. An advantage of this method is that no additives of chemicals are necessary for reduction which could influence pumping systems or columns. Poulsen et al. (109) built a photochemical reactor (consisting of a mercury vapor lamp in a specially designed housing including quartz tube and different inlets) for fluorescence detection of quinones, with a limit of detection of 20 to 30 pg per injection.

4.   Mass Selective Detection

The following sections describe a common topic which may be summarized under one title: mass selective determination of K vitamins. First, we discuss LC-MS techniques (liquid chromatography–mass spectrometry); second, we focus on mass selective detection methods hyphenated to gas chromatography, and finally we present a short view into future trends. The use of LC-MS methods has become the single most widely published analytical technique in the last 10 years (MP Balogh in LC-GC international, 11-97). Since the development of routinely working interfaces for liquid chromatography—mainly thermospray, electrospray, atmospheric pressure chemical ionization (APCI), particle beam MS, and fast atom bombardment (FAB)—LC-MS methods have claimed a leading position in common analysis, but not yet in vitamin K analysis. The development of these interfaces seems to be the "missing link" between HPLC chromatography, providing solvent flows of up to 1 mL/min, on the one hand, and the need of high vacuum in the mass analyzer, on the other hand. Most of the above-mentioned interfaces are not only a simple connection between chromatography and detection, but also an inlet system into the mass spectrometer (removing high quantities of vaporized solvent streams) and an ion source at the same time. As mass spectrometers are instruments for analysis of mixtures of ions on the basis of exact mass determination, these ionizing interfaces are important means of introduction into the vacuum system. The mass analyzer behind these interfaces is normally designed as an ion beam instrument (up to three quadrupole analyzers), with constant ion flow, or as an ion trap instrument, with pulsed ion flow. Also high-resolution "magnetic sector" instruments and "time of flight" instruments are available, but all systems own the principle of mass selective analysis and detection of ions. For technical details of the complex and changing technologies the reader is referred to basic mass spectrometry literature. In recent years different companies have developed individual instruments, with specific interfaces and ionization systems, the most important of which are described in brief in the following.

Thermospray and Plasmaspray An HPLC mobile phase is passed through a narrow capillary tube, at the end of which it is strongly heated, so that the solution vaporizes and emerges as a spray of droplets. The electrically charged droplets pass into a long, vacuum chamber, where they become smaller, and the density of electrical charge increases until ions desorb. When a plasma discharge is produced in the mist of droplets, ionization is enhanced and the procedure is called plasmaspray. Using these technologies, Bean et al. (138) reported a method for thermospray LC-MS of glutathione conjugates of menadiones.

Electrospray and APCI After HPLC elution the sample is passed through a capillary tube held at high electrical potential so that the solution vaporizes and emerges as spray of droplets. As the droplets evaporate, residual sample ions are analyzed in a mass spectrometer, after passing evaporation chambers via nozzles and skimmers. When a plasma discharge is produced in the mist of droplets, the enhanced ionization procedure is called atmospheric pressure chemical ionization (APCI). For example, Sano et al. (139) determinated MK4 and metabolites in human osteoblasts using APCI and tandem mass spectroscopy. A mass spectrum of VK1 using APCI interface and quadrupole MS technology can be seen in Figure 2A.

Particle beam MS is a system in which a stream of liquid is broken up by injecting helium at the end of a tube and nebulizing it. After removing the solvent in vacuum chambers and skimmers, a beam of solute molecules enters an extra ionization chamber. As an example, mass spectra of VK1 from vegetable samples are shown by Careri et al. (107).

In fast atom bombardment (FAB), an atom gun is used to project fast argon or xenon atoms on the surface of a matrix, consisting of compounds in a high-boiling-point solvent, which does not evaporate quickly in the mass spectrometer. The impact of the fast atoms on the solution results in desorption of molecular (or quasi-molecular) ions and neutrals. The ions are accelerated in the mass spectrometer by applying high voltages. Using glycerol as matrix, Philips et al. (140) were able to measure 1,4-quinones with electrochemically assisted FAB spectrometry.

When using the combination of HPLC with mass selective detection, one great advantage over conventional detection modes is given in the possibility of using a special system for internal standardization, which is called ''stable isotope dilution.'' Compounds labeled with stable isotopes (e.g., deuterium, ^{13}C, etc.) own nearly the same chemical and physical properties as the compounds of investigation, except the molecular weight, thus fulfilling all requirements of ideal internal standards. As MS is a mass selective detection method, where ions are distinguished from each other by different masses, these ISs can be monitored

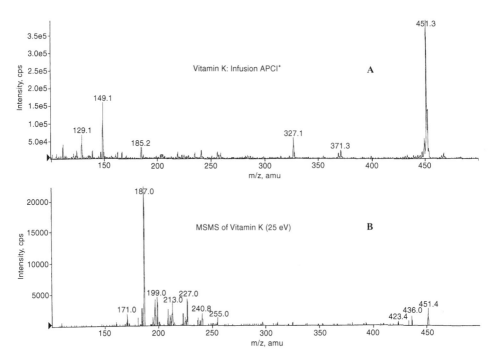

Figure 2 Mass spectra of VK1. (A) Parent ion spectrum, generated by APCI interface and quadrupole MS technology. (B) Daughter ion spectrum after fragmentation (25 eV) of parent ion m/z = 451 (MH$^+$) using quadrupole MSMS technology. (The spectra are recorded in the German PE Applied Biosystems application labs on an API365LCMSMS instrument.)

in high sensitivity together with the interesting compounds in SIM mode (single-ion monitoring) at the same time. Furthermore, losses of target compounds during sample preparation are often diminished by "dilution" with ISs of the same molecular structure. Also, eventual losses of compounds during isolation and cleanup, as well as recovery values, are of secondary importance, because the relation of added IS and natural compound is always constant. Quantitation is calculated using calibration graphs prepared with the same concentrations of ISs. A disadvantage of stable isotope dilution can arise in the case that no suitably labeled internal standards are commercially available, but very often can be prepared by scientists themselves (e.g., Leis et al. [141], is a useful review of synthetic possibilities).

Reversed-phase HPLC conditions are nearly the same as in conventional detection systems except that volatile buffer salts (e.g., NH$_4$-acetate) have to be used to prevent contamination of sensitive interfaces from depositions, a problem that sometimes also occurs by testing highly impure samples. Of course, optimal working conditions for HPLC and MS, besides individual instrument requirements (e.g., high inert gas consumption), have to be mentioned when performing LC-MS.

IV. GAS CHROMATOGRAPHY

Gas chromatography (GC) or gas liquid chromatography (GLC) as a method for analysis of K vitamins is not used as frequently as for other compounds of biochemical interest (e.g., determination of fatty acids). The main reason of this fact can be seen in a possible degradation of the vitamin K molecules on the heated GC-columns. Lefevere et al. [2] and Lambert et al. [17] reviewed early GC methods, where among other things electron capture detection was used for determination of VK1 and VK epoxide.

In recent years, through development of modern gas chromatographs, new technologies, (including special fused silica columns etc.), and the introduction of relatively low-priced, small (bench-top), simple, and stable GC-MS combinations, determination of K vitamins, using gas chromatography, reached a certain interest.

A. Internal Standardization, Isolation, and Derivatization

What we have discussed above (internal standardization in liquid chromatography) is valid in any case also for gas chromatography. Furthermore, using GC-MS, stable isotope dilution is the method of choice, performing complex analytical procedures. As an example, we have recently published (142) a method for determination of VK1 in plasma using stable isotope dilution GC-MS with a trideuterium-labeled VK1 (see also Figs. 1/2 and 3), where we reached a limit of detection of 1.0 pg per injection. It is feasible that this method could be enhanced to other homologs of the vitamin K series, after synthesis of corresponding ISs, parallel to the instructions given in (141) and (142). We believe that usual problems, arising, for example, in simultaneous HPLC detection of diverse VK homologs, could at least be diminished by the use of appropriate stable isotope labeled ISs and SIM detection mode.

Sample isolation and cleanup procedures are the same as described above for LC methods (Table 1), with the exception that a simple extraction normally is sufficient to obtain pure chromatograms. With a proper GC-MS method we were able to separate *cis/trans*-isomeres of underivatized VK1, but we have seen that the compound undergoes a nonneglectable degree of degradation during gas chromatography. This is the reason why we introduced a derivatization step after

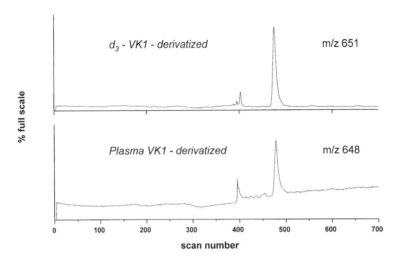

Figure 3 Typical GC-MS-SIM chromatogram obtained after analysis of VK1 in human plasma after reductive acylation with zinc and heptafluorobutyric anhydride and the use of trideuterium labeled VK1 as internal standard in stable isotope dilution (unpublished results).

isolation of the compounds. We obtained best results by esterification of the molecules with heptafluorobutyric acid (HFBA) by simultaneous reduction of the carbonyl groups (of the naphthoquinone nucleus) with zinc. This simple procedure is executed in hexane at room temperature by simultaneous addition of zinc dust and HFB anhydride and HFBA ("reductive acylation"). The hexane solution can be injected into GC-MS after washing with water. This derivatization provided us with a stable and chromatographable compound (for structure see Fig. 1/5; Refs. [2] and [143] in an early GC-MS publication of VK detection), including the further great advantage of a shift of molecular masses to m/z = 648 and m/z = 651, respectively, a region where disturbances of biological matrices are extremely low. An SIM-EI chromatogram of these compounds after isolation from plasma is shown in Figure 3. Of course, other derivatization reagents also could be used, depending on different necessities of the method and matrices etc.

B. Chromatography

Modern gas chromatography is normally performed using fused silica capillary columns with a length of 15 to 30 m, an i.d. of 0.25 mm, and a film thickness of 0.25 μm. Columns of 50 m can also be used, but retention times are increased.

For MS detection, special columns with low bleeding silicon phases should be used to diminish background noise. We normally use DB5-MS columns, consisting of 5% phenylmethylsilicon for better separation of nonpolar compounds. As carrier gas (mobile phase) nitrogen, or better helium, is used, and inlet gas pressure is set to 50 kPa.

Programmable GC oven temperatures provide analysts with fast and stable temperature programs, so that deviations in chromatography are kept at a minimum. We prefer splitless injection, using a split/splitless Grob injector (held at 260°C) and a relative high temperature program for short retention times. The initial oven temperature (160°C) is kept for 1 min, followed by an increase of 30°C per min to 290°C, an isothermal hold of 4 min, followed by another increase of 30°C per min to 310°C and a final isothermal hold of 2 min. The transfer line between GC and MS is kept at 308°C. For time-saving injection an autosampler can be used.

V. MASS SPECTROMETRY

Mass spectrometers hyphenated to gas chromatographs have the same design as those in LC-MS except that there is no need of a special interface for removal of high amounts of liquid vapors. A constant flow of carrier gas, together with the chromatographically separated compounds, is, via transfer line, directly brought into an ionization chamber. There the compounds are ionized by two different systems, and the ions are analyzed and detected by the use of different analyzer systems and electron- or photomultipliers. Turbo pumps provide a high vacuum of down to 10^{-6} mbar. Molecules can be ionized by either electron impact (EI) ionization or chemical ionization (CI).

In EI ionization, molecules of investigation interact with energetic electrons (electrically charged substances) to molecular radical cations. The resulting molecular ions often break up, resulting in ions with smaller masses (fragment ions). The total of molecular and fragment ions is separated by a mass analyzer to give a mass spectrum, relating the masses of ions with their relative abundances.

In CI mode, molecules of an reagent gas (e.g., methane, hydrogen, ammonia, etc.) and electrons interact to form ions. These ions collidate with molecules (M) of investigation, under transfer of a proton (H^+), to give quasi-molecular ions (MH^+) which do not fragment as readily as molecular ions formed by EI ionization. Therefore CI spectra contain more quasi-molecular ions and fewer fragment ions. For more detailed information regarding ion separation (quadrupole and ion trap), consult the basic GC-MS literature.

In GC-MS instruments, as well as in LC-MS-detection, mainly ion beam instruments (e.g., quadrupole or ion trap instruments) are used. In quadrupole instruments ions can be filtered along their central axis through the application of DC and RF voltages to an assembly of four parallel rods, and masses are measured to give a mass spectrum. In ion trap instruments, ions are caught in a trap through the application of DC and RF voltages to an assembly of ring electrodes; after a fixed period, defined ions are ejected by specific changes of voltages. These instruments can work in "full-scan" mode, scanning periodically the whole spectrum of ions, thus giving information of molecule mass and fragmentations. They can also be used in SIM mode (single-ion monitoring), measuring permanently only few predominant ions, as far as possible the base peak of the spectra. Using stable isotope dilution for internal standardization, mass spectrometry provides information on the concentrations of different compounds.

Up to now, quadruple instruments have gained greatest distribution worldwide, especially in routine analysis, through their stability and simplicity in handling. In the past mass selective detection mainly was used as a mean for final confirmation of hitherto unknown substances or chemical structures. Different MK_n structures from various bacteria suspensions have been confirmed with MS methods after TLC (54,58,144) and/or HPLC (56,145,146) methods. Madden et al. (53) reported confirmation of VK1 after RPTLC by GC-MS, whereas Booth et al. (70,147) confirmed the structure of HPLC-detected dihydro-VK1 in plasma following dietary intake of hydrogenated oils and foods. Kuliopulos et al. (15) used EI-MS with direct inlet or supercritical fluid chromatography and NICI-MS for determination of dioxygen transfer during VK-dependent carboxylase catalysis. Imanaka et al. (148) identified phylloquinone via GC-EI-MS as an unknown peak in ECD-GC chromatograms of pyrethroid insecticide residues. Finally, our group (142,149) used GC-EI-MS with a single-quadrupole instrument for detection of VK1 in human plasma.

VI. RECENT TECHNIQUES

Further improvements can be expected on each of the different steps of isolation, cleanup, chromatography, and detection systems for determination of VK in biological samples. TLC and direct-phase HPLC as chromatographic methods will be reduced to a minimum. Development of new materials for solid phase extraction (e.g., Oasis™, a new polymeric SPE sorbent) and improved materials for reversed-phase HPLC, including smaller, narrow-bore or capillary columns, should bring advantage to RP-HPLC, especially in connection with MS detection, but also with fluorescence and electrochemical detection and help to obtain smaller sample volumes and/or shorter analysis times. However, the greatest advance, in our opinion, will be in the field of detection instruments—in particular, in mass selective detection instruments.

A. LC-MS and LC-MS-MS

As LC-MS detectors get increasingly cheaper and smaller in size, these instruments should be used as often as alternative detectors. Due to new software programs, operation and control of these analyzers will be easy to handle. Development of simple but stable working interfaces that need less service time will make these combinations a good alternative to conventional detection modes, especially in cases of small sample volume.

A special chance for detection of VKs in heavy loaded matrices, for example, could be the combination of LC and tandem MS detection modes. Tandem MS (or two-[multi]-dimensional MS) means that monitored ions (parent ions) pass a (second) fragmentation step and the generated product ions ("daughter ions") are monitored again, either in full-scan mode or in SIM mode. As an example, a tandem MS spectrum of VK can be seen in Figure 2, where a parent ion (m/z = 451, MH^+ of VK1, generated by APCI) is fragmented to give a spectrum of daughter ions, using a quadrupole-MSMS technology. The base peak of this spectrum (m/z = 187, $[MH^+$-264]) is caused by elimination of $C_{19}H_{36}$ from the phytol side chain of the molecule (for details see 142,143). The relation of a generated daughter ion (m/z = 187) to a defined parent ion (m/z = 451) is highly specific for just one single compound (VK1) whereas background, or matrix, disturbances can be excluded. A further example is given by Sano et al. (139); they determined MK4 and its metabolites by APCI-tandem-MS.

B. GC-MS and GC-MS-MS

The same arguments that have been discussed above for LC-MS also are valid for GC-MS technology. The disadvantage of needing a derivatization step could turn to an advantage because of having no need of further sample cleanup (solid-phase extraction, semipreparative HPLC, etc.) and reaching low detection limits, through to high molecular weights. In combination with stable isotope dilution and SIM detection, GC-MS could be a method of choice for future investigations.

Also tandem MS technology is imaginable in combination with gas chromatography, especially when small sample volumes are to be analyzed. As an example, an electron impact spectrum of derivatized VK1 (Fig. 4A) is shown together with two daughter spectra (in collision-induced dissociation mode, CID; Fig. 4B: daughter spectrum of parent ion m/z = 383; Fig. 4C: daughter spectrum of parent ion m/z = 648), using GC-MS-MS and ion trap technology (unpublished results). As these instruments now are available at acceptable prices, these technologies will be a possible alternative to other analytical systems for vitamin

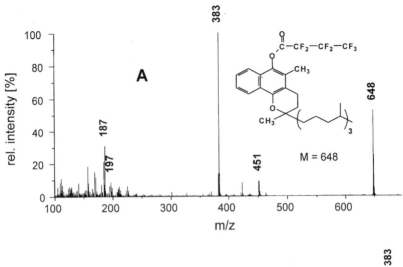

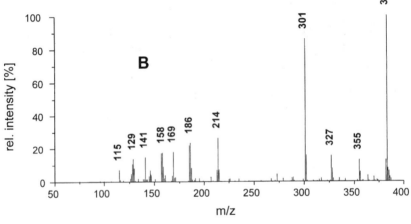

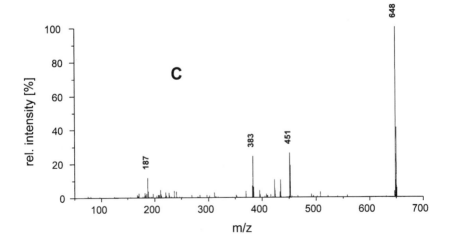

K detection, but it remains an open question whether these technologies will be applied by chromatographers in the future to a greater extent.

VII. FUTURE TRENDS

Whether other separation techniques, such as SFE, SFC, capillary electrophoresis, etc., will play a significant role depends mainly on whether instrument providers can (or find interest to) produce adequate and stable working equipment. At the moment these techniques seem not to be as handsome as others for the analysis of K vitamins. Enhanced detection sensitivity, by the use of narrow-bore or capillary-column RP-HPLC with postcolumn derivatization in simplified technologies (without a second pumping system, etc.), will be the aim of the future. Due to lowering prices of laser light sources, further development of laser technology regarding UV and fluorescence detection can be expected, but how these findings can be used in routine experiments remains to been seen.

ACKNOWLEDGMENTS

This work is dedicated to Professor Helmut Gleispach on the occasion of his 60th birthday, and was supported by the Fonds zur Förderung der wissenschaftlichen Forschung (Project No. P-10560 Med) and by grants from the Steirische Kinderkrebshilfe.

REFERENCES

1. H Dam. The antihaemorrhagic vitamin of the chick (letter). Nature 135:652–653, 1935.
2. MF Lefevere, AE Claeys, AP De Leenheer. Vitamin K: phylloquinone and menaquinones. In: AP De Leenheer, WE Lambert, MGM De Ruyter, eds. Modern Chromatographic Analysis of the Vitamins. New York: Marcel Dekker, 1985, pp 201–265.

Figure 4 Mass spectra of VK1 after reductive acylation with zinc and heptafluorobutyric anhydride, using ion trap technology. (A) EI-GC-MS-spectrum (parent ion spectrum). (B) GC-MS-MS daughter ion spectrum of parent ion m/z = 383, using collision-induced dissociation mode (CID). (C) GC-MS-MS daughter ion spectrum of parent ion m/z = 648, also in CID mode (unpublished results).

3. HR Roberts, JN Lozier. New perspectives on the coagulation cascade. Hosp Pract Off Ed 27(1): 97–105, 109–112, 1992.
4. MJ Shearer. Vitamin K and vitamin K-dependent proteins. Br J Haematol 75(2): 156–162, 1990.
5. Y Sano, K Tadano, K Kaneko, K Kikuchi, T Yuzuriha. Distribution of menaquinone-4, a therapeutic agent for osteoporosis, in bone and other tissues of rats. J Nutr Sci Vitaminol 41:499–514, 1995.
6. CC Juan, FY Wu. Vitamin K_3 inhibits growth of human hepatoma HepG2 cells by decreasing activities of both p34^{cdc2} kinase and phosphatase. Biochem Biophys Res Commun 190(3):907–913, 1993.
7. C Toxopeus, I van Holsteijn, JWF Thuring, BJ Blaauboer, J Noordhoek. Cytotoxicity of menadion and related quinones in freshly isolated rat hepatocytes: effects on thiol homeostasis and energy charge. Arch Toxicol 67(10):674–679, 1993.
8. B Bouzahzah, Y Nishikawa, D Simon, BI Carr. Growth control and gene expression in a new hepatocellular carcinoma cell line, Hep40: inhibitory actions of vitamin K. J Cell Physiol 165(3):459–467, 1995.
9. OYP Hu, CY Wu, WK Chan, FYH Wu. Determination of anticancer drug vitamin K_3 in plasma by high-performance liquid chromatography. J Chromatogr B 666(2): 299–305, 1995.
10. XQ Guo, YB Zhao, JG Xu. Study on sensitized photochemical fluorescence of menadione sodium bisulfite and its application in pharmaceutics analysis. Anal Chim Acta 343:109–116, 1997.
11. DP Morris, RD Stevens, DJ Wright, DW Stafford. Processive post-translational modification. Vitamin K–dependent carboxylation of a peptide substrate. J Biol Chem 270(51):30491–30498, 1995.
12. MJ Shearer. Vitamin K. Lancet 345:229–234, 1995.
13. JW Suttie. Recent advances in hepatic vitamin K metabolism and function. Hepatology 7:367–376, 1987.
14. SL Gardill, JW Suttie. Vitamin K epoxide and quinone reductase activities. Evidence for reduction by a common enzyme. Biochem Pharmacol 40(5):1055–1061, 1990.
15. A Kuliopulos, BR Hubbard, Z Lam, IJ Koski, B Furie, BC Furie, CT Walsh. Dioxygen transfer during vitamin K dependent carboxylase catalysis. Biochemistry 31(33):7722–7728, 1992.
16. LMT Vervoort, JE Ronden, HHW Thijssen. The potent antioxidant activity of the vitamin K cycle in microsomal lipid peroxidation. Biochem Pharmacol 54(8):871–876, 1997.
17. WE Lambert, AP De Leenheer. Vitamin K. In: AP De Leenheer, WE Lambert, HJ Nelis, eds. Modern Chromatographic Analysis of the Vitamins. 2nd ed. New York: Marcel Dekker, 1992, pp 197–233.
18. C Vermeer, KS Jie, MH Knapen. Role of vitamin K in bone metabolism. Annu Rev Nutr 15:1–22, 1995.
19. KS Jie, ML Bots, C Vermeer, JC Witteman, DE Grobbee. Vitamin K intake and osteocalcin levels in women with and without aortic atherosclerosis: a population-based study. Atherosclerosis 116(1):117–123, 1995.
20. K Hara, Y Akiyama, T Nakamura, S Murota, I Morita. The inhibitory effect of

vitamin K_2 (menatetrenone) on bone resorption may be related to its side chain. Bone 16(2):179–184, 1995.

21. SL Booth, KW Davidson, JA Sadowski. Evaluation of an HPLC method for the determination of phylloqinone (vitamin K1) in various food matrices. J Agric Food Chem 42(2):295–300, 1994.

22. SL Booth, JA Sadowski, JL Weihrauch, G Ferland. Vitamin K_1 (phylloquinone) content of foods: a provisional table. J Food Compos Anal 6:109–120, 1993.

23. Y Suzuki, M Okamoto. Production of hen's eggs rich in vitamin K. Nutr Res 17(10):1607–1615. 1997.

24. E Jakob, I Elmadfa. Application of a simplified HPLC assay for the determination of phylloqinone (vitamin K_1) in animal and plant food items. Food Chem 56(1): 87–91, 1996.

25. TJ Koivu, VI Piironen, SK Henttonen, PH Mattila. Determination of phylloquinone in vegetables, fruits, and berries by high-performance liquid chromatography with electrochemical detection. J Agric Food Chem 45(12):4644–4649, 1997.

26. SL Booth, JA Sadowski. Determination of phylloquinone in foods by high-performance liquid chromatography. Vit Coenzymes PT L 282:446–456, 1997.

27. SL Booth, JA Pennington, JA Sadowski. Food sources and dietary intakes of vitamin K-1 (phylloquinone) in the American diet: data from the FDA Total Diet Study. J Am Diet Assoc 96(2):149–154, 1996.

28. MJ Shearer, A Bach, M Kohlmeier. Chemistry, nutritional sources, tissue distribution and metabolism of vitamin K with special reference to bone health. J Nutr 126:1181S–1186S, 1996.

29. JJ Lipsky. Nutritional sources of vitamin K. Mayo Clin Proc 69(5):462–466, 1994.

30. SL Booth, HT Madabushi, KW Davidson, JA Sadowski. Tea and coffee brews are not dietary sources of vitamin K-1 (phylloquinone). J Am Diet Assoc 95(1):82–83, 1995.

31. JW Suttie. Vitamin K and human nutrition. J Am Diet Assoc 92(5):585–590, 1992.

32. LM Canfield, JM Hopkinson. State of the art vitamin K in human milk. J Pediatr Gastroenterol Nutr 8(4):430–441, 1989.

33. JM Conly, K Stein. Quantitative and qualitative measurements of K vitamins in human intestinal contents. Am J Gastroenterol 87(3):311–316, 1992.

34. JM Conly, K Stein. The absorbtion and bioactivity of bacterially synthesized menaquinones. Clin Invest Med 16:45–57, 1993.

35. JM Conly, K Stein, L Worobetz, S Rutledge-Harding. The contribution of vitamin K2 (menaquinones) produced by the intestinal microflora to human nutritional requirements for vitamin K. Am J Gastroenterol 89(6):915–923, 1994.

36. Y Usui. Assay of phylloquinone and menaquinones in human liver. Vit Coenzymes PT L 282:438–446, 1997.

37. JM Conly. Assay of menaquinones in bacterial cultures, stool samples and intestinal contents. Vit Coenzymes PT L 282:457–466, 1997.

38. J Breton, JR Burie, C Boullais, G Berger, E Nabedryk. Binding sites of quinones in photosynthetic bacterial reaction centers investigated by light-induced FTIR difference spectroscopy: binding of chainless symmetrical quinones to the QA site of *Rhodobacter sphaeroides*. Biochemistry 33(41):12405–12415, 1994.

39. SJ Hodges, J Bejui, M Leclercq, PD Delmas. Detection and measurement of vita-

mins K$_1$ and K$_2$ in human cortical and trabecular bone. J Bone Miner Res 8(8): 1005–1008, 1993.

40. MM Groenen–van Dooren, JE Ronden, BA Soute, C Vermeer. Bioavailability of phylloquinone and menaquinones after oral and colorectal administration in vitamin K–deficient rats. Biochem Pharmacol 50(6):797–801, 1995.

41. PM Loughnan, PN McDougall. Does intramuscular vitamin K$_1$ act as an unintended depot preparation? J Paediatr Child Health 32(3):251–254, 1996.

42. K Nakamura, H Toyohira, H Kariyazono, M Ishibashi, H Saigenji, S Shimokawa, A Taira. Anticoagulant effects of warfarin and kinetics of K vitamins in blood and feces. Artery 21(3):148–160, 1994.

43. R Kries, FR Greer, JW Suttie. Assessment of vitamin K status of the newborn infant. J Pediatr Gastroenterol Nutr 16(3):231–238, 1993.

44. R Kries, MJ Shearer, U Göbel. Vitamin K in infancy. Eur J Pediatr 147(2):106–112, 1988.

45. FI Clark, EJ James. Twenty-seven years of experience with oral vitamin K1 therapy in neonates. J Pediatr 127(2):301–304, 1995.

46. PM Loughnan, PN McDougall. The efficacy of oral vitamin K1: implications for future prophylaxis to prevent haemorrhagic disease of the newborn. J Paediatr Child Health 29(3):171–176, 1993.

47. PM Loughnan, PN McDougall, H Balvin, LW Doyle, AL Smith. Late onset haemorrhagic disease in premature infants who received intravenous vitamin K$_1$. J Paediatr Child Health 32(3) 268–269, 1996.

48. PM Loughnan, PN McDougall. Epidemiology of late onset haemorrhagic disease: a pooled data analysis. J Paediatr Child Health 29(3):177–181, 1993.

49. MA Brousson, MC Klein. Controversies surrounding the administration of vitamin K to newborns: a review. Can Med Assoc J 154(3):307–315, 1996.

50. J Golding, M Paterson, LJ Kinlen. Factors associated with childhood cancer in a national cohort study. Br J Cancer 62(2):304–308, 1990.

51. J Golding, R Greenwood, K Birmingham, M Mott. Childhood cancer, intramuscular vitamin K and pethidine given during labour. BMJ 305:341–346, 1992.

52. B Fried, J Sherma. Vitamins. In: Thin-Layer Chromatography, Techniques and Applications. 3rd ed. New York: Marcel Dekker, 1994, pp 337–360.

53. UA Madden, HM Stahr. Reverse phase thin layer chromatography assay of vitamin K in bovine liver. J Liquid Chromatogr 16(13):2825–2834, 1993.

54. A Das, J Hugenholtz, H van Halbeek, LG Ljungdahl. Structure and function of a menaquinone involved in electron transport in membranes of *Clostridium thermoautotrophicum* and *Clostridium thermoaceticum*. J Bacteriol 171(11):5823–5829, 1989.

55. K Hirauchi, T Sakano, S Notsumoto, T Nagaoka, A Morimoto, K Fujimoto, S Masuda, Y Suzuki. Measurement of K vitamins in animal tissues by high-performance liquid chromatography with fluorimetric detection. J Chromatogr 497:131–137, 1989.

56. CW Moss, MA Lambert-Fair, MA Nicholson, GO Guerrant. Isoprenoid quinones of *Campylobacter cryaerophila, C. cinaedi, C. fennelliae, C. hyointestinalis, C. pylori,* and "*C. upsaliensis.*" J Clin Microbiol 28(2):395–397, 1990.

57. N Sakamoto, M Kimura, H Hiraike, Y Itokawa. Changes of phylloquinone and

menaquinone-4 concentrations in rat liver after oral, intravenous and intraperitoneal administration. Int J Vitam Nutr Res 66:322–328, 1996.

58. AJ Rutherford, D Williams, RF Bilton. Isolation of menaquinone-7 from *Pseudomonas* N.C.I.B. 10590: a natural electron acceptor for steroid A-ring dehydrogenations. Biochem Soc Trans 19(1):64S, 1991.

59. EM Kirk, AF Fell. Analysis of supplemented vitamin $K_{1(20)}$ in serum microsamples by solid-phase extraction and narrow-bore HPLC with multichannel ultraviolet detection. Clin Chem 35(7):1288–1292, 1989.

60. D Blanco Gomis, VJ Escotet Arias, LE Fidalgo Alvarez, MD Gutierrez Alvarez. Simultaneous determination of vitamins D_3, E and K_1 and retinyl palmitate in cattle plasma by liquid chromatography with a narrow-bore column. J Chromatogr B, 660:49–55, 1994.

61. D Blanco Gomis, V Escotet Arias, LE Fidalgo Alvarez, MD Gutierrez Alvarez. Determination of fat-soluble vitamins by capillary liquid chromatography in bovine blood plasma. Anal Chim Acta 315:177–181, 1995.

62. MJ Winn, S Cholerton, BK Park. An investigation of the pharmacological response to vitamin K_1 in the rabbit. Br J Pharmacol 94(4):1077–1084, 1988.

63. G Schubiger, O Tönz, J Grüter, MJ Shearer. Vitamin K_1 concentration in breast-fed neonates after oral or intramuscular administration of a single dose of a new mixed-micellar preparation of phylloquinone. J Pediatr Gastroenterol Nutr 16(4):435–439, 1993.

64. M Guillaumont, M Leclercq, H Gosselet, K Makala, B Vignal. HPLC determination of serum vitamin K1 by fluorometric detection after post-column electrochemical reduction. J Micronutr Anal 4(4):285–294, 1988.

65. K Hirauchi, T Sakano, T Nagaoka, A Morimoto. Simultaneous determination of vitamin K_1, vitamin K_1-2,3-epoxide and menaquinone-4 in human plasma by high-performance liquid chromatography with fluorimetric detection. J Chromatogr 430(1):21–29, 1988.

66. K Hirauchi, T Sakano, A Morimoto. Measurement of K vitamins in human and animal plasma by high-performance liquid chromatography with fluorometric detection. Chem Pharm Bull 34(2):845–849, 1986.

67. F Moussa, L Dufour, JR Didry, P Aymard. Determination of *trans*-phylloquinone in children's serum. Clin Chem 35(5):874–878, 1989.

68. WE Lambert, AP De Leenheer, EJ Baert. Wet-chemical postcolumn reaction and fluorescence detection analysis of the reference interval of endogenous serum vitamin $K_{1(20)}$. Anal Biochem 158:257–261, 1986.

69. MJ Pettei, D Israel, J Levine. Serum vitamin K concentration in pediatric patients receiving total parenteral nutrition. J Parenter Enter Nutr 17(5):465–467, 1993.

70. SL Booth, KW Davidson, AH Lichtenstein, JA Sadowski. Plasma concentrations of dihydro-vitamin K_1 following dietary intake of a hydrogenated vitamin K_1-rich vegetable oil. Lipids 31(7):709–713, 1996.

71. JR Soedirman, EA De Bruijn, RA Maes, A Hanck, J Grüter. Pharmacokinetics and tolerance of intravenous and intramuscular phylloquinone (vitamin K_1) mixed micelles formulation. Br J Clin Pharmacol 41(6):517–523, 1996.

72. M Shino. Determination of endogenous vitamin K (phylloquinone and menaquinone-n) in plasma by high-performance liquid chromatography using plat-

inum oxide catalyst reduction and fluorescence detection. Analyst 113(3):393–397, 1988.

73. H Tamai, Z Mingci, N Kawamura, T Kuno, T Ogihara, M Mino. Fat soluble vitamins in cord blood and colostrum in the south of China. Int J Vitam Nutr Res 66: 222–226, 1996.

74. T Nakamura, K Takebe, K Imamura, Y Tando, N Yamada, Y Arai, A Terada, M Ishii, H Kikuchi, T Suda. Fat-soluble vitamins in patients with chronic pancreatitis (pancreatic insufficiency). Acta Gastroenterol Belg 59(1):10–14, 1996.

75. H Hiraike, M Kimura, Y Itokawa. Determination of K vitamins (phylloquinone and menaquinones) in umbilical cord plasma by a platinum-reduction column. J Chromatogr 430(1):143–148, 1988.

76. N Sakamoto, M Kimura, H Hiraike, Y Itokawa. Changes of K vitamins in portal and femoral venous plasma of rats after oral administration of phylloquinone and menaquinone-4. Int J Vitam Nutr Res 65(2):105–110, 1995.

77. WA MacCrehan, E Schönberger. Determination of vitamin K_1 in serum using catalytic-reduction liquid chromatography with fluorescence detection. J Chromatogr B 670(2):209–217, 1995.

78. Y Haroon, DS Bacon, JA Sadowski. Liquid-chromatographic determination of vitamin K_1 in plasma, with fluorometric detection. Clin Chem 32(10):1925–1929, 1986.

79. Y Haroon, DS Bacon, JA Sadowski. Chemical reduction system for the detection of phylloquinone (vitamin K_1) and menaquinones (vitamin K_2). J Chromatogr 384: 383–389, 1987.

80. Y Haroon, DS Bacon, JA Sadowski. Reduction of quinones with zinc metal in the presence of zinc ions: application of post-column reactor for the fluorimetric detection of vitamin K compounds. Biomed Chromatogr 2(1):4–8, 1987.

81. JA Sadowski, SJ Hood, GE Dallal, PJ Garry. Phylloquinone in plasma from elderly and young adults: factors influencing its concentration. Am J Clin Nutr 50(1):100–108, 1989.

82. NJ Kazzi, NB Ilagan, KC Liang, GM Kazzi, LA Grietsell, YW Brans. Placental transfer of vitamin K_1 in preterm pregnancy. Obstet Gynecol 75(3):334–337, 1990.

83. KE Nestor, HR Conrad. Metabolism of vitamin K and influence on prothrombin time in milk-fed preruminant calves. J Dairy Sci 73(11):3291–3296, 1990.

84. G Ferland, JA Sadowski, ME O'Brien. Dietary induced subclinical vitamin K deficiency in normal human subjects. J Clin Invest 91(4):1761–1768, 1993.

85. HN Rosen, LA Maitland, JW Suttie, WJ Manning, RJ Glynn, SL Greenspan. Vitamin K and maintenance of skeletal integrity in adults. Am J Med 94(1):62–68, 1993.

86. JN Hagstrom, EG Bovill, RF Soll, KW Davidson, JA Sadowski. The pharmacokinetics and lipoprotein fraction distribution of intramuscular vs. oral vitamin K_1 supplementation in women of childbearing age: effects on hemostasis. Thromb Haemost 74(6):1486–1490, 1995.

87. FR Greer, SP Marshall, AL Foley, JW Suttie. Improving the vitamin K status of breastfeeding infants with maternal vitamin K supplements. Pediatrics 99(1):88–92, 1997.

88. E Jakob, I Elmadfa. Rapid HPLC assay for the assessment of vitamin K_1, A, E

and beta-carotene status in children (7–19 years). Int J Vitam Nutr Res 65(1):31–35, 1995.

89. WE Lambert, AP De Leenheer, MF Lefevere. Determination of vitamin K in serum using HPLC with post-column reaction and fluorescence detection. J Chromatogr Sci 24:76–79, 1986.

90. M Sakon, M Monden, M Gotoh, K Kobayashi, T Kanai, K Umeshita, W Endoh, T Mori. The effects of vitamin K on the generation of des-gamma-carboxy prothrombin (PIVKA-II) in patients with hepatocellular carcinoma. Am J Gastroenterol 86(3):339–345, 1991.

91. Y Sano, K Tadano, K Kikuchi, K Kaneko, T Yuzuriha. Pharmacokinetic characterization of menaquinone-4 in dogs by sensitive HPLC determination. J Nutr Sci Vitaminol 39:555–566, 1993.

92. HE Indyk, VC Littlejohn, RJ Lawrence, DC Woollard. Liquid chromatographic determination of vitamin K_1 in infant formulas and milk. J AOAC Int 78(3):719–723, 1995.

93. HE Indyk, DC Woollard. Vitamin K in milk and infant formulas: determination and distribution of phylloquinone and menaquinone-4. Analyst 122:465–469, 1997.

94. H Isshiki, Y Suzuki, A Yonekubo, H Hasegawa, Y Yamamoto. Determination of phylloquinone and menaquinone in human milk using high performance liquid chromatography. J Dairy Sci 71(3):627–632, 1988.

95. LM Canfield, JM Hopkinson, AF Lima, GS Martin, K Sugimoto, J Burr, L Clark, DL McGee. Quantitation of vitamin K in human milk. Lipids 25(7):406–411, 1990.

96. LM Canfield, JM Hopkinson, AF Lima, B Silva, C Garza. Vitamin K in colostrum and mature human milk over the lactation period—a cross-sectional study. Am J Clin Nutr 53(3):730–735, 1991.

97. MM Delgado Zamarreno, A Sanchez Perez, MC Gomez Perez, MA Fernandez Moro, J Hernandez Mendez. Determination of vitamins A, E and K_1 in milk by high-performance liquid chromatography with dual amperometric detection. Analyst 120(10):2489–2492, 1995.

98. MA Schneiderman, AK Sharma, KR Mahanama, DC Locke. Determination of vitamin K_1 in powdered infant formulas, using supercritical fluid extraction and liquid chromatography with electrochemical detection. J Assoc Off Anal Chem 71(4):815–817, 1988.

99. WE Lambert, L Vanneste, AP De Leenheer. Enzymatic sample hydrolysis and HPLC in a study of phylloquinone concentration in human milk. Clin Chem 38(9):1743–1748, 1992.

100. H Ikeda, Y Doi. A vitamin-K_2-binding factor secreted from *Bacillus subtilis*. Eur J Biochem 192(1):219–224, 1990.

101. PJ Ross, MJ Shearer, AT Diplock, SA Schey. A fibroblast cell culture model to study vitamin K metabolism and the inhibition of vitamin K epoxide reductase by known and suspected antagonists. Br J Haematol 77(2):195–200, 1991.

102. Y Usui, H Tanimura, N Nishimura, N Kobayashi, T Okanoue, K Ozawa. Vitamin K concentrations in the plasma and liver of surgical patients. Am J Clin Nutr 51(5):846–852, 1990.

103. Y Usui, N Nishimura, N Kobayashi, T Okanoue, M Kimoto, K Ozawa. Measurement of vitamin K in human liver by gradient elution high-performance liquid chro-

matography using platinum-black catalyst reduction and fluorimetric detection. J Chromatogr 489(2):291–301, 1989.

104. HHW Thijssen, MJ Drittij-Reijnders. Vitamin K metabolism and vitamin K1 status in human liver samples: a search for inter-individual differences in warfarin sensitivity. Br J Haematol 84:681–685, 1993.

105. HHW Thijssen, MJ Drittij-Reijnders. Vitamin K status in human tissues: tissue-specific accumulation of phylloquinone and menaquinone-4. Br J Nutr 75(1):121–127, 1996.

106. H Hiraike, M Kimura, Y Itokawa. Distribution of K vitamins (phylloquinone and menaquinones) in human placenta and maternal and umbilical cord plasma. Am J Obstet Gynecol 158:564–569, 1988.

107. M Careri, A Mangia, P Manini, N Taboni. Determination of phylloquinone (vitamin K_1) by high performance liquid chromatography with UV detection and with particle beam–mass spectrometry. Fresenius J Anal Chem 355:48–56, 1996.

108. BE Cham, HP Roeser, TW Kamst. Simultaneous liquid-chromatographic determination of vitamin K_1 and vitamin E in serum. Clin Chem 35(12):2285–2289, 1989.

109. JR Poulsen, JW Birks. Photoreduction fluorescence detection of quinones in high performance liquid chromatography. Anal Chem 61(20):2267–2276, 1989.

110. SL Booth, JAT Pennington, JA Sadowski. Dihydro–vitamin K_1: primary food sources and estimated dietary intakes in the American diet. Lipids 31(7):715–720, 1996.

111. V Piironen, T Koivu, O Tammisalo, P Mattila. Determination of phylloquinone in oils, margarines and butter by high-performance liquid chromatography with electrochemical detection. Food Chem 59(3):473–480, 1997.

112. F Moussa, F Depasse, V Lompret, JV Hautem, JP Girardet, JL Fontaine, P Aymard. Determination of phylloquinone in intravenous fat emulsions and soybean oil by high performance liquid chromatography. J Chromatogr A 664(2):189–194, 1994.

113. C Lennon, KW Davidson, JA Sadowski, JB Mason. The vitamin K content of intravenous lipid emulsions. J Parenter Enter Nutr 17(2):142–144, 1993.

114. H Indyk. The photoinduced reduction and simultaneous fluorescence detection of vitamin K_1 with HPLC. J Micronutr Anal 4:61–70, 1988.

115. MA Schneiderman, AK Sharma, DC Locke. Determination of menadione in an animal feed using supercritical fluid extraction and HPLC with electrochemical detector. J Chromatogr Sci 26(9):458–462, 1988.

116. MA Schneiderman, AK Sharma, DC Locke. Determination of anthraquinone in paper and wood using supercritical fluid extraction and high performance liquid chromatography with electrochemical detection. J Chromatogr 409:343–353, 1987.

117. SM Billedeau. Fluorimetric determination of vitamin K_3 (menadione sodium bisulfite) in synthetic animal feed by high-performance liquid chromatography using a post-column zinc reducer. J Chromatogr 472(2):371–379, 1989.

118. I Gil Torro, JV Garcia Mate, J Martinez Calatayud. Spectrofluorimetric determination of vitamin K_3 by a solid phase zinc reactor immobilized in a flow injection assembly. Analyst 122:139–142, 1997.

119. JC Vire, V Lopez, GJ Patriarche, GD Christian. Determination of vitamin K_1 by adsorptive stripping square wave voltametry. Anal Lett 21(12):2217–2225, 1988.

120. ESM Po, JW Ho, BY Gong. Simultaneous chromatographic analysis of eight fat-soluble vitamins in plasma. J Biochem Beophys Meth 34(2):99–106, 1997.

121. PT McCarthy, DJ Harrington, MJ Shearer. Assay of phylloquinone in plasma by high-performance liquid chromatography with electrochemical detection. Vit Coenzymes PT L 282:421–433, 1997.

122. JP Langenberg, UR Tjaden. Determination of (endogenous) vitamin K_1 in human plasma by reversed phase high performance liquid chromatography using fluorimetric detection after postcolumn electrochemical reduction. J Chromatogr 305: 61–72, 1984.

123. JP Langenberg, UR Tjaden. Improved method for the determination of vitamin K_1 epoxide in human plasma with electrofluorimetric reaction detection. J Chromatogr 289:377–385, 1984.

124. K Abe, O Hiroshima, K Ishibashi, M Ohmae, K Kawabe, G Katsui. Fluorometric determination of phylloquinone and menaquinone-4 in biological materials using high performance liquid chromatography (in Japanese). Yakugaku Zasshi 99:192–200, 1979.

125. KW Davidson, JA Sadowski. Determination of vitamin K compounds in plasma or serum by high-performance liquid chromatography using postcolumn chemical reduction and fluorimetric detection. Vitam Coenzymes PT L 282:408–421, 1997.

126. HC Buitenhuis, BAM Soute, C Vermeer. Comparison of the vitamins K_1, K_2 and K_3 as cofactors for the hepatic vitamin K-dependent carboxylase. Biochim Biophys Acta 1034(2):170–175, 1990.

127. A Hiraishi. High-performance liquid chromatographic analysis of demethylmenaquinone and menaquinone mixtures from bacteria. J Appl Bacteriol 64(2):103–105, 1988.

128. Z Liu, T Li, J Li, E Wang. Detection of menadione sodium bisulfite (vitamin K_3) by reversed-phase high performance liquid chromatography with series dual-electrode amperometric detector. Anal Chim Acta 338:57–62, 1997.

129. H Wakabayashi, M Nakajima, S Yamato, K Shimada. Determination of idebenone in rat serum and brain by high-performance liquid chromatography using platinum catalyst reduction and electrochemical detection. J Chromatogr 573(1):154–157, 1992.

130. L Mandelbrot, M Guillaumont, M Leclercq, JJ Lefrere, D Gozin, F Daffos, F Forestier. Placental transfer of vitamin K_1 and its implications in fetal hemostasis. Thromb Haemost 60(1):39–43, 1988.

131. WE Lambert, AP De Leenheer. Simplified post-column reduction and fluorescence detection for the high-performance liquid chromatographic determination of vitamin $K_{1(20)}$. Anal Chim Acta 196:247–250, 1987.

132. EAM Cornelissen, AF van Lieburg, K Motohara, CG van Oostrom. Vitamin K status in cystic fibrosis. Acta Paediatr 81(9):658–661, 1992.

133. EAM Cornelissen, LAA Kollee, TGPJ van Lith, K Motohara, LAM Monnens. Evaluation of a daily dose of 25 µg vitamin K_1 to prevent vitamin K deficiency in breast-fed infants. J Pediatr Gastroenterol Nutr 16(3):301–305, 1993.

134. EAM Cornelissen, LAA Kollee, RA De Abreu, K Motohara, LAM Monnens. Prevention of vitamin K deficiency in infancy by weekly administration of vitamin K. Acta Paediatr 82(8):656–659, 1993.

135. O Amedee Manesme, WE Lambert, D Alagille, AP De Leenheer. Pharmacokinetics and safety of a new solution of vitamin $K_{1(20)}$ in children with cholestasis. J Pediatr Gastroenterol Nutr 14(2):160–165, 1992.

136. M Guillaumont, H Weise, L Sann, B Vignal, M Leclercq, A Frederich. Hepatic concentration of vitamin K active compounds after application of phylloquinone to chickens on a vitamin K deficient or adequate diet. Int J Vitam Nutr Res 62(1): 15–20, 1992.

137. M Guillaumont, L Sann, M Leclercq, L Dostalova, B Vignal, A Frederich. Changes in hepatic vitamin K1 levels after prophylactic administration to the newborn. J Pediatr Gastroenterol Nutr 16(1):10–14, 1993.

138. MF Bean, SL Pallante-Morell, DM Dulik, C Fenselau. Protocol for liquid chromatography/mass spectrometry of glutathione conjugates using postcolumn solvent modification. Anal Chem 62(2):121–124, 1990.

139. Y Sano, K Kikuchi, K Tadano, K Hoshi, Y Koshihara. Simultaneous determination of MK4 and its metabolite in human osteoblasts by high performance liquid chromatography/atmospheric pressure chemical ionization tandem mass spectrometry. Anal Sci 13:67–73, 1997.

140. LR Phillips, JE Bartmess. Electrochemically assisted fast atom bombardment negative ion mass spectrometry of quinones. Biomed Environ Mass Spectrom 18(10): 878–883, 1989.

141. HJ Leis, G Fauler, W Windischhofer. Stable isotope labeled target compounds: preparation and use as internal standards in quantitative mass spectrometry. Curr Org Chem 2(2):131–144, 1998.

142. G Fauler, HJ Leis, J Schalamon, W Muntean, H Gleispach. Method for the determination of vitamin $K_{1(20)}$ in human plasma by stable isotope dilution/gas chromatography/mass spectrometry. J Mass Spectrom 31:655–660, 1996.

143. W Vetter, M Vecchi, H Gutmann, R Rüegg, W Walther, P Meyer. Gas-chromatographische und massenspektrometrische Untersuchung von Phytylubichinon, Vitamin K_1 and Vitamin K_2. Helv Chim Acta 50(7):1866–1879, 1967. (In German.)

144. AF Yassin, KP Schaal, H Brzezinka, M Goodfellow, G Pulverer. Menaquinone patterns of *Amycolatopsis* species. Int J Med Microbiol 274(4):465–470, 1991.

145. MD Collins, F Fernandez, OW Howarth. Isolation and characterization of a novel vitamin K from eubacterium lentum. Biochem Biophys Res Commun 133(1):322–328, 1985.

146. OW Howarth, E Grund, RM Kroppenstedt, MD Collins. Structural determination of a new naturally ocurring cyclic vitamin K. Biochem Biophys Res Commun 140(3):916–923, 1986.

147. KW Davidson, SL Booth, GG Dolnikowski, JA Sadowski. The conversion of phylloquinone to 2′, 3′-dihydro-phylloquinone during the hydrogenation of vegetable oils. J Agric Food Chem 44:980–983, 1996.

148. M Imanaka, M Kadota, K Kumashiro, T Mori. Identification of phylloquinone (vitamin K_1) as an unknown peak in electron capture detection gas chromatograms of pyrethroid insecticide residues. J AOAC Int 79(2):538–543, 1996.

149. H Gleispach, HJ Leis, W Windischhofer, W Muntean. A contribution to the measurement of Vitamin $K_{1(20)}$ by mass spectrometry. J Hi Res Chrom 16:738–740, 1993.

5
Ascorbic Acid

Kristiina Nyyssönen and Jukka T. Salonen
University of Kuopio, Kuopio, Finland

Markku T. Parviainen
Kuopio University Hospital, Kuopio, Finland

I. INTRODUCTION

A. History

The necessity of vitamin C (ascorbic acid) for human health is firmly established. As humans are not able to synthesize ascorbic acid, they are dependent on their dietary intake. The dietary sources of vitamin C are fruits and vegetables, especially in uncooked forms. The historical discovery of the beneficial effects of fruits as food dates back to the Middle Ages. Scurvy was common among sailors during the long sea expeditions of the 15th and 16th centuries. The sailors suffered from symptoms of scurvy: capillary hemorrhages, bleeding gums and loosening of teeth, reduced rate of wound healing, depression, and fatigue (1). Vasco da Gama, for example, lost about 100 of his 160 seamen in his India passage between 1497 and 1499. As late as 1740, the British admiral Anson lost five of his six ships and 1165 of 1500 seamen before reaching the coast of South America. During wars in the 19th century, when food shortage was acute, scurvy was also a problem.

In 1753, James Lind published a book about scurvy. His classical study of prevention of scurvy is regarded as the first controlled clinical trial (2). He divided 12 sailors with scurvy into six groups to receive either wine, diluted sulfuric acid with ginger and cinnamon, vinegar, sea water, oranges and lemons, or nutmeg and garlic daily. The result was that only the men receiving oranges and lemons recovered from the scurvy.

In 1928, Szent-Györgyi isolated the nutritional factor with antiscorbutic activity ("ascorbutic" vitamin) and named it as "hexuronic acid" on the basis of its chemical properties. In 1933, the name of vitamin C for hexuronic acid was first introduced, and soon after that, its chemical structure was determined. It has been agreed that "vitamin C" is the generic descriptor for all compounds exhibiting qualitatively the biological activity of ascorbic acid (3).

At the present time, scurvy is very rare. Fruits and vegetables are available throughout the year in every industrial country to prevent the clinical symptoms of scurvy. An adult requires 10 mg/day of dietary ascorbic acid to avoid scurvy (4). The U.S. Recommended Daily Allowance (RDA) is 60 mg/day; however, tissue saturation appears to require an ascorbic acid intake of 100 mg/day (5). Recently published studies have been interpreted to provide evidence that vitamin C at intakes higher than the current recommendations might improve a number of functions in the human body and might reduce the risk of some chronic degenerative diseases such as cataract, cancer, and cardiovascular diseases (6,7). The mechanism of these beneficial functions of ascorbic acid has been proposed to be its ability to prevent or stop oxidative free-radical attacks in the human body (8). We found that plasma ascorbic acid concentration is the strongest determinant of serum lipid resistance to oxidation and plasma antioxidative capacity (9). Linus Pauling hypothesized that high doses of vitamin C might prevent colds and influenza (10), but this effect of ascorbic acid is still to be proved.

B. Biosynthesis and Chemical Structure of Ascorbic Acid

Ascorbic acid (2,3-endiol-L-gulonic acid-γ-lactone, L-ascorbic acid) is biosynthesized in all chlorophyll-containing plants and in the liver and kidney of most mammals, amphibians, reptiles, and most birds. In plants, two major biosynthetic pathways have been presented and named as the glucose-glucuronic- gulonic path and the galactose-galacturonate-galactonolactone path. In most animals, ascorbic acid is formed from D-glucose (Fig. 1) in the liver and kidney. The ability to synthesize ascorbic acid is lacking in the insects, invertebrates, most fish, and humans. The last enzymatic reaction for the biosynthesis of ascorbic acid in the microsomal fraction of liver, the oxidation of L-gulono-γ-lactone to 2-keto-L-gulono-γ-lactone by the enzyme L-gulonolactone oxidase (E.C. 1.1.3.8), is absent in primates and guinea pigs. It has also been suggested that L-gulonolactone oxidase could be inactivated by the active reaction byproducts. For example, hydrogen peroxide, formed in the reaction of L-gulonolactone oxidase (Fig. 1), is likely to deactivate its parent enzyme, thereby tending to inhibit this step in ascorbate biosynthesis.

L-ascorbic acid is the naturally occurring form of ascorbic acid and has the most biological activity. The D-ascorbic acid as well as D- and L-isoascorbic acids (erythorbic acid, IAA) have only marginal vitamin C activity. However, IAA is used in the food industry as an antioxidant, even though it has only about 5%

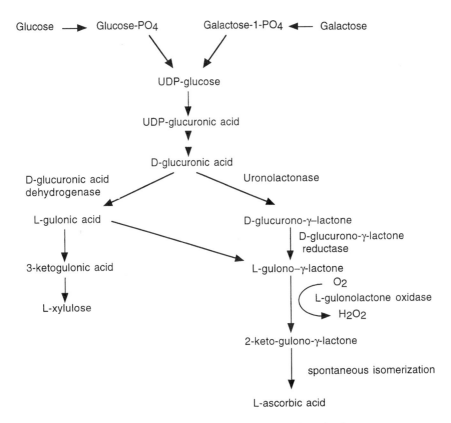

Figure 1 Biosynthesis of L-ascorbic acid and L-xylulose in animals.

of the activity of L-ascorbic acid (11). The oxidized form of L-ascorbic acid is the dehydroascorbic acid (DHA) (Fig. 2). This is very unstable in aqueous solution and is degraded by hydrolysis to 2,3-diketo-L-gulonic acid and further transformed and degraded to several compounds (12). The fatty acid esters of ascorbic acid, particularly ascorbyl palmitate, are used as antioxidants in fatty foods because of their lipophilic character (11).

The L-ascorbic acid that we use as a supplement in tablets or capsules and food industry uses as an antioxidant is synthesized by the classic Reichstein synthesis from D-glucose to D-sorbitol followed by oxidation of D-sorbitol to L-sorbose by *Acetobacter suboxydans* in submersion culture; the production of diacetone L-sorbose and its oxidation with permanganate in alkaline solution to the carbonic acid; hydrolytic cleavage of acetone to form 2-oxo-L-gulonic acid; and enolization to ascorbic acid.

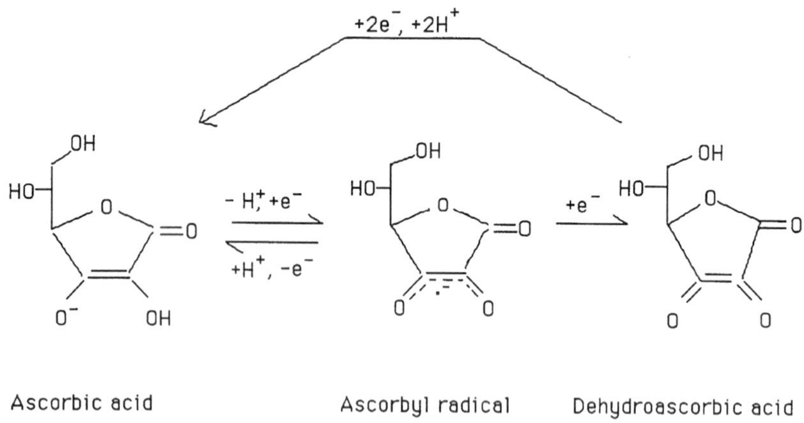

Figure 2   Oxidation of ascorbic acid to dehydroascorbic acid.

The molecular weight of ascorbic acid is 176.13 g/mol. It contains a double-bond between C-2 and C-3 carbons and a ring configuration (Fig. 2). The double bond endows it with the properties of optical absorbance and optical rotation. Ascorbic acid is ionized in two stages as the pH of its aqueous solution is raised on addition of alkali. The first pK value is 4.18, and the second pK value is 11.6. The ionizations take place at the C-2 OH and the C-3 OH sites. The double negatively charged ascorbate anion decomposes quickly to form DHA.

C.   Biological Function of Ascorbic Acid

1.   Reducing Properties of Ascorbic Acid

The oxidation of ascorbic acid is greatly favored by the presence of oxygen, oxygen radicals or metal ions, especially Fe^{3+} and Cu^{2+}, and alkaline pH. In the absence of catalytic metals ascorbate is stable at neutral pH (13). The most important reducing property of ascorbic acid in biological systems is the radical chain terminating reaction (reaction 1).

$$AH^- + \cdot OH \rightarrow A^{\cdot -} + H_2O \qquad\qquad (reaction\ 1)$$

In reaction 1, AH^- is the ascorbate anion. $A^{\cdot -}$ formed in this reaction is the ascorbyl radical (Fig. 2), and $\cdot OH$ is hydroxyl radical.

The reactive forms of oxygen, superoxide ($O_2^{\cdot -}$), $\cdot OH$, and peroxyl radicals are mostly formed in the mitochondrial electron transfer chain for the reduction of molecular oxygen to water. These reactions are necessary for all aerobic cells. In a reasonable amount, they are useful to destroy bacteria or tumoral cells in phagocytosis, to modulate vasodilatation of capillaries or to stimulate the growth of cells (14). Paradoxically, the radical forms of oxygen when produced in excess are toxic for cells. Oxygen radicals can oxidize lipids, proteins, sugars, other carbohydrates, DNA, and RNA, which then lose their physiologic function. If there are not enough protective antioxidants present to defend against "free-radical stress" in the cells, the radical chain reaction can cause harmful effects and even cell death (15). Extracellularly, radical reactions can cause defects in the functions of cell membrane (14).

$A^{\cdot-}$ is reactive and can react with another radical to yield dehydroascorbic acid (A):

$$A^{\cdot-} + {}^{\cdot}OH \rightarrow A + OH^- \hspace{3cm} \text{(reaction 2)}$$

Thus 2 moles of hydroxyl radical are reduced for every mole of ascorbate consumed (16). On the other hand, Wayner et al. (17) have found that at least in vitro, at high concentrations of ascorbic acid a mole of ascorbate can trap less than 2 moles of radicals because ascorbic acid is converted to ascorbyl radical and dehydroascorbic acid. Thus, the reducing capacity of ascorbic acid in vitro is concentration dependent: as ascorbic acid concentration increases, its reducing capacity decreases.

The existence of the ascorbyl radical in biological fluids has been proved by electron spin resonance (ESR) studies in vitro (18,19) and in vivo (20).

2. Pro-oxidant Properties

Ascorbic acid can in certain conditions act as a pro-oxidant and promote the generation of the same active oxygen species (${}^{\cdot}OH$, $O_2^{\cdot-}$, and H_2O_2) it is considered to destroy. It is generally agreed that this pro-oxidant activity is derived from the ability of ascorbic acid to reduce transition metals Fe^{3+} or Cu^{2+} by a 1-electron mechanism:

$$AH^- + Fe^{3+} \rightarrow A^{\cdot-} + Fe^{2+} \hspace{3cm} \text{(reaction 3)}$$

or by a 2-electron mechanism:

$$AH^- + O_2 + H^+ \rightarrow H_2O_2 + A \hspace{3cm} \text{(reaction 4)}$$

The formation of Fe^{2+} (ferrous ion) and H_2O_2 gives rise to the Fenton reaction (reaction 5), in which iron is oxidized and active hydroxyl radical is formed (21).

$$Fe^{2+} + H_2O_2 \rightarrow Fe^{3+} + {}^{\cdot}OH + OH^- \hspace{3cm} \text{(reaction 5)}$$

In addition to iron, also copper, titanium, cobalt, chromium, vanadium, and nickel can react similarly. This reaction in vivo is thought to require that there is free (non-protein-bound) metal available. The prominent view is that metal bound to protein cannot catalyze the reaction (22). It has also been suggested that some transition metals, e.g., iron, that are chelated to small peptides, could catalyze the Fenton reaction (23).

Halliwell (22) suggests that the possible in vivo pro-oxidant effects of ascorbate are related to the availability of catalytic transition metal ions. The content of vitamin C in meals increases nonheme iron absorption (24). In patients with iron accumulation diseases such as hemochromatosis or thalassemia, this might lead to increased iron overload and deleterious clinical effects (22). Non-protein-bound iron, as far as it exists in the human body, can induce lipid peroxidation especially if it is present together with the pro-oxidative ascorbic acid (reactions 3 and 5). According to previous reviews (6,24), vitamin C ingestion enhances the iron absorption also in individuals with iron deficiency, but may have a rather small effect in individuals with normal iron status.

3. Regeneration of Ascorbic Acid

When ascorbic acid becomes oxidized, the DHA that is formed can be reduced back to ascorbic acid in the presence of a suitable reductant (Fig. 3). Two glutathione (GSH) molecules can reduce one DHA molecule to ascorbic acid since this reaction is energetically feasible (25,26). GSH is an important antioxidant in

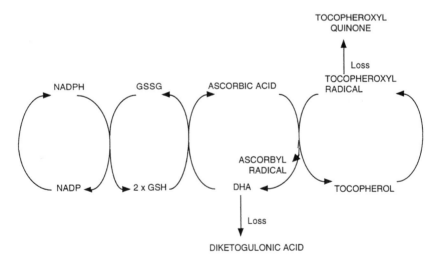

Figure 3 Cyclic reactions among glutathione, ascorbic acid, and tocopherol. NADPH, nicotinamide-adenine-dinucleotide phosphate, reduced form; NADP, nicotinamide-adenine-dinucleotide phosphate, oxidized form; GSH, glutathione; GSSG, glutathione disulphate; DHA, dehydroascorbic acid. (From Ref. 26.)

cells. It maintains high concentrations of ascorbic acid in cells by reducing DHA to ascorbic acid. In neutrophils (27) and in erythrocytes (28), DHA, but not ascorbic acid, is transported through cell membrane by glucose transporters. Intracellular GSH is mainly responsible for the reduction of DHA to ascorbic acid. This leads to the accumulation of ascorbic acid in cells. This theory also implies that oxidation of extracelluar ascorbate is essential for the high levels of intracellular concentration. The activation of neutrophils produces compounds with radical activity (27). These radicals, when released into the extracellular medium, are able to oxidize ascorbic acid to DHA that is taken up into cells and recycled to ascorbic acid.

The existence of a DHA reductase enzyme in animal or human tissue has been a theoretical possibility. Recently, Park and Levine have purified and expressed a glutaredoxin with dehydroascorbic acid–reducing activity from human neutrophils (29). They found that glutaredoxin was responsible for most of the protein-mediated DHA reduction in lysates. DHA reduction was at least fivefold greater in neutrophil lysates than in myeloid tumor cell lysates where glutaredoxin could not be detected (29). However, it may be still possible that more than one enzyme participates in the process (30). On the other hand, it seems that a nonenzymatic, direct reaction between GSH and DHA is the major physiologically relevant process underlying the reduction of DHA to ascorbic acid in mammalian tissues (26).

4. Cofactor in Enzymatic Systems

Ascorbic acid is required for many hydroxylase enzymes in the human body. Ascorbic acid is needed for conversion of tyrosine to the neurotransmitter dopamine and further hydroxylation to adrenaline and noradrenaline, for synthesis of carnitine from lysine, and probably for hydroxylation of steroid hormones. Ascorbate is also known to participate in hydroxylation of aromatic drugs and carcinogens via microsomal mono-oxygenase systems of liver endoplasmic reticulum (31,32). Its role in the formation of collagen is thought to be to maintain iron in its ferrous state for an iron-dependent proline hydroxylase, or to act as a direct source of electrons for reduction of O_2 (31).

II. THIN-LAYER CHROMATOGRAPHY

Paper and thin-layer chromatography (TLC) was applied in early analytical studies to increase selectivity in the analysis of ascorbic acid and other similar chemical compounds. Nowadays these methods have little if any qualitative or quantitative use. However, TLC has proven to have some advantages in the analysis of vitamin C compounds, especially in pharmaceuticals, which often consist of a quite few compounds. El Sadek et al. (33) described a high-performance TLC method for the determination of the components of analgesic mixtures. They used silica gel precoated plates for separating paracetamol, ascorbic acid, caffeine, and phenylephrine. The plates were scanned with a scanning spectrophotometer at the wavelength of 264 nm for ascorbic acid, 254 nm for paracetamol, and 274 nm for phenylephrine and caffeine. However, the peak height versus concentration was not linear at concentrations over 4 µg/mL of ascorbate.

III. HIGH-PERFORMANCE LIQUID CHROMATOGRAPHY

A. Introduction

At the present time, high-performance liquid chromatography (HPLC) is widely used for determining ascorbic acid and its degradation products in foods (34–37), body fluids (37–42), tissues (43–49), and pharmaceuticals (48). However, direct spectrophotometric (49–52), fluorometric (53), or electrochemical (54–57) determinations are still widely used, especially in the assays of ascorbic acid in non-biological materials such as pharmaceuticals. Biological samples often contain unknown compounds that can interfere in spectrophotometric, fluorometric, and electrochemical detection if the method includes no chromatographic separation. Recently, capillary electrophoresis has been introduced as a choice for analyzing ascorbic acid and its metabolites both in pharmaceuticals and in biological samples (58–63).

B. Sample Preparation

Ascorbic acid in biological samples is readily oxidized by oxygen, temperature, light, pH, ionic strength of solvent, oxidizing enzymes, and transition metals—e.g., iron and copper (8,27). DHA formed in the oxidation of ascorbic acid is spontaneously and irreversibly oxidized to 2,3-diketogulonic acid. It is apparent that the reduced form of ascorbic acid and DHA are both present in most biological samples. The equilibrium between the two substances is dependent on the sample source and sample handling. An analytical goal is to measure what the sample originally contains without artificially shifting the equilibrium between the two substances. Thus, it is essential to test the stability of ascorbic acid and DHA during sample handling and storage.

An ascorbic acid standard solution of 5 mg/mL has a pH value of 3. Thus, the stock standard can be made in distilled water. For dilute standard solutions, it is advisable to make the standard in acid, such as 5% metaphosphoric acid (MPA). For extracting biological samples, acids are used to precipitate proteins, to stabilize ascorbic acid and DHA and to remove interfering substances. MPA (62–64), trichloroacetic acid (TCA) (27,67), or perchloric acid (PCA) (53,68,69) are the most commonly used stabilizers and protein precipitators. The dilutions of MPA and TCA should be fresh; otherwise oxides may be formed and ascorbic acid degradation may occur. EDTA is added to many extracting solutions to chelate divalent metal cations (43,70), which otherwise could accelerate ascorbic acid oxidation and dehydroascorbic acid hydrolysis. Compounds containing sulf-hydryl groups, e.g., homocysteine (71), dithiothreitol (72), or L-cysteine (73), have been added to samples to stabilize ascorbic acid, but these compounds may also convert DHA to ascorbic acid and total ascorbic acid will be determined instead of reduced ascorbic acid. Since extractants and chelators can also interfere with assays, these agents must be selected carefully.

The recovery of ascorbic acid has been tested by adding ascorbic acid into plasma and deproteinizing plasma with either MPA, TCA, or acetic acid (74). The recoveries were 93.9% for MPA, 82.6% for TCA and 62.3% for acetic acid. In that study, 5% MPA with 0.5% thioglycolic acid was the most effective in recovering essential amounts of ascorbic acid in plasma.

Acidification should be done promptly after sampling. For blood plasma samples, we separate plasma by centrifugation and stabilize it in MPA within 30 min from blood drawing (75). Freezing the samples increases the stability of ascorbic acid, but without acidification this cannot fully prevent degradation, which is also dependent on the type of the sample (plasma, tissue, food) (27). Liau and coworkers found that ascorbic acid was more stable in whole blood than in plasma when stored below +4°C (76). At −20°C deproteinized plasma in a perchloric acid/MPA mixture was stable up to six days. Esteve et al. (77) stored heparin plasma and MPA deproteinized plasma at −18°C for 2, 4, 6, and 8 days. Heparin plasma was stable for the first 2 days at −18°C, but in the MPA extract there was no degradation of ascorbic acid during the 8 days. There are only a few reports concerning the long-term stability of ascorbic acid during freezing. Margolis and Davis (78) observed that ascorbic acid in plasma can be preserved for 88 weeks at −70°C by adding dithiothreitol. They also tested the stability of ascorbic acid in plasma samples treated with an equal volume of 10% metaphosphoric acid. During freezing at −70°C for 6 years, only 1% of ascorbic acid was degraded each year (79). We have found that plasma ascorbic acid is stable for 18 months at −80°C (80) and at least 1 day in room temperature when stabilized in 5% MPA (1 + 9).

In liver tissue, ascorbic acid has been found to be stable without any stabi-lizing agent for many hours at +5°C (69). This is possibly due to the high levels of endogenous low-molecular-weight reductants, such as glutathione, which are highly concentrated in liver.

In addition to stabilization with acid, methanol has been used (81). Dhari-wal et al. (81) extracted human neutrophils with 60/40 methanol/water containing 1 mM EDTA. However, in methanol/water, ascorbic acid was stable for only some days at −80°C, and cooling was necessary in the HPLC autosampler. DHA was reduced to ascorbic acid by 2,3-dimercapto-1-propanol which interfered in the chromatogram and had to be removed by extraction with ethyl ether before injection into HPLC.

Iwase and Ono (73) tested the stability of plasma ascorbic acid after deproteinization with small hydroxyapatite cartridges and addition of L-cysteine as stabilizator. They found that in protein-free plasma, ascorbic acid with L-cysteine was stable for 30 min at room temperature and at least for 24 hours at −20°C. Without deproteinization, stored at −20°C for 24 hours, plasma ascorbic acid decreased by 79%.

The low level of DHA in most biological samples causes difficulties in its quantitative measurement with any HPLC detector. Thus DHA is often reduced to ascorbic acid before measurement and total ascorbic acid is measured. This demands two chromatographic runs: for reduced ascorbic acid, and for total ascorbic acid. Homocysteine (82,83), dithiothreitol (66), dithioerytriol (84), β-mercaptoethanol (43), and 2,3-dimercapto-1-propanol (81) are the most usual agents used to reduce DHA before the chromatographic run. The unstable nature of DHA and effect of pH and temperature on DHA degradation should be taken into account in sample handling. It was found that decay of DHA proceeded much more rapidly at high pH (7–8) than at low pH (3–5) and was more rapid at 37° or 45°C than at 0° or 23°C (85).

C. HPLC Conditions

Ion pair reversed-phase chromatography with octadecyl (C18) columns (43,63,66,74,76,86,87) and ion exchange chromatography (37,79,82) are two practical separation techniques. For reversed-phase chromatography, tetrabutyl-ammonium hydroxide (88), tributylamine (86), myristyltrimethylammonium bromide (64,69), and cetyltrimethylammonium bromide (77) are used as ion-pairing agents in the mobile phase, which mainly consists of water and some organic modifier such as methanol or acetonitrile. The pH is usually adjusted between 2.4 and 6.5 with phosphate or acetate buffer. The choice of ion-pairing reagent is important. It must be easily soluble in mobile phase and in sample diluent. For example, the commonly used myristyltrimethylammonium bromide and n-octylamine are precipitated with samples stabilized with MPA (66,76).

Manoharan and Schiwille (65) used a triaconyl (C30) column instead of a C18 column in their reversed-phase system. They found that the C30 column gives baseline separation of ascorbate from the MPA peak, whereas on the C18 column ascorbate appears in the tailing of MPA (65,76).

A macroporous copolymer of styrene and divinylbenzene is suitable as a stationary phase for reversed-phase separations. Bilic (34) used a column with a 5-μm polystyrene/divinylbenzene copolymer packing for separation of fluorescent derivatives of ascorbic acid and DHA (see Sect. III.F). Recently, Koshiishi and Imanari introduced a poly(ethylene glycol) copolymer for the separation of ascorbic acid and DHA. They found this copolymer has characteristics of low absorption of proteins and is suitable for separating low molecular weight organic molecules. They used the column for direct injection of human plasma and urine (42).

Usually, in reversed-phase systems, ascorbic acid is eluted within 3 to 10 min. However, as extracting biological samples there may be compounds that elute far after ascorbic acid. Gradient elution may be needed to shorten the column purification time between injections. Tanishima and Kita (74) used a stepwise increasing methanol concentration to remove uric acid from the column after plasma samples. Lazzarino et al. (88) separated ascorbic acid, uric acid, xanthine, hypoxanthine, malondialdehyde, and several nucleotide derivatives with ion pairing and methanol gradient in a reversed phase system. The method was used for determining those compounds in ischemic-reperfused rat heart.

Margolis and Schapira (37) have introduced an anion exchange chromatographic method for determining IAA, DHA and ascorbic acid in biological samples, including human plasma, infant formula mixed-food diet, rat lungs, and serum and perfusate of isolated rat lungs. The mobile phase in their system consisted of monobasic potassium phosphate, water, acetonitrile, and phosphoric acid. Sometimes in silica-based aminopropyl columns used in anion exchange mode, ascorbic acid can partly be oxidized by the column matrix and broad peaks may appear (89). However, in the amino column, homocysteine used for reduction of DHA does not interfere with ascorbic acid peak, whereas in ODS columns it may disturb the separation.

D. Electrochemical Detection

Electrochemical detection of ascorbic acid is based on the oxidation of ascorbic acid to DHA (Fig. 2). Amperometry and coulometry are the measurements of current at a constant electrode potential. The main difference between these two measurements is the amount of analyte oxidized in the detector: in amperometry the oxidation and current are limited; in coulometry, the analyte is totally oxidized. The structure of an amperometric detector is usually a ''flow-by'' cell, whereas in coulometry a porous ''flow-through'' cell is used. In coulometry, a higher amount of analyte is allowed in contact with the electrode surface and sensitivity increases. Working with electrochemical detector, the components of mobile phase must allow for distinct separation of ascorbic acid and be conductive to carry the charge of the analyte. However, the mobile phase must not yield too high background signal.

The successful use of an electrochemical detector requires knowledge of the appropriate potential to affect the electrochemical reaction of the analyte. This potential depends on many factors including the nature of the electrode surface, the composition and pH of the mobile phase, and the chemical structure of the analyte. Thus a current/voltage curve, called hydrodynamic voltammogram (HDV) should be determined for ascorbic acid when the HPLC system with electrochemical detector is initially set up (43). From the curve, the adequate electrode potential for ascorbic acid detection can be determined.

The potential for detection of ascorbic acid in amperometric or coulometric detectors is between 70 and 700 mV (41,73,76,87). To eliminate high background current derived from the mobile phase, a cell between pump and injector with adequate oxidizing potential can be used (66). Lykkesfeldt et al. (66) had a coulometric detector with a ''guard'' cell operated at 200 mV for oxidizing interfering compounds of the mobile phase, and an analytical cell operated at 100 mV for detection of ascorbic acid. The mobile phase for the reversed-phase column consisted of disodium hydrogen phosphate buffer with EDTA and dodecyltrimethylammonium chloride, pH 3.0. They used dithiothreitol (DTT) for reducing plasma DHA to ascorbic acid before being injected to the HPLC system. DHA is electrochemically inactive, and cannot be measured directly by electrochemical detectors. DHA was calculated as the difference between total and reduced ascorbic acid. For confirmation of the peak identity, quantitative oxidation of ascorbic acid was performed by addition of copper(II)bromide or ascorbic acid oxidase (EC 1.10.3.3) to the sample (Fig. 4).

Daily et al. (90) described an automatically controlled system with enzyme reactor bed where in the first injection, the sample is passed through an enzyme packed bed that had been previously heat-denatured. In the second injection, ascorbic acid is eliminated by ascorbate oxidase before amperometric detection. The difference between the two signals generated at the detector is related to the concentration of ascorbic acid in the sample. The method was adapted for foodstuffs.

Tsai et al. (41) measured extracellular ascorbic acid of brain cortex or ventricular myocardium of anesthetized rats. They used an automatic continuous microdialysis system for collecting the perfusates and injecting the dialysates into the HPLC system. The mobile phase was a sodium acetate buffer with EDTA, tetrabutylammonium hydroxide, and 7.5% methanol, pH 4.75. HPLC comprised of a C18 column and an amperometric detector set to 600 mV versus Ag/AgCl electrode.

Rose and Bode (30) used a coulometric dual cell detector at a reducing potential of -100 mV and oxidation potential of 400 to 800 mV for measuring ascorbic acid, glutathione, and uric acid in a single run. The method consisted of a C18 pre- and analytical column, potassium dihydrogen phosphate buffer as the mobile phase, and a coulometric dual cell detector. They found that good sensitivity for ascorbic acid and uric acid is reached at a relatively low setting of oxidizing potential, whereas maximal sensitivity for glutathione needed higher

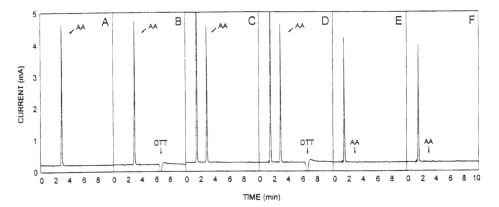

Figure 4 Chromatographic analysis with coulometric detection of ascorbic acid samples. Column was C18, 75 × 3.9 mm i.d., particle size 4 μm, fitted with a C18 precolumn. Mobile phase consisted of 0.1 M disodium hydrogen phosphate, 2.5 mM disodium-EDTA, and 2.0 mM n-dodecyltrimethylammonium chloride, pH 3.0. The detector signal is expressed in mA. (A) Aqueous sample containing 35 μM ascorbic acid (equals an actual injection of 17.5 pmol of ascorbic acid). (B) Aqueous sample reduced with DTT. (C) Typical plasma sample. (D) Plasma sample reduced by DTT. (E) plasma sample after incubation with 5 μM $CuBr_2$ (5 min, 25°C). (F) Plasma sample after incubation with ascorbic acid oxidase (5 min, 25°C). (From Ref. 66.)

voltage. However, they did not prefer to operate at voltages above 600 mV, because the background current became excessive.

Liau et al. (76) compared amperometric and UV detection for ascorbic acid analysis. Their chromatographic system consisted of a C18 column and ammonium dihydrogenphosphate buffer with 0.015% MPA. For electrochemical detection they adjusted the pH of the mobile phase to 2.55 and for UV detection to 2.95. The amperometric detector was set at 700 mV versus Ag/AgCl reference electrode. The UV detector was set at 245 nm. They preferred relatively simple mobile phase without ion-pairing reagents which tended to precipitate with MPA that was present in mobile phase as well as in plasma sample for stabilization. Ascorbic acid was well measurable with both detectors. The detection limit for electrochemical detection was 0.3 ng and for UV detection 1.2 ng per injection. The coefficient of variation (CV) for the between-day assay was <12% using electrochemical detection and <5% using UV detection. They conclude that UV detection is apparently a better choice for fast, routine measurements of ascorbic acid concentrations. Electrochemical detection takes more time to stabilize, but is more sensitive.

DHA converts nonenzymatically to a variety of transformation products in aqueous solution. Kimoto et al. (12) described an ion-pairing HPLC method with electrochemical and UV detection for analyzing electrochemically active erythro-L-ascorbic acid, 2,3-enediolgulonolactone, 5-methyl-3,4-dihydroxytetrone (MDT), 3-hydroxy-2-pyrone, and electrochemically inactive 2-furonic acid. MDT has been recognized as a product of the nonenzymatic browning reaction of ascorbic acid in food stuffs during processing (12).

E. UV Detection

Reduced ascorbic acid has its optical absorbance maximum at 245 to 270 nm depending strongly on pH. At pH 2, the absorption maximum is at 245 nm, and at pH 6.4 it is at 265 nm. Since fixed-wavelength 254-nm mercury lamp detectors are relatively inexpensive, 254-nm UV detectors are commonly used for ascorbic acid measurement. DHA absorbs at 210 to 227 nm, which limits the choice of the sample matrix, solvents, buffers, and other reagents used. Moreover, DHA degradation products may interfere with DHA in the chromatogram (12). Thus the direct UV detection of DHA is complicated, and usually DHA is reduced to ascorbic acid before chromatographic separation and measured as total ascorbic acid (43,66,81,82).

Esteve et al. described a method for determining ascorbic acid from blood plasma and serum at the wavelength of 254 nm (Fig. 5). They used 4-hydroxyacetanilide as an internal standard and deproteinized sample with 10% MPA (1 + 1). The sensitivity of the detection was 31 ng/mL. They tested the possible effect of hemolysis on the chromatogram and found no interference with the ascorbic acid peak (77). Tanishima and Kita used the wavelength of 265 nm that was also suitable for detecting maleic acid that was added as internal standard for ascorbic acid extraction and chromatography (74).

The wavelength of 254 nm was also used for detecting ascorbic acid from mice plasma and testis (44). The pH of the mobile phase was 3.1. The detection limit was carefully determined as 174 ng/mL for plasma and testis ascorbic acid.

In addition to ascorbic acid, uric acid is an important antioxidant in blood plasma (9). To assess the antioxidative capacity of plasma it is useful to determine ascorbic acid and uric acid in a single chromatographic run. Ascorbic acid and uric acid have been measured simultaneously from heparinized plasma by UV detection (64). To optimize the detection during method setup, the sample was dissolved in the mobile phase, pH 5.5, where ascorbic acid had the absorbance maximum at 262 nm and uric acid at 285 nm. The wavelength of 262 nm was chosen because ascorbic acid is present at a lower concentration than uric acid in plasma. The same mobile-phase pH (5.5) and UV detection at 280 nm for

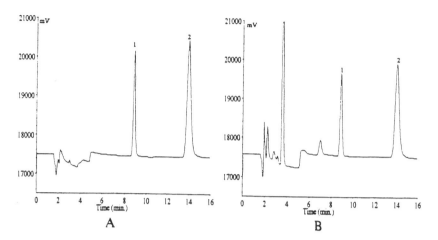

Figure 5 Chromatograms of ascorbic acid standard (A) and plasma sample (B). UV detection at 254 nm. Peak 1 = ascorbic acid; peak 2 = internal standard (4-hydroxyacetanilinide). Chromatographic conditions: C18 column 250 × 4 mm i.d., 50 mM potassium dihydrogen phosphate, and 5 mM cetyltrimethylammoniumbromide, pH 4.5, as mobile phase. Reprinted (From Ref. 77.)

simultaneous ascorbic acid and uric acid determination in animal tissues have been used by Barja de Quiroga et al. (69). The detection limit was determined as 3.3 μmol/L for ascorbic acid and 0.5 μmol/L for uric acid (69).

Malondialdehyde is an end product of lipid peroxidation. For assessing the oxidative/antioxidative state of biological samples, simultaneous determination of malondialdehyde and ascorbic acid has been described (88). The detection was at 266 nm in mobile-phase pH 5.5. The method was adapted for ischemic and reperfused rat heart and human erythrocytes after in vitro peroxidation. However, it was obvious that the method was not sensitive enough for malondialdehyde measurements without strong in vitro induction of malondialdehyde generation.

Combined UV and electrochemical detection was used for determining ascorbic acid and DHA in small animal and brain tissues. Ascorbic acid was detected amperometrically and DHA at the wavelength of 215 nm in a single run. However, quantitative measurement of DHA might be difficult with this method, because of several interfering compounds absorbing at this wavelength (86).

Yasui and Hayashi introduced a postcolumn derivatization method for simultaneous determination of ascorbic acid and DHA at a wavelength of 300 nm

(91). Sodium hydroxide together with sodium borohydride were used for oxidizing ascorbic acid and DHA to products that have the absorbance maximum at higher wavelength than the native compounds. The derivatives could not be identified due to their unstable chemical characters, but their spectra were identical. The method was adapted for determining ascorbic acid and DHA in tomato juice (91) and in fish tissues (45).

F. Fluorometric Detection

Most of the fluorometric HPLC methods are based on oxidation of ascorbic acid to DHA, which is then condensed to o-phenylenediamine to form a quinoxaline derivative, fluorophor. Precolumn oxidation of ascorbic acid yields measurement of total ascorbic acid. If the aim is to measure ascorbic acid and DHA separately, either the combination of UV and fluorescence detectors (35,48), two injections into the HPLC (92), or postcolumn derivatives (42) are needed. Vanderslice and Higgs (35) used a reversed-phase HPLC system for column separation of ascorbic acid and DHA, postcolumn oxidation with $HgCl_2$ and subsequent reaction with o-phenylenediamine to obtain the fluorophores. In their method, IAA was used as an internal standard and it behaved as an indicator for possible oxidation of ascorbic acid occurring during extraction and quantitation. The disadvantage of this postcolumn derivatization is the need of a long reaction coil (32 cm), a heating coil (45.7 m), a cooling bath, and a cooling coil. The method was adapted for the measurement of ascorbic acid and DHA contents of several food sources. Ali and Phillippo (36) used the same method for determining ascorbic acid and IAA content of multicomponent meat-based food products. As mentioned earlier, IAA is a commonly used additive in meat products as an antioxidant. Thus, for ascorbic acid analysis in meat products, IAA cannot be used as an internal standard. Recoveries were 102.5% and 83.5% for ascorbic acid and IAA, respectively, after adding known amounts of ascorbic acid and IAA to ground beef samples (36).

Quite a complicated derivatization process was introduced by Bilic (34). Two-step sequential derivatization was applied to the simultaneous assay of both ascorbic acid and DHA by HPLC. First, the sample was derivatized with 4-ethoxy-1,2-phenylenediamine, followed by isolation of the products of derivatization and removal of excess of reagent by small C18 silica gel and cation exchange columns. Second, the ascorbic acid in the sample was oxidized to DHA and then derivatized with 4-methoxy-1,2-phenylenediamine. In the first step, ascorbic acid remains constant during derivatization of DHA. This was obtained by performing the derivatization of DHA at pH 2. The methoxyquinoxaline (corresponding ascorbic acid) and ethoxyquinoxaline (corresponding DHA) derivatives were separated by reversed-phase HPLC.

Kmetec (48) determined ascorbic acid, DHA, acetylsalicylic acid, and its degradation product, salicylic acid, from pharmaceuticals. He used UV detection for the measurement of ascorbic acid, acetylsalicylic acid, and fluorescence detection with the exitation wavelength set at 350 nm and emission wavelength set at 430 nm for measurement of the quinoxaline derivative of o-phenylenediamine and DHA. The o-phenylenediamine reagent was eluted as an unretained compound in the amino column thus causing no interference with other compounds in either UV or fluorescence detection. The detection limit was about 0.002 mg/mL of sample, the injection volume was 20 μL.

Koshiishi and Imanari (42) used the postcolumn ascorbic acid oxidation with cupric acetate to DHA and o-phenylenediamine for derivatization. Ascorbic acid and DHA were assessed in a single run (Fig. 6). Their system consisted of a double-plunger pump for pumping o-phenylenediamine and cupric acetate for postcolumn reactions. The fluorescence detector was set at exitation wavelength of 345 nm and emission wavelength of 410 nm. They introduced the poly(ethylene glycol) copolymer as separation material in the analytical column (see Sect. III.C).

IV. GAS LIQUID CHROMATOGRAPHY

Gas chromatography is rarely used for ascorbate determination. However, ascorbic acid from calf brain has been characterized by mass spectral analysis (93), and Niemelä used gas liquid chromatography with mass spectrometry (GC-MS) for >50 trimethylsilylated degradation products of ascorbic acid (94). Recently, an isotope dilution assay with mass spectrometry for ascorbic acid and DHA has been developed using $[^{13}C_6]$ascorbic acid and $[^{13}C_6]$- and $[6,6-^2H_2]$dehydroascorbate (95). The method used *tert*-butyldimethyl derivatives of ascorbic acid and DHA, and was adapted for blood plasma analyses. Briefly, internal standards were added into plasma and plasma proteins were precipitated by the addition of TCA. The sample was then extracted to remove TCA, derivatized, and analyzed by GC-MS. The isotope dilution assays for unstable compounds, such as ascorbate, are useful due to the sensitivity and specificity of these techniques and also because degradation of the endogenous compound should also cause degradation of the internal standard without changing the ratio between internal standard and the endogenous substance. A major problem with the GC-MS, and probably with all other quantitative methods, is the chemical impurity of the available DHA standards (95). Although the sensitivity of GC-MS is adequate to examine small concentrations of DHA, potential error in absolute quantitations could occur when the content is related to an added impure standard. This disadvantage can be passed by using prechromatographic reduction of DHA to ascorbic acid.

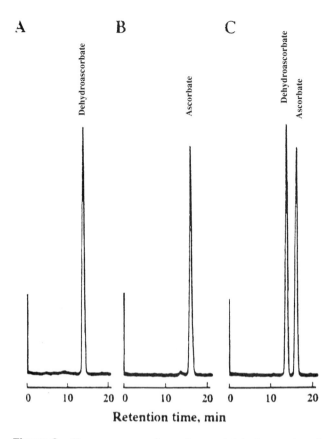

Figure 6   Chromatograms of ascorbate and dehydroascorbate in standard (A), human plasma (B), and human urine (C). Conditions: GS-320H poly(ethylene glycol) copolymer column; 0.1 M acetic acid containing 0.5 mM EDTA (1.0 mL/min) as mobile phase. The postcolumn reaction conditions were: reagent 1, 0.1 M acetate buffer (pH 4.0) containing 20 mM o-phenylenediamine; reagent 2, 0.1 M acetate buffer (pH 4.0) containing 5 mM cupric acetate; reaction temperature 55°C; reaction time 1 min; sample volume 5 µL; fluorescence detector (excitation 345 nm, emission 410 nm). (From Ref. 42.)

V.   RECENT TECHNIQUES

A.   Capillary Electrophoresis

Capillary electrophoresis (CE) is a technique that combines some characteristics of traditional polyacrylamide gel electrophoresis (PAGE) and modern HPLC. During the last 10 years, the number of CE applications for quantitation of ascorbic acid and related compounds in food, blood and tissue has increased.

This is due to the improvements in CE systems and their stability and also the prices of the instruments have decreased. CE is mostly adapted in pharmaceutical industry for checking the purity and quality control of the pharmaceutics, in addition to HPLC and gas chromatographic methods. The effective separation of compounds in CE has been found a powerful tool also in research laboratories.

Mainly, CE is based on the different electrophoretic mobility of the analytes in certain circumstances. CE employs silica capillary tubing (diameter 20 to 500 µm, length 20 to 60 cm); high electric field strengths (often >500 V/cm); UV, diode array, electrochemical, fluorescence, or mass spectrometric detection; a vacuum pump for washing the capillary; and often an autosampler. The separation of the analytes in the capillary is based on their charge and on the electro-osmotic flow that is used in zone electrophoresis (Fig. 7). The negatively charged wall attracts positively charged ions from the buffer, creating an electrical double layer. When a voltage is applied across the capillary, cations migrate in the direction of the cathode, carrying water with them. The result is a net flow of buffer solution in the direction of the negative electrode. The analytes pass the detector in the order of cations, neutrals, and anions (96). The electro-osmotic flow can be enhanced by washing the capillary with sodium hydroxide for totally filling the capillary surface with negative charges. The use of micelle-forming surfactants in CE buffers can give rise to separations that resemble reversed-phase HPLC with the benefits of CE. Micelle-forming surfactants are long chain molecules with hydrophobic tail and hydrophilic head group, for example pentanesulphonic acid or sodium dodecyl sulphonic acid (SDS), that help to control the electro-osmotic force. To perform techniques such as isoelectric focusing (IEF)

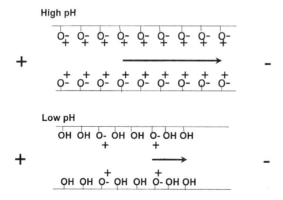

Figure 7 Capillary zone electrophoresis. The arrows indicate the direction and force of electro-osmotic flow (electroendo-osmosis).

or isotachophoresis (ITP), the electro-osmotic force must be suppressed. This is possible if the capillary is uncharged which is obtained by coating the capillary inner surface.

Capillary zone electrophoresis has been applied for measurement of ascorbic acid from fruits (58–60), serum and urine (60), cerebrospinal fluid (61), pharmaceuticals (58,62), plant tissues (63), and rat brain (97). Phosphate or borate are used as running buffers. The pH of the buffer is usually 5 to 9 to achieve the ionization of ascorbic acid. However, the degradation of ascorbic acid at alkaline pH might occur. To prevent that, L-cysteine has been added as an antioxidant for stabilization of ascorbic acid in running buffer (62,98). Sample pretreatment with either MPA or TCA was performed.

Koh et al. used tricine buffer, pH 8.8, uncoated capillary 37 cm $\times$ 75 μm, and 11 kV voltage for separating ascorbic acid and IAA (internal standard) from urine, plasma, and serum (60). The capillary was treated with NaOH before use. They used either MPA or TCA for stabilizing ascorbic acid in the sample before the CE run. However, as found later, TCA may result in the loss of nearly 40% of the ascorbic acid content (63). The between-day coefficient of variation was 1.9% to 3.3%. The detection limit was 1.6 μg/mL and the method was linear to 480 μg/mL (60).

Davey et al. (63) analyzed ascorbic acid and IAA in plant tissues with a 57 cm $\times$ 75 μm uncoated capillary at 25 kV and borate buffer, pH 9. The UV detector was set at 194 nm. Samples were extracted in MPA/EDTA. They studied the effect of the concentration of carrier electrolyte, pH, and SDS on the separation of ascorbate and IAA. Doubling the buffer concentration from 50 mM to 100 mM borate, pH 9, improved the resolution of ascorbate and IAA (Fig. 8). The increase of pH from 7.5 to 9 increased the migration times that was unexpected, because at higher pH, the ionization of the silanol groups of the capillary wall is usually increased, electro-osmotic force increases, and migration times decrease. They explained this discrepancy by the equilibrium between the free and borate-complexed forms of the analyte, which favors complex formation under alkaline conditions. Therefore, the increased migration time observed at high pH is the result of the increased proportion of anionic borate–ascorbic acid complexes that move toward anode and against electro-osmotic force. Increasing SDS

Figure 8   Capillary zone electrophoresis. Influence of carrier electrolyte concentration on the resolution of L-ascorbic acid (L-AA) and D-ascorbic acid (D-AA) dissolved in extraction buffer. Three-second (17.7 nL) hydrostatic injections of 0.05 mg/mL solution of L-AA and D-AA dissolved in 3% MPA/1 mM EDTA. UV detection carried out at 194 nm. Carrier electrolyte was 50 mm borate, pH 9 (A); and 100 mM borate, pH 9 (B). (From Ref. 63.)

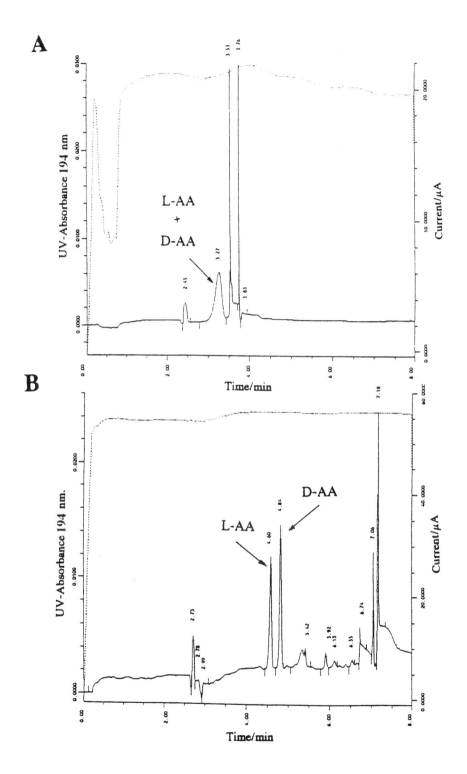

Table 1 Recently Published Detection Limits of Chromatographic Ascorbic Acid Analysis

Mode of separation	Column	Detection	Detection limit	Year (Ref.)
HPLC	Axxi-Chrom C18	Coulometric	1 nmol/L	1990 (81)
HPLC	Waters C18	Coulometric	50 nmol/L	1993 (43)
HPLC	Partisphere 5 C18	Amperometric	1.7 pmol/injection	1993 (76)
		UV 245 nm	6.8 pmol/injection	
HPLC	Nova-Pak C18	Coulometric	20 fmol/injection	1995 (66)
HPLC	Spherisorb S3NH$_2$	UV 254 nm	1.7 µmol/L	1986 (75)
HPLC	Nucleosil C18	UV 280 nm	3.3 µmol/L	1991 (69)
HPLC	Shimpack SCR-102H	UV 300 nm after postcolumn derivatization	4.54 pmol/injection	1991 (91)
HPLC	Tosoh ODS-80TM	UV 265 nm	0.57 µmol/L	1993 (74)
HPLC	µBondapak C18	UV 254 nm	0.99 µmol/L	1993 (44)
HPLC	Triaconyl C30	UV 250 nm	11.4 pmol/injection	1994 (65)
HPLC	Shimpack SCR-101H	UV 300 nm	0.57 µmol/L	1995 (45)
HPLC	Spherisorb C18 ODS	UV 254 nm	0.176 µmol/L	1997 (77)
HPLC	Polymer Lab PLRP-S	Fluorescence, quinoxaline derivative	0.28 nmol/injection	1991 (35)
				1996 (36)
HPLC	PRP-1 polystyrene divinylbenzene	Fluorescence, quinoxaline derivative	50 µmol/injection	1991 (34)
HPLC	Nucleosil NH$_2$	Fluorescence, quinoxaline derivative	11.4 µmol/L	1992 (48)
HPLC	Waters C18	Fluorescence, quinoxaline derivative	0.825 µmol/L	1996 (92)
GC-MS	Supelco SPB-1	Mass spectrometry, isotope dilution	50 fmol/injection	1993 (95)
CE	Uncoated or coated silica	UV 265 nm	1.42–2.84 µmol/L	1992 (58)
CE	Uncoated silica	UV 254 nm	9.1 µmol/L	1993 (60)
CE	Uncoated silica	UV 266 nm	6.8 µmol/L	1995 (98)
				1996 (62)
CE	Uncoated silica	UV 260 nm	85 fmol/injection	1996 (63)

concentration appeared to have significant influence on the migration of ascorbate and IAA by increasing their migration time.

Chiari et al. measured total ascorbic acid from fruits after homocysteine treatment (59). They coated the capillary inner wall with 3-(trimethoxysilyl)propyl methacrylate and acrylamide. UV detection at 254 nm and phthalic acid as an internal standard were used. The analysis was carried out with phosphate buffer, pH 7, at 6 kV in a 40 cm × 100 μm capillary.

The principle of isotachophoresis was used in a study for measurement of ascorbic acid in urine, neuroblastoma cell extracts, and cell-free system (99). HCl/β-alanine-methylcellulose was used as leading buffer and caproic acid as terminating electrolyte. The column was 23 cm × 500 μm, but the type of column coating was not mentioned. UV detection at the wavelength of 254 nm was used.

A summary of separation and detection modes and their detection limits are presented in Table 1. The limits of detection have been measured at the signal-to-noise ratio of 1.5–5.0 (or not mentioned) and published either per sample volume or per injection into chromatographic system. Thus, it is quite difficult to compare the limits between different methods.

VI. FUTURE TRENDS

A variety of methods for the assay of vitamin C in foods and biological samples are available. During the last 10 years, the number of publications dealing with HPLC combined with electrochemical detection and capillary electrophoretic methods have increased but only a few papers for HPLC with UV detection have been published. This does not mean that UV detection in ascorbic acid analysis is out-of-date, but the development of electrochemical detectors and electrophoretic methods has been stronger. However, UV detectors are common in most laboratories. Thus UV is probably the simplest and the most frequently used HPLC detector for ascorbic acid analyses, when the measurement of very low concentrations is not required. Electrochemical detection can be more sensitive, but the stabilization of the system may take more time and the between-day variation is higher than with UV (76).

CE methods will develop further. By now, only UV detectors have been used after ascorbic acid separation in capillary, but there is strong evidence from the field of HPLC that electrochemical detection will be adopted also in CE. The instrumentation of CE will improve further to become more convenient for the user, and especially the capacity will increase with modern autosamplers that allow also the increased capacity of automatic washes between samples.

Owing to the large scale of different methods for ascorbic acid analysis

Table 2 Results from the Quality Control Ring Test of European Community FLAIR Concerted Action on Micronutrient Measurement, Absorption, and Status

Laboratory	Method	μM Vitamin C in sample		
		1	2	3
1	HPLC	33.5	63.5	4.2
2	HPLC	31.0	66.4	4.5
3	HPLC	27.3	57.0	1.8
4	HPLC-Q[a]	31.4	63.6	4.4
5	HPLC-Q[a]	30.1	60.4	12.7
6	Direct-Q[a]	31.1	66.2	20.9
7	Direct-Q[a]	29.0	63.6	8.7
8	Direct-DNPH[b]	51.4	69.6	16.8
9	Direct-DNPH[b]	71.0	114.2	33.6
Mean (n = 9)		37.3	69.4	12.0
CV %		39%	25%	87%
Predicted value (μM)		33	69	4.5

[a] Q: based on determination of quinoxaline derivative.
[b] DNPH: based on dinitrophenylhydrazine method.
Source: Ref. 100.

the variation between laboratories is high. In European community FLAIR (Food Linked Agro-Industrial Research) program of interlaboratory assay method comparisons for micronutrients, nine European laboratories performed coordinated ring tests (100). For that, a sample of stored human heparin plasma with undetectable levels of ascorbic acid or DHA was subdivided and spiked with pure L-ascorbic acid. The samples were then stabilized with MPA to yield extracts of 5% in MPA and 4.5 μM, 33 μM, and 69 μM in ascorbic acid. Precipitated protein was removed by centrifugation and the supernatants were frozen and sent in dry ice to participating laboratories. As presented in Table 2, the between-laboratory CV varied from 25% to 87% depending on the concentration of the ascorbate in the control sample. However, HPLC-based methods seemed more sensitive and accurate than other procedures. It should also be noted that in Table 2, laboratories 1–3 measure only the reduced form of ascorbic acid, laboratories 4–7 ascorbic acid plus DHA, and laboratories 8 and 9 ascorbic acid, DHA, and diketogulonic acid together. These results show that there is need for generally available quality control materials with verified and assigned vitamin C contents for interlaboratory verification and method development. At the least, the laboratories performing ascorbate analysis should exchange samples for checking their analytical level and calibration. Also, there is a lack of an internationally accepted reference method for ascorbic acid, DHA, or other related compounds.

REFERENCES

1. HE Sauberlich. A history of scurvy and vitamin C. In: L Packer, J Fuchs, eds. Vitamin C in Health and Disease. New York: Marcel Dekker, 1997, pp. 1–24.
2. W Friedrich. Vitamin C. In: Friedrich W, ed. Vitamins, Berlin: W. De Gruyter, 1988, pp 929–1001.
3. GF Combs. The Vitamins. Fundamental Aspects in Nutrition and Health. San Diego: Academic Press, 1992.
4. BJ Burri, RA Jacob. Human metabolism and the requirement for vitamin C. In: L Packer, J Fuchs, eds. Vitamin C in Health and Disease. New York: Marcel Dekker, 1997, pp 341–366.
5. M Levine, C Conry-Cantilena, Y Wang, RW Welch, PW Washko, KR Dhariwal, JB Park, A Lazarev, JF Graumlich, J King, LR Cantilena. Vitamin C pharmacokinetics in healthy volunteers: evidence for a recommended dietary allowance. Proc Natl Acad Sci USA 93:3704–3709, 1996.
6. A Bendich, L Langseth. The health effects of vitamin C supplementation: a review. J Am Coll Nutr 14:124–136, 1995.
7. JT Salonen, K Nyyssönen, MT Parviainen. Vitamin C, lipid peroxidation, and the risk of myocardial infarction: epidemiological evidence from eastern Finland. In: L Packer, J Fuchs, eds. Vitamin C in Health and Disease. New York: Marcel Dekker, 1997, pp 457–469.
8. E Niki. Action of ascorbic acid as a scavenger of active and stable oxygen radicals. Am J Clin Nutr 54:1119S–1124S, 1991.
9. K Nyyssönen, E Porkkala-Sarataho, J Kaikkonen, JT Salonen. Ascorbate and urate are the strongest determinants of plasma antioxidative capacity and serum lipid resistance to oxidation in Finnish men. Atherosclerosis 130:223–233, 1997.
10. L Pauling L. Vitamin C and the common cold and the flu. San Francisco: W.H. Freeman, 1970.
11. LJ Machlin. Handbook of vitamins: nutritional, biochemical, and clinical aspects. New York: Marcel Dekker, 1991, pp 195–232.
12. E Kimoto, H Tanaka, T Ohmoto, M Choami. Analysis of the transformation products of dehydro-L-ascorbic acid by ion-pairing high-performance liquid chromatography. Anal Biochem 214:38–44, 1993.
13. GR Buettner. In the absence of catalytic metals ascorbate does not autoxidize at pH 7: ascorbate as a test for catalytic metals. J Biochem Biophys Methods 16:27–40, 1988.
14. A Favier. Defence of the body against oxygen free radicals. In: Training in Free Radical Methodologies: Production, Damage, Repair. Grenoble: Cerlib, 1996.
15. B Halliwell, JMC Gutteridge. Free Radicals in Biology and Medicine. 2nd ed. Oxford: Clarendon Press, 1989.
16. ER Stadtman. Ascorbic acid and oxidative inactivation of proteins. Am J Clin Nutr 54:1125S–1128S, 1991.
17. DDM Wayner, GW Burton, KU Ingold. The antioxidant efficiency of vitamin C is concentration-dependent. Biochem Biophys Acta 884:119–123, 1986.
18. VA Roginsky, HB Stegmann. Ascorbyl radical as natural indicator of oxidative stress: quantitative regularities. Free Rad Biol Med 17:93–193, 1994.

19. D Bartlett, DF Church, PL Bounds, WH Koppenol. The kinetics of the oxidation of L-ascorbic acid by peroxynitrite. Free Rad Biol Med 18:85–92, 1995.

20. A Mori, X Wang, J Liu. Electron spin resonance assay of ascorbate free radicals in vivo. Methods Enzymol 233:149–154, 1994.

21. B Halliwell, JMC Gutteridge. Role of free radicals and catalytic metal ions in human disease: an overview. Methods Enzymol 186:1–85, 1990.

22. B Halliwell. Vitamin C: antioxidant or pro-oxidant in vivo? Free Rad Res 25:439–454, 1996.

23. JMC Gutteridge. Iron and free radicals. In: L Hallberg, N-G Asp, eds. Iron Nutrition in Health and Disease. London: John Libbey, 1996, pp 239–246.

24. L Rossander-Hulthén, L Hallberg. Dietary factors influencing iron absorption—an overview. In: L Hallberg, N-G Asp, eds. Iron Nutrition in Health and Disease. London: John Libbey, 1996, pp 105–115.

25. BS Winkler. Unequivocal evidence in support of the nonenzymatic redox coupling between glutathione/glutathione disulfide and ascorbic acid/dehydroascorbic acid. Biochim Biophys Acta 1117:287–290, 1992.

26. BS Winkler, SM Orselli, TS Rex. The redox couple between glutathione and ascorbic acid: a chemical and physiological perspective. Free Rad Biol Med 17:333–349, 1994.

27. PW Washko, RW Welch, KR Dhariwal, Y Wang, M Levine. Ascorbic acid and dehydroascorbic acid analyses in biological samples. Anal Biochem 204:1–14, 1992.

28. JM May, Z-C Qu, RR Whitesell, CE Cobb. Ascorbate recycling in human erythrocytes: role of GSH in reducing dehydroascorbate. Free Rad Biol Med 20:543–551, 1996.

29. JM Park, M Levine. Purification, cloning and expression of dehydroascorbic acid–reducing activity from human neutrophils: identification as glutaredoxin. Biochem J 315:931–938, 1996.

30. RC Rose, AM Bode. Tissue-mediated regeneration of ascorbic acid: is the process enzymatic? Enzyme 46:196–203, 1992.

31. MC Linder. Nutrition and metabolism of vitamins. In: MC Linder, ed. Nutritional Biochemistry and Metabolism with Clinical Applications. New York: Elsevier Science, 1991, pp 111–189.

32. SK Gaby, A Bendich, VN Singh, LJ Machlin. Vitamin Intake and Health, a Scientific Review. New York: Marcel Dekker, 1991.

33. M El Sadek M, A El Shanawany, A Aboul Khier. Determination of the components of analgesic mixtures using high-performance thin-layer chromatography. Analyst 115:1181–1184, 1990.

34. N Bilic. Assay for both ascorbic and dehydroascorbic acid in dairy foods by high-performance liquid chromatography using precolumn derivatization with methoxy- and ethoxy-1,2-phenylenediamine. J Chromatogr 543:367–374, 1991.

35. JT Vanderslice, DJ Higgs. Vitamin C content of foods: sample variability. Am J Clin Nutr 54:1323S–1327S, 1991.

36. MS Ali, ET Phillippo. Simultaneous determination of ascorbic, dehydroascorbic, isoascorbic and dehydroascorbic acids in meat-based food products by liquid chromatography with postcolumn fluorescence detection: a method extension. J AOAC Int 79:803–808, 1996.

37. SA Margolis, RM Schapira. Liquid chromatographic measurement of L-ascorbic acid and D-ascorbic acid in biological samples. J Chromatogr 690:25–33, 1997.

38. KR Dhariwal, WO Harzell, M Levine M. Ascorbic acid and dehydroascorbic acid measurements in human plasma and serum. Am J Clin Nutr 54:712–716, 1991.

39. SA Margolis, RG Ziegler, KJ Helzlsouer. Ascorbic acid and dehydroascorbic acid measurement in human serum and plasma. Am J Clin Nutr 54:1315S–1518S, 1991.

40. DG Watson, Z Iqbal, JM Midgley, H Pryce-Jones, L Morrison, GN Dutton, S Karditsas, W Wilson. Measurement of ascorbic acid in human aqueous humour and plasma and bovine aqueous humour by high-performance liquid chromatography with electrochemical detection. J Pharm Biomed Anal 11:389–392, 1993.

41. P-J Tsai, J-P Wu, N-N Lin, J-S Kuo, C-S Yang. In vivo, continuous and automatic monitoring of extracellular ascorbic acid by microdialysis and on-line liquid chromatography. J Chromatogr 686:151–156, 1996.

42. I Koshiishi, T Imanari. Measurement of ascorbate and dehydroascorbate contents in biological fluids. Anal Chem 69:216–220, 1997.

43. DA Schell, AM Bode. Measurement of ascorbic acid and dehydroascorbic acid in mammalian tissue utilizing HPLC and electrochemical detection. Biomed Chromatogr 7:267–272, 1993.

44. RS Harapanhalli, RW Howell, DV Rao. Testicular and plasma ascorbic acid levels in mice following dietary intake: a high-performance liquid chromatographic analysis. J Chromatogr 614:233–243, 1993.

45. T Ito, H Murata, Y Yasui, M Matsui, T Sakai, K Yamauchi. Simultaneous determination of ascorbic acid and dehydroascorbic acid in fish tissues by high-performance liquid chromatography. J Chromatogr 667:355–357, 1995.

46. T Sakai, H Murata, T Ito. High-performance liquid chromatographic analysis of ascorbyl-2-phosphate in fish tissues. J Chromatogr 685:196–198, 1996.

47. RC Rose, AM Bode. Analysis of water-soluble antioxidants by high-pressure liquid chromatography. Biochem J 306:101–105, 1995.

48. V Kmetec. Simultaneous determination of acetylsalicylic, salicylic, ascorbic and dehydroascorbic acid by HPLC. J Pharm Biomed Anal 10:1073–1076, 1992.

49. H Goldenberg, L Jirovetz, P Krajnik, W Mosgöller, T Moslinger, E Schweinzer. Quantitation of dehydroascorbic acid by the kinetic measurement of a derivatization reaction. Anal Chem 66:1086–1089, 1994.

50. T Moeslinger, M Brunner, I Volf, PG Spieckermann. Spectrophotometric determination of ascorbic acid and dehydroascorbic acid. Clin Chem 41:1177–1181, 1995.

51. IFF Benzie. An automated, specific, spectrophotometric method for measuring ascorbic acid in plasma (EFTSA). Clin Biochem 1996;29:111–116.

52. D Every. Enzymatic method to determine dehydroascorbic acid in biological samples and in bread dough at various stages of mixing. Anal Biochem 242:234–239, 1996.

53. W Lee, SM Roberts, RF Labbe. Ascorbic acid determination with an automated enzymatic procedure. Clin Chem 43:154–157, 1997.

54. S Uchiyama, Y Kobayashi, S Suzuki, O Hamamoto. Selective biocoulometry of vitamin C using dithiothreithol, N-ethylmaleimide, and ascorbate oxidase. Anal Chem 63:2259–2262, 1991.

55. K Matsumoto, JJB Baeza, HA Mottola. Simultaneous kinetic-based determination of fructose and ascorbate with a rotating bioreactor and amperometric detection: application to the analysis of food samples. Anal Chem 65:1658–1661, 1993.

56. PS Cahill, RM Wightman. Simultaneous amperometric measurement of ascorbate and catecholamine secretion from individual bovine adrenal medullary cells. Anal Chem 67:2599–2605, 1995.

57. Z Gao, KS Siow, A Ng, Y Zhang. Determination of ascorbic acid in a mixture of ascorbic acid and uric acid at a chemically modified electrode. Anal Chim Acta 343:49–57, 1997.

58. B LinLing, WRG Baeyens, P Van Acker, C Dewaele. Determination of ascorbic acid and isascrobic acid by capillary zone electrophoresis: application to fruit juices and to a pharmaceutical formulation: J Pharm Biomed Anal 10:717–721, 1992.

59. M Chiari, M Nesi, G Carrea, PG Righetti. Determination of total vitamin C in fruits by capillary zone electrophoresis. J Chromatogr 645:197–200, 1993.

60. EV Koh, MG Bissell, RK Ito. Measurement of vitamin C by capillary electrophoresis in biological fluids and fruit beverages using a stereoisomer as an internal standard. J Chromatogr 633:245–250, 1993.

61. A Hiraoka, J Akai, I Tominaga, M Hattori, H Sasaki, T Arato. Capillary zone electrophoretic determination of organic acids in cerebrospinal fluid from patients with central nervous system diseases. J Chromatogr 680:243–246, 1994.

62. R Neubert, J Schiewe. Vitamin analysis using capillary zone electrophoresis. Am Biotechnol Lab 14:12–14, 1996.

63. MW Davey, G Bauw, M Van Montagu. Analysis of ascorbate in plant tissues by high-performance capillary zone electrophoresis. Anal Biochem 239:8–19, 1996.

64. MA Ross. Determination of ascorbic acid and uric acid in plasma by high-performance liquid chromatography. J Chromatogr 657:197–200, 1994.

65. M Manoharan, PO Schwille. Measurement of ascorbic acid in human plasma and urine by high-performance liquid chromatography. Results in healthy subjects and patients with idiopathic calcium urolithiasis. J Chromatogr 654:134–139, 1994.

66. J Lykkesfeldt, S Loft, HE Poulsen. Determination of ascorbic acid and dehydroascorbic acid in plasma by high performance liquid chromatography with coulometric detection. Are they reliable biomarkers of oxidative stress? Anal Biochem 229: 329–335, 1995.

67. CS Tsao, PY Leung, M Young. Effect of dietary ascorbic acid intake on tissue vitamin C in mice. J Nutr 117:291–297, 1987.

68. W Lee, P Hamernyik, M Hutchinson, VA Raisys, RF Labbe. Ascorbic acid in lymphocytes: cell preparation and liquid-chromatography assay. Clin Chem 28:2165–2169, 1982.

69. G Barja de Quiroga, M López-Torres, R Pérez-Campo, C Rojas. Simultaneous determination of two antioxidants, uric and ascorbic acid, in animal tissue by high-performance liquid chromatography. Anal Biochem 199:81–85, 1991.

70. KL Khanduja, A Koul. Stabilization of ascorbic acid in plasma. Clin Chim Acta 174:351–352, 1988.

71. CJ Bates. Use of homocysteine to stabilize ascorbic acid, or to reduce dehydroascorbic acid, during HPLC separation of large volumes of tissue extracts. Clin Chim Acta 205:249–252, 1992.

72. SA Margolis, RC Paule, RG Ziegler. Ascorbic and dehydroascorbic acids measured in plasma preserved with dithiothreitol or metaphosphoric acid. Clin Chem 36: 1750–1755, 1990.

73. H Iwase, I Ono. Determination of ascorbic acid in human plasma by high-performance liquid chromatography with electrochemical detection using a hydroxyapatite cartridge for precolumn deproteinization. J Chromatogr 655:195–200, 1994.

74. K Tanishima, M Kita. High-performance liquid chromatographic determination of plasma ascorbic acid in relationship to health care. J Chromatogr 613:275–280, 1993.

75. MT Parviainen, K Nyyssönen, IM Penttilä, R Rauramaa, JT Salonen, C-G Gref. A method for routine assay of plasma ascorbic acid using high-performance liquid chromatography. J Liq Chromatogr 9:2185–2197, 1986.

76. LS Liau, BL Lee, AL New, CN Ong. Determination of plasma ascorbic acid by high-performance liquid chromatography with ultraviolet and electrochemical detection. J Chromatogr 612:63–70, 1993.

77. MJ Esteve, R Farré, A Frigola, JM Garcia-Cantabella. Determination of ascorbic acid and dehydroascorbic acids in blood plasma and serum by liquid chromatography. J Chromatogr 688:345–349, 1997.

78. SA Margolis, TP Davis. Stabilization of ascorbic acid in human plasma, and its liquid chromatographic measurement. Clin Chem 34:2217–2223, 1988.

79. SA Margolis, DL Duewer. Measurement of ascorbic acid in human plasma and serum: stability, intralaboratory repeatability, and interlaboratory reproducibility. Clin Chem 42:1257–1262, 1996.

80. K Nyyssönen. Ascorbic acid: analysis in biological samples and role in lipid peroxidation. PhD dissertation, Kuopio University, Kuopio, Finland, 1997.

81. KR Dhariwal, PW Washko, M Levine. Determination of dehydroascorbic acid using high-performance liquid chromatography with electrochemical detection. Anal Biochem 189:18–23, 1990.

82. K Nyyssönen, S Pikkarainen, MT Parviainen, K Heinonen, I Mononen. Quantitative estimation of dehydroascorbic acid and ascorbic acid by high-performance liquid chromatography—application to human milk, plasma, and leukocytes. J Liq Chromatogr 11:1717–1728, 1988.

83. CJ Bates. Use of homocysteine to stabilize ascorbic acid, or to reduce dehydroascorbic acid, during HPLC separation of large volumes of tissue extracts. Clin Chim Acta 205:249–252, 1992.

84. L Ødum. pH optimum of the reduction of dehydroascorbic acid by dithioerytriol. Scand J Lab Invest 53:367–371, 1993.

85. AM Bode, L Cunningham, RC Rose. Spontaneous decay of oxidized ascorbic acid (dehydro-L-ascorbic acid) evaluated by high-pressure liquid chromatography. Clin Chem 36:1807–1809, 1990.

86. J Cammack, A Oke, RN Adams. Simultaneous high-performance liquid chromatographic determination of ascorbic acid and dehydroascorbic acid in biological samples. J Chromatogr 565:529–532, 1991.

87. H Iwase. Determination of ascorbic acid in elemental diet by high-performance liquid chromatography with electrochemical detection. J Chromatogr 606:277–280, 1992.

88. G Lazzarino, DD Pierro, B Tavazzi, L Cerroni, B Giardina. Simultaneous separation of malondialdehyde, ascorbic acid, and adenine nucleotide derivatives from biological samples by ion-pairing high-performance liquid chromatography. Anal Biochem 197:101–196, 1991.

89. RC Rose, MJ Koch. Liquid chromatographic behaviour of ascorbate on amine columns. Anal Biochem 143:21–24, 1984.

90. S Daily, SJ Armfield, BGD Haggett, MEA Downs. Automated enzyme packed-bed system for the determination of vitamin C in foodstuffs. Analyst 116:569–572, 1991.

91. Y Yasui, M Hayashi. Simultaneous determination of ascorbic acid and dehydroascorbic acid by high performance liquid chromatography. Anal Sci 7(suppl):125–128, 1991.

92. F Tessier, I Birlouez-Aragon, C Tjani, JC Guilland. Validation of a micromethod for determining oxidized and reduced vitamin C in plasma by HPLC-fluorescence. Int J Vitam Nutr Res 66:166–170, 1996.

93. D Knaack, T Podleski. Ascorbic acid mediates acetylcholine receptor increase induced by brain extract on myogenic cells. Proc Natl Acad Sci USA 82:575–579, 1985.

94. K Niemelä. Oxidative and non-oxidative alkali-catalyzed degradation of L-ascorbic acid. J Chromatogr 399:235–243, 1987.

95. JC Deutsch, JF Kolhouse. Ascorbate and dehydroascorbate measurements in aqueous solutions and plasma determined by gas chromatography–mass spectrometry. Anal Chem 65:321–326, 1993.

96. MJ Gordon, X Huang, SL Pentoney Jr, RN Zare. Capillary Electrophoresis. Science 242:224–228, 1988.

97. MW Lada, RT Kennedy. Quantitative in vivo measurements using microdialysis on-line with capillary zone electrophoresis. J Neurosci Methods 63:147–152, 1995.

98. J Schiewe, Y Mrestani, R Neubert. Application and optimization of capillary zone electrophoresis in vitamin analysis. J Chromatogr 717:255–259, 1995.

99. S Gebhardt, K Kraft, N HN Lode, D Niethammér. Determination of ascorbic acid by isotachophoresis with regard to its potential in neuroblastoma therapy. J Chromatogr 638:235–240, 1993.

100. CJ Bates, Members of EC FLAIR Concerted Action No. 10. Micronutrient measurement, absorption and status. Plasma vitamin C assays: a European experience. Int J Vitam Nutr Res 64:283–287, 1994.

6

Chromatographic Determination of Folates

Liisa Vahteristo
University of Helsinki, Helsinki, Finland

Paul Michael Finglas
Institute of Food Research, Colney, Norwich, Norfolk, England

I. INTRODUCTION

The parent compound of folates, one of the B-vitamins, is pteroyl-L-glutamic acid (folic acid). Folic acid does not occur naturally but is used for food fortification purposes because it is more stable than naturally occurring folates. The latter forms are tetrahydrofolates which usually exist as in the polyglutamate form with up to 12 molecules of glutamic acid. One-carbon units are attached to the N_5 or N_{10} positions or linked across both positions (Fig. 1) to give a large group of compounds exhibiting biological activity of folic acid. Folates are coenzymes required in many metabolic bathways, including purine and pyrimidine biosynthesis and amino acid interconversions (1).

The potential of folate in preventive medicine against neural tube defects, cardiovascular disease, and various cancers (colon) (reviewed in Ref. 2) has greatly attracted attention to this vitamin in recent years. It is now established that folic acid can prevent neural tube defects such as spina bifida. More recently, it has been shown that folic acid, along with vitamin B_6 and B_{12}, can reduce plasma homocysteine concentration, an amino acid that has been linked to possible increased risk of atherosclerosis (3–5).

Folates are sensitive to heat, acids, oxidation, and light, and their concentration in foods and biological samples is usually very low. Therefore, the extraction and determination of folates presents a difficult analytical problem. Folates vary

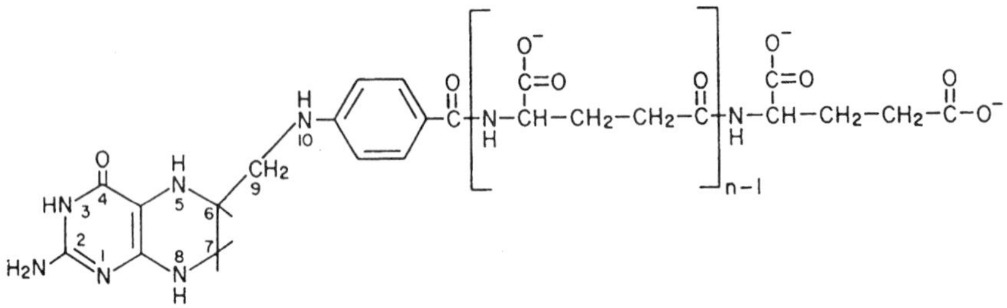

Figure 1 Structure of reduced polyglutamyltetrahydrofolate.

in their stability and to some extent in their bioavailability. As there is increased research aimed at understanding folate absorption and metabolism in man, chromatographic methods capable of separating and quantifying various folate forms, both mono- and polyglutamates, are required.

II. METHODS OF ANALYSIS

Suitable techniques for the determination of folates in food and biological samples include typically microbiological assays, high-performance liquid chromatographic techniques (HPLC), and competitive-binding radioassay procedures using radiolabeled and enzyme labels (1,6), and combined gas chromatographic–mass spectrometric techniques using stable isotopes.

   The microbiological method for total folate determination is still the most widely used and accepted procedure, and competitive-binding assays are most typically used for blood folate analyses. Separation, identification, and quantification of natural folates using chromatographic procedures, particularly HPLC, can be achieved but is not straightforward and requires careful selection of chromatographic conditions for the required separations (7).

   In this chapter chromatographic folate methods for various applications are presented. Sample treatments including extraction, cleanup, and deconjugation steps are also discussed.

III. EXTRACTION

A. Foods and Biological Samples

Folates are usually extracted from complex matrix using heat treatment in dilute buffer with added antioxidants. These buffered solutions are of pH close to neutral, or mildly acidic/alkaline. For fresh samples usually 5 to 8 volumes of buffer, or at least 10 times the dry weight of the sample in grams, are used to extract folates from the matrix. It has been recommended to use a second extraction step to improve extraction efficiency (8). The sample matrix is often disrupted by homogenizing in buffered solutions, and together with heat, bound folates are released into solution. Heat treatment denatures folate-binding and other proteins, and prevents interconversion of folates (reviewed in 1, 9, 10). Freeze-drying prior to extraction is sometimes performed (11–13). Heat extraction of folates is typically carried out at 100°C or 121°C (autoclaving) for 5 to 60 min (6,9). Ascorbic acid is the most commonly used antioxidant but thiols, such as 2-mercaptoethanol and dithiothreitol, can be used as well. For the extraction of mycelial and conidiospores of *Neurospora crassa*, Kruschwitz et al. (14) used HEPES buffer (pH 7.6) with EDTA, 2-mercaptoethanol (50 mM), and sodium ascorbate (2%; w:v) as antioxidants. Hot buffer and 5 min boiling was used to extract H_4-folates. Moran et al. (15) extracted cellular folates using hot buffer (3 mL) at pH 6.0 containing ascorbate (1%) and 2-mercaptoethanol (1%) and boiling the mixture in a water bath for 3 min. Wilson and Horne (16) recommended use of both ascorbic acid (2%) and 2-mercaptoethanol (0.2 M) at pH 7.85 to overcome possible stability problems with 10-HCO-H_4-folate and other, more labile folates, but also to protect excessive losses of ascorbic acid during extraction. The combination of those two antioxidants has been used at various concentrations (8,13,17–20), but adequate stability has also been reported with the use of only one antioxidant (21–24). Some methods do not use antioxidants for the extraction step at all (25–27). However, the use of antioxidants is essential if total folates are to be determined.

The effects of pH in the oxidative destruction of folates has been reviewed by several authors (6,10,28). Each vitamer has its own unique pH stability, and it is therefore sometimes difficult to optimize extraction conditions for all folate forms in a single extraction. The pH of the extractant, presence of oxygen, the temperature used, and in addition, buffer type, can all affect the efficiency of folate extraction (8). Several folate derivatives can be altered during extraction. The presence of phosphate accelerates conversion of 10-HCO-H_4-folate to (1) 5,10-CH^+-H_4-folate and (2) 5-HCO-H_4-folate at pH values below pH 7 during extraction. Also, 5,10-CH_2-H_4 folates can be converted to H_4 folate during extraction (29). As the use of ascorbic acid reduces 5-CH_3-H_2-folate to 5-CH_3-H_4-folate, the former can be quantified only in the 5-CH_3-H_4-folate pool (30). Thus, extraction conditions can affect the vitamer distribution and therefore need always to be carefully considered and fully reported. The chosen extraction conditions are dependent on the purpose of the analysis to be carried out.

Some methods for the determination of added folic acid, and even endogenous 5-CH_3-H_4-folate, do not use heating for extraction (31–33). For example, in the analysis of bile folates, Shimoda et al. (34) first stored the samples in sodium ascorbate (0.4%; 1:1, v:v), and after thawing and centrifugation, the supernatants were diluted with sodium ascorbate (0.2%) prior to HPLC injection. In addition to heat extraction, the use of amylase and protease can be important in maximizing measurable folate from certain foods. Using a microbiological assay, Martin et al. (35) reported significant increases in folate concentration with the combined use of chicken pancreas, α-amylase and protease compared to chicken pancreas alone. For several vegetables and wheat flour, the increase was 20% to 25%, for canned tuna as high as 50%; the mean increase for all foods was 19%.

DeSouza and Eitenmiller (36) used different enzyme treatments including conjugase, α-amylase and protease. They observed significant increases in measurable folate by using the triple enzyme combination prior to microbiological and radioassay. More recently, Rader et al. (37), determined total folate in 56 enriched foods by microbiological assay following trienzyme treatment (α-amylase and chicken pancreas followed by protease, pH 7.8). These workers concluded that for many foods, total folate was about 25% higher with the trienzyme digestion than with conjugase treatment alone. Similarly, Lim et al. (38) reported significantly higher (85%) folate content of human milk using trienzyme extraction at pH 4.1 and microbiological assay. Neither pH of extraction buffer, source of folate conjugase, nor method/length of heat treatment was found to affect folate content. Tamura and coworkers (39,40) concluded the trienzyme treatment to be essential for determining food folate content accurately in food samples, although the extent of the increase was dependent on the extraction pH, and the type of food analyzed.

Claradiastase and trypsin in phosphate buffer have been used for hydrolysis of sample matrix for the determination of supplemental folic acid (27). Jacoby and Henry (26) further modified the method of Hoppner and Lampi (41) for folic acid by HPLC, in which the folic acid added to infant formulas and liquid medical nutritionals is quantitatively extracted with the aid of bacterial protease and papain. One disadvantage of the enzymatic extraction method was the large number of UV-absorbing compounds that are formed during enzymatic hydrolysis. These can interfere in the quantitation of the folate peaks. Thus, addition of α-amylase and protease to enhance extraction has not been common practice for methods involving HPLC. However, Pfeiffer et al. (13) successfully used triple-enzyme treatment prior to HPLC analysis with a purification method based on affinity chromatography.

B. Blood Plasma and Red Blood Cells

Extraction of plasma folates, which are present as folylmonoglutamates, can be carried out with several extraction methods. Kohashi and Inoue (42) used acetone extraction of serum protected with 2-mercaptoethanol for HPLC determination of 5-CH$_3$-H$_4$-folate with electrochemical detection. Acetonitrile extraction, evaporation, and reconstitution of the residue with sodium ascorbate could also be used when coupled chiral-achiral HPLC system with postchiral column peak compression was applied (43). This hyphenated technique for the determination of stereoisomers of 5-CH$_3$-H$_4$-folate and 5-HCO-H$_4$-folate required three columns, and a well-equipped and timed switching system. Schleyer et al. (44) applied acetonitrile extraction for similar type of hyphenated chromatography separation of plasma and urine samples.

For deproteinatization of plasma samples, perchloric acid with or without ascorbic acid is often used. The centrifuged extract can then be injected after pH adjustment (45,46). Etienne et al. (47) reported significant background interference from all organic extraction procedures and therefore recommended an extraction method based on solid-phase extraction (SPE) cartridges. The use of reversed-phase SPE columns is commonly reported (47–49) but phenyl cartridges (50,51), in addition to strong anion exchange cartridges (52–54), can also be used for plasma folate extraction. Using permanganate as a fluorogenic reagent, plasma could be injected directly after permanganate oxidation (55). Despite of its simplicity, this method has very little use due to its very low sensitivity.

Mild heat extraction was applied by Bohlman and Nau (56) in the analysis of rat plasma folates. Plasma was mixed with 0.1 M acetate buffer (pH 5.5) containing ascorbic acid (1%) and 2-mercaptoethanol (0.2M), after flushing with argon, the mixture was heated for 5 min at 80°C. Kelly et al. (45) described an assay for serum folate involving HPLC fractionation of deproteinized serum prior to *Lactobacillus rhamnosus* assay of the collected peaks. The method was validated particularly for the determination of unmetabolized folic acid in human plasma. The limit of detection for folic acid using *L. rhamnosus* microtitration plate assay was 1 ng/mL serum.

Erythrocyte folates can be extracted by mixing with ascorbic acid, and polyglutamylfolates are hydrolyzed with endogenous plasma γ-glutamylhydrolase before HPLC separation and analysis. Lucock et al. (57) injected this extract after vortex mixing with perchloric acid and subsequent pH adjustment with sodium hydroxide. Leeming et al. (46) added ascorbic acid before the perchloric acid extraction. The blood extract with ascorbic acid added had a pH of 4.5, and a 1-hour incubation at room temperature was sufficient for the conversion of red blood cell polyglutamyl folates to monoglutamates. After perchloric acid extraction and centrifugation, the supernatant was mixed with ascorbic acid (3%) prior to HPLC analysis. Instead, Hoppner and Lampi (41) used a combination of DEAE-cellulose and Bio-Beads SM-2 resin for the purification of the blood extract. Their combined use was reported to aid extraction.

It has been recently argued (58) that as deoxyhemoglobin (which can bind red cell folate electrostatically) picks up oxygen and switches quaternary structure, any bound folate can become physically "trapped." Venous blood taken for folate analysis is 65% to 75% saturated with oxygen, and pro-rata trapping may lead to serious underestimation of red cell folate. These workers suggested that as ascorbic acid can protect hemolysates that are destined to be stored prior to analysis from oxidation, 10-fold dilution of deoxygenation blood (to promote complete lysis) should probably be carried out with a solution of ascorbic acid (1%), provided that the diluent is deoxygenated immediately before use. To avoid reoxygenation during the subsequent deconjugation step, hemolysates could be placed in containers, flushed with an appropriate gas, and sealed with a gas-tight closure.

IV. DECONJUGATION OF FOLATE POLYGLUTAMATES

Folates in plant and animal tissues are mostly present as poly-γ-glutamates. Methods of analysis are usually that these are either hydrolyzed to mono-, di-, or triglutamates before quantitative measurement by microbiological assay, or monoglutamates for HPLC quantitation. The glutamyl residues are linked by a unique peptide bond which is not hydrolyzed by usual peptidases. The hydrolysis is achieved with γ-glutamylhydrolase (EC 3.4.22.12), the most common sources of which are chicken pancreas, hog kidneys, and human and rat plasma. The conjugases from different sources differ in their mode of action, type of end product, and pH optima. The type and distribution of the end products depend on the conditions used (59,60). In general, chicken pancreas deconjugation produces mainly folate diglutamates, whereas hog kidney and plasma conjugases give folate monoglutamates as their end products. For microbiological analysis, chicken pancreas is widely used as it is readily available and highly active, and the end products exhibit equal response with the growth organism. For HPLC methods where separated individual folate monoglutamates are determined, chicken pancreas cannot be used and one of the other enzymes must be employed. In some chromatographic methods, folate polyglutamates can be separated according to their chain length, rendering the deconjugation step unnecessary. For electrophoretic methods using ternary complexes of glutamate residues, deconjugation of the polyglutamyl forms is again unnecessary (61,62).

It is important to ensure that, for whatever deconjugase enzyme is used, the deconjugation procedure is optimized and completed (63). Generally, short incubation times (1 to 4 hours), are favorable, as prolonged incubation times (24 hours) can cause degradation of some folate forms. Large surplus of conjugase to the theoretically sufficient amount is useful in overcoming any effects of enzyme inhibitors present in some foods, but also the presence of amylases in chicken pancreas seem to be beneficial in liberating bound folates from some starchy samples with a high amylose content (39).

V. HIGH-PERFORMANCE LIQUID CHROMATOGRAPHY

Many separation systems have been developed for the determination of folate vitamers using HPLC including reversed-phase, ion pair, and anion exchange types. The developments in the liquid chromatographic analysis of folates have been comprehensively reviewed by several workers (6,28,64,65).

The water-soluble nature of the folates, together with differences in ionic properties and hydrophobicity, make these compounds well suited for either ion exchange, or reversed-phase HPLC (64). Analysis of various naturally occurring folate polyglutamate derivatives can be accomplished on hydrolysed extracts (as the corresponding folate monoglutamates), or as intact poly-γ-glutamyl folates. Monoglutamyl folates are usually separated by reversed-phase HPLC, or by ion pair techniques. In reversed-phase separations, suppression or enhancement of the ionization of functional groups by pH can effectively be used to regulate retention on the column. The pH, ionic strength, and polarity of solvents are used to optimize the separation. Usually low pH with or without gradient of the organic phase is used with C_{18} or phenyl supports (65). In Figure 2, separation of five reduced and oxidized folate monoglutamate derivatives using acetonitrile phosphate buffer (pH 2.3) gradient is given. The method used (13) was applied to the analysis of cereal foods, which, due to the large number of vitamin active compounds, as well as the low concentrations present, poses a difficult and demanding matrix to work with. Separation with ion pairing is accomplished at near neutral pH with cationic surfactants designed for acids such as tetrabutylammoniumphosphate (TBAP). Typically, separations involve methanol and TBAP (18,22,66–68), or gradient consisting of acetonitrile or ethanol with TBAP (69,70). Separation of seven or eight folate vitamers can typically be achieved with ion pair HPLC (Figs. 3, 4).

The chain lengths of folate polyglutamates can be determined after cleavage of folates to p-aminobenzoylpolyglutamates followed by HPLC analysis. Shane (71) quantitatively cleaved folates in mammalian tissues into p-aminobenzoylpolyglutamates, which after derivatizations and purification, could be separated utilizing HPLC. However, only data on the chain lengths, without any information of the substituents, or the reduction state of those forms, could then be obtained. Conversion of p-aminobenzoylpolyglutamates to azodyes prior to HPLC analysis at 560 nm has also been reported (72). Excellent baseline resolution was obtained and the method was used to analyse rat liver polyglutamates. Separation of folate poly-γ-glutamate derivatives can be fully completed using methods of Selhub and coworkers (19,73,74), although very specific and laborious purification is required. Reversed-phase ion pair chromatography and diode array detection were applied for quantification of isolated folates from tissue extract (Fig. 5). Separation of 35 derivatives, into seven clusters, was arranged in order of increasing number of glutamate residues. Although some important

STANDARD MIXTURE

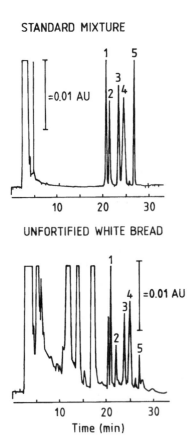

UNFORTIFIED WHITE BREAD

Time (min)

Figure 2   Reversed-phase HPLC separation of the main folate forms found for (top) standard mixture, and (bottom) cereal grain products using acetonitrile gradient in 30 mM phosphoric acid; UV absorption at 280 nm; flow rate 1 mL/min; and Phenomenex Ultremex C_{18} column (5 μm, 250 × 4.6 mm i.d.). Peaks: (1) 5-CH_3-H_4-folate; (2) 10-HCO-H_2-folate; (3) 10-HCO folic acid; (4) 5-HCO-H_4-folate; (5) folic acid. (From Ref. 13.)

derivatives (folic acid and 5-CH_3-H_4-folate) eluted in coinciding peaks, identification was reported based on differential spectral properties. This elegant and rather rapid method represents a major advance in folate polyglutamate research techniques. It has been applied to several foods and biological tissues (19,74).

Several methods for separating single folate compounds such as folic acid from matrix interference are available. Akhtar et al. (75) applied paired-ion chromatography for HPLC determination of folic acid and its photodegradation

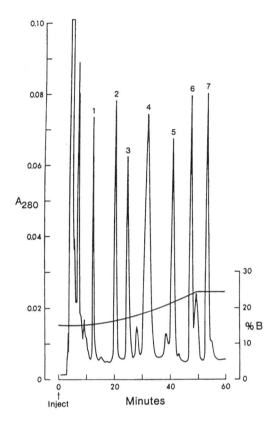

Figure 3 Chromatogram of folate derivatives with UV absorbance detection at 280 nm; ethanol gradient in 10mM TBAP reagent (pH 7.55); μ-Bondapak C_{18} column (3.9 × 300 mm, 10 μm); flow rate 1 mL/min. Peaks: (1) p-aminobenzoylglutamic acid; (2) 10-HCO-H_4-folate; (3) H_4-folate, (4) 5-HCO-H_4-folate; (5) H_2-folate; (6) 5-CH_3-H_4-folate; (7) folic acid (3–5 nmol of each derivative). (From Ref. 70.)

products. μ-Bondapak C_{18} column with TBAP and methanol as isogratic solvent system was used with UV detection at 254 nm. Hoppner and Lampi (41) determined added folic acid from fortified infant formulas with a Spherisorb ODS column (10 μm, 250 × 4.6 mm) using a linear gradient of acetonitrile and 0.1 M acetate buffer (pH 4.0) and UV detection at 280 nm. Schieffer et al. (25) used reversed-phase HPLC with absorbance detection at 365 nm employing isogratic elution with acetonitrile (6%) and 0.1 M sodium acetate buffer at pH 5.7 (94%) for the determination of folic acid in commercial diets. Detection of folic acid at 360 nm has been used with acetonitrile-water (pH 2.1 adjusted

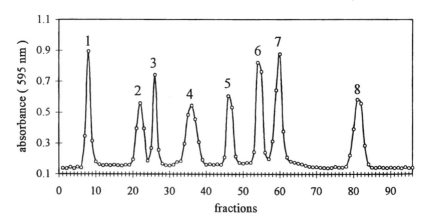

Figure 4 Ion pair chromatogram of eight folate monoglutamates (standard mixture); mobile phase (1 mL/min) consisted of 100 mM sodium acetate buffer with 5 mM TBAP (pH 6.0) (81%) and methanol (19%); Spherisorb ODS-2, 3 μm (125 × 4.6 mm i.d.), and run time = 25 min. Peak identification: (1) 5,10-CH$^+$-H$_4$-folate; (2) 10-HCO-H$_4$-folate; (3) H$_4$-folate; (4) 5-HCO-H$_4$-folate; (5) H$_2$-folate; (6) 5,10-CH$_2$-H$_4$-folate; (7) folic acid; (8) 5-CH$_3$-H$_4$-folate. (From Ref. 68.)

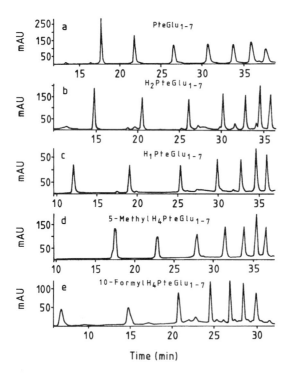

Figure 5 Ion pair chromatography of pteroylmono- to heptaglutamates (PteGlU$_{1-7}$) before and after conversion to reduced forms. Gradient of A (5 mM TBAP, 0.5 mM dithioerythritol in 25 mM phosphate/Tris buffer, pH 7.4, in water), and B (as A but in acetonitrile [64%], ethanol [9%], and water [27%]); UV absorption at 280 nm; Econosphere column (5 μm, 100 × 4.6 mm i.d.), flow rate 1 mL/min. (From Ref. 73.)

with phosphoric acid) eluent containing Na_2EDTA (1 mM) from elemental diet samples (32). A detection wavelength of 360 nm was chosen to eliminate coexisting compounds from the chromatogram. The authors reported that addition of Na_2EDTA to the mobile phase was necessary to obtain sharp, well-resolved peaks.

VI. SAMPLE CLEANUP FOR HPLC ANALYSIS

Many HPLC techniques have had only limited application because they lacked selectivity and sensitivity. In various European interlaboratory studies with several food matrices, the use of sample cleanup was considered essential to maintain high reliability (7). Several sample preparation methods have been designed to overcome these problems. In particular, cartridges packed with modified silica sorbents have provided a simple alternative to traditional sample preparative procedures, such as ion exchange resins (76).

Most common purification procedures for biological matrices include strong or weak anion exchange phases with automation of the purification stage being easily accomplished. Folates are retained on the basis of their charges and eluted using suitable buffers. Various food types including Brussels sprouts, powered diets, milk and milk products, and several other foods of plant and animal origin have successfully been purified using these techniques (25,77–81,83,84). Jacoby and Henry (26) described strong anion exchange (SAX) purification for folic acid determination in infant formulas using automated switching valves followed by separation in C_8 column. Reversed-phase chromatography can also be applied as a purification step as shown by Belz et al. (48). White et al. (67) automated a reversed-phase purification procedure using a switching 10-port valve. This hyphenation allowed direct injection of filtered orange juice into the HPLC system and $5\text{-}CH_3\text{-}H_4\text{-}folate$ was then backflushed from the C_{18} precolumn to the analytical column.

Affinity chromatography with folate-binding protein (FBP) provides a powerful mean for folate extract purification. Folate-binding protein from milk has high affinity for both mono- and polyglutamate folates at neutral pH, and this property can be utilized in affinity chromatography. FBP is fixed covalently to activated gel and this preparation can then be used for retaining biological folates. After elution, these folates can be chromatographed and assayed microbiologically, or, if labeled with radio tracers or stable isotopes, the labeling specificity can be determined (82). This approach has been applied as a purification step to the determination of polyglutamate folates (19,74) with adequate repeatability, reproducibility, and correlation with microbiological assay. The multiplicity of fluorescing compounds in urine can effectively be purified with FBP affinity columns prior to the determination of urinary folates by HPLC (85).

Pfeiffer et al. (13) combined affinity and reversed-phase liquid chromatography for determination of folates as monoglutamates in cereal grain products (Fig. 2). The main drawbacks of this very promising approach for purifying food extracts are that columns prepared with FBP are not commercially available, and that the application of a sample, which can only be done at low pressure, is time-consuming. In addition, the presence of suitable antioxidants such as ascorbic acid in eluting solutions is essential in order to protect the more labile folates such as 5-CH_3-H_4-folate and H_4-folate.

VII. DETECTION

Following HPLC separation of different folate forms, a number of detection systems can be used including UV, fluorometric, or electrochemical activity (65). Due to its sensitivity and selectivity, fluorescence detection is most commonly used, particularly for reduced folate forms (86). Quantitation is usually based on external standard methods, although an internal method using a stable isotope label has been recently reported (see below) (87). More cumbersome identification and quantifications are needed in methods in which folates are eluted in clusters (69). It is important that the actual concentrations of folate calibration solutions are determined using known extinction factors (30) because most commercially available folate calibrants are of variable purity (7).

Postcolumn derivatization can be used to enhance the fluorescence of reduced folates, or to produce fluorescence from some oxidized forms. For example, postcolumn oxidation with hypochlorite can be used to assay low concentrations of folic acid in foods as the parent molecule is cleaved to a more highly fluorescent pterin (77). Hahn et al. (88) studied the influence of flow rate of the postcolumn derivatization reagent [1% (w:v) $K_2S_2O_8$] to the peak areas obtained. Higher flow rates resulted in smaller peak areas. Precolumn derivatization of folic acid with permanganate oxidation has been used in plasma analysis to permit fluorometric detection (55). When optimized, the sensitivity of this approach was comparable to that of other methods. On-line postcolumn irradiation with UV at 254 nm has been applied to enhance the fluorescence intensity of 5-HCO-H_4-folate peak (89), resulting in a fourfold increase in fluorescence intensity after 90-sec irradiation in a Teflon capillary. However, applicability of the method to biological samples had not been tested, nor was the resulting compound structure or composition known.

Detection limits reported for several HPLC methods are given in Table 1. With fluorometric (F) or electrochemical (EC) detection, it is possible to achieve much better sensitivity than with UV detection. However, UV detection has sufficient sensitivity for the determination of folic acid in fortified foods and vitamin

Table 1 Comparison of Detection Limits for Various HPLC and Microbiological Methods Used for Determination of Folate Monoglutamates in Foods and Biological Samples (ng per Injection)

Detection[a]	H_4-folate	5-CH$_3$H$_4$-folate[b]	5-HCO-H$_4$-folate[b]	Folic acid	Others[b]	Reference
MA	0.0013	0.0014	0.0014	0.0009	(1) 0.0014 (2) 0.00456 (3) 0.0091 (4) 0.0044	68
F	0.89	0.18	9.4			24
F	4.9	0.3	18.8	—	(3) 52.4	107
F	—	0.01	2	—	—	48
F				1.5		55
F	—	0.01	0.2			89
F	0.07	0.014	0.15	1.0		77
F	0.003	0.002	0.05		(5) 0.04	83
UV				1.4		83
UV	—	0.9–1.8	0.95–1.9	0.9–1.8	(5) 0.9–1.9 (6) 3.3	13
UV	2	2	1.5	1.5	(1) 2.5 (4) 1.0	17
UV	0.79	1.7	3.8	3.8	(3) 9.6 (4) 2.7 (7) 1.2	107
UV	8.9	0.92	0.71	0.66	(3) 0.89	108
EC	0.01	0.01			(1) 0.01 (3) 0.02	109
EC	0.013	0.01			(1) 0.011	34
EC	0.44	0.24	0.8	—	(2) 2.4–7.4 (4) 0.18	107
EC	3	1	150	80	(1) 125	42

[a] MA, microbiological assay; F, fluorometric; EC, electrochemical detection.
[b] Folate forms detected: (1) 10-HCO-H$_4$-folate; (2) 5,10-CH$^+$-H$_4$-folate; (3) 5,10-CH$_2$-H$_4$-folate; (4) H$_2$-folate; (5) 10-HCO folic acid; (6) 10-HCO-H$_2$-folate; (7) 5-CH$_3$-H$_2$-folate.

premixes. Applicability of EC is more limited as smaller number of folate derivatives or antioxidants can be used, and results are only available for a limited number of matrices compared to the other detection methods. Microbiological assay of chromatographed folates seems to be the most sensitive way to detect separated derivatives, although it is not very commonly used to date, and the need for such an approach is questionable in most situations. Large differences in the detection limits within one detection system may be caused by the differences in instrument and chromatographic responses, including eluent pH and other variables.

VIII. HPLC METHODS FOR MULTIVITAMIN ANALYSIS

Folic acid can be determined in methods designed for multivitamin analysis either from multivitamin preparations (90–93) or from fortified samples such as infant milk (94). In these methods trichloroacetic acid extraction of liquid and powdered infant milk followed by ion pair chromatography with reversed-phase C_{18} column was applied. Satisfactory separation was achieved with octanesulfonic acid (5 mM) with triethylamine (0.5%) in methanol-water (15:85; v:v) at pH 3.6 (Fig. 6). UV detection and wavelength switching were used for six vitamins giving sensitivity of 1 ng folic acid per injection (282 nm for folic acid). Total run time of this isogratic separation was 55 min. This approach has not been reported for nonfortified samples.

IX. OTHER CHROMATOGRAPHIC METHODS

Other chromatographic approaches to separate folates include thin-layer chromatography of pteroylmonoglutamates as fluorescing compounds (95). Anion exchange and gel filtration column chromatography were used in the early work on folate separation of biological samples, as reviewed by Gregory (64). For example, separation of polyglutamate folates in meat using DEAE cellulose anion

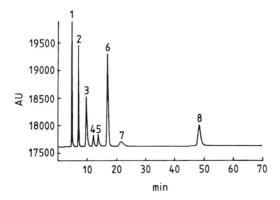

Figure 6   Chromatogram of water-soluble vitamins from standard solution. Ion pair chromatography with a reversed-phase C_{18} column (Tracer Spherisorb ODS 2, 250 × 4.6 mm i.d., 5 μm); UV detection at different wavelengths; mobile phase (1 mL/min) contained octanesulfonic acid (5 mM), triethylamine (0.5%), glacial acetic acid (2.4%), and methanol (15%) at pH 3.6. Peak identities: (1) nicotinamide; (2) pyridoxal; (3) pyridoxine; (4) pyridoxamine; (5) folic acid; (6) riboflavin; (7) cyanocobalamin; (8) thiamin. (From Ref. 94.)

exchange resin has been reported (96) which uses oxidative degradation of folate polyglutamates to the corresponding p-aminobenzoylpolyglutamates prior to chromatographic separation and quantitation. DEAE cellulose was used by Moran et al. (15) for separation of several folate fractions with gradient elution and verified by absorbance measurements at 285 nm radiolabel tracers. Additionally, peaks were identified with UV scanning and by microbiological growth.

X. MASS SPECTROMETRIC PROCEDURES

Folate compounds are not generally suited for gas chromatographic (GC) determination due to their ionic properties. However, derivatives of folate active compounds can be quantified with combined GC-MS method (97,98). For example, the quantification of red blood cell folates has recently been achieved with mass spectrometric (MS) detection of p-aminobenzoic acid (98). In this method, the C_9-N_{10} linkage of pteridine and p-aminobenzoylglutamic acid moiety of the folate molecule is cleaved to p-aminobenzoic acid via p-aminobenzoylglutamic acid. The quantitation is performed using internal standard with stable isotope label. The methodology appears complex but is highly specific and sensitive, and could ultimately form the basis of reference procedure for the determination of folate in blood and other matrices.

Combined GC-MS can also be used in the quantitation of nanogram or lower quantities of various folates in other biological matrices following chemical cleavage of the folate molecule to p-aminobenzoylglutamic acid, and derivatization to the respective lactam (97). This approach has been used in bioavailability studies in humans (99,100) but not for the direct determination of folates in biological samples. However, this type of approach would appear to be the most promising for a reference procedure for folates. The application of combined liquid chromatographic–mass spectrometric techniques for either direct analysis of folates, or cleavage fragments, has not been reported to date due to problems associated with the transition of the sample from its acidic solution, as eluted from the HPLC column, into the alkaline environment, which is essential for ESI-MS (101). However, this approach would alleviate the need for derivatization as various p-aminobenzoylglutamic acid fragments could be determined directly by LC-MS.

Human folate metabolism has recently been studied using ^{14}C accelerator mass spectrometry (AMS) by measuring the ^{14}C-folate levels in plasma, urine and feces samples taken over a 150-day period after dosing a healthy adult volunteer (102). AMS avoids the need for high specific activities and pharmacological doses as needed with scintillation counting. The volunteer receives an effective radiation dose of only 1.1 mrem.

XI. CONCLUSION

To increase our understanding about folate absorption and metabolism, identification and quantification of a range of folate derivatives need to be achieved with properly validated and detailed methods. Recent interlaboratory studies on methods for the determination of folates in foods (7,103), and blood (104,105) found high between-laboratory variation in the measurement of individual folates using microbiological, radioassay, and HPLC methods. Future research in improving the selectivity of cleanup and purification of sample extracts (especially using affinity columns) and instrumentation will enable the production of more reliable and more comparable folate data. The preparation of suitably stable and homogeneous food certified reference materials for folate analysis in foods (106) and whole blood (WHO 1st International Standard for Folate, National Institute for Biological Standards and Control, South Mimms, U.K.) will greatly assist in the quality control of the measurements performed.

REFERENCES

1. GFM Ball. Bioavailability and Analysis of Vitamins in Foods. 1st ed. London: Chapman & Hall, 1998, pp 439–496.
2. K Pietrzik, A Brönstrup. Folate in preventive medicine: a new role in cardiovascular disease, neural tube defects and cancer. Ann Nutr Metab 41:331–343, 1997.
3. J Selhub, PF Jacques, AG Bostom, RB S'Agostino, PWF Wilson, AJ Belanger, DH O'Leary, PA Wolf, EJ Schaefer, IH Rosenberg. Association between plasma homocysteine concentrations and extracranial carotid-artery stenosis. N Engl J Med 332:286–291, 1995.
4. P Verhoef, MJ Stampfer, JE Buring, JM Gaziano, RH Allen, SP Stabler, RD Reynolds, FJ Kok, CH Hennekens, WC Willett. Homocysteine metabolism and risk of myocardial infarction: relation with vitamins B_6, B_{12}, and folate. Am J Epidemiol 143:845–859, 1996.
5. EB Rimm, WC Willett, FB Hu, L Sampson, GA Colditz, JE Manson, C Hennekens, MJ Stampfer. Folate and vitamin B_6 from diet and supplements in relation to risk of coronary heart disease among women. JAMA 279(5):359–364, 1998.
6. JF Gregory. Chemical and nutritional aspects of folate research: analytical procedures, methods of folate synthesis, stability, and bioavailability of dietary folates. Adv Food Nutr Res 33:1–101, 1989.
7. PM Finglas, K Wigertz, L Vahteristo, C Witthöft, S Southon, I de Froidmont-Görtz. Standardisation of HPLC techniques for the determination of naturally-occurring folates in food. Food Chem 64:245–255, 1998.
8. JF Gregory, R Engelhardt, SD Bhandari, DB Sartain, SK Gustafson. Adequacy of extraction techniques for determination of folate in foods and other biological materials. J Food Comp Anal 3:134–144, 1990.
9. ID Lumley. Vitamin analysis in foods. In: PB Ottaway, ed. The Technology of Vitamins in Foods. Bishopbriggs, U.K.: Chapman & Hall, 1993, pp 172–232.

10. JF Gregory. Vitamins. In: OR Fennema, ed. Food Chemistry. 3rd ed. New York: Marcel Dekker, 1996, pp 531–616.

11. WJ Mullin, DF Wood, SG Howsam. Some factors affecting folacin content of spinach, Swiss chard, broccoli and Brussels sprouts. Nutr Rep Int 26:7–16, 1982.

12. K Hoppner, B Lampi. Folate retention in dried legumes after different methods of meal preparation. Food Res Int 26:45–48, 1993.

13. CM Pfeiffer, LM Rogers, JF Gregory. Improved analysis of folate in cereal-grain food products using tri-enzyme extraction and combined affinity and reverse-phase liquid chromatography. J Agric Food Chem 45:407–413, 1997.

14. HI Kruschwitz, D McDonald, EA Cossins, V Schirch. 5-Formyltetrahydropteroyl-polyglutamates are the major folate derivatives in *Neurospora crassa* conidiospores. J Biol Chem 269(46):28757–28763, 1994.

15. RG Moran, WC Werkheiser, SF Zakrzewski. Folate metabolism in mammalian cells in culture. I. Partial characterization of the folate derivatives present in L1210 mouse leukaemia cells. J Biol Chem 251(12):3569–3575, 1976.

16. SD Wilson, DW Horne. High-performance liquid chromatographic determination of the distribution of naturally occurring folic acid derivatives in rat liver. Anal Biochem 142:529–535, 1984.

17. DS Duch, SW Bowers, CA Nichol. Analysis of folate cofactor levels in tissues using high-performance liquid chromatography. Anal Biochem 130:385–392, 1983.

18. DL Holt, RL Wehling, MG Zeece. Determination of native folates in milk and other dairy products by high-performance liquid chromatography. J Chromatogr 449:271–279, 1988.

19. E Seyoum, J Selhub. Combined affinity and ion pair column chromatographies for the analysis of food folate. J Nutr Biochem 4:488–494, 1993.

20. L Vahteristo, V Ollilainen, P Koivistoinen, P Varo. Improvements in the determination of reduced folate monoglutamates and folic acid in food using high-performance liquid chromatography. J Agric Food Chem 44:477–482, 1996.

21. CK Clifford, AJ Clifford. High pressure liquid chromatographic analysis of food for folates. J Assoc Off Anal Chem 60:1248–1251, 1977.

22. A Schultz, K Wiedemann, I Bitsch. Stabilization of 5-methyltetrahydrofolic acid and subsequent analysis by reversed-phase high-performance liquid chromatography. J Chromatogr 328:417–421, 1985.

23. S DeSouza, R Eitenmiller. Effects of processing and storage on the folate content of spinach and broccoli. J Food Sci 51:626–628, 1986.

24. JC Gounelle, H Ladjimi, P Prognon. A rapid and specific extraction procedure for folates determination in rat liver and analysis by high-performance liquid chromatography with fluorometric detection. Anal Biochem 176:406–411, 1989.

25. GW Schieffer, GP Wheeler, CO Cimino. Determination of folic acid in commercial diets by anion-exchange solid-phase extraction and subsequent reversed-phase HPLC. J Liq Chromatogr 7:2659–2669, 1984.

26. BT Jacoby, FT Henry. Liquid chromatographic determination of folic acid in infant formula and adult medical nutritionals. J AOAC Int 75:891–898, 1992.

27. R Gauch, U Leuenberger, U Müller. Die Bestimmung von Folsäure (Pteroyl-L-

glutaminsäure) in Lebensmitteln mit HPLC. Mitt Geb Lebensmittelunters Hyg 84: 295–302, 1993.

28. JG Hawkes, R Villota. Folates in foods: reactivity, stability during processing, and nutritional implications. Crit Rev Food Sci Nutr 28(6):438–538, 1989.

29. V Schirch. Enzymatic determination of folylpolyglutamate pools. Methods Enzymol 281:77–81, 1997.

30. RL Blakley. The Biochemistry of Folic Acid and Related Pteridines. Amsterdam: North Holland, 1969.

31. G Farrar, DH Buss, J Loughridge, RJ Leeming, K Huges, JA Blair. Food folates and the British total diet study. J Hum Nutr Dietet 5:237–249, 1992.

32. H Iwase. Determination of folic acid in an elemental diet by high-performance liquid chromatography with UV detection. J Chromatogr 609:399–401, 1992.

33. AOAC. Official Methods of Analysis. 16th ed. Washington: Association of Official Analytical Chemists, 1995.

34. M Shimoda, H-C Shin, E Kokue. Simultaneous determination of tetrahydrofolate, 10-formyltetrahydrofolate and 5-methyltetrahydrofolate in rat bile by high-performance liquid chromatography with electrochemical detection. J Vet Med Sci 56: 701–705, 1994.

35. JI Martin, WO Landen Jr, A-GM Soliman, RR Eitenmiller. Application of a tri-enzyme extraction for total folate determination in foods. J Assoc Off Anal Chem 73:805–808, 1990.

36. S DeSouza, R Eitenmiller. Effects of different enzyme treatments on extraction of total folate from various foods prior to microbiological assay and radioassay. J Micronutr Anal 7:37–57, 1990.

37. JI Rader, CM Weaver, G Angyal. Use of microbiological assay with tri-enzyme extraction for measurement of pre-fortification levels of folates in enriched cereal-grain products. Food Chem 62:451–465, 1998.

38. H-S Lim, AD Mackey, T Tamura, SC Wong, MF Picciano. Measurable human milk folate is increased by treatment with α-amylase and protease in addition to folate conjugase. Food Chem 63:401–407, 1999.

39. T Tamura, Y Mizuno, KE Johnston, RA Jacob. Food folate assay with protease, α-amylase, and folate conjugase treatments. J Agric Food Chem 45:135–139, 1997.

40. T Tamura. Determination of food folate. J Nutr Biochem 9(5):285–293, 1998.

41. K Hoppner, B Lampi. Reversed phase high pressure liquid chromatography of folates in human whole blood. Nutr Rep Int 27:911–919, 1983.

42. M Kohashi, K Inoue. Microdetermination of folate monoglutamates in serum by liquid chromatography with electrochemical detection. J Chromatogr 382:303–307, 1986.

43. L Silan, P Jadaud, LR Whitfield, IW Wainer. Determination of low levels of stereoisomers of leucovorin and 5-methyltetrahydrofolate in plasma using a coupled chiral-achiral high-performance liquid chromatographic system with post-chiral column peak compression. J Chromatogr 532:227–236, 1990.

44. E Schleyer, J Reinhardt, M Unterhalt, W Hiddemann. Highly sensitive coupled-column high-performance liquid chromatographic method for the separation and quantitation of the diastereomers of leucovorin and 5-methyltetrahydrofolate in serum and urine. J Chromatogr B 669:319–330, 1995.

45. P Kelly, J McPartlin, J Scott. A combined high-performance liquid chromato-graphic–microbiological assay for serum folic acid. Anal Biochem 238:179–183, 1996.

46. RJ Leeming, A Pollock, LJ Melville, CGB Hamon. Measurement of 5-methyltet-rahydrofolic acid in man by high-performance liquid chromatography. Metabolism 39:902–904, 1990.

47. M-C Etienne, N Speziale, G Milano. HPLC of folinic acid diastereoisomers and 5-methyltetrahydrofolate in plasma. Clin Chem 39(1):82–86, 1993.

48. S Belz, C Frickel, C Wolfrom, H Nau, G Henze. High-performance liquid chro-matographic determination of methotrexate, 7-hydroxymethotrexate, 5-methyltet-rahydrofolic acid and folinic acid in serum and cerebrospinal fluid. J Chromatogr B: Biomed Appl 661(1):109–118, 1994.

49. O van Tellingen, HR van der Woude, JH Beijnen, CJT van Beers, WJ Nooyen. Stable and sensitive method for the simultaneous determination of N^5-methyltet-rahydrofolate, leucovorin, methotrexate and 7-hydroxymethotrexate in biological fluids. J Chromatogr 488:379–388, 1989.

50. MD Lucock, R Hartely, RW Smithells. A rapid and specific HPLC-electrochemical method for the determination of endogenous 5-methyltetrahydrofolic acid in plasma using solid phase sample preparation with internal standardization. Biomed Chro-matogr 3(2):58–63, 1989.

51. MD Lucock, J Wild, RW Smithells, R Hartley. In vivo characterization of the ab-sorption and biotransformation of pteroylmonoglutamic acid in man: a model for future studies. Biochem Med Metab Biol 42:30–42, 1989.

52. C Witthöft, I Bitsch. Analytische Grundlagen zur Bestimmung der Bioverfügbar-keit von Lebensmittelfolaten für den Menschen. Lebensmittelchemie 49:22–23, 1995.

53. I Pölönen, LT Vahteristo, EJ Tanhuanpää. Effect of folic acid supplementation on folate status and formate oxidation rate in mink (*Mustela vision*). J Anim Sci 75: 1569–1574, 1997.

54. K Wigertz, M Jägerstad. Comparison of a HPLC and radioprotein-binding assay for the determination of folates in milk and blood samples. Food Chem 54(4):429–436, 1995.

55. N Ichinose, T Tsuneyoshi, M Kato, T Suzuki, S Ikeda. Fluorescent high-perfor-mance liquid chromatography of folic acid and its derivatives using permanganate as a fluorogenic reagent. Fresenius J Anal Chem 346:841–846, 1993.

56. AM Bohlmann, H Nau. Folate metabolite pattern in mice following treatment with the antiepileptic drug valproic acid (VPA) and 2-en-VPA: relationship to neural tube defect induction. In: W Pfleiderer, H Rokos, eds. Chemistry and Biology of Pteridines and Folates 1977. Berlin: Blackwell Science, 1997, pp 313–315.

57. MD Lucock, I Daskalakis, CJ Schorah, MI Levene, R Hartley. Analysis and bio-chemistry of blood folate. Biochem Mol Med 58:93–112, 1996.

58. AJA Wright, PM Finglas, S Southon. Erythrocyte folate analysis: a cause for con-cern. Clin Chem 44(9):1886–1891, 1998.

59. R Engelhardt, JF Gregory. Adequacy of enzymatic deconjugation in quantification of folate in foods. J Agric Food Chem 38:154–158, 1990.

60. DM Goli, JT Vanderslice. Investigation of the conjugase treatment procedure in the microbiological assay of folate. Food Chem 43:57–64, 1992.

61. DG Priest, KK Happel, M Mangum, JM Bendarek, MT Doig, CM Baugh. Tissue folypolyglutamate chain-length characterization by electrophoresis as thymidylate synthtase–fluorodeoxyuridylate ternary complexes. Anal Biochem 115:163–169, 1981.

62. DG Priest, MA Bunni, RJ Mullin, DS Duch, J Galivan, MS Rhee. A comparison of HPLC and ternary complex-based assays of tissue reduced folates. Anal Lett 25(2):219–230, 1992.

63. JC Pedersen. Comparison of gamma-glutamyl hydrolase (conjugase; EC 3.4.22.12) and amylase treatment procedures in the microbiological assay for food folates. Br J Nutr 59:261–271, 1988.

64. JF Gregory. Folacin. Chromatographic and radiometric assays. In: J Augustin, BP Klein, DA Becker, PB Venugopal, eds. Methods of Vitamin Assay. 4th ed. New York: John Wiley & Sons, 1985, pp 473–496.

65. RJ Mullin, DS Duch. Folic acid. In: AP de Leenheer, WE Lambert, HJ Nelis, eds. Modern Chromatographic Analysis of Vitamins. 2nd ed. New York: Marcel Dekker, 1992, pp 261–283.

66. RN Reingold, MF Picciano. Two improved high-performance liquid chromatographic separations of biologically significant forms of folate. J Chromatogr 234: 171–179, 1982.

67. DR White, HS Lee, RE Krüger. Reversed-phase HPLC/EC determination of folate in citrus juice by direct injection with column switching. J Agric Food Chem 39: 714–717, 1991.

68. S Belz, H Nau. HPLC coupled with a microbiological assay for the determination of folates. In: W Pfleiderer, H Rokos, eds. Chemistry and Biology of Pteridines and Folates 1997. Berlin: Blackwell Science, 1997, pp 341–344.

69. PJ Bagley, J Selhub. Analysis of folates using combined affinity and ion-pair chromatography. Methods Enzymol 281:16–25, 1997.

70. DW Horne, WT Briggs, C Wagner. High-pressure liquid chromatographic separation of the naturally occurring folic acid monoglutamate derivatives. Anal Biochem 116:393–397, 1981.

71. B Shane. High performance liquid chromatography of folates: identification of poly-γ-glutamate chain lengths of labelled and unlabelled folates. Am J Clin Nutr 35:599–608, 1980.

72. I Eto, CL Krumdieck. Determination of three different pools of reduced one-carbon-substituted folates. III. Reversed-phase high-performance liquid chromatography of the azo dye derivatives of p-aminobenzoylpolyglutamates and its application to the study of unlabeled endogenous pteroylpolyglutamates of rat liver. Anal Biochem 120:323–329, 1982.

73. J Selhub, B Darcy-Vrilloin, D Fell. Affinity chromatography of naturally occurring folate derivatives. Anal Biochem 168:247–251, 1988.

74. J Selhub. Determination of tissue folate composition by affinity chromatography followed by high-pressure ion pair liquid chromatography. Anal Biochem 182:84–93, 1989.

75. MJ Akhtar, MA Khan, I Ahmad. High performance liquid chromatographic determination of folic acid and its photodegradation products in the presence of riboflavin. J Pharm Biomed Anal 16:95–99, 1997.

76. T Rebello. Trace enrichment of biological folates on solid-phase adsorption cartridges and analysis by high-pressure liquid chromatography. Anal Biochem 166: 55–64, 1987.

77. JF Gregory, DB Sartain, BPF Day. Fluorometric determination of folacin in biological materials using high performance liquid chromatography. J Nutr 114:341–353, 1984.

78. GL Case, RD Steele. Determination of reduced folate derivatives in tissue samples by high-performance liquid chromatography with fluorometric detection. J Chromatogr 487:456–462, 1989.

79. H Müller. Bestimmung der Folsäure-Gehalt von Gemüse und Obst mit Hilfe der Hochleistungsflüssigchromatographie (HPLC). Z Lebensm-Unters Forsch 196: 137–141, 1993.

80. H Müller. Die Bestimmung der Folsäure-Gehalte von Lebensmitteln tierisher Herkunft mit Hilfe der Hochleistungsflüssigchromatographie (HPLC). Z Lebensm-Unters Forsch 196:518–521, 1993.

81. H Müller. Bestimmung der Folsäure-Gehalte von Getreide, Getreideprodukten, Backwaren und Hülsenfrüchten mit Hilfe der Hochleistungsflüssigchromatographie (HPLC). Z Lebensm-Unters Forsch 197:573–577, 1993.

82. J Selhub, O Ahmad, IR Rosenberg. Preparation and use of affinity columns with bovine milk folate-binding protein (FBP) covalently linked to Sepharose 4B. Methods Enzymol 66:686–690, 1980.

83. L Vahteristo, K Lehikoinen, V Ollilainen, P Varo. Application of an HPLC assay for the determination of folate derivatives in some vegetables, fruits and berries consumed in Finland. Food Chem 59:589–597, 1997.

84. LT Vahteristo, V Ollilainen, P Varo. Folate monoglutamates is some fish, meat, egg and milk products consumed in Finland: liquid chromatographic determination. J AOAC Int 80(20):373–378, 1997.

85. LM Rogers, CH Pfeiffer, LB Bailey, JF Gregory. A dual-label stable-isotopic protocol is suitable for determination of folate bioavailability: evaluation of urinary excretion and plasma folate kinetics of intravenous and oral doses of $^{13}C_5$ and $^2H_2$ folic acid. J Nutr 127:2321–2327, 1997.

86. RR Eitenmiller, WO Landen Jr. Vitamins. In: IJ Joen, WG Ikins, eds. Analyzing Food for Nutrition Labelling and Hazardous Contaminants. New York: Marcel Dekker, 1995, pp 195–281.

87. CR Santhosh-Kumar, JC Deutsch, KL Hassell, NM Kolhouse, JF Kolhouse. Quantitation of red blood cell folates by stable isotope dilution gas chromatography-mass spectrometry utilizing a folate internal standard. Anal Biochem 225:1–9, 1995.

88. A Hahn, J Stein, U Rump, G Rehner. Optimized high-performance liquid chromatographic procedure for the separation and quantification of the main folacins and some derivatives. I. Chromatographic system. J Chromatogr 540:207–215, 1991.

89. A Mandl, W Lindner. Improved detection of leucovorin in mixed folates and antifolates by reversed-phase liquid chromatography and on-line post-column UV irradiation. Chromatographia 43:327–330, 1996.

90. M Amin, J Reusch. High-performance liquid chromatography of water-soluble vitamins. Part 3. Simultaneous determination of vitamins B_1, B_2, B_6, B_{12} and C, nicotin-

amide and folic acid in capsule preparations by ion-pair reversed-phase high-performance liquid chromatography. Analyst 112:989–991, 1987.

91. E Wang, W Hou. Determination of water-soluble vitamins using high-performance liquid chromatography and electrochemical or absorbance detection. J Chromatogr 447:256–262, 1988.

92. SM El-Gizawy, An Ahmed, NA El-Rabbat. High performance liquid chromatographic determination of multivitamin preparations using a chemically bonded cyclodextrin stationary phase. Anal Lett 24(7):1173–1181, 1991.

93. IN Papadoyannis, GK Tsioni, VF Samanidou. Simultaneous determination of nine water and fat soluble vitamins after SPE separation and RP-HPLC analysis in pharmaceutical preparations and biological fluids. J Liq Chromatogr Rel Technol 20(19):3203–3231, 1997.

94. S Albala-Hurtado, MT Veciana-Nogues, M Izquierdo-Pulido, A Marine-Font. Determination of water-soluble vitamins in infant milk by high-performance liquid chromatography. J Chromatogr A 778:247–253, 1997.

95. JM Scott. Thin-layer chromatography of pteroylmonoglutamates and related compounds. Methods Enzymol 66:437–443, 1980.

96. B Reed, D Weir, J Scott. The fate of folate polyglutamates in meat during storage and processing. Am J Clin Nutr 29:1393–1396, 1976.

97. JP Toth, JF Gregory. Analysis of folacin by GC/MS: Structures and mass spectra of fluorinated derivatives of *para*-aminobenzoyl glutamic acid. Biomed Environ Mass Spectrom 17:73–79, 1988.

98. CR Santhosh-Kumar, NM Kolhouse. Molar quantitation of folates by gas chromatography–mass spectrometry. Methods Enzymol 281:26–38, 1997.

99. JF Gregory, SD Bhandari, LB Bailey, JP Toth, JJ Cerda. Bioavailability of deuterium-labeled monoglutamyl forms of folic acid and tetrahydrofolates in human subjects. Am J Clin Nutr 55:1147–1153, 1992.

100. CM Pfeiffer, LM Rogers, LB Bailey, JF Gregory. Absorption of folate from fortified cereal-grain products and of supplemental folate consumed with or without food determined using a dual-label stable-isotope protocol. Am J Clin Nutr 66:1388–1397, 1997.

101. MA Razzaque, PM Finglas, C Witthoft, B Ridge, P Maunder, AI Mallet. The development of mass spectrometric techniques for the determination of folates. Proceedings 22nd Annual British Mass Spectrometric Society Meeting, Swansea, U.K., 1996.

102. AJ Clifford, A Arjomand, SR Dueker, H Johnson, PD Schneider, RA Zulim, BA Buchholz, JS Vogel. Human folate metabolism using ^{14}C-accelerator mass spectrometry. In: Synthesis and Applications of isotopically labelled compounds 1997. Proceedings of the 6th International Symposium, Philadelphia, 1997. Chichester, U.K.: John Wiley & Sons, 1998, pp 605–608.

103. L Vahteristo, PM Finglas, C Witthöft, K Wigertz, R Seale, I de Froidmont-Görtz. Third EU MAT intercomparison study on food folate analysis using HPLC procedures. Food Chem 57:109–111, 1996.

104. H van den Berg, PM Finglas, C Bates. FLAIR intercomparison on serum and red cell folate. Int J Vitam Nutr Res 64(4):288–293, 1994.

105. EW Gunter, BA Bowman, SP Caudill, DB Twite, MJ Adams, EJ Sampson. Results

of an international round robin for serum and whole blood folate. Clin Chem 42(10): 1689–1694, 1996.

106. PM Finglas, KJ Scott, CM Witthoft, H van den Berg, I de Froidmont-Gortz. The certification of the mass fraction of vitamins in four reference materials: wholemeal flour (CRM 121), milk powder (CRM 421), mixed vegetables (CRM 485) and pig's liver (CRM 487). EUR-report 18320. Luxembourg: Commission of the European Union, 1999.

107. MD Lucock, M Green, M Priestnall, I Daskalakis, MI Levene, R Hartley. Optimisation of chromatographic conditions for the determination of folates in foods and biological tissues for nutritional and clinical work. Food Chem 53:329–338, 1995.

108. M Fawas, M Novovitch, J Alary. Identification et dosage de l'acide folique et des folates par chromatographie liquide haute performance. Validation de la methode. Ann Pharm Fr 46:121–128, 1988.

109. H-C Shin, M Shimoda, E Kokue. Identification of 5,10-methylenetetrahydrofolate in rat bile. J Chromatogr B 661:237–244, 1994.

7
Nicotinic Acid and Nicotinamide

Katsumi Shibata
The University of Shiga Prefecture, Hikone, Shiga, Japan

Hiroshi Taguchi
Mie University, Tsu, Mie, Japan

I. INTRODUCTION

A. History

Pellagra (which is now recognized as the set of symptoms related to niacin deficiency) was first reported as "Mal de la Rosa" by Casal in 1735. The term "pellagra" consisting of a synthesis of the Italian word from *pelle* (skin) and *agra* (rough) was introduced by Frapolli in 1771. This disorder persisted in southern Europe and the southern United States until the early 1900s. The occurrence of pellagra was associated with corn consumption, although pellagra was initially believed to be an infectious disease. Goldberger (1) showed in 1916 that pellagra could not be transmitted to healthy people, and it was postulated that the disease was attributable to a deficiency of a nutrient, the so-called pellagra-preventing factor. In 1937 Elvehjem et al. (2) showed that nicotinic acid cured the black tongue of dogs (similar symptom of human pellagra) and isolated nicotinamide as a pellagra preventing factor from liver. In 1938, Spies et al. (3) showed that administration of nicotinic acid would cure pellagra, and it was gradually accepted that pellagra was caused by a nicotinic acid or nicotinamide deficiency. The pellagragenic effect of corn is now considered to be due to nicotinic acid in a form not released by digestion. It is bound with glucose, imbedded in a glycopeptide (4). Alkali, or even ammonia vapor, will release the vitamin in a form useful to man and animals (4). The fact that Central American diets do not cause pellagra, in spite of a very high corn content, has been attributed to cooking the corn in lime as part of tortilla preparation.

Pyridine-3-carboxylic acid (nicotinic acid) was first discovered as an oxidation product of nicotine by Huber in 1867. In 1873 Weidel described the elemental analysis and crystalline structure of salts and other derivatives of nicotinic acid in some detail. Nicotinic acid was first isolated from rice bran by Suzuki in 1912 and from yeast by Funk in 1913. But, these investigators did not know the anti-pellagra activity of nicotinic acid. Pyridine-3-carboxamide (nicotinamide) was first isolated from liver as an anti-pellagra factor in 1937 by Elvehjem et al. (2) as described above. NAD (NAD^+ + NADH) and NADP ($NADP^+$ + NADPH) were discovered as cozymase of alcohol fermentation and as a hydrogen-transporting coenzyme by Harden and Young in 1904 and by Warburg and Christian in 1934, respectively. Subsequently, nicotinamide was found to be an integral part of NAD^+ (5) and $NADP^+$ (6). Over 350 enzymes require NAD and NADP as coenzyme.

B. Chemical Properties

1. Nicotinic Acid

Nicotinic acid has been designated as pyridine-3-carboxylic acid, pyridine-β-carboxylic acid, PP (pellagra-preventing) factor, vitamin PP, or antipellagra vitamin. The structure of nicotinic acid is given in Figure 1. Nicotinic acid ($C_6H_5O_2N$, molecular weight 123.11) forms colorless, nonhygroscopic needles and is stable in air. It sublimes without decomposition at 234–237°C. It is amphoteric, the pKa's are 4.9 and 2.07. The pH of a saturated aqueous solution is 2.7. One gram dissolves in 60 mL water and in 80 mL ethanol. It is freely soluble in boiling water, alcohol, alkali hydroxides, and carbonates, and soluble in propylene glycol. It is insoluble in diethyl ether. Aqueous nicotinic acid can be autoclaved for 10 min at 120°C without decomposition. Furthermore, nicotinic acid is stable even when autoclaved in 1–2 N mineral acid or alkali. The molar absorptivity is 2800 M^{-1} cm^{-1} at 260 nm in 50 mM potassium phosphate buffer, pH 7.0.

2. Nicotinamide

Nicotinamide is known as pyridine-3-carboxamide, nicotinic acid amide, pellagramine, pyridine-β-carboxylic acid amide, and vitamin PP. The structure of nicotinamide is given in Figure 1. Nicotinamide ($C_6H_6ON_2$, molecular weight 122.12) has a melting point of 129–131°C. It is a colorless crystalline compound that crystallizes as needles from benzene. One gram dissolves in 1 mL water, in 1.5 mL ethanol, and in 10 mL glycerol. Nicotinamide dissolves in acetone, chloroform, and butanol, but is only slightly soluble in diethyl ether and benzene. In neutral solution it is very stable. However, nicotinamide is converted into nicotinic acid in 1 N mineral acid and alkali when heated at 100°C. A 10% (w/v) solution in water has a neutral pH and can be autoclaved for 10 min at 120°C

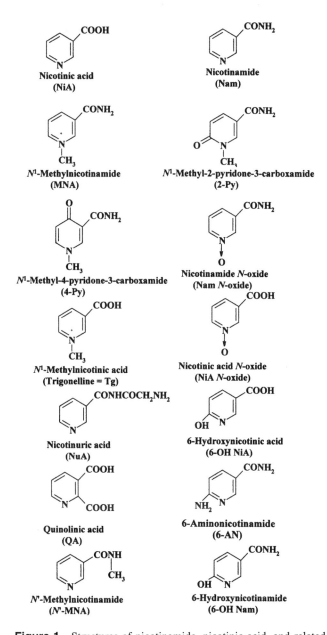

Figure 1 Structures of nicotinamide, nicotinic acid, and related compounds.

without decomposition. The molar absorptivity is 3300 M^{-1} cm^{-1} at 260 nm in water.

3. Nicotinamide Adenine Dinucleotide (NAD$^+$)

NAD$^+$ has been designated as diphosphopyridine nucleotide (DPN), coenzyme I, factor V, codehydrogenase I, Harden's coferment, cozymase I, and codihydrase. NAD$^+$ ($C_{21}H_{27}O_{14}P_2$, molecular weight 663.4) (inner salt) is a very hydroscopic white powder and is freely soluble in water. A 1% solution has a pH of about 2.

4. Nicotinamide Adenine Dinucleotide Phosphate (NADP$^+$)

NADP$^+$ is known as triphosphopyridine nucleotide (TPN), coenzyme II, Warburg's coferment, codehydrogenase II, cozymase II, and phosphocozymase. NADP$^+$ ($C_{21}H_{28}N_7O_{17}P_3$, molecular weight 743.44) (inner salt) is a grayish-white powder. It is soluble in water and in methanol, but much less soluble in ethanol and practically insoluble in diethyl ether and ethyl acetate.

5. Stability of NADH (Reduced NAD$^+$) and NADPH (Reduced NADP$^+$)

NADH and NADPH are rapidly destroyed in acid (99% loss in 0.6 min in 0.02 N HCl (7). However, NADH and NADPH are very stable in alkali (7). Although directly stable to alkali, NADH and NADPH tend to oxidize after long periods of time (7). NADH could be stored for nearly 3 months without decomposition at pH 9.1 either at 4°C or −20°C.

6. Stability of NAD$^+$ and NADP$^+$

Alkali accelerates the rate of destruction of NAD$^+$ and NADP$^+$ (7). However, NAD$^+$ and NADP$^+$ are stable in acidic conditions. Aqueous solutions are stable for about 1 week. When these solutions are neutralized, they are stable for about 2 weeks at 0°C.

C. Biochemistry

1. Biosynthesis

NAD$^+$ is biosynthesized via four pathways as shown in Figure 2: (I) nicotinic acid → NaMN → NaAD → NAD$^+$; (II) nicotinamide → NMN → NAD$^+$; (III) nicotinamide → nicotinic acid → NaMN → NaAD → NAD$^+$; (IV) quinolinic acid → NaMN → NaAD → NAD$^+$. In the four NAD$^+$ biosynthetic pathways, pathways II and IV are physiologically important. Quinolinic acid is synthesized

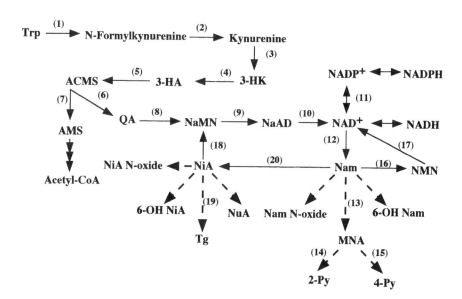

Figure 2 NAD metabolism. Trp = tryptophan, 3-HK = 3-hydroxykynurenine, 3-HA = 3-hydroxyanthranilic acid, ACMS = α-amino-β-carboxymuconate-ε-semialdehyde, AMS = α-aminomuconate-ε-semialdehyde, NaMN = nicotinic acid mononucleotide, NMN = nicotinamide mononucleotide, NaAD = nicotinic acid adenine dinucleotide. For other abbreviations, see Figure 1. (1) tryptophan oxygenase [EC 1.13.11.11], (2) formydase [EC 3.5.1.9], (3) kynurenine 3-hydroxylase [EC 1.14.13.9], (4) kynureninase [EC 3.7.1.3], (5) 3-hydroxyanthranilic acid oxygenase [EC 1.13.11.6], (6) nonenzymatic, (7) aminocarboxymuconate-semialdehyde decarboxylase [EC 4.1.1.45], (8) quinolinate phosphoribosyltransferase [EC 2.4.2.19], (9) NaMN adenylyltransferase [EC 2.7.2.18], (10) NAD$^+$ synthetase [EC 6.3.5.1], (11) NAD$^+$ kinase [EC 2.7.1.23], (12) NAD$^+$ glycohydrolase [EC 3.2.2.5], (13) nicotinamide methyltransferase [EC 2.2.1.1], (14) 2-Py-forming MNA oxidase [EC 1.2.3.1], (15) 4-Py-forming MNA oxidase [EC number not given], (16) nicotinamide phosphoribosyltransferase [EC 2.4.2.12], (17) NMN adenylytransferase [EC 2.7.71], (18) nicotinate phosphoribosyltransferase [EC 2.4.2.11], (19) nicotinate methyltransferase [EC 2.7.1.7], and nicotinamidase [EC 3.5.1.19]. Solid line, biosynthesis; dotted line, catabolism.

from an essential amino acid tryptophan. Table 1 shows the [NAD$^+$ + NADH] (8) and total nicotinamide levels (9) in various tissues of rats.

2. Catabolism

NAD$^+$ is catabolized mainly through five pathways as shown in Figure 2: (I) NAD$^+$ → nicotinamide → MNA → 2-Py; (II) NAD$^+$ → nicotinamide → MNA

Table 1 NAD (NAD$^+$ + NADH) and Total
Nicotinamide (Free Nicotinamide + NAD$^+$ +
NADH + NADP$^+$ + NADPH) Levels in Various
Tissues of Rats

Tissue	NAD	Total nicotinamide
Liver	753 ± 12	1259 ± 42
Kidney	616 ± 30	1061 ± 35
Small intestine	219 ± 24	453 ± 15
Spleen	144 ± 13	504 ± 13
Heart	599 ± 26	1047 ± 16
Brain	271 ± 26	457 ± 13
Testis	154 ± 9	241 ± 6
Muscle	574 ± 20	677 ± 14
Lung	106 ± 9	391 ± 22
Pancreas	233 ± 17	352 ± 7
Blood	95 ± 3	136 ± 4

Each value is expressed as nmol/g tissue wet weight and
means ±SEM for 5 rats.
Source: Refs. 8 and 9.

→ 4-Py; (III) NAD$^+$ → nicotinamide → nicotinamide *N*-oxide; (IV) NAD$^+$ →
nicotinamide → nicotinic acid → nicotinuric acid; (V) NAD$^+$ → nicotinamide
→ nicotinic acid → N^1-methylnicotinic acid (trigonelline). In rats, pathway II
mainly functions, and in humans pathway I. Pathway V functions in mushrooms,
shellfish, and plants, but not in mammals.

D. Physiological Function

1. Pellagra-Preventing Factor

Pellagra is a typical hypovitaminosis disorder. A sufficient supply of niacin to
humans or animals totally prevents the symptoms of pellagra. This disorder is
characterized by three symptoms: dermatitis, diarrhea, and dementia. A number
of nonspecific symptoms precede the actual clinical manifestation, such as an-
orexia, weakness, and maldigestion. The minimum requirement of niacin is about
4.5 mg/1000 kcal.

2. Biochemical Function

The biochemical function of nicotinic acid and nicotinamide actually comes down
to that of NAD$^+$ and NADP$^+$ as coenzymes of dehydrogenases. The latter func-

tion in oxidation reduction systems is well known. Furthermore, NAD^+ is a substrate in nonoxidation/reduction reactions, such as mono-ADP-ribosylation (10), poly-ADP-ribosylation (10), and the formation of cyclic ADP-ribose (11).

E. Pharmacological Function

1. Reduction of Serum Cholesterol and Triglycerides by Nicotinic Acid

The first report that large doses of nicotinic acid reduce serum cholesterol levels in man appeared in 1955 by Altschul (12). Several subsequent studies indicated that nicotinic acid lowers plasma triglycerides as well. However, its use has been limited by the well-known flushing of the face that results from large doses. The conclusions reached by a collaborative research group (13) indicated that it is a safe drug and that it causes a 22% lowering of plasma cholesterol and a 52% lowering of plasma triglycerides. The mechanism by which nicotinic acid lowers plasma cholesterol and triglycerides is not known. Some authors suggest that the synthesis of lipoprotein is reduced in the liver as a result of reduced lipolysis in the adipose tissue (14). Nicotinamide does not possess this pharmacological function.

2. Vasodilatation by Nicotinic Acid

Nicotinic acid is used as a vasodilator. The flush reaction to nicotinic acid which occurs very rapidly is highly characteristic. The biochemical mechanism of the vascular response to nicotinic acid is not well understood. However, some reports suggest that nicotinic acid somehow enhances the biosynthesis or release of prostaglandin E_1, which stimulates adenylate cyclase and raises the level of cAMP (15). Wilson and Douglass (16) reported that the amount of nicotinic acid per injection should be <200 mg because the injection of 250 mg nicotinic acid to humans causes very severe flushing.

3. Nicotinamide in the Prevention of Diabetes

The intraperitoneal injection of streptozotocin to mouse, rat, monkey, and dog induces diabetes. However, when a large amount of nicotinamide is administered before the administration of streptozotocin, the streptozotocin-induced diabetes is inhibited. This is also effective when nicotinamide is administered within 2 hr after the streptozotocin administration (17,18). The NAD content in β-cells of Langerhans islands in the pancreas is reduced by the administration of streptozotocin. However, this decrease does not occur after prior administration of a large amount of nicotinamide (19,20). The decrease in NAD content by streptozotocin is attributable to the stimulation of poly(ADP-ribose) synthetase in the β-cells [21]. Nicotinic acid does not possess this preventive effect.

II. THIN-LAYER CHROMATOGRAPHY AND ION EXCHANGE CHROMATOGRAPHY

A. Introduction

Thin-layer chromatography has not been used for separation of unlabeled nicotinic acid and nicotinamide because of a low sensitivity and a low resolution. These compounds are rather determined by HPLC and GC. However, these chromatographic techniques are still useful for the separation of labeled compounds.

Ion exchange chromatography is still used for preparative purification of nicotinic acid and nicotinamide and their related compounds. Ion exchange systems can separate a large number of substances, but require long analysis times.

B. Thin-Layer Chromatography

Table 2 shows R_f values of nicotinic acid and nicotinamide and their derivatives in different eluents. The detection is performed by illumination under short-wavelength (257.3 nm) UV light.

C. Ion Exchange Chromatography

1. Separation of the Intermediates of Niacin-NAD$^+$ Biosynthetic Pathway on a Dowex 1-Formate Column

Figure 3 shows the elution profile of NMN, nicotinic acid, NAD$^+$, NaMN, NaAD, quinolinic acid, NADP$^+$, and ADP-ribose on a Dowex 1-X2 formate column (29 cm $\times$ 0.8 cm i.d.) (22,23). The elution was carried out by application of a formic acid concave–concentration gradient 0.01 N $\rightarrow$ 0.25 N $\rightarrow$ 2 N $\rightarrow$ 6 N. The mixing chamber initially contained 250 mL of water. The detection was at 260 nm.

2. Separation of Niacin Metabolites on a Dowex 1-Formate Column

Figure 4 shows the elution profile of MNA, N^1-methylnicotinic acid, nicotinamide N-oxide, 4-Py, 2-Py, nicotinamide, 6-hydroxynicotinamide, nicotinic acid, nicotinuric acid, and 6-hydroxynicotinic acid on a Dowex 1-X4 formate column (73.6 cm $\times$ 0.9 cm i.d.) (24). Gradient elution was done with ammonium formate–formic acid of various pHs and ionic strengths. Detection was performed by UV absorption at 254 nm.

Table 2 Paper and Thin-Layer Chromatography of Nicotinic Acid and Nicotinamide and Their Metabolites

	Paper chromatography								Thin-layer chromatography			
	(I)[a]	(II)[a]	(III)[a]	(IV)[a]	(V)[a]	(VI)[a]	(VII)[a]	(VII)[b]	(I)[c]	(II)[c]	(VIII)[c]	(IX)[d]
NADP+	—	0.02	0.24	—	0	—	—	0	0.03	0.50	0.70	—
NAD+	0.13	0.15	0.40	0.62	0	—	—	0	0.13	0.61	0.58	—
NaAD	0.11	0.16	0.26	0.64	0	—	—	0	0.15	0.52	0.57	—
NMN	0.27	0.18	0.42	0.28	0	—	—	0	0.11	0.63	0.73	—
NaMN	0.23	0.20	0.23	0.48	0	—	—	0	0.13	0.47	0.75	—
Nam	0.80	0.81	0.94	0.84	0.61	0.70	—	0.25	0.87	0.88	0.45	0.32
NiA	0.75	0.64	0.89	0.74	0.27	0.11	—	0.49	0.77	0.82	0.55	—
QA	0.41	0.33	—	0.65	0.36	—	0.08	0.20	—	—	—	—
MNA	—	0.67	0.81	—	0.10	0.04	—	0.39	—	—	—	—
Tg	—	0.63	0.66	—	0.06	—	—	—	—	—	—	0.53
Nam N-oxide	—	0.70	0.76	—	0.29	0.30	—	—	—	—	—	0.56
NiA N-oxide	—	0.62	0.52	—	0.10	—	—	—	—	—	—	0.67
2-Py	—	0.80	0.85	—	0.52	0.52	—	—	—	—	—	0.81
4-Py	—	0.70	0.85	—	0.46	0.44	—	—	—	—	—	0.44
NuA	—	0.62	0.66	—	0.12	0.04	0.55	—	—	—	—	—
Reference	72	73	73	72	73	74	74	75	76	76	76	77

Solvent system: (I) 1 M ammonium acetate:95% ethanol (3:7, pH 5.0); (II) isobutyric acid:ammonia:water (66:1.7:33); (III) pyridine:water (2:1); (IV) upper layer of n-butanol:acetone:water (45:5:50); (V) upper layer of n-butanol saturated with water:pyridine (60:1); (VI) upper layer of n-butanol saturated with 3% ammonia; (VII) upper layer of ethyl acetate:formic acid:water (60:1); (VIII) the solvent consists of 600 g ammonium sulfate in 0.1 M sodium phosphate:2% n-propanol (pH 6.8); (IX) isopropanol:conc. HCl:water (70:15:15)

[a] Whatman No. 1 paper, ascending chromatography
[b] Toyo No. 50 paper, ascending chromatography
[c] MN 300G cellulose plate, ascending chromatography
[d] silica gel 60/Kieselguhr F254, ascending chromatography

Abbreviations: NaAD, nicotinic acid adenine dinucleotide; NMN, nicotinamide mononucleotide; NaMN, nicotinic acid mononucleotide; Nam, nicotinamide; NiA, nicotinic acid; QA, quinolinic acid; MNA, N^1-methylnicotinamide; Tg, trigonelline = N^1-methylnicotinic acid; Nam N-oxide, nicotinamide N-oxide; NiA N-oxide, nicotinic acid N-oxide; 2-Py, N^1-methyl-2-pyridone-5-carboxamide; 4-Py, N^1-methyl-4-pyridone-3-carboxamide; NuA, nicotinuric acid.

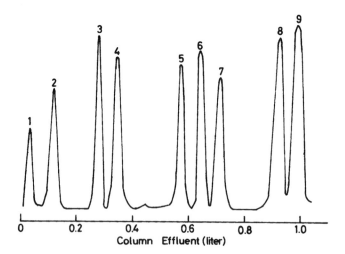

Figure 3 Ion exchange chromatographic separation of the substances involved in bio-synthesis of NAD$^+$ on a conventional column of Dowex 1-X2 formate. Elution was conducted under gravity. Peaks: 1, Nam; 2, NMN; 3, NiA; 4, NAD$^+$; 5, NaMN; 6, NaAD; 7, QA; 8, NADP$^+$; 9, adenosine diphosphate-ribose. For abbreviations, see Figures 1 and 2. The elution was carried out by application of a formic acid concave concentration gradient 0.01 N(150 mL) → 0.25 N(250 mL) → 2 N(400 mL) → 6 N(400 mL). The mixing chamber initially contained 250 ml of water. The detection was at 260 nm. (From Refs. 22 and 23.)

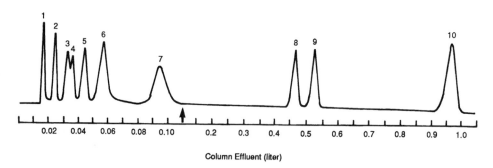

Column Effluent (liter)

Figure 4 Ion exchange chromatographic separation of the substances involved in the catabolism of NAD$^+$ on a conventional column of Dowex 1-X4 formate. Elution was conducted at 0.2 mL/min for the first 110 mL (arrow) and at 1 mL/min with ammonium formate-formic acid of various pH and ionic strengths thereafter. Peaks: 1, MNA; 2, Tg; 3, Nam N-oxide; 4, 4-Py; 5, 2-Py; 6, Nam; 7, 6-OH Nam; 8, NiA; 9, NuA; 10, 6-OH NiA. For abbreviations, see Figure 1. (From Ref. 24.)

III. HIGH-PERFORMANCE LIQUID CHROMATOGRAPHY

A. Introduction

Total niacin (nicotinic acid + nicotinamide) in food is still measured by the microbiological method using *Lactobacillus planturum* or by the colorimetric method. However, these methods suffer from several disadvantages: nicotinic acid and nicotinamide are not distinguished because of hydrolysis before the determination, the microbiological method is time-consuming, and the reproducibility is rather low, whereas the colorimetric method has low sensitivity and the reagents required are harmful and unstable. Reversed-phase chromatographic systems are now the most popular, simple and precise methods.

B. Nicotinic Acid and Nicotinuric Acid

Nicotinic acid is used as a drug; however, nicotinuric acid is not. Nicotinuric acid is a detoxified metabolite of nicotinic acid and is detected in urine only when a large amount of nicotinic acid is administered.

1. Biosamples

In 1978 Hengen et al. (25) measured nicotinic acid and nicotinuric acid in plasma and urine with reversed-phase high-performance liquid chromatography (HPLC). The preparation of plasma and urine extract is time-consuming (two extractions with organic solvents and evaporation to dryness are included), and losses during these procedures are likely. This method requires fairly large amounts of plasma (0.5 mL), and the analytical methods for urine and blood are different. In 1982, McKee et al. (26) measured nicotinamide, MNA, 2-Py nicotinic acid and nicotinuric acid in urine with HPLC using a linear ion pair mobile-phase gradient. This method was fascinating at first glance, but was impractical for the measurement of nicotinic acid and nicotinuric acid in biological samples owing to interfering peaks. In 1984 Tsuruta et al. (27) reported a sensitive method for the determination of nicotinic acid in serum, in which nicotinic acid is reacted with N,N'-dicyclohexyl-O-(7-methoxycoumarin-4-yl) methylisourea (DCCI) in acetone to give the corresponding fluorescent 4-hydroxymethyl-7-methoxycoumarin esters. The compound is separated by reversed-phase HPLC on LiChrosorb RP-18 with isocratic elution using aqueous acetonitrile containing a small amount of sodium 1-hexanesulphonate as the mobile phase. This method requires 100 μL of serum. This method might be superior but DCCI is not commercially available and phosgene is needed for the synthesis of DCCI. In 1988, Shibata (28) reported a simultaneous measurement of nicotinic and nicotinuric acid in blood and urine by a reversed-phase HPLC. This method is the most reliable and practical. A chromatogram of a reference mixture of nicotinic acid and nicotinuric acid is described in Figure 5. The detection limits of nicotinic acid and nicotinuric acid were 10 pmol (1.23 ng) and 10 pmol (1.80 ng), respectively, at a signal-to-noise ratio of 5 : 1.

The chromatograms of extracts of blood samples before and after nicotinic acid administration to rats (25 mg injected intraperitoneally) are shown in Figure 6A and 6B, respectively. Nicotinic acid and nicotinuric acid were not detected in blood (obtained from tail vein) before the nicotinic acid injection. After the nicotinic acid injection, only nicotinic acid was detected in blood. The chromatograms of extracts of urine samples before and after nicotinic acid injection are shown in Figure 7A and 7B, respectively. Nicotinic acid and nicotinuric acid were not detected in urine before the nicotinic acid injection.

2. Food

Nicotinic acid is contained mainly in plant food such as coffee, cereals, and seeds, but not in animal foods, such as fish, pork, beef, and chicken (animal food contains nicotinamide). There are no reports about the determination of nicotinic acid in animal food. Coffee contains appreciable amounts of nicotinic acid. Nicotinic acid contents in roasted coffees and instant coffees in the range of about 24 mg/100 g (29). In 1985, Trugo et al. (30) reported the measurement of nicotinic acid in instant coffees. Chromatograms of reference nicotinic acid and the extract of instant coffee are shown in Figure 8. The detection limit of nicotinic acid was 0.8 nmol (0.1 μg). Tyler and Shargo (31) reported a determination of

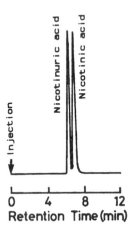

Figure 5 Chromatogram of a reference mixture of nicotinic acid and nicotinuric acid. Conditions: column, Chemcosorb 5-ODS-H (150 mm × 4.6 mm i.d.); mobile phase, 10 mM potassium phosphate buffer (pH 7.0) containing 5 mM tetra-*n*-butylammonium bromide-acetonitrile (100:9, v/v); flow-rate, 1 mL/min; detection wavelength, 260 nm; column temperature, 25°C. (From Ref. 28.)

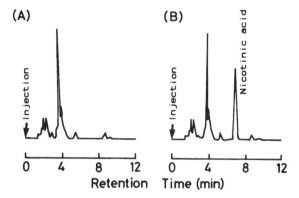

Figure 6 (A) Chromatogram of the extract of rat blood before the nicotinic acid injection. (B) Chromatogram of the extract of rat blood, 0.5 hr after the nicotinic acid injection (25 mg nicotinic acid per rat dissolved in 1 mL sterilized saline was intraperitoneally injected). For conditions, see Figure 5. (From Ref. 28.)

nicotinic acid in cereal samples by reversed-phase HPLC with a UV method using a reversed-phase column.

Hirayama and Maruyama (32) reported the HPLC measurement of a small amount of nicotinic acid in foodstuffs such as rice vinegar, crane vinegar, strawberry jam, orange jam, and apple jam. This method is applicable for variety of foods and performs excellent. The detection limit of nicotinic acid is about 0.01 mg/100 g food. A chromatogram is shown in Figure 9. The other HPLC methods for analysis of nicotinic acid are summarized in Table 3.

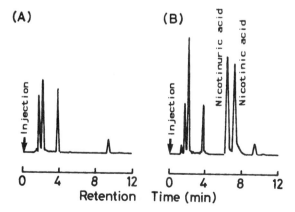

Figure 7 (A) Chromatogram of the extract of rat urine before the nicotinic acid injection. (B) Chromatogram of the extract of rat urine excreted 0 to 5 hr after the nicotinic acid injection (25 mg/rat). For conditions, see Figure 5. (From Ref. 28.)

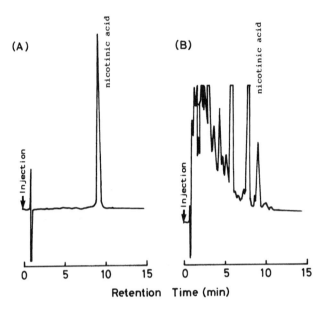

Figure 8 (A) Chromatogram of a reference nicotinic acid. (B) Chromatogram of the extract of instant coffee. Conditions: column, Spherisorb ODS-2 (150 mm × 5 mm i.d.); mobile phase, 5 m*M* tetra-*n*-butylammonium hydroxide (pH adjusted to 7.0 with 4 *M* sulphuric acid)-methanol (92:8, v/v); flow rate, 1.5 mL/min; detection wavelength, 254 nm; column temperature, ambient. (From Ref. 30.)

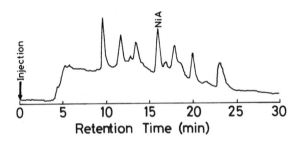

Figure 9 Chromatogram of niacin extracted from strawberry jam (niacin content; 0.07 mg/100 g) after clean-up. Conditions: column; Asahipak NH₂ P-50 (250 × 4.6 mm i.d.); mobile phase, acetonitrile:water (6:4) containing 75 m*M* sodium acetate; flow rate, 0.5 mL/min; detection wavelength 261 nm; column temperature, ambient. (From Ref. 32.)

Table 3 HPLC Determination of Nicotinic Acid in Biosamples

Reference	Source	Pretreatment	Column	Mobile phase	Detection	Detection limit	Other compounds detected simultaneously
Chase (78)	Food	H_2SO_4 extraction, florisil column	Hamilton PRP-X 100 (250 × 4.1 mm)	2% acetic acid	254 nm	0.11 μg/mL	
Pelzer (79)	Plasma	Acetonitrile extraction	IB-SIL CN (150 × 4.6 mm)	acetonitrile:methanol: water: acetic acid (700:150:150:1)	263 nm	20 ng/mL	6-methylnicotinc acid (IS)
Iwaki (80)	Urine	Bond Elut SCX column	Inertsil ODS-2 (250 × 4.6 mm)	10 mM potassium phosphate buffer (pH 7.0) containing tetra-n-ammonium phosphate: acetonitrile (9:1)	254 nm	40 ng/injection	nicotinuric acid 6-Methylnicotinic acid (IS)

Table 4 HPLC Determination of Nicotinamide in Biosamples

Reference	Source	Pretreatment	Column	Mobile phase	Detection	Detection limit	Other compounds detected simultaneously
De Vries (77)	Plasma Urine	Sep Pak C_{18} cartrige	μBondpack C_{18} (250 × 4 mm) LiChrosorbRP-18 (300 × 4 mm)	4.446 g Dioctylsulfosuccinate/ 1450 mL water + 1050 mL methanol	254 nm	0.1 mg/L (plasma) 1.0 mg/L (urine)	Isonicotinic acid (IS) Nicotinic acid Nicotinuric acid
McKee (26)	Urine	Filtration through 0.2-μm membrane filter	Ultrasphere-ODS (150 × 4.6 mm)	Linear gradient: A, 10 mM KH_2PO_4 containing 10 mM PSA and 10 mM TMA (pH 3.3)	254 nm	0.1 μg/injection	Nicotinic acid MNA Nicotinuric acid 2-Py
Takatsuki (81)	Meat	Water extraction, deproteinization with zinc sulfate	μBondpack C_{18} (300 × 3.9 mm)	10 mM PIC-B7	263 nm	1 mg/100 g meat	
Shibata (34)	Urine Blood Tissues Meat Drug	Diethyl ether extraction	Chemcosorb 7-ODS-L (250 × 4.6 mm) Tosoh ODS 80_{TS} (250 × 4.6 mm)	10 mM KH_2PO_4 (pH 3.0):acetonitrile (96:4)	260 nm	1.22 ng/injection	Isonicotinamide (IS) 2-Py 4-Py

Reference	Sample	Extraction	Column	Mobile phase	Detection	Amount	Compounds
Hamano (82)	Meat	Boiling water extraction	Partisil SCX (250 × 4.6 mm)	50 mM KH$_2$PO$_4$ (pH 3.0)	260 nm	4 ng/injection	Ascorbic acid, Sorbic acid, Nicotinic acid
Balschukat (83)	Food	HCl extraction	Precolumn: RP-18 (17 × 4.6 mm), Column A: RP-18 (250 × 4.6 mm), Column B: Nucleosil 5 SA cation-exchange column (250 × 4.6 mm)	Column-switching method, 10 mM KH$_2$PO$_4$, 10 mM KH$_2$PO$_4$:acetonitrile (6:4)	264 nm	0.2 mg/100 g food	
Mawatari (84)	Serum	Perchloric acid extraction	Capcell pak C18 (250 × 4.6 mm)	70 mM KH$_2$PO$_4$ (pH 4.5) containing 75 mM H$_2$O$_2$ and 5 μM CuSO$_4$	Exi: 322 nm, Emi: 380 nm	0.6 ng/injection	Nicotinic acid, MNA
Iwase (85)	Elemental diet	Hot water extraction	Capcell pak C$_{18}$ (250 × 4.6 mm)	Acetonitrile:water (pH 2.1, adjusted with phosphoric acid) (1.5:98.5)	260 nm	5.5 ng/injection	Pyridoxine
Miyauchi (86)	Plasma	Bakerbond spe 7090-03	Shim-pack CLC-ODS (150 × 6	10 mM phosphate buffer (pH 2.1) with 10 mM sodium octanesulfonic acid and 3% acetonitrile	254 nm	Not described	Isonicotinic acid (IS), Nicotinic Acid, Nicotinuric acid

Abbreviations: PSA, 1-pentanesulfonic acid sodium salt; TMA, tetramethylammmonium chloride; PIC-B7, heptane sulfonic acid, adjusted to pH 3.5 with phosphoric acid.

C. Nicotinamide, N¹-Methyl-2-pyridone-5-carboxamide (2-Py), and N¹-Methyl-4-pyridone-3-carboxamide (4-Py)

Nicotinamide is present in pork, beef, chicken, and fish. 2-Py and 4-Py are the major metabolites of nicotinic acid and nicotinamide excreted in urine. The contents of 2-Py and 4-Py in blood are below the limit of detection (33). The HPLC methods for analysis of nicotinamide and the related compounds are summarized in Table 4. Shibata et al. (34) reported the simultaneous measurement of nicotinamide, 2-Py, and 4-Py. This method is commonly applicable not only to urine and pharmaceutical preparations but also biological materials and foods. Chromatograms of a reference mixture of isonicotinamide (used as an internal standard), nicotinamide, 2-Py, and 4-Py and of extracts of rat urine, human urine, rat liver, and of the extract of multivitamin preparations are shown in Figure 10. The detection limits for nicotinamide, 2-Py, and 4-Py were 4 pmol (552 pg), 10 pmol (1220 pg), and 2 pmol (304 pg), respectively, at a signal-to-noise ratio of 5:1. Daily urinary excretion of nicotinamide, 2-Py and 4-Py in rats, mice, guinea pigs, hamsters and humans is given in Table 5.

D. NAD⁺, NADH, NADP⁺, and NADPH

Chromatographic analysis of NAD^+, NADH, $NADP^+$, and NADPH is not practical because these compounds are not stable. In biological materials they are mea-

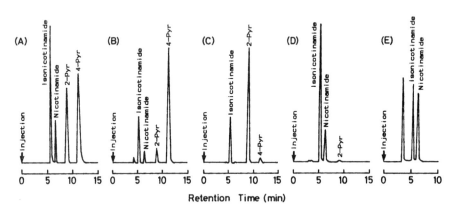

Figure 10 (A) Chromatogram of a reference mixture of isonicotinamide, nicotinamide, 2-Py, and 4-Py. (B) Chromatogram of an extract of rat urine. (C) Chromatogram of an extract of human urine. (D) Chromatogram of an extract of rat liver. (E) Chromatogram of an extract of a multivitamin complex. Conditions: column, Chemcosorb 7-ODS-L (250 mm × 4.6 mm i.d.); mobile phase, 10 mM KH_2PO_4 (pH adjusted to 3.0 by addition of conc. phosphoric acid), plus acetonitrile (96:4, v/v); flow-rate, 1 mL/min; detection wavelength, 260 nm; column temperature, 25°C. (From Ref. 34.)

Table 5 Daily Urinary Excretion of Nicotinic Acid and Nicotinamide and Their Catabolic Metabolites

	Human[a]	Rat[b]	Mouse[c]	Hamster[c]	Guinea pig[c]
Nicotinamide (μmol/day)	N.D.	0.51 ± 0.03	0.18 ± 0.03	0.58 ± 0.04	0.10 ± 0.03
MNA (μmol/day)	31 ± 12	0.85 ± 0.09	0.16 ± 0.04	0.13 ± 0.02	0.35 ± 0.01
2-Py (μmol/day)	60 ± 27	0.78 ± 0.09	0.20 ± 0.06	0.27 ± 0.06	2.60 ± 0.09
4-Py (μmol/day)	7 ± 3	9.78 ± 9.27	0.10 ± 0.04	0.13 ± 0.03	0.10 ± 0.01
Nam N-oxide (μmol/day)	N.D.	0.13 ± 0.01	0.35 ± 0.04	0.20 ± 0.04	0.10 ± 0.03
NiA (μmol/day)	N.D.	N.D.	N.D.	N.D.	N.D.
NuA (μmol/day)	N.D.	N.D.	N.D.	N.D.	N.D.

[a] Values are means ± SD (n = 84). From Ref. 87.
[b] Values are means ± SEM (N = 5). From Ref. 88.
[c] Values are means ± SEM (n = 4–5). From Ref. 44.

sured usually by enzymatic methods (35,36). However, HPLC analysis of pyridine nucleotide coenzymes in rat liver has been reported. Chromatograms of an extract of rat liver are shown in Figures 11 and 12 (37).

Klaidman et al. (38) reported a fluorometric ultrasensitive HPLC method for NAD^+, NADH, $NADP^+$, and NADPH. The oxidized nucleotides were transformed to stable fluorescent products with cyanide in a basic solution. A chromatogram of a reference mixture of NAD^+ and the related compounds is shown in Figure 13. In our laboratory, standard NAD, NADPH, NADP, and NADPH have been reported using HPLC as shown in Figure 14.

E. N^1-Methylnicotinamide (MNA)

Kutnink et al. (39) reported an isocratic reversed-phase HPLC-UV method for the determination of MNA in human urine. A method for urinary MNA and 2-Py using isocratic reversed-phase HPLC-UV detection after a simple anion exchange cleanup procedure has been described by Carter (40). However, UV detection suffers from sensitivity. In 1987, Shibata (41) developed a highly sensitive HPLC method with fluorescence detection. MNA was reacted with acetophenone in a strongly alkaline medium at 0°C in the presence of a large amount of isonicotinamide. After 10 min, formic acid was added and the mixture was kept at 0°C for another 15 min. The mixture was heated in a boiling water bath for 5 min.

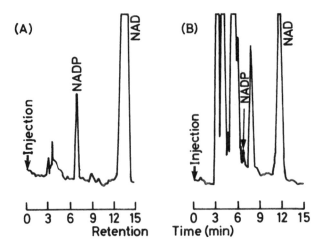

Figure 11 (A) Chromatogram of a reference mixture of NAD$^+$ and NADP$^+$. (B) Chromatogram of an extract of rat liver. Conditions: column, µBondpak C$_{18}$ (150 × 4.6 mm i.d.); mobile phase, 0.2 M ammonium phosphate buffer (pH 5.25): methanol (97:3, v/v); flow rate, 1.1 mL/min; detection wavelength, 254 nm; column temperature, ambient. (From Ref. 37.)

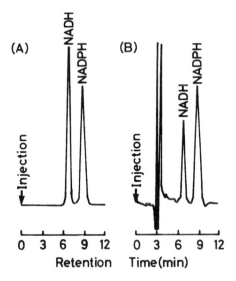

Figure 12 (A) Chromatogram of a reference mixture of NADH and NADPH. (B) Chromatogram of an extract of rat liver. Conditions: column, µBondpak C$_{18}$ (150 × 4.6 mm); mobile phase, 0.2 M ammonium phosphate buffer (pH 6.0)-methanol-trimethylamine (82: 17.87:0.13); flow rate, 1.1 mL/ min; detection wavelength, 340 nm; column temperature, ambient. (From Ref. 37.)

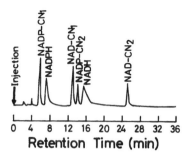

Figure 13 Chromatograms of a reference mixture of NAD$^+$ and related compounds. Conditions: column, Rexchrome-ODS (250 × 4.6 mm i.d., particle size, 5 μm); mobile phase, initially 96% 0.2 M ammonium acetate and 4% methanol. At 1 min, the methanol was set to increase by 0.2% per min and allowed to increase for 25 min; flow rate, 1 mL/min; excitation and emission wavelengths of 330 nm and 460 nm; column temperature, ambient. NAD(P)-CN$_1$, NAD(P)$^+$ added cyanide in the α-configuration at the 4-position of the Nam ring; NAD(P)-CN$_2$, NAD(P)$^+$ added cyanide in the β-configuration at the 4-position of the Nam ring. (From Ref. 38.)

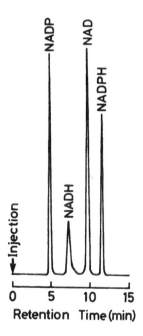

Figure 14 Chromatogram of a reference mixture of NAD$^+$, NADP$^+$, NADH, and NADPH. Conditions: column, Chemcosorb 5-ODS-H (150 mm × 4.6 mm i.d.); mobile phase, acetonitrile in 25 mM KH$_2$PO$_4$ (pH 4.5) increasing linearly from 1% to 7% at 0.4%/min; flow rate, 1 mL/min; detection wavelength, 260 nm, column temperature, 25°C.

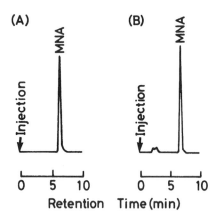

Figure 15   (A) Chromatogram of 1-methyl-7-phenyl-1,5-dihydro-5-oxo-1,6-naphthyridine from reference MNA. (B) Chromatogram of 1-methyl-7-phenyl-1,5-dihydro-5-oxo-1,6-naphthyridine from MNA in rat urine. Conditions: column, Chemcosorb 300-7SCX (150 mm × 4.6 mm i.d.); mobile phase, 31.25 mM KH_2PO_4 (pH 4.5): acetonitrile (80: 20, v/v); flow rate, 1 mL/min; excitation and emission wavelengths of 382 nm and 440 nm; column temperature, 25°C. (From Ref. 41.)

The reaction product—1-methyl-7-phenyl-1,5-dihydro-5-oxo-1,6-naphthyridine—was analyzed by HPLC. When acetophenone is omitted from the mixture, the reaction product is not formed. A large amount of isonicotinamide was added to protect MNA from the deamidation reaction yielding N^1-methylnicotinic acid. N^1-Methylnicotinic acid did not react with acetophenone. This approach can be applied to liver and blood extracts as well as to urine (42). The detection limit was 0.01 pmol (1.37 pg as MNA) at a signal-to-noise ratio of 5:1. Chromatograms of reference MNA and MNA in urinary extracts are shown in Figure 15. Daily urinary excretion of MNA in rats, mice, guinea pigs, hamsters, and humans is given in Table 5.

F. Nicotinamide *N*-oxide

Nicotinamide *N*-oxide is synthesized from nicotinamide, probably in the liver and excreted into the urine. Nicotinamide *N*-oxide is detected only in rodents such as rats, mice, guinea pigs, and hamsters (43,44). The nicotinamide *N*-oxide content of rat and mouse liver was below the limit of detection (unpublished data). A method for the determination of nicotinamide *N*-oxide has been reported by Shibata (45). Chromatograms of reference nicotinamide *N*-oxide and an extract of rat urine are shown in Figure 16. Chloroform was used to extract nicotinamide N-oxide from urine, saturated with potassium carbonate. The detection limit for nicotinamide N-oxide was 2 pmol (276 pg) at a signal-to-noise ratio of 5:1. 2-Py and 4-Py could also be extracted completely, the two compounds elute at around 8 min under the present HPLC conditions (Figure 16B). When this sample was subjected to the HPLC system for simultaneous analysis of nicotinamide, 2-Py and 4-Py (Fig. 10), the peak of nicotinamide N-oxide overlapped partly with an unknown peak. However, under the present analytical conditions, nicotinamide N-oxide was eluted without interfering peaks.

Nicotinamide was also partly extracted in this process and eluted at about 10 min (Fig. 16B). Nicotinamide N-oxide eluted at the same time even when sodium 1-octanesulfonate was omitted from the mobile phase, although identification of the peak of nicotinamide N-oxide was easier in the presence of sodium 1-octanesulfonate than in its absence, because no interfering peaks were eluted at the elution time of nicotinamide N-oxide. Daily urinary excretion of nicotinamide N-oxide in rats, mice, guinea pigs, and hamsters is given in Table 5.

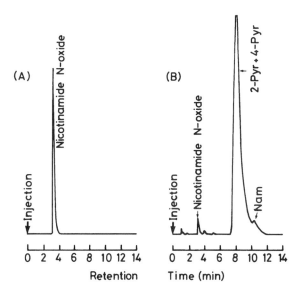

Figure 16 (A) Chromatogram of reference nicotinamide N-oxide. (B) Chromatogram of an extract of rat urine. Conditions: column, Chemcosorb 5-ODS-H (150 mm × 4.6 mm i.d.); mobile phase, 10 mM KH$_2$PO$_4$ (pH adjusted to 3.0 by the addition of conc. phosphoric acid) containing 0.1 g/L 1-octanesulfonate-methanol (100:4, v/v); flow rate, 1 mL/min; detection wavelength, 260 nm; column temperature, 25°C. (From Ref. 45.)

G. N¹-Methylnicotinic Acid (Trigonelline)

Trigonelline (N^1-methylnicotinic acid) was named after the leguminous plant *Trigonella foenum graecum* L. (fenugreek), from which the compound was first isolated and characterized (46). In plant cells, N^1-methylnicotinic acid functions as a "G2 factor" which promotes cell arrest in the G2 stage of the cell cycle (47). However, N^1-methylnicotinic acid is inactive in mammalian cells; e.g., when a large amount of N^1-methylnicotinic acid was fed to weanling rats, the body weight gain was not affected compared with the control (48). All of the administered N^1-methylnicotinic acid was recovered unchanged from urine (48).

The HPLC determination of N^1-methylnicotinic acid has been reported by Trugo et al. (49). Chromatograms of a reference mixture of N^1-methylnicotinic acid, theobromine, theophylline, and caffeine and of an extract of instant coffee are shown in Figure 17.

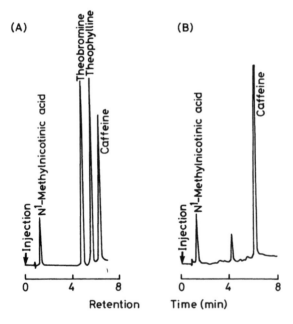

Figure 17   (A) Chromatogram of a reference mixture of N^1-methylnicotinic acid, theobromine, theophylline, and caffeine. (B) Chromatogram of an extract of instant coffee. Conditions: column, Spherisorb ODS (250 mm × 5 mm i.d.); mobile phase, methanol in 15 mM tripotassium citrate buffer (pH 6.0) increasing linearly from 0% to 60% at 10%/min; flow rate, 2 mL/min; detection wavelength, 272 nm; column temperature, ambient. (From Ref. 49.)

H. Quinolinic acid

Quinolinic acid is synthesized not only from tryptophan in mammals (50) but also from aspartic acid and dihydroxyacetone phosphate in microorganisms (51). Therefore, quinolinic acid is the key intermediate in the de novo NAD biosynthethic pathway. Furthermore, quinolinic acid has been shown to excite nerve cells in rodents and primates on iontophoretic application (52) and intracerebral injection of quinolinic acid in rats results in selective "axon-sparing" neuronal lesions (53). The quinolinic acid content in biological materials is very low, so its content cannot be measured by the HPLC-UV method described under (28) in Section III.B. Recently, Mawatari et al. (54) found that quinolinic acid induces fluorescence by photoirradiation in the presence of hydrogen peroxide and established a system based on HPLC with detection by the fluorescence reaction. The detection limit was 1.1 pmol. Chromatograms of reference quinolinic acid and of quinolinic acid in urine are shown in Figure 18.

I. Others

1. 6-Aminonicotinamide

6-Aminonicotinamide causes a disease similar to human pellagra (55) and the lethal effect of 6-aminonicotinamide is prevented by simultaneous injection of nicotinic acid or nicotinamide (56). Therefore, it is one of the putative pellagragenic compounds. This antiniacin compound can be measured by using the

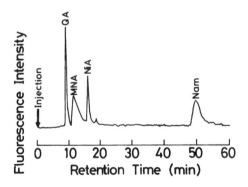

Figure 18   Chromatogram of a reference mixture of QA, MNA, NiA, and Nam. Conditions: column, Unisil Q C18 (250 × 4.6 mm i.d.); mobile phase, 35 mM KH_2PO_4 (adjusted to pH 3.8 with 0.2 M citric acid) containing 350 mM hydrogen peroxide and 0.05 mM TMA; flow rate, 0.6 mL/min; excitation and emission wavelength of 326 nm and 380 nm; column temperature, ambient. (From Ref. 54.)

same HPLC conditions of Figure 10 and eluted at around 4 min (57). A chromatogram is shown in Figure 19.

2. Nicotinic Acid N-oxide

Standard nicotinic acid *N*-oxide can be determined by the HPLC system described in Section III.B (28). Nicotinic acid *N*-oxide was eluted at around 3.5 min, as shown in Figure 20. Nicotinic acid *N*-oxide is not detected in biological materials.

3. 6-Hydroxynicotinic Acid and 6-Hydroxynicotinamide

6-Hydroxynicotinic acid and 6-hydroxynicotinamide were found as urinary excretion metabolites in rats following intraperitoneal injection of nicotinic acid-7-^{14}C and nicotinamide-7-[14]C (58). The acid was eluted after about 6 min in the HPLC system as in Figure 10 (unpublished data). 6-Hydroxynicotinic acid is available from commercial sources, but 6-hydroxynicotinamide is not. Therefore, the determination of 6-hydroxynicotinamide using HPLC has not been reported.

4. NaMN, NMN, and NaAD

NaMN, NMN, and NaAD are intermediates in the biosynthesis of NAD. The contents of these substances in biological materials are very low. Determinations

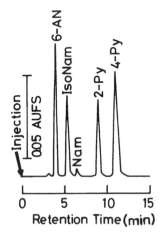

Figure 19   Chromatogram of an extract of rat urine fed with an NiA-free and 20% casein diet being added 0.005% 6-AN. For conditions, see Figure 10. (From Ref. 57.)

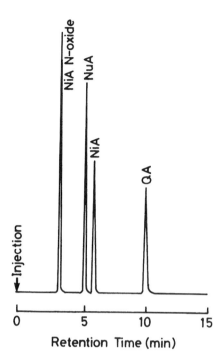

Figure 20 Chromatogram of a reference mixture of nicotinic acid *N*-oxide, nicotinuric acid, nicotinic acid, and quinolinic acid. For conditions, see Figure 5.

of these substances have not been reported. However, in our laboratory, the purity of standard NaMN, NMN, and NaAD is routinely checked by HPLC, as demonstrated in Figure 21.

Rocchingiani et al. (59) reported the separation of standard NMN, NaMN, nicotinamide, NAD$^+$, nicotinic acid, NADP$^+$, and NaAD (Fig. 22).

5. Cyclic Adenosine Diphosphate Ribose (cADPR) and Adenosine Diphosphate Ribose (ADPR)

cADPR is an endogenous metabolite of NAD$^+$ and involves mobilization of intracellular Ca^{2+} (60) and ADPR is the product of cADPR. These reactions of NAD$^+$ → cADPR → ADPR are catalyzed by ADP-ribosyl cyclase. Casabona et al. (61) reported the separation of these compounds by HPLC with an amperometric detector. A chromatogram is shown in Figure 23.

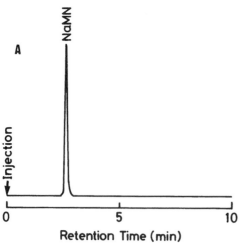

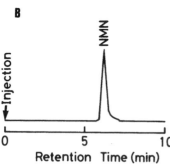

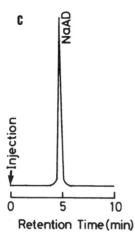

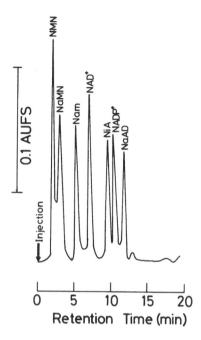

Figure 22 Chromatogram of a reference mixture of pyridine compounds. Conditions: column, ODS-2 Spherisorb (125 × 4.6 mm, i.d.); mobile phase, gradient elution with 0.1 M KH$_2$PO$_4$, containing 6 mM tetra-n-butylammonium hydrogen sulfate, at pH 5.5 (eluant A), and methanol (eluant B). Initial conditions 96% A and 4% B, increasing stepwise to 10% B in 5 min, 12% B at 10 min, and 30% B from 13 to 15 min, followed by a return to initial conditions (96% A and 4% B) from 19 to 20 min. Flow rate, 1 mL/min; detection wavelength, 254 nm; column temperature, ambient. (From Ref. 59.)

Figure 21 (A) Chromatogram of reference nicotinic acid mononucleotide (NaMN). Conditions: column, Tosoh TSKgel ODS-80$_{τM}$ (150 mm × 4.6 mm i.d.); mobile phase, 10 mM KH$_2$PO$_4$ (pH adjusted to 3.0 by the addition of conc. phosphoric acid): acetonitrile (98:2, v/v); flow rate, 1 mL/min; detection wavelength, 260 nm; column temperature, 25°C. (B) Chromatogram of reference nicotinamide mononucleotide (NMN). Conditions: column, Shimadzu PNH$_2$-10/S2504 (250 mm × 4.0 mm i.d.); mobile phase, 10 mM KH$_2$PO$_4$ (pH 4.5); flow rate, 1 mL/min; detection wavelength, 260 nm; column temperature, 35°C. (C) Chromatogram of reference nicotinic acid adenine dinucleotide (NaAD). Conditions: column, Shimadzu PNH$_2$-10/S2504 (250 mm × 4.0 mm i.d.); mobile phase, 250 mM KH$_2$PO$_4$ (pH 4.5); flow rate, 1 mL/min; detection wavelength, 260 nm; column temperature, 45°C.

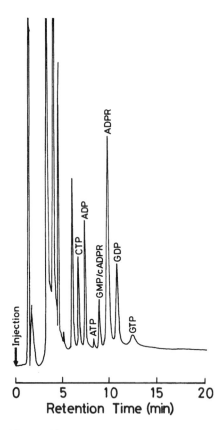

Figure 23 Chromatogram of a crude extract of synaptic hippocampal membranes incubated for 20 min in the presence of NAD^+, GTP, and ATP. CTP was added as internal standard. Conditions: column, Bio-Sil O 90-5S strong anion exchange (SAX) (150 × 4.6 mm i.d.); mobile phase, buffer A (7 mM KH_2PO_4 containing 7 mM KCl at pH 4.0), buffer B (0.25 M KH_2PO_4 containing 0.5 M KCl at pH 5.0), a linear gradient from 100% A to 100% B starting from 5 to 40 min, after 5 initial min of isocratic conditions with 100% A. Flow rate, 1 mL/min; detection wavelength, 254 nm; column temperature, ambient. (From Ref. 61.)

IV. GAS CHROMATOGRAPHY

A. Introduction

Nicotinamide and nicotinic acid are insufficiently volatile to be analyzed directly by GC. However, after suitable derivatization their GC analysis becomes feasible.

B. Nicotinamide

Tanaka et al. (62) reported that the determination of nicotinamide as 3-cyanopyridine after dehydration by heptafluorobutyric anhydride was performed by GC with flame ionization detection and using a column of 5% OV-17 on Chromosorb W AW DMCS at 130°C. The minimum detectable amount of 3-cyanopyridine was ~2.0 µg. Nicotinamide in foods was extracted with acetonitrile without the need for a cleanup stage.

C. N'-Methylnicotinamide

N'-Methylnicotinamide was first identified from human urine by Holmen et al. (63) in 1981. The identification and quantification of N'-methylnicotinamide in the urine samples was performed by GC and GCMS. However, it is not clarified whether N'-methylnicotinamide is a metabolite of nicotinic acid and nicotinamide or not (e.g., it derives from dietary, drinking, or smoking habits).

We measured N'-methylnicotinamide in urine using Japanese healthy women (20 to 22 years old), rat, and mouse by a HPLC method (64); however, we could not detect N'-methylnicotinamide.

V. MASS SPECTROMETRY

Quinolinic acid was identified as a metabolite of tryptophan in the rat by Henderson and Hirsch in 1949 (65). Quinolinic acid has been known to be a key intermediate of the tryptophan-NAD and aspartate + dihydroxyacetone phosphate-NAD pathways since the clear demonstration in vitro by Nishizuka and Hayaishi in 1963 (50). It has been proven that quinolinic acid is transformed into nicotinic acid mononucleotide in the presence of 5-phosphoribosyl-1-pyrophosphate by quinolinate phosphoribosyltransferase (50). In 1978 Lapin (52) first reported that quinolinic acid induced seizures when injected directly into the brains of mice. Later, quinolinic acid was shown to increase neuronal activity when ionophoretically applied to neurons in rats cerebral cortex, striatum, and hippocampus (53,66,67). Now, quinolinic acid is known as a potent neurotoxin and as a precursor of NAD in the liver.

GC analysis of nonvolatile compounds such as quinolinic acid requires a derivatization step. Heyes and Markey (68) have reported quantification of quinolinic acid in rat brain, whole blood, and plasma after quinolinic acid derivatization to its hexafluoroisopropyl (HFIP) ester and GC electron capture negative ionization (CI) MS with [^{18}O]quinolinic acid as an internal standard. Negative CI efficiently forms a characteristic molecular anion and consequently negative CI GC/

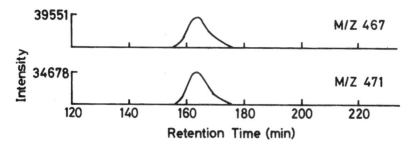

Figure 24   Selected ion recording of QA-HFIP and [^{18}O]QA-HFIP at m/z 467 and 471. Sample: rat cerebral cortex including the [^{18}O]QA internal standard. HFIP = hexafluoroisopropanol. Conditions: column, SP-2401 (1.5 m × 1.5 mm i.d.); column temperature, 115°C; carrier gas, methane; flow rate, not described. (From Ref. 68.)

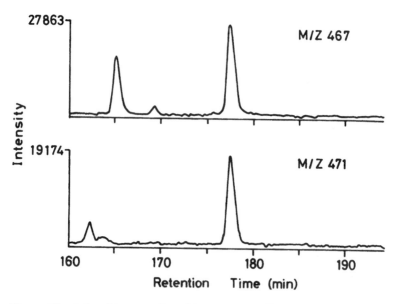

Figure 25   Selected ion recording of QA-HFIP and [^{18}O]QA-HFIP at m/z 467 and 471. Sample, rat frontal cortex including the [^{18}O]QA internal standard. Conditions: column, DB-5 (30 m × 0.25 mm i.d.; column temperature, 95°C for 1 min → 170°C, 30°C/min; carrier gas, helium; flow rate, not described; sample injection, splitless injection mode. (From Ref. 68.)

MS has higher sensitivity than electron impact (EI) ionization GC/MS. The chromatograms of selected ion recording of QA-HFIP and [^{18}O]QA-HFIP at m/z 467 and 471 with packed and capillary columns are shown in Figures 24 and 25, respectively. The minimum detectable limit of quinolinic acid for standards at signal-to-noise ratio of 10:1 was 30 fmol on packed columns and 3 fmol on capillary columns.

Quinolinic acid concentrations in whole blood and plasma were about 170 and 362 nmol/mL. Two hours after tryptophan (0.370 mmol/kg) administration, quinolinic acid increased in whole blood 135-fold, in plasma 74-fold, and in frontal cortex 23-fold.

VI. RECENT TECHNIQUES

Capillary zone electrophresis (CZE), one of the separation modes of capillary electrophoresis (CE), is a powerful separation technique for many ionic substances. CZE has a high resolution for ionic species but not for non-ionic species. For the separation of nonionic species, micellar electrokinetic current chromatography (MECC) can be applied. These CE methods are attractive tools for the determination of pyridine coenzymes because of their high separation efficiency, easy operation, and low running costs.

In 1994 Nesi et al. (69) reported a CZE separation of NAD$^+$, NADP$^+$, NADH, NADPH, and related compounds. The CZE profile of a mixture of seven NAD(P)$^+$ and NAD(P)H derivatives is shown in Figure 26. In 1995, Zarycki

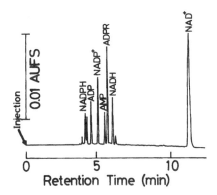

Figure 26 Electropherogram of a mixture of seven NAD$^+$ and NADH derivatives. Conditions: applied voltage, 15 kV; sample injection, 5 s by hydrostatic pressure; capillary, coated capillary of 45 cm × 100 μm i.d.; buffer, 50 mM MOPS (pH 7.0); detection wavelength, 254 nm; anodic migration (reverse polarity). (From Ref. 69.)

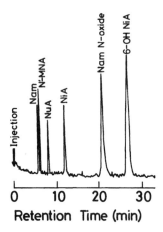

Figure 27 Electropherogram of a mixture of nicotinic acid and related compounds. Conditions: applied voltage, 25 kV; sample injection, 7 s pneumatic injection; capillary, unmodified fused-silica (57 cm × 50 μm i.d.); buffer, acetonitrile-buffer [10 mM KH$_2$PO$_4$ (pH 2.50), titrated with phosphoric acid]; detection wavelength, 254 nm. N'-MNA, N^1-methylnicotinamide (see Fig. 1). (From Ref. 70.)

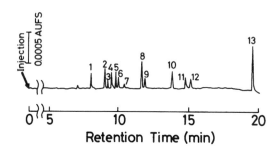

Figure 28 Electrokinetic separation of the niacin derivatives in the presence of SDS. Conditions: applied voltage, 15 kV; sample injection, injected by raising the positive end of the capillary about 4 cm higher than the other end; capillary, a polyimide-coated fused-silica (70 cm × 50 μm i.d.); micellar solution, 0.15 M SDS in 0.02 M borate-0.01 M KOH (pH 9.1); detection wavelength, 210 nm. 1, Isonicotinic acid hydrazide; 2, Nam; 3, pyridine-3-methanol; 4, 6-AN; 5, MNA; 6, pyridine-3-aldehyde; 7, pyridine; 8, 3-acetylpyridine; 9, thionicotinamide; 10, NiA; 11, pyridine-3-sulfonic acid; 12, β-picoline; 13, nicotinic acid ethyl ester. (From Ref. 71.)

et al. (70) reported CZE measurement of nicotinic acid in human plasma. The electropherogram of a standard solution containing nicotinic acid, nicotinamide, nicotinuric acid, 6-hydroxynicotinic acid, nicotinamide N-oxide, and N^1-methyl-nicotinamide is shown in Figure 27. The migration behavior of niacin derivatives such as β-picoline, 3-pyridinemethanol, 6-aminonicotinamide, and pyridine-3-aldehyde was investigated by CZE and MECC by Tanaka et al. (71), which electropherogram is shown in Figure 28.

VII. FUTURE TRENDS

The quantification of the acids such as quinolinic and nicotinic acids proceeds into two stages. The first stage consists of extraction and concentration of the compounds from biomaterials. This procedure is so complicated that the reproducibility of the recovery of the target compound is poor in some cases. The progress of computer technology will make it possible to use automated sample pretreatment systems (e.g., auto injector with pretreatment function for HPLC and/or advanced automated sample processor) or by a laboratory robot (e.g., a robotic system for HPLC or an automatic preparation system for HPLC) with high reproducibility.

The second stage is the instrumental analysis of target compounds. Quantitation of low amounts of quinolinic acid and nicotinic acid continues to be a challenge. In HPLC derivatization is required to increase the detector sensitivity. Evidently, for GC purposes the derivative should be volatile and thermostable. Many derivatizing agents which react with carboxyl groups have been produced for both HPLC and GC; the application of these agents will be further and thoroughly investigated. These developments will enable trace analysis of quinolinic acid and nicotinic acid by the use of HPLC with a fluorescence detector, electrochemical detector, or MS, and by GC with electron capture, thermionic, and negative CI-MS detection devices.

For quantification of the radioactive compounds, HPLC with a radiochemical detector has already been developed. Undoubtedly, this technique also has the potential to become popular.

As for separation columns, capillary columns will more frequently be used in HPLC, as is already the case in GC. In addition, multidimensional column chromatographic systems combined with different columns for pretreatment and separation of target compounds appear to be useful.

For the analysis of small samples, CE seems powerful as a complementary method to HPLC. Finally, supercritical fluid chromatography (SFC) might become the third powerful method for the analysis of bioorganic compounds in addition to HPLC, GC, and CE.

REFERENCES

1. J Goldberger. Pellagra: Causation and a method of prevention. A summary of some of the recent studies of the United States Public Health Service. JAMA 66:471–476, 1916.
2. CA Elvehjem, RJ Madden, FM Strong, DW Wooley. Relation of nicotinic acid and nicotinic acid amide to canine black tongue. J Am Chem Soc 59:1767–1768, 1937.
3. TD Spies, C Cooper, MA Blankenhorn. The use of nicotinic acid in the treatment of pellagra. JAMA 110:622–627, 1938.
4. E Kodicek, PW Wilson. The isolation of niacytin, the bound form of nicotinic acid. Biochem J 76:27p–28p, 1960.
5. H von Euler, H Ahlers, F Schlenk. Chemical investigations on highly purified cozymase. Z Physiol Chem 124:113–126, 1936.
6. O Warburg, W Christian. Coenzyme problem. Biochem Z 275:464, 1935.
7. OH Lowry, JV Passonneau, MK Rock. The stability of pyridine nucleotides. J Biol Chem 236:2756–2759, 1961.
8. K Shibata. The metabolism of niacin in each organ and the biological method for assessing the nutritional status of niacin in the rats. Vitamins (Japan) 61:39–56, 1987.
9. K Shibata, H Matsuo, K Iwai. Non-uniform decrease of nicotinamide in various tissues of rats fed on a niacin-free and tryptophan-limited diet. Agric Biol Chem 51:3429–3430, 1987.
10. O Hayaishi, K Ueda. ADP-Ribosylation Reactions. Biology and Medicine. New York: Academic Press, 1982.
11. N Rusinko, HC Lee. Widespread occurrence in animal tissues of an enzyme catalyzing the conversion of NAD^+ into a cyclic metabolite with intracellular Ca^{2+}-mobilizing activity. J Biol Chem 264:11725–11731, 1989.
12. R Altschul, H Hoffer, JD Stephen. Influence of nicotinic acid on serum cholesterol in man. Arch Biochem Biophys 54:558–559, 1955.
13. Coronary Drug Project Research Group. Clofibrate and niacin in coronary heart disease. JAMA 231:360–381, 1975.
14. SM Grundy, GL Vega. Fibric acid: effects on lipids and lipoprotein metabolism. Am J Med 83(suppl 5B):9–30, 1987.
15. WS Phillips, SL Lightman. Is cutaneous flushing prostaglandin mediated? Lancet 8223:754–756, 1981.
16. DWS Wilson, AB Douglass. Niacin skin flush is not diagnostic of schizophrenia. Biol Psychiatry 21:974–977, 1986.
17. WE Dulin, BM Wyse. Reversal of streptozotocin diabetes with nicotinamide. Proc Soc Exp Biol Med 130:992–994, 1969.
18. W Stauffacher, I Burr, A Gutzeit, D Veleminsky, AE Renold. Streptozotocin diabetes: time course of irreversible β-cell damage; further observations on prevention by nicotinamide. Proc Soc Exp Biol Med 133:194–200, 1970.
19. PS Schein, DA Cooley, MG Anderson, T Anderson. Streptozotocin diabetes—further studies on the mechanism of depression of nicotinamide adenine dinucleotide

concentrations in mouse pancreatic islets and liver. Biochem Pharmacol 22:2625–2631, 1973.

20. T Anderson, PS Schein, MG McMenamin, DA Cooney. Streptozotocin diabetes. Correlation with extent of depression of pancreatic islet nicotinamide adenine dinucleotide. J Clin Invest 54:672–677, 1974.

21. H Yamamoto, H Okamoto. Protection by picolinamide, a novel inhibitor of poly (ADP-ribose) synthetase, against both streptozotocin-induced depression of proinsulin synthesis and reduction of NAD content in pancreatic islets. Biochem Biophys Res Commun 95:474–481, 1980.

22. H Ijichi, A Ichiyama, O Hayaishi. Studies on the biosynthesis of nicotinamide adenine dinucleotide. III. Comparative in vivo studies on nicotinic acid, nicotinamide, and quinolinic acid as precursors of nicotinamide adenine dinucleotide. J Biol Chem 241:3701–3707, 1966.

23. T Negishi, A Ichiyama. Studies of the metabolism of nicotinamide adenine dinucleotide. Vitamins (Japan) 40:38–45, 1969.

24. C Bernofsky. New synthesis of the 4- and 6-pyridones of 1-methylnicotinamide and 1-methylnicotinic acid (trigonelline). Anal Biochem 96:189–200, 1979.

25. N Hengen, V Seiberth, M Hengen. High-performance liquid-chromatographic determination of free nicotinic acid and its metabolites, nicotinuric acid, in plasma and urine. Clin Chem 24:1740–1743, 1978.

26. RW McKee, YA Kang-Lee, M Panaque, ME Swedseid. Determination of nicotinamide and metabolic products in urine by high-performance liquid chromatography. J Chromatogr 230:309–317, 1982.

27. Y Tsuruta, K Kohashi, S Ishida, Y Ohkura. Determination of nicotinic acid in serum by high-performance liquid chromatography with fluorescence detection. J Chromatogr 309:309–315, 1984.

28. K Shibata. Simultaneous measurement of nicotinic acid and its major metabolite, nicotinuric acid in blood and urine by a reversed-phase high-performance liquid chromatography. Agric Biol Chem 52:2973–2976, 1988.

29. P Okungbowa, MC Ma, AS Truswell. Niacin in instant coffee. Proc Nutr Soc 36:26A, 1977.

30. LC Trugo, R Macrae, NMF Trugo. Determination of nicotinic acid in instant coffee using high-performance liquid chromatography. J Micronutr Anal 1:55–63, 1985.

31. TA Tyler, RR Shrago. Determination of niacin in cereal samples by HPLC. J Liq Chromatogr 3:269–277, 1980.

32. S Hirayama, M Maruyama. Determination of a small amount of niacin in foodstuffs by high-performance liquid chromatography. J Chromatogr 588:171–175, 1991.

33. K Shibata, H. Matsuo. Levels of NAD, NADP and their related compounds in human blood. Vitamins (Japan) 63:569–572, 1989.

34. K Shibata, T Kawada, K Iwai. Simultaneous micro-determination of nicotinamide and its major metabolites, N^1-methyl-2-pyridone-5-carboxamide and N^1-methyl-4-pyridone-3-carboxamide, by high-performance liquid chromatography. J Chromatogr 424:23–28, 1988.

35. K Shibata, K Murata. Blood NAD as an index of niacin nutrition. Nutr Int 2:177–181, 1986.

36. K Shibata, K Tanaka. Simple measurement of blood NADP and blood levels of NAD and NADP in humans. Agric Biol Chem 50:2941–2942, 1986.

37. TF Kalhorn, KE Thummel, SD Nelson, JT Slattery. Analysis of oxidized and reduced pyridine dinucleotides in rat liver by high-performance liquid chromatography. Anal Biochem 151:343–347, 1985.

38. LK Klaidman, AC Leung, JD Adams Jr. High-performance liquid chromatography analysis of oxidized and reduced pyridine dinucleotides in specific brain regions. Anal Biochem 228:312–317, 1995.

39. MA Kutnink, H Vannucchi, HE Sauberlich. A simple high performance liquid chromatography procedure for the determination of N^1-methylnicotinamide in urine. J Liq Chromatogr 7:969–977, 1984.

40. EG Carter. Quantitation of urinary niacin metabolites by reversed-phase liquid chromatography. Am J Clin Nutr 36:926–930, 1982.

41. K Shibata. Ultramicro-determination of N^1-methylnicotinamide in urine by high-performance liquid chromatography. Vitamins (Japan) 61:599–604, 1987.

42. K Shibata. Micro-determination of nicotinamide and its metabolites by high-performance liquid chromatography. Vitamins (Japan) 62:225–233, 1988.

43. K Shibata, H Taguchi, Y Sakakibara. Comparison of the urinary excretion of niacin and its metabolites in various mammals. Vitamins (Japan) 63:369–372, 1989.

44. K Shibata, H Kakehi, H Matsuo. Niacin metabolism in rodents. J Nutr Sci Vitaminol 36:87–98, 1990.

45. K Shibata. High-performance liquid chromatographic measurement of nicotinamide N-oxide in urine after extracting with chloroform. Agric Biol Chem 53:1329–1331, 1989.

46. E Jahns. Uber die Alkaloides des Bockshornsamens. Ber Deut Chem Ges 18:2518–2523, 1885.

47. DG Lynn, K Nakanishi, SL Patt, JL Occolowitz, S Almedia, LS Evans. Isolation and characterization of the first mitotoc cycle hormone that regulates cell proliferation. J Am Chem Soc 100:7759–7760, 1978.

48. K Shibata, H Taguchi. Effect of dietary N^1-methylnicotinamide or trigonelline on the growth and the niacin metabolism in weanling rats. Vitamins (Japan) 61:493–499, 1987.

49. LC Trugo, R Macrae, J Dick. Determination of purine alkaloids and trigonelline in instant coffee and other beverages using high performance liquid chromatography. J Sci Food Agric 34:300–306, 1983.

50. Y Nishizuka, O Hayaishi. Enzymic synthesis of niacin nucleotides from 3-hydroxy-anthranilic acid in mammalian liver. J Biol Chem 238:PC483–485, 1963.

51. S Nasu, F Diwicks, S Sakakibara, RK Gholson. L-Aspartate oxidase, a newly discovered enzyme of Escherichia coli, is the B protein of quinolinate synthetase. J Biol Chem 257:626–632, 1982.

52. IP Lapin. Stimulant and convulsive effects of kynurenes injected into brain ventricles in mice. J Neural Trans 32:37–43, 1978.

53. R Schwarcz, WO Whetsell Jr, RM Mangano. Quinolinic acid: an endogenous metabolite that produces axon-sparing lesions in rat brain. Science 219:316–318, 1983.

54. K Mawatari, K Oshida, F Iinuma, M Watanabe. Determination of quinolinic acid in human urine by liquid chromatography with fluorimetric detection. Anal Chim Acta 302:179–183, 1995.
55. N Horita, S Oyanagi, T Ishii, Y Izumiyama. Ultrastructure of 6-aminonicotinamide-induced lesions in the central nervous system of rats. Acta Neuropathol 44:111–119, 1978.
56. K Shibata. Reparative effect of nicotinamide and nicotinic acid on the aggravation of tryptophan-nicotinamide metabolism caused by 6-aminonicotinamide. Biosci Biotech Biochem 58:1729–1730, 1994.
57. K Shibata. Measurement of 6-aminonicotinamide, an antagonist of niacin, by high-performance liquid chromatography. Vitamins (Japan) 67:493–497, 1993.
58. YC Lee, RK Gholson, N Raica. Isolation and identification of two new nicotinamide metabolites. J Biol Chem 244:3277–3282, 1969.
59. M Rocchingiani, V Michaeli, JA Duley, HA Simmonds. Determination of nicotinamide phosphoribosyltransferase activity in human erythrocytes: high-performance liquid chromatography-linked method. Anal Biochem 205:334–336, 1992.
60. HC Lee, TF Walseth, GT Bratt, RN Hayes, DL Clapper. Structural determination of a cyclic metabolite of NAD$^+$ with intracellular Ca^{2+}-mobilizing activity. J Biol Chem 264:1608–1615, 1989.
61. G Casabona, L Sturiale, MR L'Episcopo, G Raciti, A Fazzio, MG Sarpietro, AA Genazzani, A Cambria, F Nicoletti. HPLC analysis of cyclic adenosine diphosphate ribose and adenosine diphosphate ribose: determination of NAD$^+$ metabolites in hippocampal membranes. Ital J Biochem 44:258–268, 1995.
62. A Tanaka, M Iijima, Y Kikuchi, Y Hoshino, N Nose. Gas chromatographic determination of nicotinamide in meats and meat products as 3-cyanopyridine. J Chromatogr 466:307–317, 1989.
63. H Holmen, H Egsgaard, J Funck, E Larsen. N'-Methylnicotinamide in human urine. Biomed Mass Spectrom 8:122–124, 1981.
64. K Shibata, T Kawada, K Iwai. High-performance liquid chromatographic determination of nicotinamide in rat tissue samples and blood after extraction with diethylether. J Chromatogr 422:257–262, 1987.
65. LM Henderson, HM Hirsch. Quinolinic acid metabolism. I. Urinary excretion by the rat following tryptophan and 3-hydroxynanthranilic acid administration. J Biol Chem 181:667–675, 1949.
66. TW Stone, MN Perkins. Quinolinic acid: a potent endogenous excitant at amino acid receptors in CNS. Eur J Pharmacol 72:411–412, 1981.
67. MN Perkins, TW Stone. Pharmacology and regional variations of quinolinic acid-evoked excitations in the rat central nervous system. J Pharmacol Exp Ther 226:551–557, 1983.
68. MP Heyes, SP Markey. Quantification of quinolinic acid in rat brain, whole blood, and plasma by gas chromatography and negative chemical ionization mass spectrometry: effects of systemic L-tryptophan administration on brain and blood quinolinic acid concentrations. Anal Biochem 174:349–359, 1988.
69. M Nesi, M Chiari, G Carrea, G Ottolina, PG Righetti. Capillary electrophoresis of nicotinamide-adenine dinucleotide and nicotinamide-adenine dinucleotide

phosphate derivatives in coated tubular columns. J Chromatogr 670:215–221, 1994.

70. PK Zarzycki, P Kowalski, J Nowakowska, H Lamparczyk. High-performance liquid chromatographic and capillary electrophoretic determination of free nicotinic acid in human plasma and separation of its metabolites by capillary electrophoresis. J Chromatogr 709:203–208, 1995.
71. S Tanaka, K Kodama, T Kaneta, H Nakamura. Migration behavior of niacin derivatives in capillary electrophoresis. J Chromatogr 718:233–237, 1995.
72. Y Nishizuka, O Hayaishi. Studies on the biosynthesis of nicotinamide adenine dinucleotide. J Biol Chem 238:3369–3377, 1968.
73. Y Hagino, SJ Lan, CY Ng, LM Henderson. J Biol Chem 243:4980–4986, 1968.
74. S Chaykin, M Dagani, L Johnson, M Samli. J Biol Chem 240:932–938, 1965.
75. H Taguchi, H Takamizawa, M Muto, Y Shimabayashi, K Iwai. Vitamins (Japan) 52:363–370, 1978.
76. K Shibata, unpublished data.
77. JX DeVries, W Gunthert, R Ding. J Chromatogr 221:161–165, 1980.
78. GW Chase Jr, WO Landen Jr, AM Soliman, RR Errenmiller. J AOAC Int 76:390–393, 1993.
79. M Pelzer, S Northcott, G Hanson. J Liq Chromatogr 16:2563–2570, 1993.
80. M Iwaki, T Ogiso, H Hayashi, ET Lin, LZ Benet. J Chromatogr 661:154–158, 1994.
81. K Takatsuki, S Suzuki, M Sato, K Sakai, I Ushizawa. J Assoc Off Anal Chem 70:698–702, 1987.
82. T Hamano, Y Mitsuhashi, N Aoki, S Yamamoto. J Chromatogr 457:403–408, 1989.
83. D Balschukat, E Kress. J Chromatogr 502:79–85, 1990.
84. K Mawatari, F Iinuma, M Watanabe. Anal Sci 7:733–736, 1991.
85. H Iwase. J Chromatogr 625:377–381, 1992.
86. Y Miyauchi, N Sano, T Nakamura. Internal J Vit Nutr Res 63:145–149, 1993.
87. K Shibata, H Matsuo. Am J Clin Nutr 50:114–119, 1989.
88. K Shibata, H Matsuo. Agric Biol Chem 53:2031–2036, 1989.

8
Thiamine

Takashi Kawasaki
Hiroshima University School of Medicine, Hiroshima, Japan

Yoshiko Egi
Suzugamine Women's College, Hiroshima, Japan

I. INTRODUCTION

A. Chemical Properties

The chemical structure of thiamine is shown in Figure 1. A pyrimidine moiety (2-methyl-4-amino-5-hydroxymethylpyrimidine) and a thiazole moiety (4-methyl-5-hydroxyethylthiazole) are connected by a methylene group. The double-salt form of thiamine with hydrochloric acid ($C_{12}H_{17}N_4OSCl$-HCl; molecular weight 337.28) is readily soluble in water, less soluble in methanol and glycerol, nearly insoluble in ethanol, and insoluble in ether and benzene (1).

Thiamine in water is most stable between pH 2 and 4 and unstable at alkaline pH; it is heat labile with its decomposition dependent on pH and exposure time to heat.

The structures of the phosphate esters of thiamine are also shown in Figure 1. Thiamine monophosphate (TMP), thiamine pyrophosphate (TPP), and thiamine triphosphate (TTP) are commonly found in organisms. About 80% to 90% of the total thiamine content in cells is TPP, the coenzyme form of thiamine. In some animal tissues, especially pig skeletal muscle (2) and chicken white skeletal muscle (3), TTP is present in an extremely high amount (70% to 80% of total thiamine—i.e., thiamine plus thiamine phosphate esters). However, TTP has no coenzyme activity. Thiamine pyrophosphate in the dried state is stable for several months when stored at a low temperature in the dark. In solution, TPP is unstable and partially decomposes to TMP and/or thiamine when stored for several months at pH 5 and 38°C. However, TPP in solution at pH 2 to 6 and at 0°C is

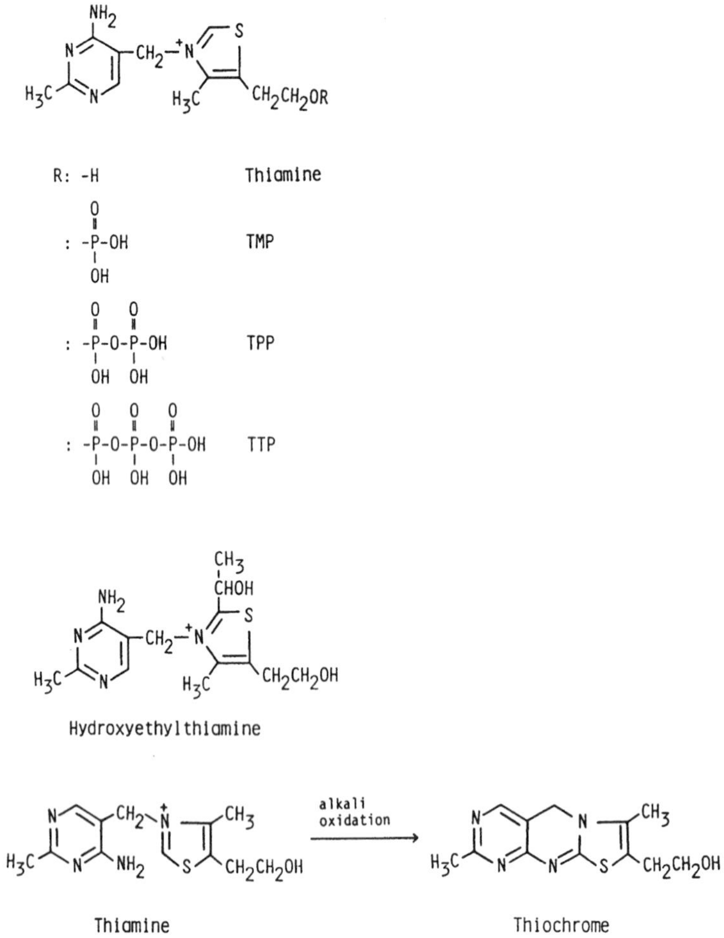

R: -H   Thiamine

: -P-OH   TMP

: -P-O-P-OH   TPP

: -P-O-P-O-P-OH   TTP

Hydroxyethylthiamine

Thiamine   Thiochrome

Figure 1 Structures of thiamine and its related compounds.

stable for 6 months. In an aqueous solution, TTP is stable for at least 6 months when stored at −80°C. Aqueous TPP solutions stored at −20°C have an even better stability.

For the chemical determination of thiamine, thiochrome (Thc) is the most important compound. Its structure is also shown in Figure 1. Thiochrome is quantitatively formed from thiamine by alkaline oxidation with cyanogen bromide or potassium ferricyanide. It is a highly fluorescent compound. Thiamine phosphate esters are also quantitatively converted to thiochrome phosphate esters without affecting the phosphate bond: thiochrome mono-, pyro-, and triphosphate are referred to as ThcMP, ThcPP, and ThcTP, respectively.

Thiochrome and its phosphates are fluorescent at a pH > 8, and all of these have identical excitation maxima at 375 nm and nearly identical fluorescence maxima at 432 to 435 nm (4), as shown in Figure 2.

Alkaline solutions (pH > 9) of thiochrome and its phosphates are stable for at least 3 days at room temperature. ThcPP and ThcTP in a 0.1 N HCl solution are converted quantitatively to ThcMP upon heating above 95°C for 10 min (H. Sanemori and T. Kawasaki, unpublished observation).

Hydroxyethylthiamine, a derivative of thiamine, is found in cells in the form of hydroxyethyl-TPP. It is converted to the corresponding thiochrome deriv-

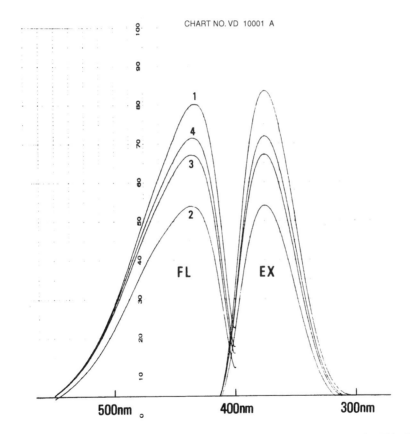

Figure 2 Excitation and fluorescence spectra of thiochrome compounds. Thiochrome (Thc) derivatives were prepared from corresponding thiamine derivatives by alkaline BrCN oxidation: (1) Thc; (2) ThcMP; (3) ThcPP; and (4) ThcTP. (From Ref. 4.).

ative by alkaline potassium ferricyanide oxidation, but not by alkaline BrCN oxidation (5). This difference in oxidation property is useful for its determination. Factors affecting a cyanogen bromide–based assay of thiamine include drugs and phosphatases used for treating biological samples (6).

B. Biochemistry

1. Metabolism

Thiamine is metabolized to TPP by thiamine pyrophosphokinase (EC 2.7.6.2) in animal cells including red and white blood cells. This enzyme is also present in plants, yeast, and a bacterium (*Paracoccus denitrificans*) (7). However, in some bacteria, for example in *Escherichia coli*, thiamine is metabolized to TPP by a two-step reaction catalyzed by thiamine kinase (EC 2.7.1.89) and TMP kinase (EC 2.7.4.10). Thiamine pyrophosphate is further metabolized to TTP in yeast, animal tissues, and human red blood cells. Evidence has been obtained which indicates that cytosolic adenylate kinase (EC 2.7.4.3) catalyzes TTP synthesis from TPP in vitro (8) and in vivo (3). The enzyme system involved in thiamine metabolism to TTP in human red blood cells was recently identified, purified, and reconstituted (9).

2. Physiological Functions

Thiamine has to be converted to TPP before exerting its physiological function as a coenzyme in cells. Thiamine pyrophosphate functions as a coenzyme for several important enzymes in the carbohydrate and amino acid metabolism (10), including pyruvate dehydrogenase (EC 1.2.4.1), oxoglutarate dehydrogenase (EC 1.2.4.2), transketolase (EC 2.2.1.1), branched-chain α-acid dehydrogenase (EC 1.2.4.4), pyruvate decarboxylase (EC 4.1.1.1), and carboxylate carboligase (EC 4.1.1.47).

In the oxidative decarboxylation of pyruvate and oxoglutarate, as shown in Figure 3, TPP functions as a carrier of ''active aldehyde'' to form the intermediates, hydroxyethyl-TPP and α-hydroxy-β-carboxypropyl-TPP, which are finally transferred to coenzyme A (CoA) to form acetyl-CoA and succinyl-CoA, respectively.

Thiamine is known to affect nerve functions as an antipolyneuritis factor. There are many reports on the effects of thiamine on nerve conductance or neurotransmission (11). However, the exact mechanism of neurological function is not known. Neither has the role of TTP been elucidated as yet. Schoffeniels' group suggested that (1) TTP is involved in some kind of anion transport mechanism (12), (2) TTP has a regulatory role on chloride permeability of rat brain (13), and (3) TTP is an activator of chloride channels having a large unit conductance (14).

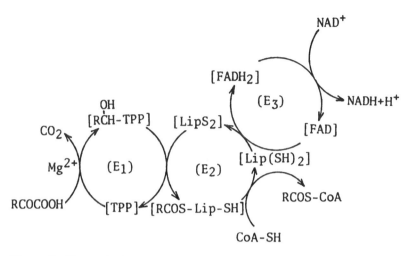

Figure 3 The mechanism of oxidative decarboxylation of α-keto acids. E_1, α-keto acid dehydrogenase; E_2, lipoate acyltransferase; E_3, lipoamide dehydrogenase. R : CH_3^-, HOOC-$CH_2CH_2^-$; R-CH(OH)-TPP, "active aldehyde"; $LipS_2$, lipoate; $Lip (SH)_2$, dihydrolipoate; RCOS-Lip SH, 6-acyldihydrolipoate; [ ], coenzymes bound to the enzyme proteins.

II. PARTITION CHROMATOGRAPHY AND ION EXCHANGE CHROMATOGRAPHY

Paper partition chromatography (PPC) was the first used for separating thiamine phosphates in biological materials. A good PPC separation of thiamine phosphates was reported by using several different solvent systems. Photometry at 270 nm of eluted, individual spots allowed quantitation of each thiamine compound down to the 10-μg level.

A combined microbiological assay for thiamine phosphates after separation by PPC was reported to yield a detection limit of approximately 0.02 μg (60 pmol) of thiamine, which is still at least three orders of magnitude higher than that obtained by high-performance liquid chromatography (HPLC) with fluorescence detection.

Two-dimensional thin-layer chromatography (TLC) was applied to separate thiamine and its metabolites and precursor compounds (15). Most of the TLC procedures have been used to analyze thiamine in pharmaceutical preparations at levels of 20 μg to 10 mg.

Ion exchange chromatography was also used for the quantitative assay of thiamine and its phosphates, not only in pharmaceutical materials but also in biological samples. An anion exchange column and a Sephadex cation exchange column (16) have been used. Both approaches resulted in the coelution of thiamine and TMP in the same fraction, but a good separation of TPP and TTP in micromolar amounts could be achieved.

III. HIGH-PERFORMANCE LIQUID CHROMATOGRAPHY

A. Introduction

High-performance liquid chromatography (HPLC) methods have been developed in recent years to allow the rapid, sensitive, and specific analysis of thiamine and its phosphates in drugs, biological materials, and metabolites. To determine total thiamine in biological materials, samples must first be treated with Taka-diastase or acid phosphatase to hydrolyze thiamine phosphate esters to free thiamine (17,18). The complete separation of thiamine and its phosphate esters at subpicomolar level by use of HPLC was first reported by Kawasaki's group (4,17) and then continued to be developed until recent years (19,20).

B. Analytical Systems

1. Stationary Phase

There are many kinds of stationary phases used for the analysis by HPLC. Porous silica particles, chemically bound to a monomolecular octadecyl layer, such as μBondapak C_{18}, are most frequently used in reversed-phase HPLC. As an improved reversed-phase sorbent, a poly(styrene-divinylbenzene) resin can be used within an extended pH range of 1 to 13 because of its chemical stability (17). LiChrosorb NH_2, a straight-phase sorbent, has found applications for thiamine (21) as well as thiamine phosphate esters (17).

2. Mobile Phase

The mobile phase used in HPLC is dependent on the hydrophobicity of the compounds. Appropriate mixtures of methanol (or an appropriate organic solvent such as acetonitrile) and water (or buffer solution), with or without ion pair chromatography (IPC) reagents, are used.

3. Detection

The detection of thiamine compounds in the eluate is carried out either spectrophotometrically, usually at 254 nm, or fluorometrically.

In fluorometric detection, two different procedures are used. First, samples containing thiamine compounds are converted to thiochrome compounds using reagents for alkaline oxidation and then chromatographed (precolumn derivatization). In the second procedure, samples are first directly chromatographed, and the thiamine compounds in the eluate are then converted to thiochrome compounds (postcolumn derivatization). The chemical principle of postcolumn derivatization procedure is the same: the effluent from the column is mixed with a potassium ferricyanide-NaOH solution at an appropriate ratio, which is sent to a mixing coil by a proportioning pump at the rate of 0.25 to 1.0 mL/min (17,18). Thiochrome fluorescence in the mixture is then measured.

The excitation wavelength for the fluorometric detection is usually 365 to 375 nm, while the emission wavelength is usually 425–435 nm.

The detection limit depends on the method used. Fluorometric detection is much more sensitive than its spectrophotometric counterpart. The latter method is suitable for analysis of large quantities of thiamine in pharmaceutical preparations and foods: the detection limit is approximately 2 ng or 6 pmol as thiamine hydrochloride. On the other hand, the detection limit by the fluorometric method is <17 pg or 0.05 pmol as thiamine hydrocholoride. The lowest detection limit so far reported for thiamine is 5 fmol, using fluorescence (21,22). Fluorescence detection is therefore more suitable for the analysis of thiamine in biological materials such as cells, blood, and urine.

The time required for completion of the chromatographic separation is usually <10 min although in some cases up to 25 min (17).

4. Precision

In all analytical systems listed above, the recovery of thiamine added to the sample is 100 ± 10%, and the coefficient of variation is 1% to 5%.

5. Internal Standard

As internal standard, either salicylamide or anthracene was used (17). These compounds, which are not related to thiamine in structure but have fluorescence characteristics comparable to those of thiochrome, were used only to compensate for variability in the volume injected into the column. In other studies, thiochrome compounds derived from authentic thiamine and its phosphates are used as calibration standards for the analysis of thiamine phosphates (4,17), since the fluorescence intensity of thiochrome phosphates varies considerably (see Fig. 2).

C. Analysis of Thiamine

1. Pharmaceutical Preparations

Reversed-phase HPLC systems are often used for the simultaneous assay of thiamine and other water-soluble vitamins in multivitamin preparations. No special treatment of the sample is required before chromatography and UV detection at 270 nm or 254 nm is usually employed.

With a mobile phase of 0.2 M ammonium phosphate buffer (pH 5.1) in a reversed-phase system, folate, pyridoxine, nicotinamide, and thiamine could be separated, in this order, within 20 min, followed by vitamin B_{12} and riboflavin (17). The latter two compounds eluted after a step gradient to 30% aqueous methanol. This procedure has the advantage of completely separating at least nine coenzyme forms of water-soluble vitamins, including TPP. A similar mobile phase, composed of methanol-water (50:50) was used in connection with a LiChrosorb RP-18 column to separate thiamine, pyridoxine, vitamin B_{12}, and riboflavin in 3 min, as shown in Figure 4 (23). The detection limit at 254 nm is 5 ng (15 pmol) for thiamine and 10 to 20 ng for the other three vitamins, with a coefficient of variation of <4%.

2. Foods

Thiamine was analyzed by HPLC in rice, cereals, meat, and other byproducts. In these systems, thiamine could be determined simultaneously with riboflavin and niacin.

Samples, in either 0.1 N HCl or 0.1 N H_2SO_4, were autoclaved for 30 min and then subjected to enzymatic hydrolysis with either Taka-diastase or papain. The filtrate was then chromatographed.

The HPLC system for cereals was: column, μBondapak C_{18}; detection, UV

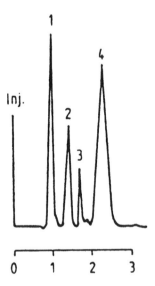

Figure 4 Typical separation of four vitamins from capsules. Peaks 1 to 4 are thiamine, pyridoxine, vitamin B_{12}, and riboflavin, respectively. (From Ref. 23.)

254 nm; mobile phase, a mixture of 12.5% acetonitrile and 87.5% 10 mM phosphate buffer (pH 7.0) containing 5 mM heptane sulfonate. The system used for rice is quite similar to that for cereals except for the mobile phase (17).

Thiamine in meat and meat products was determined by the HPLC system with precolumn derivatization. The thiochrome derived from thiamine in the sample filtrate, treated with acid phosphatase, was extracted with isobutyl alcohol, chromatographed with chloroform-methanol (90:10, v/v), and then determined fluorometrically (17). The recovery values ranged between 84.4% and 94.2%, and the detection limit for thiamine was 0.05 ng (0.15 pmol).

A simultaneous determination of thiamine and riboflavin in a wide range of foods by HPLC with postcolumn derivatization was described (24). Extractions of thiamine and riboflavin from foods, and enzymatic digestion of the extract, were carried out according to the AOAC methods (1980). After filtration, an aliquot (50 to 100 µL) was injected onto a µBondapak C$_{18}$ column and eluted with methanol-water (40:60, v/v) containing 5 mM PIC B6. Riboflavin fluorescence in the eluate was detected directly at 360 nm (excitation) and 500 nm (emission). Thiamine in the eluate was then passed into a reaction coil, reacted with an alkaline ferricyanide solution (to convert thiamine to thiochrome), and determined at 360 nm (excitation) and 425 nm (emission). This procedure could be successfully used for determining the two vitamins in a wide range of foods, including raw meats, processed meats, cereal products, fruits, and vegetables (24).

HPLC determination of total thiamine in biological and food products, with or without simultaneous determination of riboflavin, pyridoxine, or niacin, was recently reviewed by Fayol (18). This review concludes that, first, for pharmaceutical preparations, methodologies based on UV detection can be performed. Although these techniques are less sensitive, sample preparation is simpler. Secondly, methodologies based on thiochrome formation (precolumn or postcolumn derivatization) are needed for biological and food products. The advantage or disadvantage of precolumn derivatization versus postcolumn derivatization is also described. For example, the advantage of precolumn derivatization is the decreased peak broadening and hence better resolution, and the method requires no additional equipment. The disadvantage of precolumn derivatization is the instability of thiochrome or its phosphate esters at pH values below 8.0 and hence the elution solvent must have a pH of at least 8.0 (4), which affects the lifetime of the column.

3. Clinical Specimens

The determination of thiamine in blood by HPLC with either precolumn or postcolumn derivatization is a valuable way to assess the thiamine status in humans because of its sensitivity, specificity, reproducibility, speed, and simplicity. Most procedures determine total thiamine in blood after hydrolysis of the phosphates with an appropriate phosphatase (17,18). Other HPLC methods were used to quantitate thiamine and its phosphates in blood separately, total thiamine being the sum of the individual levels (17).

Schrijver et al. (25) described a reliable postcolumn derivatization HPLC method for total thiamine in whole blood. Two milliliters of whole blood was deproteinized with trichloroacetic acid, neutralized with sodium acetate buffer to a final pH of 4.5, and then treated with Taka-diastase for 2 h at 45°C. After centrifugation, the clear supernatant was used for direct HPLC analysis. A LiChrosorb Si-100 column (250 × 4.6 mm; 10μm) was used and 240 μL of the extract was injected onto the column, eluted with a mobile phase composed of 40 mM Na_2HPO_4-30 mM KH_2PO_4 and ethanol (87:13, v/v), at pH 6.8.

The effluent was mixed with the thiochrome reagent [12 mM $K_3Fe(CN)_6$-1.8 M NaOH] in a reaction coil, followed by fluorescence detection (excitation, 367 nm; emission, 430 nm). The concentration of thiamine in the original sample was calculated from the peak area versus a thiamine standard.

The within-assay and between-assay coefficients of variation for the determination of total thiamine in whole blood were 4.2% and 4.4%, respectively. The between-assay analytical recovery of TPP added to blood samples was 99.9 ± 11.7% (mean ±SD). The analysis of samples from 98 normal volunteers, not matched for age and sex, revealed an overall range for total thiamine of 70 to 185 nmol/L blood, with a mean value of 115 nmol/L blood. The frequency distribution of total thiamine in these blood samples is shown in Figure 5; a ''normal'' range of total thiamine 95 to 155 nmol/L blood was obtained (25). Schrijver extended the number of analyses of total thiamine in whole blood to 598 normal volunteers and obtained a value of 129.2 ± 21.9 nmol/L blood (personal communication). This HPLC method could also be applied to the analysis of thiamine in plasma and erythrocytes.

A similar HPLC method with a postcolumn derivatization system was used for the analysis of total thiamine in human whole blood as well as in serum, cerebrospinal fluid, and milk (26). The HPLC system consisted of a μBondapak column and the mobile phase was a mixture of methanol-50 mM sodium citrate buffer pH 4.0 (45:55, v/v) plus 10 mM sodium 1-octanesulfonate. Two milliliters of blood was needed. The minimum detectable amount was 60 fmol of thiamine. The intra-assay and inter-assay coefficients of variation were 2.3% and 3.9%, respectively. The recovery of TPP added to blood samples was 98.7%.

The total thiamine content in whole blood of 56 healthy volunteers determined by this method ranged from 71 to 185 nmol/L with a mean value of 117 nmol/L. The reference range was found to be 88–157 nmol of total thiamine per liter blood (26).

Weber and Kewitz (21) reported a sensitive HPLC method for thiamine in human plasma, based on precolumn oxidation of thiamine to thiochrome followed by HPLC separation and fluorescence detection. The method includes a disadvan-

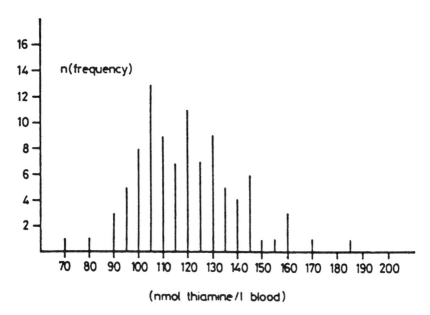

Figure 5 Frequency distribution of total thiamine in whole-blood samples of normal healthy Dutch volunteers (n = 98). (From Ref. 25.)

tageous step to extract the thiochrome into isobutanol but, on the other hand, has a detection limit as low as 5 fmol. Plasma isolated from blood samples was deproteinized and the supernatant obtained after centrifugation was neutralized before hydrolysis with acid phosphatase. The free thiamine formed was oxidized to thiochrome with $HgCl_2$ and then alkalinized. The thiochrome was extracted with isobutanol and then analyzed on the HPLC system: column, LiChrosorb NH_2 (120 × 4.6 mm; 5μm); mobile phase, methanol-diethyl ether (25:75, v/v); detection, fluorescence (excitation, 365 nm; emission, 440 nm). The minimum sample volume required for the assay was 0.3 mL of plasma, and the minimum detectable plasma concentration was 0.5 nmol of thiamine per liter (0.17 μg/L). Plasma levels of thiamine in 91 volunteers ranged from 6.6 to 43 nmol/L, showing a mean of 11.6 nmol/L. This HPLC method was used to study the pharmacokinetics of thiamine excretion into urine after intravenous or oral dosing.

D. Analysis of Thiamine Phosphate Esters

1. Analytical Systems

To elucidate the physical role of thiamine phosphates, especially of TTP, in cells and tissues, it is primarily necessary to establish a quantitative analytical method for thiamine and its phosphates present in such low levels as nM concentration or even less than that.

The first complete separation and determination of thiamine and thiamine phosphate esters at the 1 pmol level was reported by Kawasaki's group (17) using the prederivatization procedure in either straight-phase HPLC (4) or reversed-phase HPLC (27). Subsequent improvement in instrumentation succeeded in reducing the detection limit to 50 fmol (28).

The analytical system in the straight-phase HPLC consisted of LiChrosorb NH$_2$ (250 × 4.6 mm ID; 10 μm) as the stationary phase, and acetonitrile-90 mM potassium phosphate buffer, pH 8.4 (60:40, v/v), as the mobile phase. Fluorometric detection was carried out at 375 nm for excitation and 430 nm for emission. Thiamine, TMP, TPP, and TTP in the sample were detected in this order as the corresponding thiochromes within 7 min.

A chromatogram for Thc, ThcMP, ThcPP, and ThcTP (1 pmol each) obtained with the improved straight-phase HPLC system is illustrated in Figure 6. The resolution factors (Rs) calculated from the widths and retention times are 2.4 for Thc-ThcMP, 1.6 for ThcMP-ThcPP, and 2.1 for ThcPP-ThcTP, respectively, indicating a complete separation of each peak. The detection limit of thiamine and its phosphates is 50 fmol.

This isocratic straight-phase HPLC system has been successfully employed to determine thiamine phosphates, especially TTP, in biological materials (2–4) as well as TPP or TTP after enzymatic synthesis (8,9).

More recently, determination of thiamine and its phosphate esters by use of a straight-phase HPLC system was described by Tollaksen et al. (19), based on their original work (29). They used a straight-phase (Supelcosil NH$_2$) column (250 mm × 4.6 mm ID) and phosphate buffer (85 mM, pH 7.5)-acetonitrile as the solvent system. The eluent consisted of buffer-acetonitrile in the ratio of 90:10 (v/v) for the elution of thiamine and 60:40 (v/v) for the thiamine phosphate esters. Thiamine and its phosphate esters are prederivatized with cyanogen bromide and separated within 15 min. The detection limit is 13 to 16 fmol. The elution profile for a standard solution, a whole-blood sample, and a plasma sample is shown in Figure 7.

A reversed-phase HPLC system for thiamine and its phosphates, with pre-column derivatization to thiochromes, was first described by Kawasaki's group (17,27). In this system TTP elutes first, followed by TPP, TMP, and finally thiamine. Since then, several reversed-phase HPLC systems were described with either precolumn (30–32) or postcolumn derivatization (33,34).

In our HPLC system, thiamine compounds were converted to thiochromes by alkaline BrCN oxidation prior to separation. An ODS column was used as the stationary phase and 2.5% N,N-dimethylformamide (DMF)-25 mM potassium phosphate buffer (pH 8.4) as the mobile phase. The thiochromes were detected fluorometrically. After completion of ThcMP elution, the mobile phase was

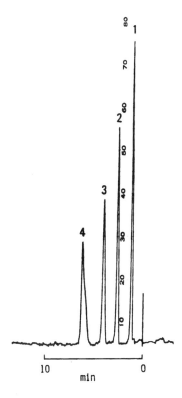

Figure 6 Chromatogram of thiochrome and its phosphate esters analyzed by the improved straight-phase LC system. Thiochrome (Thc) and its phosphates were prepared from thiamine and its phosphates by alkaline BrCN oxidation: (1) Thc; (2) ThcMP; (3) ThcPP; and (4) ThcTP (1 pmol each).

changed to 25% DMF-25 mM potassium phosphate buffer (pH 8.4) for elution of Thc. The detection limit of thiamine and its phosphates was 50 fmol (28).

A reversed-phase HPLC of thiamine phosphates followed by postcolumn fluorogenic oxidation was described by Kimura et al. (33). A μBondapak C_{18} column was used with a mobile phase of a 0.2 M sodium phosphate–phosphoric acid buffer (pH 4.3). The effluent was oxidized and detected fluorometrically.

Schoffeniels et al. described a reversed-phase HPLC system for thiamine and its phosphates with precolumn derivatization by alkaline $K_3Fe(CN)_6$ (30). They used an ODS column as the stationary phase and gradient elution with 25 mM phosphate buffer (pH 8.4)/methanol. The gradient program was as follows: after being kept for 1 min at 10%, the methanol concentration was raised to 100% in 3 min, and then 6 min after the injection of the sample the initial 10% was

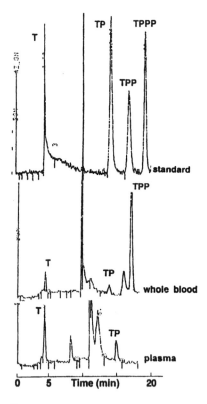

Figure 7 Elution profile for a standard solution, a whole-blood sample, and a plasma sample. Peaks: T, thiamine; TP, thiamine monophosphate; TPP, thiamine pyrophosphate; TPPP, thiamine triphosphate. Injection at time 0. (From Ref. 19.)

restored. Detection was carried out fluorometrically (excitation 390 nm, emission 475 nm). ThcTP, ThcPP, ThcMP, and Thc eluted in this order, within 14 min, and with a detection limit of 50 fmol each.

The same group (22) subsequently modified and improved this HPLC method by using a styrene-divinylbenzene reversed-phase column (PRP-1), eluted isocratically. The main advantage of the PRP-1 reversed-phase compared to silica-based reversed-phase is its stability over the pH range 1 to 13. The determination of thiamine phosphates by precolumn derivatization must be carried out above the mobile phase pH of 8.0 (4). Under these conditions only, the thiochrome phosphates are fluorescent. At alkaline pH, however, silica-based sorbents rapidly deteriorate. The isocratic elution is carried out in two different modes: the mobile phase contains either 10% methanol for the determination of thiochrome phosphates, or 10% tetrahydrofuran (THF) for thiochrome. The elution was completed within 14 min. As compared to previous methods, the detection limit was lowered to 10 fmol for thiochrome phosphates and 5 fmol for thiochrome.

Another reversed-phase HPLC system with precolumn derivatization has been reported by Iwata and his group (31). Thiamine and its phosphates were converted to fluorophores by alkaline cyanogen bromide. The thiochromes were injected, in a neutralized solution, on an ODS silica column. The mobile phase, containing 100 mM Na$_2$HPO$_4$-H$_3$PO$_4$ buffer (pH 2.5) and methanol (92:8), was mixed postcolumn for alkalinization with 0.2 M NaOH–70% methanol, followed by fluorescence detection. The elution of these four thiochromes was completed in 30 min, and the detection limit was 0.1 pmol for thiamine phosphate esters and 0.05 pmol for thiamine.

Gerrits et al. (20) presented a simple, inexpensive, and sensitive method with automated precolumn derivatization of thiamine and its phosphate esters using reversed-phase HPLC. The method is optimized for the simultaneous determination of thiamine and its phosphates. It has previously been described in detail (32).

They used a Microspher C$_{18}$ column and two solvent systems that are principally composed of a phosphate buffer (pH 7.0) and methanol (final concentration, 12%) and used in a partial gradient manner. The beneficial point of the system is to use a buffer with pH 7.0, which extends column life. Total analysis time is 15 min, and the minimum detectable amount was 0.5 nmol/L for thiamine as well as its phosphates. A typical chromatogram of a standard sample repeatedly injected is shown in Figure 8, which indicates the stability of the system.

2. Tissue Contents

To determine thiamine and its phosphates at low concentrations, in cells and tissues, the straight-phase and reversed-phase HPLC systems are preferred because of the sensitivity, specificity, and relative simplicity of the procedure. An example of a procedure to determine the tissue content of thiamine and its phosphates is described.

Male Wistar rats of an appropriate body weight were sacrificed by decapitation, and the organ to be analyzed was immediately removed. Approximately 200 mg tissue was weighed and homogenized in 5 volumes of 10% trichloroacetic acid (TCA) by use of a Polytron 20 ST at 4°C. After centrifugation for 15 min at 16,000g, the supernatant was extracted, at least three times, with the same volume of ether to remove TCA. The water layer obtained, 0.4 mL, was mixed with 50 µL 0.3 M BrCN and then with 50 µL 1 M NaOH. In blank experiments, 50 µL 1 M NaOH was first added to 0.4 mL of the water layer and mixed, followed by the addition of 50µL 0.3 M BrCN. An aliquot of both the oxidized

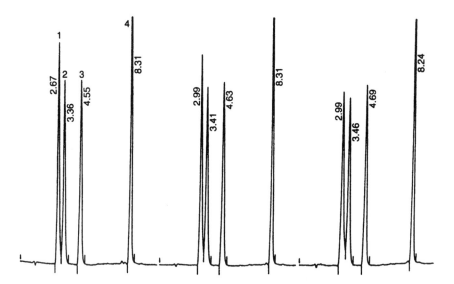

Figure 8   Elution profile for a standard solution injected three times. Peaks: 1, TTP; 2, TPP; 3, TMP; and 4, thiamine. (From Ref. 20.)

sample and the blank was injected onto the column and chromatographed in the improved straight-phase HPLC system, as described above.

Chromatograms of thiamine and its phosphates in rat liver analyzed by the improved straight-phase HPLC system are shown in Figure 9. Good separation of peaks 1, 2, 3, and 4, representing thiamine, TMP, TPP, and TTP, respectively, is evident. Although unidentified peaks can be seen in the sample, no peaks appear in the region of TTP. Blank sample peaks are subtracted from test sample peaks and cellular concentrations of thiamine and its phosphates can then be calculated (Table 1).

When the contents of thiamine phosphates in guinea pig and pig tissues were determined with the same HPLC system, an extremely high concentration of TTP, averaging 70% of total thiamine (26.1 nmol/g wet weight), was detected in adult pig skeletal muscle (2). In one extreme case 88.7% of the total thiamine (19.6 nmol/g wet weight) was present as TTP. The chromatogram is shown in Figure 10. Chicken white skeletal muscle was also found to contain a high TTP to total thiamine ratio (70% at a total level of 1.9 nmol/g wet weight), while in chicken red muscle this ratio averaged 30% for the same amount of total thiamine (3).

The reversed-phase HPLC was also successfully applied to the analysis of animal tissues (27). Cellular concentrations of TTP, TPP, and TMP in different

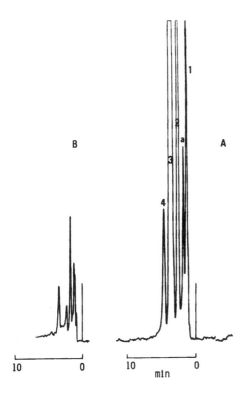

Figure 9 Elution pattern of thiamine and its phosphates in rat liver analyzed by the improved straight-phase HPLC system (see text): (A) sample; (B) blank; (1) thiamine; (2) TMP; (3) TPP; (4) TTP; and (a) unidentified peak.

Table 1 Contents of Thiamine and Its Phosphate Esters in Rat Tissues[a]

Compound	Liver (nmol/g wet weight)	Brain (nmol/g wet weight)
Thiamine	1.45 ± 0.72 (4.8)	0.34 ± 0.13 (5.3)
TMP	3.98 ± 1.87 (13.2)	1.27 ± 0.60 (19.6)
TPP	24.02 ± 4.49 (79.7)	4.83 ± 0.49 (74.5)
TTP	0.68 ± 0.01 (2.3)	0.04 ± 0.02 (0.6)
Total thiamine	30.13	6.48

[a] Values represent the mean ± SD for five rats. The numbers in parentheses refer to percentages of total thiamine.

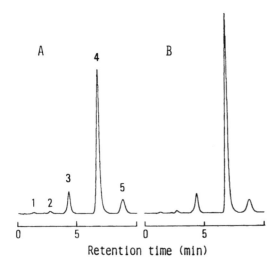

Figure 10 Typical chromatogram of thiamine phosphates in an extract of pig skeletal muscle: (A) skeletal muscle extract; (B) A plus 1 pmol authentic TTP converted to ThcTP; 1, thiamine; 2, TMP; 3, TPP; 4, TTP; 5, unknown. (From Ref. 2.)

tissues determined by reversed-phase HPLC agree with those determined by the straight-phase HPLC.

With the reversed-phase HPLC method described above (31), Matsuda et al. determined the contents of thiamine and its phosphates in various tissues of different animals (35). Total thiamine and TPP levels in the brain and kidney were significantly higher in the particulate fraction than in the soluble one, while those in the liver, soleus muscle (red muscle), and EDL (extensor digitorum longus: white muscle) muscle were markedly lower in the former fraction than in the latter.

TTP was localized in the particulate fraction in the brain, heart, and kidney, while in the liver, soleus, and EDL muscles it occurred mainly in the soluble fraction. In the pig skeletal muscle, especially in the white muscle, the soluble fraction contained TTP to the extent of 77% of total thiamine (Table 2), correlating well with the results reported elsewhere (2). Thiamine phosphates in these muscles were localized solely in the soluble fraction, and the TPP concentration was higher in the soleus than in the EDL (35).

3. Blood and Other Clinical Specimens

As described above (section III.C), the determination of total thiamine in blood specimens is important to assess thiamine status in humans as an indicator for

Table 2 Thiamine and Its Phosphate Esters in Soluble and Particulate Fractions of Pig Red and White Muscles[a]

	Content (pmol/mg protein)				
	Total	TTP	TDP	TMP	Thiamine
Red					
Soluble	330.2	164.7	132.6	23.7	9.2
Particulate	15.24	4.63	8.84	1.61	0.16
White					
Soluble	116.9	89.7	24.2	1.7	1.3
Particulate	12.27	1.80	9.72	0.72	ND

[a] Values are means of two determinations.
Source: Ref. 35.

thiamine deficiency. Most of the HPLC methods include a hydrolysis step of the blood sample with an appropriate phosphatase, before the HPLC step (17,19). Some of the other methods determine thiamine and its phosphates in blood specimens, directly without enzymatic hydrolysis (17,19,20). Total thiamine is obtained by summing the levels of these thiamine forms.

Warnock (36) determined TPP in erythrocyte hemolysates by HPLC with precolumn derivatization using alkaline BrCN and fluorescence detection. The stationary phase was Bondapak-NH$_2$ and the mobile phase consisted of methanol–100 mM potassium phosphate buffer (pH 7.5) (50:50, v/v). A mean TPP value of 123 $\pm$ 27 ng/mL packed normal human erythrocytes was obtained, which is consistent with results obtained previously using an enzymatic method.

Brunnekreeft et al. (32) also reported the determination of thiamine and its phosphates in human whole blood. Sixty-five samples were obtained from healthy volunteers. The mean level ($\pm$SD) of TPP was 120 $\pm$ 17.5 nmol/L versus 4.1 $\pm$ 1.6 nmol/L for TMP, and 4.3 $\pm$ 1.9 nmol/L for thiamine. For TTP, levels were all <4.0 nmol/L. Total thiamine calculated was 132 nmol/L (mean), which is in good agreement with the results reported by Schrijver et al. (25; also personal communication).

We also analyzed the content of thiamine and its phosphates in human blood by straight-phase HPLC with precolumn derivatization (28). Blood samples were obtained from 12 healthy volunteers after a 12-h fasting period. The mean ($\pm$SD) values in nmol/L were 2.23 $\pm$ 1.24 for TTP, 51.6 $\pm$ 8.01 for TPP, 4.77 $\pm$ 1.24 for TMP, 5.39 $\pm$ 2.58 for thiamine, and 64.0 $\pm$ 7.90 for total thiamine (17). The mean total thiamine is lower than the one that has been described by others (25,32), which is probably due to the fasting conditions before blood sampling.

More recently, determination of thiamine and its phosphate esters in human

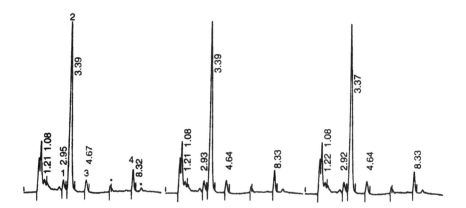

Figure 11 Elution profile for a human blood sample, injected three times. Peaks: 1, TTP; 2, TPP; 3, TMP; and 4, thiamine. Unknown impurities are indicated by an asterisk. (From Ref. 20.)

blood, plasma, and urine was summarized by Tollaksen et al. (19), based on their original work (29). By use of the straight-phase HPLC system with precolumn derivetization, the values (n*M*, mean ±SD) obtained for 30 healthy adults (15 females/15 males) are as follows. In serum: thiamine, 10.9 ± 2.9/16.9 ± 3.3; TMP, 8.3 ± 1.5/3.7 ± 1.5. In whole blood: thiamine, 29.6 ± 10.0/33.4 ± 10.4; TMP, 9.7 ± 2.3/10.9 ± 5.1; TPP, 121 ± 29.6/165 ± 40.4. Trace amounts of TTP were found in a few whole blood samples only. The same group reported similar results for the concentrations of thiamine, TMP, and TPP in plasma and whole blood or cerebrospinal fluid of healthy subjects (19).

Gerrits et al. (20) determined thiamine and its phosphates in whole blood by reversed-phase HPLC with precolumn derivatization. Good resolution of the elution profile for thiamine and its phosphate esters was obtained for a standard sample (Fig. 8) and for a human blood sample (Fig. 11). Reference values obtained from the whole blood samples of 65 healthy volunteers were in n*M* (mean ± SD): thiamine, 4.3 ± 1.9; TMP, 4.1 ± 1.6; TPP, 120 ± 17.5. For TTP, all measurements were <4.0 n*M*. However, a few patients with low TPP and relatively high TTP were found, and in most cases these patients were alcoholics. The precise role of TTP in a clinical setting remains to be elucidated.

IV. GAS CHROMATOGRAPHY

A. Introduction

Gas chromatography (GC) is a fast, selective, and sensitive analytical method that is useful for relatively volatile compounds. The polar nature and low volatility of thiamine prohibit its direct analysis at temperatures at which it is stable. Derivatization of thiamine, prior to GC analysis, is therefore required. GC analysis of thiamine and TPP in biological materials has been briefly reviewed (17). That study concluded that the method remains qualitative, poor in sensitivity, and its application therefore seems to be limited in the assay of total thiamine in meat or cereals (37).

B. Procedures for Determination

A procedure used for the determination of thiamine in meat (37) is briefly described.

Meat was homogenized in 0.1 M HCl, heated at 100°C for 1 h, cooled, and adjusted to a pH of 4.5 to 5. The mixture was then subjected to enzymatic hydrolysis with mylase 100 and papain for 4 h at 45 to 50°C. To cleave thiamine to 4-methyl-5-(2-hydroxyethyl)thiazole, $NaHSO_3$ or Na_2SO_3 was added, the pH was readjusted to 4 to 5, and the mixture was again heated at 100°C for 2 h. After the addition of trichloroacetic acid (TCA), the mixture was chilled and filtered. The filtrate was adjusted to a pH of 11 to 12 and then extracted with chloroform, which was evaporated to dryness. The residue was taken up in 1 mL internal standard solution and the latter was concentrated to 0.5 mL. The resulting sample was stored for analysis.

The internal standard used was 2-(hydroxyethyl)pyridine (I), dissolved in methanol (2 mg/mL) and diluted appropriately. The calibration standard was 4-methyl-5-(2-hydroxyethyl)thiazole (II), which was obtained commercially or by cleavage of thiamine with $NaHSO_3$, as described above. The calibration was done by use of aliquots of the standard solution of compound II, which were mixed with an aliquot of compound I and then chromatographed by GC at an appropriate temperature. The difference in the retention times of compounds I and II is quite clear and can vary from less <1 up to 5 min, depending on the temperatures. The concentration of a given standard obtained by the cleavage of thiamine agreed within 5% with a calibration standard of the same concentration prepared from compound II.

Thiamine contents in meats using the GC methods are in μg/5 g (±SD): calf liver, 25.3 ± 4.0; pork chops, 54.6 ± 1.6; chicken, 7.2 ± 1.5; and ground beef, 10.5 ± 0.9. These values are in good agreement with those obtained by the official AOAC method (1980). Deviation from the experimental mean of 10 trials was usually within ±10%. The detection limit of this GC method for thiamine is 0.8 μg/mL, corresponding to 2.4 pmol injected.

The advantages of the GC method are its simplicity, ease of standardization, and lack of cleanup steps in the sample preparation. Disadvantages are the relatively large amount of sample required and the inability to separate thiamine phosphate esters.

V. MASS SPECTROMETRY AND RECENT TECHNIQUES

Mass spectrometry and capillary electrophoresis seem to attract attension of analysts in the assay field of vitamins. However, neither mass spectrometry nor capillary electrophoresis is practically used in the quantitative determination of thiamine. Mass spectrometry, either stand-alone or in combination of HPLC/capillary gas chromatography, is often used to determine the structure of intermediates of thiamine degradation or biosynthesis. Actually only one paper (38) reported, as far as we know, the use of quantitative mass spectrometry in the past 10 years. It used particle beam HPLC/mass spectrometry with operational parameters optimized for the sensitive analysis of several drugs, including thiamine, in agricultural products, such as milk, and in tissue extracts. Under full-scan conditions, detection limits were in the 100 ng range (0.3 nmol of thiamine) for most drugs, including α-lactams, tetracyclines, furosemide, and others. Another paper (39) descried the application of a HPLC/atmospheric pressure chemical ionization–mass spectrometric (APCI-MS) system to the analysis of thiamine in dried yeast. Thiamine in the extract of dried yeast was first analyzed by HPLC on a reversed-phase ODS column with UV detection at 254 nm. Response was linear in the range of 25 to 300 ng of thiamine. Identification of the thiamine peak was confirmed by its mass spectrum, using the HPLC/APCI-MS system.

VI. FUTURE TRENDS

High-performance liquid chromatography is still a very useful technique for the quantitative determination of thiamine and its phosphate esters, not only in pharmaceutical preparations but also in biological materials, especially clinical specimens.

HPLC systems have rapidly proliferated since the first edition of this book was published. The detection limit for thiamine, as well as thiamine phosphates, as obtained with fluorescence detection is now in the femtomole range. This allows accurate determination of these compounds in human blood samples.

However, the analytical techniques for thiamine and its phosphate esters seem to be in a steady state during the recent past for nearly 10 years. No marked developments in practical techniques and/or methodology in the theoretical basis for the analysis of thiamine and its phosphates have been reported. For example, mass spectrometry and capillary electrophoresis seem to increase in importance in the whole field of vitamin analysis other than thiamine. As described in the text, however, these techniques were applied only in a limited number of thiamine analyses.

The analysis of total thiamine in human whole blood is important to assess the status of thiamine deficiency. This can be successfully carried out by HPLC either with or without enzymatic hydrolysis. Reversed-phase HPLC and straight-phase HPLC are both useful for this purpose (19,31). Concentrations of total thiamine or thiamine phosphates in whole blood are in a good agreement among different HPLC systems used (22,23,28).

Future improvement of stationary-phase materials in HPLC systems is necessary to maintain the stability and reliability of the column, especially to treat the many clinical specimens, mainly blood samples.

REFERENCES

1. S Budavari, ed. Merck Index. 12th ed. Rahway, NJ: Merck & Co., 1996 p 1586.
2. Y Egi, S Koyama, H Shikata, K Yamada, T Kawasaki. Content of thiamin phosphate esters in mammalian tissues—an extremely high concentration of thiamin triphosphate in pig skeletal muscle. Biochem Int 12:385–390, 1986.
3. K Miyoshi, Y Egi, T Shioda, T Kawasaki. Evidence for in vivo synthesis of thiamin triphosphate by cytosolic adenylate kinase in chicken skeletal muscle. J Biochem 108:267–270, 1990.
4. K Ishii, K Sarai, H Sanemori, T Kawasaki. Analysis of thiamin and its phosphate esters by high-performance liquid chromatography. Anal Biochem 97:191–195, 1979.
5. M Morita, T Kanaya, T Minesita. Simultaneous determination of thiamine and 2-(1-hydroxyethyl) thiamine in biological materials. J Vitaminol 15:116–125, 1969.
6. DT Wyatt, M Lee, RE Hillman. Factors affecting a cyanogen bromide–based assay of thiamin. Clin Chem 35:2173–2178, 1989.
7. H Sanemori, T Kawasaki. Purification and properties of thiamine pyrophosphokinase in *Paracoccus denitrificans*. J Biochem 88:223–230, 1980.
8. H Shikata, Y Egi, S Koyama, K Yamada, T Kawasaki. Properties of the thiamin triphosphate-synthesizing activity catalyzed by adenylate kinase (isoenzyme 1). Biochem Int 8:943–949, 1989.
9. Y Egi, S Koyama, T Shioda, K Yamada, T Kawasaki. Identification, purification and reconstitution of thiamin metabolizing enzymes in human red blood cells. Biochim Biophys Acta 1160:171–178, 1992.
10. LO Krampitz. Catalytic functions of thiamin diphosphate. Annu Rev Biochem 38: 213–240, 1969.
11. L Eder, Y Dunant. Thiamine and cholinergic transmission in the electric organ of *Torpedo*. I. Cellular localization and functional changes of thiamine and thiamine phosphate esters. J Neurochem 35:1278–1286, 1980.
12. L Bettendorff, P Wins, E Schoffeniels. Regulation of ion uptake in membrane vesicles from rat brain by thiamine compounds. Biochem Biophys Res Commun 171: 1137–1144, 1990.
13. L Bettendorff, M Peeters, P Wins, E Schoffeniels. Metabolism of thiamine triphosphate in rat brain: correlation with chloride permeability. J Neurochem 60:423–434, 1993.

14. L Bettendorff, B Hennuy, A DeClarck, P Wins. Chloride permeability of rat brain membrane vesicles correlates with thiamine triphosphate content. Brain Res 652: 157–160, 1994.

15. ZZ Ziporin, PP Waring. Thin-layer chromatography for the separation of thiamine, N^1-methylnicotinamide, and related compounds. Methods Enzymol 18A:86–87, 1970.

16. JM Parkhomenko, AA Rybina, AC Khalmuradov. Separation of thiamine phosphoric esters on Sephadex cation exchanger. Methods Enzymol 62:59–62, 1979.

17. T Kawasaki. Vitamin B_1: Thiamine. In: AP DeLeenheer, WE Lambert, HJ Nelis, eds. Modern Chromatographic Analysis of Vitamins. 2nd ed. New York: Marcel Dekker, 1992, pp 319–354.

18. V Fayol. High-performance liquid chromatography determination of total thiamin in biological and food products. Methods Enzymol 279:57–66, 1997.

19. CME Tallaksen, T Bøhmer, J Karlsen, H Bell. Determination of thiamin and its phosphate esters in human blood, plasma, and urine. Methods Enzymol 279:67–74, 1997.

20. J Gerrits, H Eidhof, JWI Brunnekreeft, J Hessels. Determination of thiamin and thiamin phosphates in whole blood by reversed-phase liquid chromatography with precolumn derivatization. Methods Enzymol 279:74–82, 1997.

21. W Weber, H Kewitz. Determination of thiamine in human plasma and its pharmacokinetics. Eur J Clin Pharmacol 28:213–219, 1985.

22. L Bettendorff, C Grandfils, C DeRycker, E Schoffeniels. Determination of thiamine and its phosphate esters in human blood serum at femtomole levels. J Chromatogr 382:279–302, 1986.

23. N Amin, J Reusch. High-performance liquid chromatography of water-soluble vitamins. II. Simultaneous determinations of vitamins B_1, B_2, B_6 and B_{12} in pharmaceutical preparations. J Chromatogr 390:448–453, 1987.

24. P Wimalasiri, RBH Wills. Simultaneous analysis of thiamin and riboflavin in foods by high-performance liquid chromatography. J Chromatogr 318:412–416, 1985.

25. J Schrijver, AJ Speek, JA Klosse, HJM VanRijn, WHP Schreurs. A reliable semiautomated method for the determination of total thiamine in whole blood by the thiochrome method with high-performance liquid chromatography. Ann Clin Biochem 19: 52–56, 1982.

26. JPM Wielders, CJK Mink. Quantitative analysis of total thiamine in human blood, milk and cerebrospinal fluid by reversed-phase ion-pair high-performance liquid chromatography. J Chromatogr 277:145–156, 1983.

27. H Sanemori, H Ueki, T Kawasaki. Reversed-phase high-performance liquid chromatographic analysis of thiamine phosphate esters at subpicomole levels. Anal Biochem 107:451–455, 1980.

28. T Kawasaki. Determination of thiamin and its phosphate esters by high-performance liquid chromatography. Methods Enzymol 122:15–20, 1986.

29. CME Tallaksen, T Bøhmer, H Bell. Concomitant determination of thiamin and its phosphate esters in human blood and serum by high-performance liquid chromatography. J Chromatogr 564:127–136, 1991.

30. J Bontemps, P Phillippe, L Bettendorff, J Lombet, G Dandrifosse, E Schoffeniels, J Crommen. Determination of thiamine and thiamine phosphates in excitable tissues

as thiochrome derivatives by reversed-phase high-performance liquid chromatography on octadecyl silica. J Chromatogr 307:283–294, 1984.

31. H Iwata, T Matsuda, H Tonomura. Improved high-performance liquid chromatographic determination of thiamine and its phosphate esters in animal tissues. J Chromatogr 450:317–323, 1988.

32. JWI Brunnekreeft, H Eidhof, J Gerrits. Optimized determination of thiochrome derivatives of thiamine and thiamine phosphates in whole blood by reversed-phase liquid chromatography with precolumn derivatization. J Chromatogr 491:89–96, 1989.

33. E Kimura, B Panijpan, Y Itokawa. Separation and determination of thiamin and its phosphate esters by reversed-phase high-performance liquid chromatography. J Chromatogr 245:141–143, 1982.

34. M Kimura, Y Itokawa. Determination of thiamine and its phosphate esters in human and rat blood by high-performance liquid chromatography with postcolumn derivatization. J Chromatogr 332:181–188, 1985.

35. T Matsuda, H Tonomura, A Baba, H Iwata. Tissue difference in cellular localization of thiamin phosphate esters. Comp Biochem Physiol 94B:405–409, 1989.

36. LG Warnock. The measurement of erythrocyte thiamin pyrophosphate by high-performance liquid chromatography. Anal Biochem 126:394–397, 1982.

37. RE Echols, J Miller, W Winzer, DJ Carmen, YR Ireland. Gas chromatographic determination of thiamine in meats, vegetables and cereals with a nitrogen-phosphorus detector. J Chromatogr 262:257–263, 1983.

38. RD Voyksner, CS Smith, PC Knox. Optimization and application of particle beam high-performance liquid chromatography/mass spectrometry to compounds of pharmaceutical interest. Biomed Environ Mass Spectrom 19:523–534, 1990.

39. K Yamanaka, S Horimoto, M Matsuoka. K Banno. Analysis of thiamine in dried yeast by high-performance liquid chromatography and high-performance liquid chromatography/atomospheric pressure chemical ionization–mass spectrometry. Chromatographia 39:91–96, 1994.

9
Flavins

Peter Nielsen
University Hospital Eppendorf, Hamburg, Germany

I. INTRODUCTION

A. History

More than 100 years ago a fluorescent compound was isolated first from whey, and later from different biological materials. When it became clear that the isolated yellow pigments, named lactochrome, ovoflavin, or lactoflavin, had a common structure, the new compound was named riboflavin (vitamin B_2) (for historical review see 2). In the years between 1933 and 1935 the structure and the main chemical reactions of riboflavin were studied and the chemical synthesis was performed. Soon afterward, the coenzyme forms, flavin mononucleotide (FMN) and flavin adenine dinucleotide (FAD), were isolated in pure form, and the structures were determined. In the last 50 years many flavoproteins were isolated and their physicochemical properties were studied. Succinate dehydrogenase was the first enzyme found with the prosthetic group (FAD) covalently bound to the protein. About 20 flavoproteins are now known to contain covalently bound coenzyme (mainly via carbon atom 8α) (3). In mammalian tissue, the number of covalently bound flavoproteins appears to be limited.

Another milestone in flavin research was the characterization of the flavosemiquinones, the first example of a stable flavin radical. In addition to riboflavin, FMN, and FAD, a number of flavin analogs with biological activities have been found in micro-organisms and plants. Among these, the coenzyme factor F_{420} isolated from methanogenic bacteria should be mentioned, which has 5-deazaflavin as its chromophore.

B. Chemical Properties

Riboflavin 7,8-dimethyl-10-(1′-D-ribityl)isoalloxazine (RF), a yellow-green, light-sensitive compound, is widely distributed in animal and plant cells. In biological systems, the two coenzyme forms "flavin mononucleotide" (FMN) and "flavin adenine dinucleotide" (FAD) are predominant (Fig. 1). These terms are formally incorrect, because FMN is no nucleotide and FAD is no dinucleotide. However, these names are still accepted. Some authors use also the name riboflavin to designate the coenzyme form of the vitamin and take vitamin B_2 as a general term. Some basic physicochemical properties of these flavins are outlined

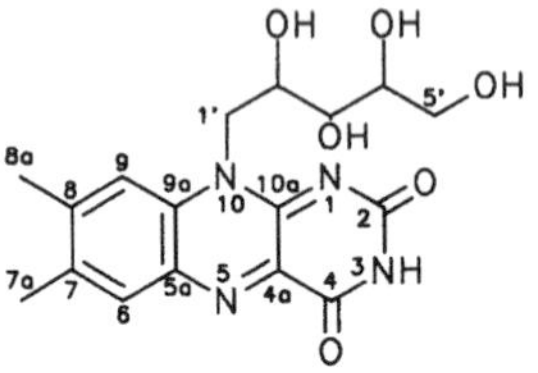

1

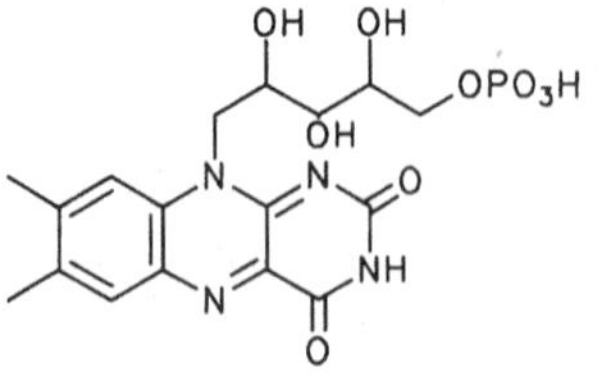

2

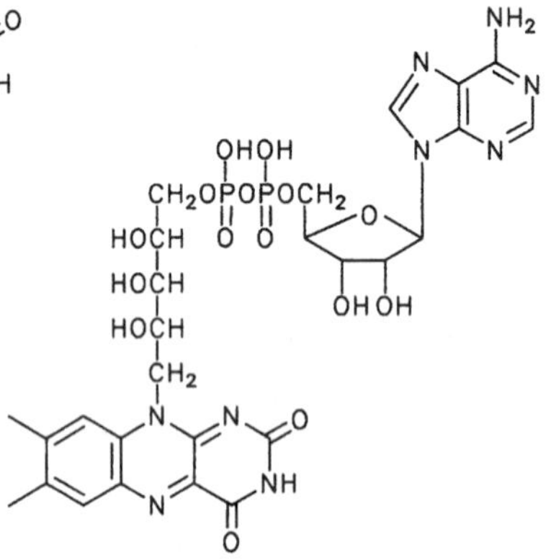

3

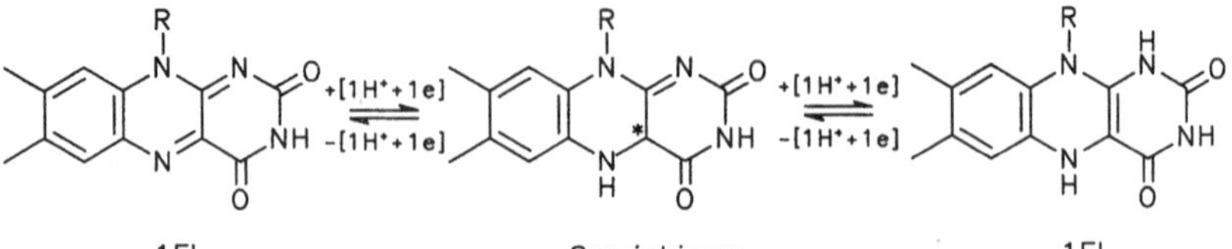

1Fl$_{ox}$ Semichinon 1Fl$_{red}$

Figure 1 Structural formulae of riboflavin, FMN, and FAD: (1) riboflavin in oxidized (FL_{ox}) or reduced form (FL_{red}): (2) FMN, "flavin mononucleotide"; (3) FAD, "flavin adenine dinucleotide."

in Table 1. FAD can be hydrolyzed to form FMN and finally riboflavin, whereas the nonglycosidic bonding between the isoalloxazine ring and the ribityl side chain withstands hydrolysis.

Based on their unique tricyclic isoalloxazine structure (benzene, pyrimidine, and pyrazine ring), the reactivity of flavin compounds in chemical and biochemical reactions is very complex (for review see 2,4), and only some reactions that are important for the analysis of flavin compounds can be mentioned in this text. Riboflavin and FMN are very sensitive to UV light (5). The effective photolysis of riboflavin is a complex, pH-dependent reaction, which results in the formation of 60% to 70% lumiflavin in basic solutions or lumichrome in neutral or slightly acid solutions as the main component (Fig. 2). In contrast to riboflavin, lumiflavin can be easily extracted from biological samples with chloroform and measured photometrically (at 450 nm). The lumiflavin method has been widely used for analytical riboflavin determination in various sources. When protected from light, riboflavin is rather stable in neutral and slightly acidic solutions.

In alkaline solution, the pyrazine ring is opened to form 1,2-dihydro-6,7-dimethyl-2-oxo-D-ribityl-3-chinoxalincarbone acid and urea (6). In acidic solution, a migration of the phosphate group in the 5′ position of FMN has to be considered that, starting from pure riboflavin 5′-phosphate, yields the formation of significant amounts of riboflavin, 4′-, 3′-, and 2′-phosphate (7). In addition, the phosphoric ester group of FMN is hydrolyzed in acidic solution with a pH maximum at about pH 4.0 (8). The spectral properties of flavins in different oxidized and reduced states have been studied in detail (9). In the oxidized state (Fl_{Ox}), flavins and flavoproteins are yellow pigments with two characteristic absorption bands at about 370 nm and 450 nm (Fig. 3).

In solution, flavins show a green-yellow fluorescence with an emission maximum around 520 to 530 nm. The fluorescence of riboflavin and FMN occurs with the same quantum yield, whereas the quantum yield is about 10 times less in FAD.

Table 1 Physicochemical Properties of Riboflavin, FMN, and FAD

Compound	Structure	Relative molecular mass	Melting point (°C)	Solubility (g/L)
Riboflavin	$C_{17}H_{20}N_4O_6$	376.4	278–282 dec.	Water: 0.10–0.13; ethanol: 0.045; insoluble: ether, $CHCl_3$, acetone
FMN (Na salt)	$C_{17}H_{20}N_4NaO_9P$	478.4	>280–290	Water: 30
FAD (Na_2 salt)	$C_{27}H_{31}N_9Na_2O_{15}P$	829.5	>250–290	Soluble in water

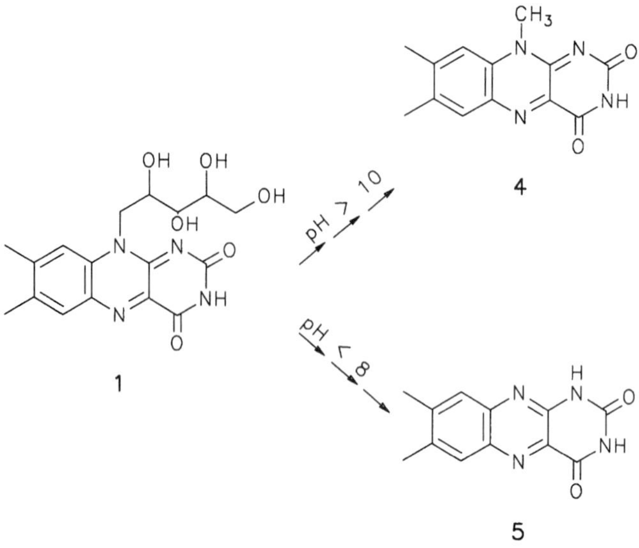

Figure 2   Photolysis of riboflavin: (1) riboflavin; (4) lumiflavin; (5) lumichrome.

Chemical modifications of the ring structure can change the emission spectrum drastically. The same is true for interactions with the protein shell in flavoproteins. The "optic" variability of flavoproteins is exemplary documented in a
recently found association of FAD with the blue light receptor CRY1 in *Arabidopsis thaliana* (10). To mediate responses to blue, UV-A, and green light, the
CRY1-bound flavin oscillates among its different redox states.

C.   Biochemistry

Flavin coenzymes are the most versatile catalysts in biological redox systems.
As they can participate both in two- and one-electron processes, flavins can act as
a redox switch between two-electron donors (NADH, succinate) and one-electron
acceptors (heme proteins, iron-sulfur proteins). In addition, flavins can react with
molecular oxygen. As a consequence, flavoproteins catalyze a large number of
different chemical reactions (for review see 2,4). The specific catalytic function
of a given flavoprotein is based on and regulated by the interaction between
coenzyme and apoprotein. A simplified classification of the biological function
of flavoproteins according to the different reaction types was given by Müller et
al. (4) (Table 2). The problem of this and other proposals of flavoprotein classification is the fact that a given enzyme can catalyze quite different redox reactions.

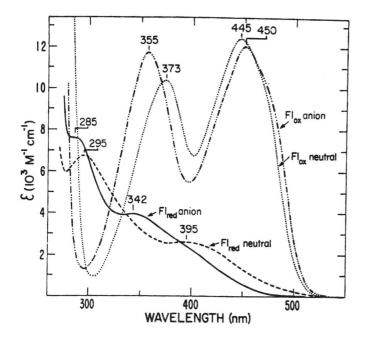

Figure 3 Absorption spectrum of FMM in water in the oxidized (FL_{ox}), reduced (FL_{red}), neutral, and anionic states. (From Ref. 9.)

Table 2 Functions of Flavoproteins

Function	Typical reaction	Typical enzyme
I. Dehydrogenation	$-CXH \rightarrow C{=}X + 2H^+ + 2e^-$ $(X{=}CH_2, NH, O)$	Acyl-CoA dehydrogenase
II. O_2 activation (hydroxylation, monooxygenation)	$O_2 + 4e^- + 4H^+ \rightarrow 2H_2O$	P-Hydroxybenzoate hydroxylase
III. Electron transfer	$+e^- - e^- \rightarrow 2e^- \rightarrow 1e^- + 1e^-$	Flavodoxin transhydrogenases
IV. Light emission		Bacterial luciferase
V. Photo(bio)chemistry	Phototropism	"Bluelight receptors"
VI. Regulation (?)		Oxynitrilase; carboligase

Source: Ref. 4.

D. Physiological Function

1. Biosynthesis of Riboflavin

Riboflavin is synthesized in plants and various micro-organisms (for a review see 11). The biosynthesis starts from guanosine 5'-triphosphate (compound 6) by the removal of C8 from the purine ring and pyrophosphate from the ribosyl side chain (Fig. 4). In yeast, the next steps are rearrangement and reduction of the sugar residue followed by a deamination in position 2 of the pyrimidine ring. The 5-amino-6-ribitylamino-2,4(1H,3H)-pyrimidinedione (compound 7) formed, reacts with 3,4-dihydroxy-2-butanone-4-phosphate (compound 8, derived from ribulose 5'-phosphate) to form 6,7-dimethyl-8-ribityllumazine (compound 9). Dismutation of two molecules of the lumazine results in the formation of riboflavin by the action of riboflavin synthase. In bacteria, the deamination in position 2 of the pyrimidine precedes the reduction of the ribosyl side chain.

Riboflavin also plays a role in the biosynthesis of vitamin B_{12}. The 5,6-dimethylbenzimidazole moiety of vitamin B_{12} is formed from riboflavin in aerobic and some aerotolerant bacteria (12).

2. Flavin Metabolism in Humans

In humans, riboflavin is absorbed in the upper jejunum by a fast and saturable mechanism (apparent K_m 0.4 µM) (13). FMN and FAD are hydrolyzed prior to intestinal absorption by the action of an FMN phosphatase and an FAD pyrophosphatase (14). Evidence indicates that a specific carrier exists in the brush border as well as in the basolateral cell membrane (13). Proteolytic digestion of food proteins containing covalently bound flavins results in the liberation of S-cysteinylriboflavin and N-histidylriboflavin. These compounds are absorbed in rats but seem to have no vitamin activity (15). In plasma, most of the riboflavin is weakly bound to albumin and more tightly bound to a subclass of immunoglobulins (13). A class of riboflavin-binding proteins (RBP) has been found in liver, blood, and eggs of hens (16). These proteins are induced by estrogen, have no catalytic activity but are involved in riboflavin storage and transport to the ovarial follicles. Especially, the structure and properties of RBP from hen egg white has been studied in some detail. RPB is also of interest in the field of affinity chromatography of flavin compounds (see Sect. II.C). Similar transport proteins have been found in pregnant mammals, including humans (17).

Within the mammalian cells, most of the riboflavin is converted to FAD and FMN (Fig. 5). Flavins are mainly excreted in human urine in the form of riboflavin and some catabolites, including 8α- and 7α-hydroxyriboflavin (18).

Tissues that receive the greatest exposure to light, that is, skin and ocular tissue, manifest the earliest clinical symptoms of a riboflavin deficiency. These include pellagralike symptoms (*pellagra sine pellagra*): lesions around the corner

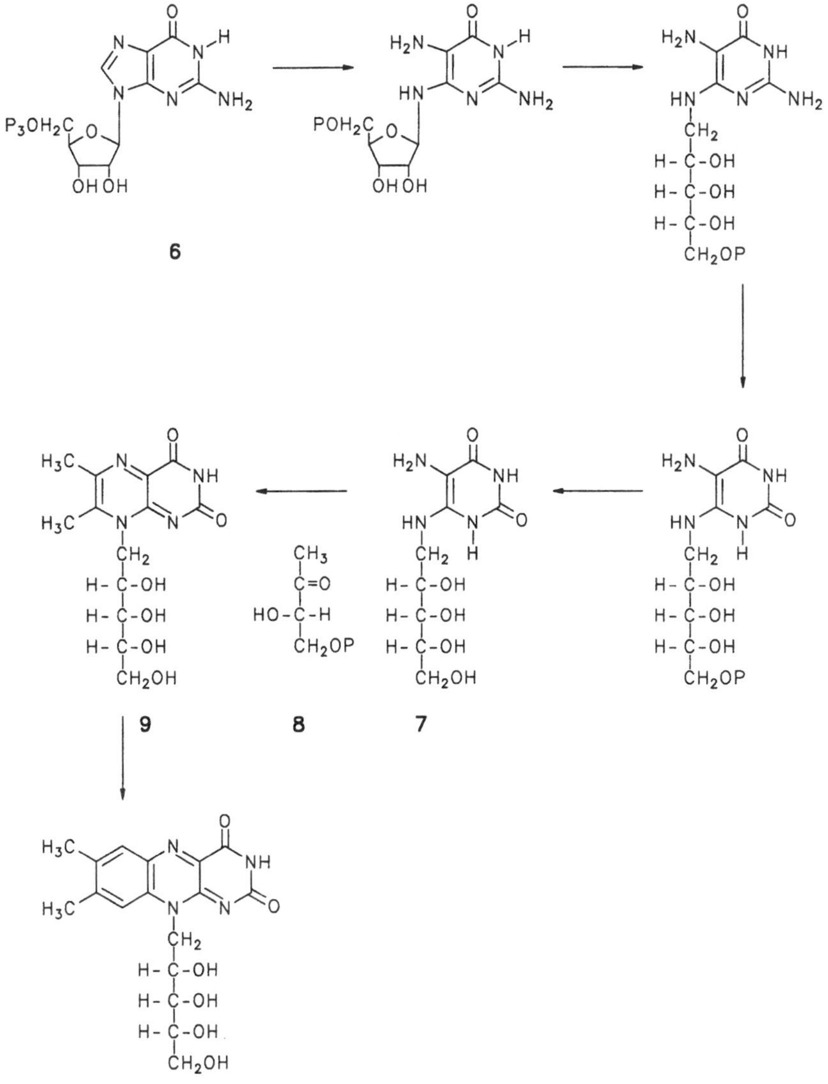

Figure 4 Biosynthesis of riboflavin in yeasts: (6) GTP; (7) 5-amino-6-ribitylamino-2,4(*1H,3H*)pyrimidine dione; (8) 3,4-dihydroxy-2-butanone 4-phosphate; (9) 6,7-dimethyl-8-ribityllumazine; (1) riboflavin.

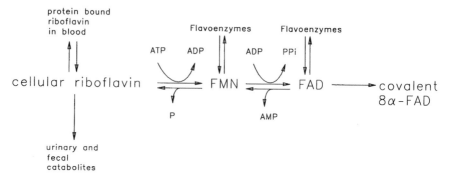

Figure 5 Metabolic pathway of riboflavin in mammals.

of the mouth (stomatitis angularis), cheilosis, glossitis, and seborrheic changes at the nose and ears. Insufficient dietary intake is the main cause of riboflavin deficiency in undeveloped countries, whereas alcoholism accounts for vitamin B_2 deficiency in economically strong countries. Besides the general malnutrition in alcoholics, ethanol seems to reduce the utilization of the vitamin, especially from FAD-containing sources. Furthermore, riboflavin deficiency is found in newborn infants with hyperbilirubinemia, which is treated with phototherapy (19).

According to the *Recommended Dietary Allowances*, 10th ed. (1989), a daily intake of 1.7 mg in males and 1.3 mg in females is desirable for an adult (20). Animal proteins (e.g., milk) and some vegetables (e.g., broccoli) are good sources of riboflavin, even when cooked. However, exposure to daylight, for example, of milk, can destroy most of the riboflavin (70% loss in 4 h).

E. Pharmacological Function

Riboflavin and synthetic FMN are used in numerous pharmaceutical preparations for the prophylactics and therapy of riboflavin deficiency. The pharmacokinetics of orally and intravenously administered riboflavin has been studied in some detail in healthy humans (21). The maximum amount of riboflavin from a single dose is 27 mg per adult.

In Western countries a clear indication for vitamin B_2 treatment is rare at the moment. When newborn infants are given 0.5 mg FMN/day during phototherapy, a decrease in riboflavin status can be avoided (19). FMN and FAD are cofactors for complex I and II in the respiratory chain. In some cases of respiratory chain encephalomyopathy, high doses of riboflavin resulted in clear clinical improvement (22).

Recently, evidence indicated that flavins may also play a role as natural antioxidant (23). For instance, ferrylmyoglobin has been implicated in cardiac reoxygenation damage. Because flavin reductase can reduce the oxidated myoglobin in the presence of sufficient flavin concentration, a potential role of riboflavin supplementation in decreasing reoxygenation injury is discussed following results in experimental animals (24). It is speculated that this cardioprotective effect may be the result of a direct reaction of reduced flavins with hydroxyl radicals.

As the measurement of riboflavin fluorescence in the urine after oral dosage is very easy to perform, riboflavin is given in some studies to test the compliance of other oral drugs in patients (25).

II. THIN-LAYER, COLUMN, AND AFFINITY CHROMATOGRAPHY

Owing to the fascinating yellow color of flavin compounds, the history of paper and column chromatography is closely connected with flavins. For the analytic determination of flavins in different matrices, paper chromatography has only historical interest, mainly owing to its rather qualitative character. To a lesser degree, the same is true for thin-layer chromatography. However, the visual impression of a flavin chromatography is preserved in TLC, and maybe, therefore, in contrast to the analysis of other vitamins, this technique is still used for the identification of flavin compounds (10,26–28).

For preparative applications, conventional column chromatography (ion exchange, gel, affinity chromatography) are still valuable tools for isolation and purification of flavin compounds.

A. Thin-Layer Chromatography

The benefit of TLC or its advanced form HPTLC (high-performance thin-layer chromatography) for identification of unknown flavin compounds by comparing R_f values of the unknown flavins with authentic material is beyond doubt (26–28). The quantification of the separated flavin compounds can be performed by the use of modern densitometry. Various high-quality precoated plates with low and uniform particle diameters are available for HPTLC, including stationary phases of silica gel, cellulose, or various reversed phases (TLC-RP). For special applications, TLC may have some advantage, for example, in separation capability (multidimensional development), even when compared with HPLC. Depending on the mobile phase, TLC can clearly differentiate riboflavin from FMN and flavin analogs, as well as riboflavin phosphates from riboflavin bisphosphates (7,26,27).

TLC was used for both qualitative (26–28) and quantitative analysis (29–31) of flavins in different matrices. The separation and quantitative determination by fluorescence quenching of the B-complex vitamins in some pharmaceutical products was performed using a newly synthesised carbamide formaldehyde polymer ''aminoplast'' with solvent systems 1-butanol/water/acetone (25:9:5, v/v) and 1-butanol/methanol/benzene/water (20:10:10:8, v/v) (29).

The quantitation of riboflavin together with thiamine and niacin using HPTLC silica gel plates and methanol/water (70:30, v/v) as mobile phase was described by Diaz et al. (30). For riboflavin the native fluorescence was used and a ''preplate'' derivatization was applied for the other two vitamins (addition of a fluorescent tracer to label nicotinic acid; conversion of thiamine into thiochrome). The developed plates were scanned by a commercially available bifurcated fiber-optic-based instrument that transferred the excitation and emission energies between the plate and the fluorescence spectrometer. Calibration curves for the determination of riboflavin 48 to 320 ng, thiamine 300 to 750 ng, and niacin 10 to 100 ng were established. The advantages of this method are that no elimination of excess oxidation reagent is necessary and that the simultaneous determination of vitamins with only one detector is possible.

A simultaneous detection of all water-soluble vitamins in multivitamin solution by only one TLC analysis was described by Postaire et al. in 1991 (31). The method used HPTLC plates with silica gel as stationary phase and n-butanol/pyridine/water (50:35:15, v/v) as mobile phase. The quantitation was carried out by photodensitometric detection without derivatization (B_1, B_2, B_6, C, folic acid, nicotinamide) or after spraying ninhydrin reagent (calcium panthothenate) or 4-dimethylaminocinnamaldehyde (B_{12}, biotin). The use of an over-pressured layer chromatograph improved the resolution of the HPTLC plates. Good reproducibility, satisfactory standard deviations, and recoveries were obtained for all the vitamins.

B. Conventional Column Chromatography

Conventional ion exchange chromatography is a useful technique for the preparation of substantial amounts of purified FMN and FAD. The separation of FMN from riboflavin and riboflavin bisphosphates can be achieved efficiently on DEAE-Sephacel (acetate form) with a discontinuous gradient of triethyl ammonium acetate pH 7.0 in 30% 2-propanol as mobile phase (8). This has some practical relevance because commercial preparations of FMN contain significant amounts of riboflavin and riboflavin bisphosphates (7). The riboflavin monophosphate fraction obtained by this chromatographic technique still consists of an isomeric mixture of riboflavin 5'-, 4'-, and 3'-phosphate. Milligram amounts of almost pure FMN can be obtained by ion exchange chromatography on DEAE-cellulose (chloride form) eluted with ammonium carbonate, pH 7.8 (8), or by preparative HPLC (7,8,27). Pure FAD in 500-mg amounts (98%) can be prepared from the culture broth of *Sarcina lutea* by the method of Chibata et al. (32). This method involves adsorption of the crude FAD on Florisil followed by ion exchange chromatography on Amberlite IRA-401 and a final purification step on p-acetoxymercurianiline-agarose.

Urinary riboflavin was determined by reversed-phase open-column chromatography (33,34). A column (290 × 11.5 mm) was packed with 50 μm C18 material (Separation Technology, Wakefield, RI) and equilibrated with a solvent system consisting of methanol/water/acetic acid (37:63:0.1, v/v/v). Riboflavin standard solutions and urine specimens were passed through the column and 5-mL fractions were collected. It was possible to separate FMN together with FAD from riboflavin as well as from lumiflavin and lumichrome, respectively. The precision of this method was determined by within-day and day-to-day reproducibility using frozen urinary pools with high (>1 μg/mL) and low (<1 μg/mL) riboflavin concentration. Results obtained from the traditional AOAC fluorimetric method and from a HPLC method were highly correlated. The reversed-phase open-column chromatography was found to be a precise and simple procedure which, unlike the HPLC technique, is not dependent on expensive equipment.

Nekoflavin (7α-hydroxyriboflavin, NF) was isolated from cat's choroids using a combination of column chromatography and a droplet countercurrent distribution method (28). A stock of choroids was prepared from the eyes of cats. After homogenization in a Waring Blendor, the homogenate was heated to 80°C and perchloric acid was added. The precipitate was centrifuged and the supernatant was added to a Florisil column. The column was washed with dilute perchloric acid and water and developed with 5% pyridine. The flavin eluate was concentrated and applied to a Sephadex G-10 column. The crude NF fraction is fractionated by a droplet countercurrent method. A final chromatography on Biogel P-2 (eluant, water) yielded about 26 μmol NF from 30,000 eyeballs. The identification of NF as 7α-hydroxyriboflavin was performed by comparison with chemically synthesized reference material, using mass spectrometry and thin-layer chromatography (silica gel 70 plates; isoamyl alcohol/methylethylketone/acetic acid/water, 40:35:7:13, w/w/w/w).

C.   Affinity Chromatography

Riboflavin-binding apoprotein from egg white has a high affinity for riboflavin (k_m about 1.3 nM). Several analytical techniques are available which are based on riboflavin-binding protein.

A nonchromatographic method for the sensitive determination of riboflavin in human plasma and urine by enzyme-linked ligand-sorbent assay (ELLSA) was recently described (35). Using standard microtiter plate formate, a conjugate of flavin with bovine serum albumin is immobilized on a plastic surface. The RF analyte competes with the immobilized flavin for soluble RBP that bears a biotin marker (B-RBP). As a result, B-RBP is absorbed on the plastic in an amount reciprocally related to RF content in the analyzed sample. A final enzyme labeling occurs via the avidine anchor. The optimized method has a detection limit of 0.8 pmol riboflavin and should work in the range of 20 nM to 4 μM riboflavin and is thus sufficient for routine determination of riboflavin in human plasma and urine. In serum samples from human volunteers, a good correlation between the ELLSA method and an HPLC technique was found.

A simple determination of riboflavin from urine and plasma based on a fluorometric titration was described by Kodentsova et al. (36). This method is based on the fact that the formation of riboflavin-apoprotein complex is accompanied by a complete loss of fluorescence peculiar to free riboflavin. A plasma sample of 1 mL was acidified by the addition of trichloroacetic acid. After centrifugation and neutralization, the fluorescence intensity was measured in a 3-mL cuvette (ex. 465 nm; em. 525 nm). The measurement was repeated after addition of riboflavin standard and riboflavin-binding apoprotein, respectively. An amount of 1.5 ng riboflavin/mL plasma can be detected using this method.

A chiral stationary phase for HPLC based on hen egg yolk riboflavin-binding protein was developed (37). The purified protein was immobilized on activated 5 NH_2-Nucleosil silica. Several chiral drugs were chromatographed on this material, and the influence of mobile-phase factors was studied. Thirteen of 20 drug enantiomers were partially or baseline separated, including nicardipine, carprofen, oxazepam, and warfarin. It is not clear whether the drugs interact with the riboflavin binding site; however, the very encouraging results deserve further investigations.

III. HIGH-PERFORMANCE LIQUID CHROMATOGRAPHY

With the development of high-performance (high-pressure) liquid chromatography (HPLC), the classical column chromatography technique has undergone a renaissance. In the field of flavin (bio)chemistry, HPLC is the analytical method of choice and has superseded other chromatographic methods more or less completely. The determination of riboflavin and flavin coenzyme as well as flavin degradation products, catabolites, or analogs from various sources, including pharmaceutical preparations, foods, and blood, has been described (see below).

A. Analytical Systems

1. Stationary Phases

Many different kinds of stationary phases are available for HPLC. Among these, reversed-phase columns (e.g., 18-C or 8-C alkyl chain length) are frequently used for chromatography of the water-soluble, ionic, or nonionic flavin compounds.

2. Mobile Phase

For reversed-phase HPLC, mixtures of methanol (or acetonitrile) and aqueous buffer solutions are preferred for the chromatography of flavin compounds. Addition of ion pair compounds (such as triethylammonium chloride or hexanesulfonates) can modify the retention behavior of ionic compounds (riboflavin phosphates, FAD) significantly (ion pair HPLC).

3. Detection

For routine HPLC analysis, the detection of flavins is carried out either spectrophotometrically, using variable- or fixed-wavelength HPLC detectors in the ultraviolet (e.g., 254 nm) or visible (e.g., 405 nm) region, or fluorimetrically. For riboflavin, the excitation wavelength for fluorimetric detection is usually 440 to 450 nm, and the emission wavelength 530 nm. The detection limit for fluorescence detectors is >1 pmol (0.38 ng) riboflavin, whereas <30 pmol (11 ng) can be detected spectrophotometrically at 254 nm. Photodiode array detectors are significantly less sensitive than normal HPLC spectrophotometers (38).

4. Internal Standard

A variety of compounds such as p-hydroxybenzoic acid (39) or acetanilide (40) have been used as internal standards in the HPLC determination of riboflavin. These compounds have solubilities in water similar to riboflavin; however, their chemical structures are quite different. Lambert et al. suggested isoriboflavin as a better internal standard for the liquid-chromatographic determination of riboflavin in human urine because of its closer structural similarity (41). More recently, sorboflavin, which has glucityl on the isoalloxazine ring, in contrast to the ribityl chain in riboflavin, was suggested as a good internal standard, which guarantees a high yield of copurification together with riboflavin during the different extraction and isolation procedures (42).

B. Riboflavin

Numerous methods for the determination of riboflavin in pharmaceutical multivitamin preparations and foods have been described. In biological samples, protein-bound FAD and FMN are the predominant flavin species. Significant amounts of free (noncoenzymic) riboflavin are found only in the urine and in the retina of the eye.

1. Pharmaceutical Preparations

Riboflavin is an ingredient of many pharmaceutical multivitamin preparations. Methods have been described for the determination of riboflavin and other, water-soluble vitamins (B_1, B_6, B_{12}, niacin, niacinamide, folic acid, ascorbic acid). For an efficient separation of the different compounds, ion pair HPLC seems to be favored because this technique results in better peak shapes, especially of thiamine, pyridoxine, and folic acid, than classical reversed-phase HPLC. Details of some methods are outlined in Table 3.

No special pretreatment procedure is necessary. The liquid or finely powdered material is dissolved directly in the mobile phase (containing internal standard). After centrifugation or filtration, appropriate aliquots are injected directly onto the column. Reversed-phase C8 or C18 columns and mixtures of methanol or acetonitrile with aqueous buffer solutions (containing different ion-pairing reagents) are used as stationary and mobile phases, respectively. Dong et al. studied the factors affecting the ion pair chromatography of water-soluble vitamins in some detail (43). Following comparison of six different reversed-phase columns (C8 and C18), it was found that the influence of the stationary phase on the separation quality is only limited. Riboflavin and folic acid have significantly higher k′ values on C18-bonded phases, but baseline separation was obtained on all the columns studied. Smaller particle size (3 and 5 µm) generated a higher resolution of the seven water-soluble vitamins. The mobile phase should contain 12.5% to 20% methanol as an organic modifier for optimum separation. Acetonitrile gave slightly sharper peaks but yielded inadequate separation of ascorbic acid and niacin. A concentration of 4 to 7 mM of hexanesulfonate as ion-pairing reagent and a pH of 2.8 to 3.2 gave the best separation of all seven compounds. At 0.10% to 0.13%, triethylamine as an additive resulted in symmetrical peaks and the best separation of thiamine from riboflavin. The influence of different columns (C8 vs. C18) on the separation of water-soluble vitamins was also studied by Blanco et al. using hexanesulfonate as ion interaction reagent (38). The best resolution was obtained using a Lichrosorb RP-8 column. An increase in sensitivity as well as lower cost was found using microcolumns packed with 3-µm particles.

2. Foods

Since the availability of the HPLC techniques in the 1980s, numerous HPLC methods for the determination of riboflavin in foods have been published (for earlier review, see 46). These methods describe the analysis of flavins alone or together with other water-soluble vitamins such as thiamine and/or niacin, ascorbic acid, or pyridoxine. Isocratic classical reversed-phase chromatography or ion pair chromatography was used for the determination of vitamin B_2 in meat or meat products (47), cow milk (26,48–51), infant milk or formulas (52–54), dairy products (42,55), cereals (51,57–61), fruits and vegetables (58,62), and beverages (40). As far as the chromatographic conditions are concerned (column, mobile phase), the variations among the different methods are only small (Table 4).

Table 3 HPLC Determination of Riboflavin in Pharmaceutical Preparations

Reference	Column	Mobile phase	Detection	Other compounds detected simultaneously	Detection limit
Dong, 1988 (43)	RP 8-3 μm Pecosphere-3CR 8.3 × 0.46 cm	Methanol/water (15:85), 5 mM Na hexanesulf/1% acetic acid/0.10–0.13%	UV spectrometry photodiode array	B_2, B_1, C, B_6, niacin, niacinamide, folic acid	0.2 ng (d.l.)
Raju, 1989 (44)	Presample cleanup Sep-Pak C18; Novapak C18	ion pair eluant Na hexanesulfonate		simult. det. of B_1, B_2, B_6, niacinamide	
Chase, 1990 (39)	NovaPak C18 (150 × 3.9 mm)	acetonitrile/phosphate buffer pH 3.6 ion pair Na hexanesulfate	UV detection	thiamine, pyridoxine, niacin	0.2 μg/mL were detected
Blanco, 1994 (38)	Sperisorb ODS-2 (100 × 2.1 mm i.d., 3 μm)	10 mM KH_2PO_4 pH 2.8 5 mM hexanesulf gradient 0–50% MeOH	UV detection 272 nm or photodiode array	niacin, niacinamide, B_6, B_1, folic acid	1.8 ng UV 33 ng diode array
Akiyama, 1991 (45)	Methoxy(3-morpholinopropyl)silanediyl gel (150 × 6 mm)	0.1 M phosphate buffer pH 2.3	UV detection 270 nm	FAD, B_1, B_{12}, chlorpheniramine, naphazoline, neostigmine in eye lotion	

Table 4 HPLC Determination of Riboflavin in Foods

Reference	Source	Pretreatment	Column	Mobile phase	Detection	Results/remarks
Maeda, 1989 (40)	Liquid tonics	no	Nucleosil 7 C$_{18}$ (250 × 4.6 mm)	acetonitrile/0.01 M KH$_2$PO$_4$/triethylamine (8 + 91.5 + 0.5 v/v/v) + 5 mM Naoctanesulfonate pH 2.8	UV 254	simult. detection of RF, FMN, B$_1$, B$_6$, caffeine, Na-benzoate det. limit: <200 µg B$_2$/100 mL
Kanno, 1991 (48)	Milk, milk fat membrane	1. boiling for 3 min to inactivate pyrophosphates 2. pronase digestion to release bound flavins 3. centrifugation 12,000 g	Capcell Pak C18; 40°C	(A) 90% aqueous MeOH (B) 10 mM NaH$_2$PO$_4$ pH 5.5 linear gradient from 35% to 95% A	fluorescence det. ex: 462 nm em: 520 nm	separation of RF, FMN, FAD bovine milk contained 202 µg flavin/100 g (14% FAD, 4.4% FMN, 83% riboflavin)
Munari, 1991 (49)	Milk, cheese	1. precipitation with MeOH/acetic acid 2. centrifugation 3. 0.45-µm pore filter	Supelcosil LC-18 (5 µm) (250 × 4.6 mm)	water:MeOH (68:32 v/v)	UV 444 nm, 267 nm	whole milk 127 µg B$_2$/100 g
Gauch, 1992 (50)	Milk	1. centrifugation 2. trypsin, clara-diastase treatment	Nucleosil 100 C18-5 µm (150 mm)	phosphate buffer 0.2%, pH 5.5/water/acetonitrile (9:78.5:12.5 v/v/v)	fluorescence det. ex: 455 nm em: 525 nm	det. limit 50 µg B$_2$/L milk
Roughead, 1990 (26)	Milk	1. TCA precipitation 2. sat. with NH$_4$-sulfate 3. phenol extraction	µBondpak RP-18 (300 × 3.9 mm)	Methanol/50 mM ammonium acetate buffer, pH 6.0 linear gradient	fluorescence ex: 305–395 nm em: 475–650 nm	total flavin in milk: 900 µg/L: 60% RF, 26% FAD, 11% 10(2'-hydroxyethyl)flavin
Bilic, 1990 (42)	Dairy products	1. sample homogenized 2. incubated in 6% formic acid, 2 M urea for 30 min 3. 0.75 mL C18 columns	Supelco LC-18 (75 × 4.6 mm) 3 µm	100 mM phosphate buffer pH 2.9/acetonitrile (86: 14, v/v)	fluorescence det. ex: 450 nm em: 530 nm	total flavin 483 pM (88% RF, 4.7% FMN, 6.8% FAD)
Greenway, 1994 (51)	Milk, cereals	on-line system 1. microwave extract. 2. dialysis 3. trace enrich. cartr. (Asorbospheres HS C18)	5 µm ODS-2 Spherisorb (250 × 4.6 mm)	A. Na-acetate buffer pH 4.8 B. H$_2$O/acetonitrile/ MeOH (50/40/ 10, v/v) linear gradient 6% to 100% B	fluorescence det. ex: 450 nm em: 520 nm	during the microwave extraction, FAD was converted completely into FMN; 15% of FMN was hydrolyzed into riboflavin; analysis time: 20 min

Reference	Food	Sample preparation	Column	Mobile phase	Detection	Remarks
Albala-Hurtado, 1997 (52)	Infant milk	1. TCA precipitation 2. 0.45 μm filter	Spherisorb ODS-2 C18-5 μm (250 × 5.6 mm)	Methanol/0.5% triethylamine, 2.4% acetic acid (15:85, v/v) 5 mM octanesulfonic acid	UV detection	simult. detect. of B_1, B_2, B_6, B_{12}, folic acid, niacin detec. limit: 0.05 μg B_2/mL; total flavin in whole milk: 1.4 mg/L
Chase, 1992 (53)	Infant formula	1. perchlorate incubation for 1 h 2. pH 3.3 filtration	NovaPak C18 (150 × 3.9 mm)	phosphate buffer pH 3.6/acetonitril (90:10), 5 mM hexanesulfonate	UV or fluorescence det. ex: 295 nm em: 395 nm	simult. determ. of B_1, B_2, B_6 detect. limit: 0.09 μg B_2/mL comparison with different AOAC methods, incl. fluoresc. and microbiolog. method
Li, 1993 (54)	Fortified milk powder	extraction with 5% TCA	Radial-Pak C_{18}	MeOH/water (35:65), 15 mM butylamin, pH 4	UV detection 254 nm	simult. detect. of B_2 and B_1; recovery of B_2: 91.8% comparison with AOAC method
Hewavitharana, 1996 (55)	Casein	1. comparison of different extraction methods: acid, hot water or pepsin extraction 2. takadiastase	Spherisorb ODS C18-5 μm (250 × 4.6 mm)	50 mM acetate buffer/MeOH (65:35, v/v) pH 5.0	fluorescence det. ex: 447 nm em: 517 nm	0.06 mg B_2/kg
Barna, 1994 (47)	Meat, liver	1. homogenized sample autoclaved in 0.01 M HCl 2. takadiastase/claradiastase/papain 3. Nucleosil C18 cartridge	Nucleosil C18 (150 × 4.6 mm) 3 μM with guard column 10 μm (20 × 4.6 mm)	10 mM phosphate buffer pH 3.0/acetonitrile (84:16 v/v) containing 5 mM Na-heptanesulfonate	UV detector 254 nm	HPLC method yields lower B_2 conc. in meat than AOAC microbiological method detect. limit: 0.03 μg B_2/mL
Otles, 1993 (56)	Eggs		μBondpak C18	water/MeOH (78:22 v/v)	UV detection	simult. determ. for B_2, B_6, niacin, folic acid methods for B_1 and B_2 described; pretreatment with different enzyme
Hägg, 1994 (57)	Various foods	1. acid autoclavation 2. enzyme treatment 3. ferricyanide oxidation 4. Sep-Pak C_{18}	μBondpak C18 Radial-Pak (100 × 8 mm)	0.005 M phosphate buffer pH 7.0/MeOH (65:35 v/v)	separate runs B_1: ex: 360 nm, em: 425 nm; B_2: ex: 440 nm em: 520 nm	preparation strongly affected determination of riboflavin from food

Table 4 Continued

Reference	Source	Pretreatment	Column	Mobile phase	Detection	Results/remarks
Sims, 1993 (58)	Various foods, cereals, flour, broccoli	1. autoclaved: 0.1 N HCl 2. Ferricyanide oxidation 3. Sep-Pak C$_{18}$	RP18 μBondpak Waters Assoc. 300 × 3.9 mm	0.005 M NH$_4$OAc/ MeOH (72/28 v/v; pH 5.0)	0–10 min: es: 370, em: 435 nm >10 min: ex: 370, em: 520 nm	simult. detec. of B$_1$ and B$_2$; detect. limit: 0.05 ng on column linear up to 100 ng total flavin in cereal samples: 1.69–2.04 mg/100 g
Ollilainen, 1991 (59)	21 food items	1. acid hydrolysis 2. clara-diastase	Spherisorb S5 ODS 2 (5 μm) 250 × 4.6 mm	MeOH/water (35:65, v/v)	fluorescence det. ex: 445 nm em: 525 nm	comparison with AOAC fluorescence method detect. limit: 20 pg B$_2$
Han, 1993 (60)	Cereals, vegetables	1. 0.4 M HCl autoclaving 2. centrifugation	μBondpak C18 (300 × 3.9 mm)	MeOH/water (30:70, v/v), 5 mM 1-heptanesulfonic acid	UV detection 254 nm	detect. limit: 5 ng, linear range 5–50 ng simultaneous analysis of B$_2$ with B$_1$, B$_5$, B$_6$, folic acid comparison with a referenced method
Hasselman, 1989 (61)	Dietetic food	enzyme treatment (takadiastase/ β-amylase)	μBondpak C18	MeOH/0.05 M Na acetate pH 4.5	fluorescence ex: 366 nm em: 522 nm	2 μg/100 g
Fernando, 1990 (62)	Soybeans, tofu	acid hydrolysis, autoclavation, centrifugation	Ultrasphere C18 5 μm (150 × 4.6 mm)	acetonitril/0.01 M acetate buffer pH 5.5 (13:87, v/v)	fluorescence det.	simult. determ. of B$_2$ and B$_1$ detect. limit: 2 ng B$_2$/mL soybeans, 0.92–1.10 μg/g
Agostini, 1997 (63)	Vitamin-enriched Brazilian foods (50 products)	1. ground to flour 2. acid hydrolysis boiling water bath 3. + methanol storage −18°C, filtration	Spherisorb ODS-2 RP-18 5 μm (150 × 4.6 mm)	(A) acetonitrile; (B) 5 mM hexanesulf. 0.15%; triethylamine pH 2.8; (C) MeOH linear gradient: 2% A, 98% B to 2% A, 41% B, 57% C	photo diode array det.	simult. determ. of B$_1$, B$_2$, B$_6$, nicotinamide, nicotinic acid; comparison with the declared values
Barna, 1992 (64)	Ready-to-eat baby food	1. acid autoclavation 2. clara-diastase/Papain pH 4.5, 20 h, 37°C 3. C$_{18}$ cartridges (Sep-Pak)	Nucleosil C18 5 μm (250 × 4.6 mm)	10 mM K-phosphate buffer pH 7.0/acetonitrile (89.5:10.5, v/v)	UV detection 268 nm	comparison with AOAC fluorescence method (correlation coefficient: 0.987) det. limit: 30 ng/mL RF
Ayranci, 1993 (65)	Macaroni	1. acid autoclavation 2. filtration	μBondpak-NH$_2$ (300 × 3.9 mm)	MeOH/water (20:80, v/v, pH 3.5)	UV 254 nm	kinetic analysis of loss of flavin content of macaroni during cooking

Some methods have been applied only to selected foods and their general validity has not been demonstrated, whereas other methods were successfully used with a wide variety of different food matrices (58–61,63). For simultaneous determination of riboflavin and thiamine in foods, various workers have proposed HPLC methods, which analyze riboflavin either as riboflavin or lumiflavin and thiamine as thiochrome by fluorescence detection. The thiochrome can be formed either before or after column separation, while lumiflavin was formed precolumn. A strong argument against the precolumn oxidation of thiamine to thiochrome is the fact that some of the riboflavin may be destroyed during the oxidation step in alkaline solution. Therefore, a state-of-the-art procedure for this special separation is the postcolumn derivatization of thiamine (see chapter on vitamin B_1).

3. Sample Pretreatment

Most of the methods involve a pretreatment procedure to remove the protein-bound coenzymes from the respective flavoproteins and to hydrolyze phosphoric ester bonds. The total flavin content is determined in these methods in the form of riboflavin, and no information is available about the concentrations of FMN and FAD. Usually, homogenized samples in 0.1 N HCl are first autoclaved for 20 to 30 min followed by an enzymatic treatment (Takadiastase or papain) to obtain a complete cleavage of the phosphate esters. Additional precipitation of proteins by the addition of trichloroacetic acid can be performed to extend the lifetime of the HPLC column. After pH adjustment and filtration or centrifugation, the resulting riboflavin solution is chromatographed. To concentrate the samples and to remove some of the interfering materials, the riboflavin-containing solution can be passed through reversed-phase cartridges (C18, Sep-Pak, or RP-8 Baker) (42,47,51,57,58). Usually, an aliquot (2 to 10 mL) of the respective solution is placed onto one of these short disposable columns. After washing the column with water or a water/methanol mixture to remove salts and some impurities, the vitamins are eluted with a small volume of pure methanol.

Two studies have looked for the influence of alternative extraction procedures in some detail. Three different approaches—acid digestion, hot water extraction, and pepsin digestion—were compared for the efficient extraction of low concentrations of riboflavin from casein (55). The pepsin extraction gave about 17% higher riboflavin values than did acid or hot water extraction. Experiments to optimize the extraction procedure clearly showed the considerable influence of enzyme concentration and the incubation time both for the pepsin-catalyzed peptide hydrolysis and the takadiastase dephosphorylation. According to the authors of this study, the new approach seems to be superior to the traditional acid extraction.

The effect of various commercially available enzyme preparation in the HPLC determination of riboflavin and thiamine in foods was studied by M. Hägg (57). Different enzymes (takadiastase, clara-diastase, α-amylase), as well as the same enzyme produced by different manufacturers, strongly affected the determination of both vitamins. The recoveries for different foods ranged from 85% to 100% for thiamine and from 80% to 100% for riboflavin. Fluka clara-diastase (6%) was the best enzyme in the study.

Sensitivity. It has been pointed out by different authors that UV detection techniques suffer from a lack of sensitivity and from UV-absorbing interference, especially when low amounts of riboflavin in some nonenriched foods have to be determined.

Validity. The HPLC results for the riboflavin content in foods have been found to agree well with those obtained by the Association of Official Analytical Chemists (AOAC) microbiological assay and fluorimetric assay. However, lower values for the HPLC method compared to the AOAC method have also been reported in some foods. This might indicate the presence of interfering material in the AOAC fluorimetric assay and the higher selectivity of the HPLC method.

To give some information on the reproducibility of the determination of water-soluble vitamins in foods between laboratories using their own routine methods, an intercomparison of methods involving 18 European laboratories was organized in 1993 to assess the state of the art of vitamin determination in foods (66). Each laboratory received identical samples. In contrast to other vitamins (B_1, niacin) but similar to vitamin B_6 and C, the differences for the determination of riboflavin content of milk powder, pork muscle, and haricots verts beans with HPLC, fluorimetric and microbiological methods were rather high, with relative standard deviation of reproducibility (RSD_{Reprod}) ranging from 28% to 74% (Fig. 6). The extraction and hydrolysis procedures were probably the most important sources of variation. The presented data indicated the need for a rigorous hydrolysis procedure including an effective enzyme treatment. Takadiastase was frequently used by most of the laboratories. However, it seemed that the performance of takadiastase varied among suppliers.

Stability of Riboflavin in Foods. Increasing information is available about the degradation of vitamin B_2 in different foodstuffs that are stored under adverse environmental conditions or that are processed. The kinetic of losses of riboflavin (together with thiamine and niacinamide) were investigated during cooking of macaroni at different temperatures (65). After acid digestion, riboflavin was analyzed by reversed-phase HPLC. A total loss of 50% of the flavin content was found after a cooking period at 90°C for 30 min. As the thermal decomposition of flavin takes place only at much higher temperature, the vitamin loss in this study results most probably from leaching into the cooking water.

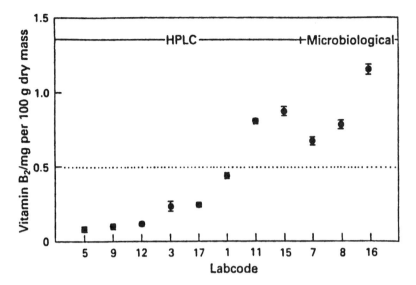

Figure 6 State-of-the-art determination of riboflavin in foods. Results of individual European laboratories for vitamin B_2 determination (HPLC methods, except 7,8,16) in pork muscle (identical samples). Data represent the mean $\pm$ standard deviation of at least three separate determinations for each laboratory. (From Ref. 66.)

Leaching is also the cause for riboflavin loss from tofu. Using HPLC analysis, a 20% loss of vitamin from tofu was found after 3 to 7 days of storage at 4°C (62). The quantitative assessment of flavins in cow milk and the effect of pasteurization on these compounds were studied by Roughead and McCormick (26) using the phenol extraction technique and reversed-phase HPLC (Table 4; Fig. 7). Riboflavin (61%), FAD (26%), and 10-(2'-hydroxyethyl)flavin (11%) were the predominant flavin compounds in all milk samples studied. The distribution of flavins in raw and pasteurized milk reveals little difference, with the exception of a decrease in the FAD content with pasteurization.

The effect of low-dose gamma radiation on the B vitamins in pork chops and chicken breasts was investigated by Fox et al. (67). Over the range of dose (0.5 to 6.7 kGy) and temperature (-20 to $+20$°C) studied, the loss of riboflavin was relatively small. These radiation conditions are appropriate for both trichina elimination in pork and salmonella control in chicken.

What is more important in respect to a loss of flavin compounds from various sources is the possibility of photochemical degradation. It is quite clear that the light intensity is the rate-determining factor for riboflavin loss, whereas increased temperature has only little effect. March et al. studied the effect of light

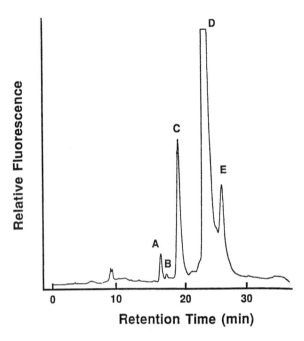

Figure 7   HPLC chromatogram of phenol-extracted flavins from pooled cow's milk. (A) 7a-hydroxyriboflavin; (B) 8a-hydroxyriboflavin; (C) FAD; (D) riboflavin; (E) 10-(2'-hydroxyethyl)flavin. (From Ref. 26.)

on the vitamin B_2 content of cheese (68). The loss of riboflavin when exposed to sunlight was shown to be primarily a surface effect whereas loss of vitamin A was similar throughout the cheese. It is speculated that hydroxyl radicals are formed in the continuous aqueous phase of the cheese structure which play a significant part in the riboflavin destruction. The protection afforded riboflavin by vacuum packaging may thus be attributed to the lack of oxygen.

4. Biosamples

As compared with the concentration of FAD and FMN in most biological materials (blood, tissue, organs), a significant amount of riboflavin is found only in urine.

Urine.   The sensitive determination of riboflavin excreted into urine is of interest because it seems to be a good means for monitoring dietary intake of the vitamin. The flavin content in urine can establish the vitamin B_2 status of an individual and it can be used as an index of the relative bioavailability of vitamin formulations. About 200 μg/day riboflavin is renally excreted by a normal person. Excretion of 40 to 70 μg/d (or <17 μg/g creatinine) indicates a riboflavin deficiency (69). Besides riboflavin, significant amounts of different RF metabolites were found in human urine, amounting up to 28% to 39% of total urinary flavins. These are in the following quantitative order: 7α-hydroxyriboflavin, 8α-sulfonylriboflavin, lumiflavin, 8α-hydroxyriboflavin, and 10-hydroxyethylflavin (18,21).

Several HPLC methods for the assessment of flavins in urine are available, the majority being published before 1989 (for review, see 1) (18,26,33,70,71). HPLC has been shown to be a fast, sensitive, simple, and reproducible analytical method for urinary flavins (Table 5). Using fluorescence detection, flavin concentrations as low as 0.05 to 0.30 μg/mL and as high as 2 to 12 μg/mL RF can be measured without concentration or dilution, respectively. No special pretreatment procedure is necessary for HPLC of urine samples. Fresh urine samples are injected either directly or after centrifugation, to remove most of the sediment. Due to bacterial degradation, the stability of riboflavin in fresh urine is limited at room temperature, even in the dark. Consequently, all flavin-containing urine samples should be stabilized by acidification (acetic or oxalic acid) and/or adding a small amount of toluene.

For special investigations, flavin compounds can be concentrated from large volumes of urine by extraction procedures. According to Chastain et al. (18), a phenol extraction is performed by saturating urine samples with ammonium sulfate followed by centrifugation to remove solids. The supernatant fluid is shaken twice with 0.1 volume of 80% aqueous phenol and centrifuged, and the upper phenolic layers are combined. An equal volume of distilled water is added and the solution is extracted with diethyl ether to remove the phenol from the aqueous flavin-containing solution.

Miscellaneous. Riboflavin in serum of athletes was determined by Rokitzki using a reversed-phase technique (Table 5) (72). Proteins were denatured by an enzymatic treatment followed by a TCA precipitation. Contaminating interferents were removed by a solid-phase extraction (Baker 10-SPE). The normal reference value for riboflavin detected by HPLC in serum was 6.27 ± 2.46 ng/mL. Corresponding values in athletes of different sport disciplines were much higher, indicating the widespread use of vitamin B-complex medication in sports.

C. Flavin Mononucleotide and Riboflavin Phosphates

Riboflavin 5'-phosphate (FMN) is synthetically prepared on a large scale for use in pharmaceutical preparations, especially for parenteral administration. It has been demonstrated by reversed-phase HPLC that the chemical phosphorylation of riboflavin invariably yields a complex mixture of various riboflavin phosphates besides FMN. The HPLC chromatogram of a commercial FMN sample is shown

Table 5 HPLC Determination of Flavins in Biosamples

Reference	Source	Pretreatment	Column	Mobile phase	Detection	Results/remarks
Huang, 1993, 1997 (33,34)	Urine	oxalic acid is added to urine samples as a preservative; centrifugation at 1500 g	μBondpak 10 μm (300 × 3.9 mm)	Methanol/water/acetic acid (34:66:0.1, v/v/v)	fluorescence det. ex: 440 nm em: 500–700 nm	comparison with results from open column RP-18 chromatography as well as AOAC fluorescence method
Zempleni, 1996 (21)	Urine	centrifugation	μBondpak (300 × 3.9 mm)	Methanol/water (34:66)	fluorescence det. ex: 450 nm em: 530 nm	rate constants for renal RF excretion were determined
Zempleni, 1996 (21)	Blood plasma after oral or IV riboflavin dosage	1. TCA precipitation 2. centrifugation 3. hot water incub. 85°C, 10 min 4. centrifugation	RP-18 10 μm (250 × 4.6 mm)	Methanol/ammonium acetate pH 4.0 (35:65, v/v)	fluorescence det. ex: 450 nm em: 520 nm	pharmacokinetic variables were determined after oral or parenteral dosage of riboflavin
Rokitzki, 1989 (72)	Serum, whole blood in athletes	enzymatic hydrolysis (?) TCA precipitation Baker 10-SPE	Novapak C18	MeOH/1% acetic acid 3 mM 1-hexanesulfonic acid	UV detection 254 nm	simul. det. of B_1, B_2, B_6 in various disciplines normal B_2 6.2 ng/mL serum
Zempleni, 1995 (73)	Blood plasma	1. TCA precipitation 2. centrifugation 3. hot water extraction (85°C, 10 min)	Gynkotek RP18 10 μm (250 × 4.6 mm)	Methanol/ammonium acetate pH 4.0 (35:65, v/v)	fluorescence det. ex: 450 nm em: 520 nm	blood plasma: 16.4 nM riboflavin, 55.4 nm flavocoenzymes (FAD + FMN); det. limit 3 nM RF, 9 nM FMN
Zempleni, 1992 (74)	Maternal blood, cord blood	1. TCA precipitation 2. centrifugation 3. hot water extraction (85°C, 10 min)	RP-18 10 μm (250 × 4.6 mm)	Methanol/ammonium acetate pH 4.0 (35:65, v/v)	fluorescence ex: 450 nm em: 520 nm	blood plasma: 8.7 nM RF, 84.5 nM FAD + FMN; venous cord plasma: 40.6 nM RF, 49.1 nM FAD + FMN
Zempleni, 1995 (75)	Placenta tissue	1. homogenization (Potter-E.) 2. TCA precipitation 3. hot water extraction (85°C, 10 min)	RP-18 5 μm (250 × 4.6 mm)	Methanol/ammonium acetate pH 4.0 (35:65, v/v)	fluorescence det. ex: 450 nm em: 565 nm	preterm: 0.24 nmol/g RF, 3.36 nmol/g FAD + FMN; fullterm; 0.27 nmol/g RF, 10.8 nmol/g FAD + FMN wet weight, respectively

Reference	Sample	Sample preparation	Column	Mobile phase	Detection	Comments
Batey, 1991 (76)	Ocular tissue rabbit	1. ocular tissue preparation 2. homogenization 3. hot water incub. 80°C, 15 min 3. centrifugation, TCA precipitation	Analytichem Sepralyte CH column 5 μm (150 × 4.6)	Acetonitrile/ ammonium phosphate pH 5.5 (10:90, v/v)	fluorescence det. ex: 447 nm em: 530 nm	among ocular tissues, retina contained the highest flavin concentration with FAD as the primary flavin
Batey, 1990 (77)	Retina of rat (+ liver, plasma)	1. retina preparation (6–9) 2. homogenization (Potter-E.); hot water incub. 80°C, 15 min 3. centrifugation, TCA precipitation	Analytichem Sepralyte CH column 5 μm (150 × 4.6 mm)	Acetonitrile/ ammonium phosphate pH 5.5 (10:90, v/v)	fluorescence det. ex: 447 nm em: 530 nm	retina: 4.8 RF, 17.6 FMN, 46.5 FAD (pmol/mg protein)
Batey, 1992 (78)	Rat retina		Novapak C18	MeOH/1% acetic acid 3 mM 1-hexanesulfonic acid	UV detection 254 nm	simul. det. of B_1, B_2, B_6 in various disciplines normal B_2, 6.2 ng/mL serum
Ryll, 1991 (79)	Animal cells	1. centrifugation of cell suspension 2. pellet resusp. in ice-cooled perchloric acid (2 times) 3. neutralized, 0.45 μm filtration	Supelcosil LC-18T 3 μm (150 × 4.6 mm)	(A) 100 mM phosphate buffer pH 6.0, 8 mM TBA sulfate gradient of A with MeOH	UV 254 nm	quantitation of nucleotides incl. FAD in animal cells
Hohl, 1992 (80)	Fungus (Phycomyces blakesleeanus)	1. homogenization (Ultra Turrax) 2. extraction with MeOH at different temperatures	Nucleosil 10 C8 (250 × 4.6 mm)	(A) THF/acetonitrile/ water (0.1:0.5:99.4, v/v/v) (B) MeOH multilinear gradient A:B	fluorescence ex: 430 nm em: 540 nm	wild-type: 5.5 μM RF, 4 μM FMN, 1.4 μM FAD; photobehavioral mutants; (madA) has altered flavin concentrations
Susin, 1993 (81)	Roots of sugar beets	1. homogenization (Potter) in ice-cold 0.1 M ammonium acetate, pH 6.1 2. centrifugation	Novapak C18 4 μm (100 × 8 mm)	methanol/0.1 M ammonium acetate pH 6.0 (30:70, v/v)	UV 375 or 445 nm	identification of riboflavin 3' and 5' sulfate as naturally occurring flavins

in Figure 8. 5'-FMN accounts only for 75% of the total flavin content. Significant amounts of isomeric monophosphates (4'-FMN, 3'-FMN) and riboflavin bisphosphates (3',5'-RBP, 4',5'-RBP, and 3',4'-RBP) are detectable in all commercial FMN samples analyzed (7,8). Thus, the HPLC methods for the detection of FMN in pharmaceutical preparations have to take into account that ~25% of the total flavin content is present in the form of isomeric riboflavin phosphates. A purity control of flavin phosphates can also be performed using capillary zone electrophoreses (see Sect. IV.A).

The isomeric riboflavin phosphates can also be separated by ion pair chromatography on reversed-phase HPLC columns (7). However, the reversed-phase technique was found to be more reproducible. Milligram amounts of pure 5'-FMN (>95%) can be obtained by preparative HPLC, which may be significant in work with rare riboflavin analogues and with isotope-labeled samples (7,27).

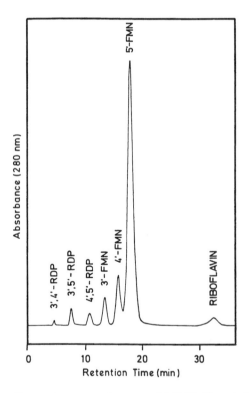

Figure 8 HPLC of commercial FMN (Sigma). Column, Nucleosil 100-10 C_{18} 25 × 0.4 cm; injection volume, 3 μL (40 μg); eluant, 100 mM ammonium formate and 100 mM formic acid in 17% methanol. 3',4'-RDP, riboflavin 3',4'-bisphosphate. (From Ref. 7.)

Pure 4'-FMN was used in an amplified colorimetric assay of alkaline phosphatase. This system allows the detection of very low amounts of enzyme in the range of 4 amol (82).

A second remark concerns the rapid isomerization of riboflavin phosphates in acid solution (pH < 2), especially at elevated temperature. The thermodynamic equilibrium is characterized by the presence of about 65% 5'-FMN, 11% 4'-FMN, 8% 3'-FMN, and 15% 2'-FMN. The rate constants for the various isomerization reactions under a variety of experimental conditions have been determined by reversed-phase HPLC (83). According to the official method of the American Association of Analytical Chemists, protein-bound flavins are extracted from biological samples by treatment with 0.1 M HCl at 121°C for 15 to 30 min. It is known that this treatment leads to hydrolysis of FAD. As shown in Figure 9, it should also be noted that a substantial fraction of 5'-FMN is converted to other isomeric phosphates under these conditions.

1. Foods

The separate determination of riboflavin and the flavin coenzymes FMN and FAD has been described in dairy products (26,43,48) and liquid tonics (40) (Table 4).

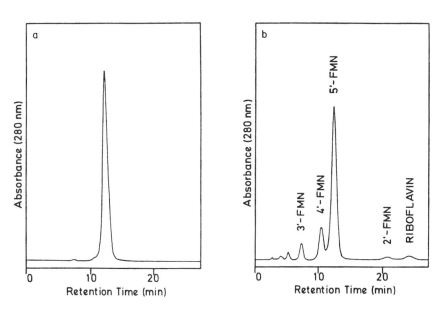

Figure 9 HPLC of (a) pure 5'-FMN and (b) a 5'-FMN sample heated at 121°C for 15 min in 0.1 M HCl. Column, Nucleosil, 100-10 C18: eluant, 100 mM ammonium formate in 20% methanol, pH 6.3. (From Ref. 8.)

Milk samples were prepared by acid hydrolysis but without an enzymatic hydrolysis step in order not to convert FMN into free riboflavin. Relatively low amounts of flavocoenzymes compared to free riboflavin were found in dairy products.

2. Biosamples

The HPLC determination of FMN in biological samples is described together with that of FAD.

D. Flavin Adenine Dinucleotide

Commercially available preparations of FAD have been found to be 94% pure. The remaining 6% are composed of four or five minor components (Fig. 10), which can be separated from FAD by reversed-phase HPLC (27).

    In most tissues and organs, protein-bound FAD is the predominant flavin species. HPLC methods have been developed for the determination of flavocoenzymes in whole blood, plasma, and different tissues, including ocular tissue. A

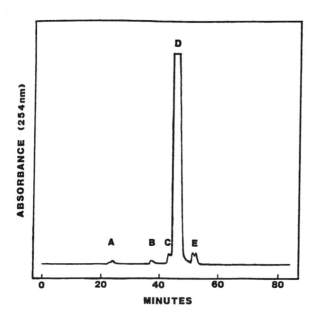

Figure 10 Reversed-phase HPLC separation of commercial FAD. Column, Dynamax Macro HPLC C18 (250 × 21.5 mm); eluant, linear gradient of 5 mM ammonium acetate pH 6.0 to 70% methanol. (D) FAD; (C) riboflavin 5′-pyrophosphate; (E) FMN. (From Ref. 27.)

significant hydrolysis of FAD due to (1) the pretreatment procedure (i.e., acid hydrolysis), and (2) an enzymatic degradation (e.g., in plasma) has to be considered.

1. Biosamples

Using reversed-phase HPLC methods, it has been demonstrated that FAD in human plasma is hydrolyzed almost completely to FMN within 60 min at 37°C in the dark, whereas FMN is quite stable under the same conditions (84). Therefore, artificially enhanced FMN and reduced FAD have to be considered in attempts to quantitate flavocoenzymes separately. To avoid this complication, methods have been developed that converted FAD completely into FMN. The flavocoenzymes were thus analyzed as total FMN. In this way, Zempleni developed HPLC methods for the separate detection of flavins in blood and tissue (Table 5) (26,73–75). Blood plasma was deproteinized by addition of 20% trichloroacetic acid. After centrifugation, the supernatant was heated for 10 min in a water bath at 85°C. By this procedure, FAD is completely hydrolyzed to yield FMN.

Using reversed-phase HPLC, the pharmacokinetics of orally and intravenously administered riboflavin was studied in healthy humans (26). After riboflavin dosage, release of flavocoenzymes into plasma was low compared with the increase of riboflavin concentration. In another study of this author using the same HPLC technique, the concentrations of thiamine, riboflavin and pyridoxal 5′-phosphate in blood plasma of pregnant women and venous and arterial cord plasma were determined (74). In maternal plasma the concentration was 8.7 nmol/L (free riboflavin) and 84.5 nmol/L (FAD + FMN). In venous cord plasma the concentration was 40.6 nmol/L (free riboflavin) and 49.1 nmol/L (FAD + FMN). The gradients of concentration between maternal plasma and venous cord plasma were 1:10 for thiamine, 1:4.7 for free riboflavin, and 1:5 for pyridoxal 5′-phosphate. For the coenzyme forms of vitamin B_2 the maternal circulation showed the higher concentration (1.7:1). Therefore an active transplacentar transport mechanism was assumed. The vitamin concentrations in the cord artery were significantly lower than those in the cord vein, indicating a massive retention by the fetus. In placental tissue, flavocoenzymes were more abundant than free riboflavin with no significant difference in placentas from a group of preterm infants as compared to a group of full-term infants (75).

The presence of substantial amounts of flavins in retina has been known for some time. The detailed function of the light-sensitive flavins in the eye as well as pathophysiological consequences of riboflavin deficiency or overload conditions are not known to date. Batey and coworkers have analyzed the flavin content in retinal tissue from rats and rabbits in some detail (76–78). Retinas from two to three rats were isolated and homogenized. Flavins were extracted by a hot water treatment and analyzed by reversed-phase HPLC (Table 5). The flavin content in retina (FAD, 46.5; FMN, 17.6; riboflavin, 4.80 pmol/mg protein) was found to be five to 10 times smaller than in liver, but 1 to 2 orders of magnitude higher than in plasma. In liver and retina, FAD was the predominant flavin, followed by FMN and riboflavin.

E. Flavin Metabolites and Analogs

Flavin analogs are of special use in the study of structure/function relationship within different flavin dependent enzymes. Even before 1989, a variety of flavin analogs at the riboflavin, FMN, and FAD level were separated, quantitated, or isolated by HPLC, including 5-deazaflavin, 1-deazaflavin, and malonylriboflavin (85). A simultaneous separation of 12 different flavin analogs by reversed-phase HPLC is demonstrated in Figure 11 (85).

More recently, riboflavin 3'- and 5'-sulfates were found in root extracts of iron-deficient sugar beets using reversed-phase HPLC (81). These novel, naturally occurring flavins account for 82% and 15%, respectively, of the total flavin concentration in the plants and cause a flavinlike yellow color as well as autoflu-

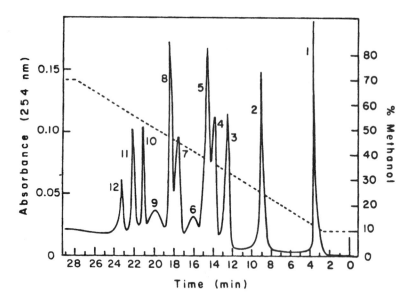

Figure 11 HPLC separation of flavin analogs: (1) ATP; (2) 8-hydroxy-FAD; (3) FAD; (4) 1-deaza-FAD; (5) 5-deaza-FAD; (6) FMN; (7) 1-deaza-FMN; (8) riboflavin; (9) 5-deaza-FMN; (10) 1-deaza-riboflavin; (11) methyl-riboflavin; (12) 5-deaza-riboflavin. Column, Lichrosorb RP18, 25 × 1 cm; linear gradient between 5 mM ammonium acetate buffer pH 6.0, and methanol. (From Ref. 85.)

orescence. Root tips were excised, ground, and homogenized using ice-cold 0.1 M ammonium acetate, pH 6.1. The extracts were centrifuged and the supernatant was directly applied onto a 100×8 mm Waters Novapak C_{18} radial compression column, equilibrated with 0.1 M ammonium acetate/methanol (pH 6.0; 95:5, v/v). Two peaks were developed from the column using water/methanol (70:30) as mobile phase and were unequivocally identified as flavin sulfates. As iron-sufficient plants exhibit no yellow color, it is suggested that theses flavins are integrated in the response of the plants to iron deficiency.

Using a preparative C18 Dynamax Macro HPLC column (21.4×250 mm), Hartman et al. described the isolation and identification of riboflavin 5'-pyrophosphate (RPP) as a minor contaminant of commercial FAD preparations (27). The sample was washed onto the column with 5 mM ammonium acetate pH 6.0 and developed from the column using a linear gradient of this ammonium acetate buffer and 70% of methanol (Fig. 10). RPP was found to have biological activity as a weak cofactor for two FAD-requiring enzymes—porcine D-amino acid oxidase and fungal glucose oxidase.

IV. GAS CHROMATOGRAPHY

As indicated by the lack of appropriate references in the literature, gas chromatography seems to have no practical relevance for the analysis of flavin compounds.

V. MASS SPECTROMETRY

Combined liquid chromatography–mass spectrometry (LC-MS) has become one of the most useful and powerful techniques for the analysis of nonvolatile and thermolabile organic compounds, which cannot be analyzed by GC-MS. However, the high costs of the hardware and the necessity of expert knowledge limit the widespread use of this ''high-end'' technique. A technical problem is still the interface technique which has to deal properly with the nonvolatile buffer solutions used for the HPLC part of the analysis. In many cases, the optimum mobile phase for the separation of the compounds of interest cannot be utilized. To overcome this problem, a new column-switching technique has been described using tocopherol or riboflavin as test compounds (86). The peak of interest was cut from the effluent, placed into sampling loops, and applied to a trapping column. Buffer constituents were washed out and riboflavin was eluted using a suitable mobile phase for LC-MS. Using a fast atom bombardment MS 0.3- to 1-µg amounts of riboflavin could be detected. The MS spectrum showed a protonated molecular ion peak $[(M+2H)^+]$ at m/z 378.

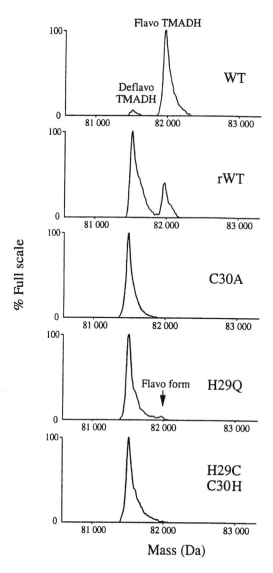

Figure 12 Electrospray mass spectra, deconvoluted to the mass scale, of native wild-type (WT) and recombinant wild-type (rWT) trimethylamine dehydrogenase. (From Ref. 88.)

An on-line coupling of capillary electrophoresis or reversed anionic capillary isotachophoresis (CITP) with electrospray ionization mass spectrometry has been described for the separation of monophosphate nucleosides, pyridine, and flavin dinucleotides (87). The combination with CITP gives an enhancement of sample loadability and concentration sensitivity in capillary zone electrophoresis/ mass spectrometry (CZE/MS). MS-compatible buffer systems were developed (trailing electrolyte, 10 mM caproic acid pH 3.4; leading electrolyte, 7 mM HCl/ 13 mM β-alanine pH 3.9). This technique seems to be valuable for the trace analysis of DNA and RNA, e.g., to study radiation-induced DNA damage.

Electrospray mass spectrometry has been used to characterize the flavinylation reaction products of wild-type and mutant forms of trimethylamine dehydrogenases purified form *Methylophilus methylottrophus* (88). Purified protein samples from wild-type and mutant trimethylamine dehydrogenase were run on reverse phase C_4 column HPLC with acetonitrile/0.1% trifluoroacetic acid as mobile phase. Fractions were injected directly into the VG BioQ triple quadrupole instrument. For the native enzyme, a major peak of mass 81,956 ± 5 Da and a minor peak of 81500 ± 6 Da were obtained. The major peak corresponded to a mass of 81,951 Da expected for a subunit coupled with FMN linked via the 6S-cysteinyl FMN bond. The minor peak (about 5% of total enzyme) corresponded to a mass expected (81,498) for deflavo trimethylamine dehydrogenase. The relation of flavo/deflavo enzyme was also analyzed in mutant enzymes altered in the FMN binding site (Fig. 12). These findings illustrate the power of electrospray mass spectrometry in analysis of flavoproteins, since it was previously thought that the enzyme purified from *M. methylotropus* contained a full complement of flavin.

VI. RECENT TECHNIQUES: CAPILLARY ZONE ELECTROPHORESIS

Capillary electrophoresis (CE) or capillary zone electrophoresis (CZE) has recently emerged as a highly promising technique consuming an extremely small amount of sample and capable of a rapid, high-resolution separation, characterization, and quantification of analytes. Separation in CZE is achieved mainly via differences in mobilities of analytes under a high electric field. Both electroosmotic flow of the bulk solution and electrophoretic mobilities contribute to the observed migration behavior of each analyte in the capillary. Since the electrophoretic mobility of an ion is directly proportional to its charge, which in turn is strongly affected by pH, manipulation of the buffer pH becomes one of the key strategies in optimizing a separation.

CE has been used to determine impurities in commercial riboflavin 5'-phosphate preparations (89). The instrumentation used consisted of a fused-silica separation capillary (100 cm × 100 μm i.d.) which was mounted in an HPLC fluorescence detector. Using a constant voltage of 17.5 kV, variations of buffer systems at pH values between 6.0 and 9.1 were studied. Corresponding electropherograms are shown in Figure 13. Without experimental proof, the different peaks were identified and quantitated as 5'-FMN and other riboflavin mono- as well as bisphosphates. The authors expressed that CE is superior to all other methods in analyzing isomeric flavin phosphates. However, this statement is simply wrong,

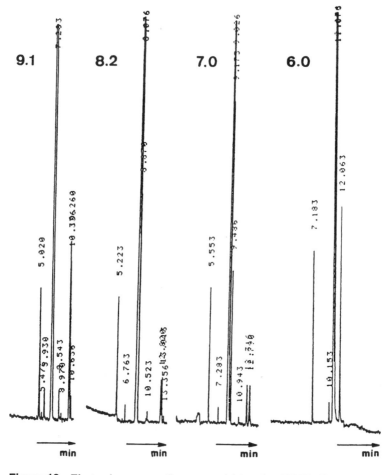

Figure 13 Electropherograms of a commercial sample of FMN. The main peak is riboflavin 5'-phosphate; the first eluting peak is riboflavin; components with longer migration times than that of FMN are other riboflavin mono- and diphosphates (?). (From Ref. 89.)

because HPLC clearly allows the separation of particular flavin monophosphates and bisphosphates (7,8,83).

The simultaneous determination of the B vitamins, thiamine, riboflavin, pyridoxal, pyridoxine, and pyridoxamine in a pharmaceutical product using CZE was described by Huopalahti and Sunell (90). Hydrochlorid acid was used for the extraction of the vitamins from the multivitamin-multimineral tablet. The applied potential was 6.0 kV, and a 75-μm fused-silica capillary tubing was used. The electrolyte used was a 20-mM sodium phosphate buffer pH 9.0. A clear separation of standards as well as of the pharmaceutical sample is shown in Figure 14. This method appears to be a fast and simple technique for the simultaneous determination of water-soluble vitamins in pharmaceutical products, where the

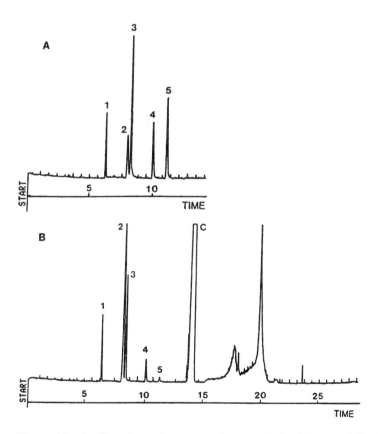

Figure 14 Capillary electropherograms of (A) standard solutions, and (B) the pharmaceutical sample. (1) Thiamine; (2) pyridoxamine; (3) riboflavin; (4) pyridoxine; (5) pyridoxal. C, ascorbic acid. (From Ref. 90.)

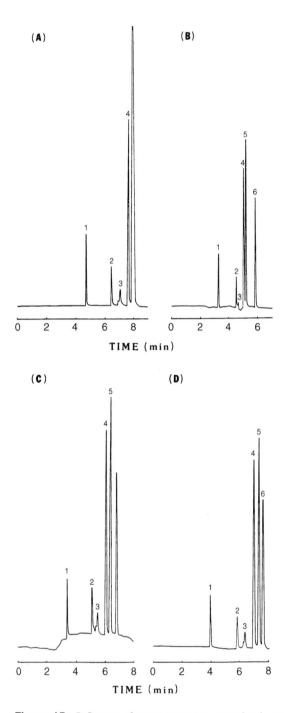

Figure 15 Influence of temperature programming in capillary zone electrophoresis on the separation of fluorescent dyes in 0.01 M Tris buffer (initial pH 8.3). (A) 20°C isothermal; (B) 70°C isothermal; (C) 70°C for 2 min, then step to 20°C; (D) temperature gradient from 53°C to 20°C. (1) Riboflavin; (2) coumarin; (3) impurities; (4) fluorescein-5-isothio-

concentrations of these vitamins are high. In the author's opinion, the quantification of lower amounts of vitamins, e.g., in foodstuffs, might be still a problem in CZE because of the low concentrations of the analytes in the detector cell.

A similar quantitative analysis of six water-soluble vitamins (B_1, B_2, B_6, C, nicotinamide, and pantothenic acid) in a pharmaceutical formulation using CZE in uncoated fused silica capillaries with UV detection was described by Fotsing et al. (91). For the B-group vitamins, a good compromise among resolution, analysis time, and analyte stability was obtained by use of 50 mM borax buffer pH 8.5. A capillary wash with sodium hydroxide was necessary between successive runs to minimize absorption of excipients from the pharmaceutical formulation to the capillary surface, otherwise giving rise to a progressive decrease of the electro-osmotic flow.

A new type of selectivity in CZE based on temperature programming was presented using a mixture of fluorescent organic compounds, including riboflavin (92). Using a buffer system with a large temperature coefficient (dpH/dT), e.g., Tris buffer, a pH step or gradient can be generated in situ simply by varying the temperature of the capillary as a function of time. The influence of temperature programming on the fluorescent dyes is shown in Figure 15. In comparison with the reported pH gradient techniques in CZE, for example, flow-generated pH gradients, problems associated with inhomogeneity in mixing do not exist. This method could be useful for the analysis of complex samples, where inadequate separation is obtained at a single buffer pH.

VII. FUTURE TRENDS

In recent years, modern chromatographic methods were introduced including capillary zone electrophoresis (CZE), capillary isotachophoresis (CITP), or supercritical fluid analysis (SCFA). As far as the analysis of flavin compounds is concerned, only a few studies are available at the moment using CZE. CZE may become an alternative or a supplement of HPLC, especially for the separation of the charged flavin coenzymes FMN and FAD.

For the moment, high-performance liquid chromatography is still the most versatile tool that can be used for the determination and the small-scale isolation of various flavin compounds. In the analytical field, HPLC has already completely superseded paper chromatography and more or less completely superseded thin-layer chromatography. A modern HPLC apparatus, including a fluorescence detector, leaves little to be desired in regard to sensitivity, reliability, and reproducibility. The on-line combination with mass spectrometry which needs expensive equipment and special expertise, is, however, a powerful technique for the separation and unequivocal identification of unknown flavin compounds from different sources. Various approved reversed-phase or ion pair HPLC systems are available that allow the separation and determination of riboflavin in different matrices. It can be expected that future studies will bring only minor technical modifications of the existing HPLC methods and will probably concentrate more on the careful and reliable extraction of various flavins from the respective sample.

The sensitive and reliable determination of flavocoenzymes or new flavin analogs or metabolites is still the basis of a full understanding of the role of flavins in health and disease. As the biological function of flavin compounds in prokaryotic and eukaryotic cells is still not known in detail, there is much room for future work in the chromatographic analysis of flavins.

REFERENCES

1. P Nielsen. Flavins. In: AP De Leenheer, WE Lambert, HJ Nelis, eds. Modern Chromatographic Analysis of the Vitamins. 2nd ed. New York: Marcel Dekker, 1992, pp 355–398.
2. P Hemmerich. The present status of flavin and flavocoenzyme chemistry. Fortschr Chem Org Naturst 33:451–527, 1971.
3. K Decker, R Brandsch. Determining covalent flavinylation. Methods Enzymol 280: 413–423, 1997.
4. F Müller, S Ghisla, A Bacher. Vitamin B_2 and natural flavins. In: O Isler, G Brubacher, S Ghisla, B Kräutler, eds. Vitamine II. Stuttgart: Thieme, 1988, pp 50–159.
5. WL Chairns, DE Metzler. Photochemical degradation of flavins VI. A new photoproduct and its use in studying the photolytic mechanism. J Am Chem Soc 93: 2772–2777, 1971.
6. AR Surrey. Alkaline hydrolysis of riboflavin. J Am Chem Soc 73:2336–2340, 1951.
7. P Nielsen, P Rauschenbach, A Bacher. Phosphates of riboflavin and riboflavin analogs: a reinvestigation by high-performance liquid chromatography. Anal Biochem 130:359–368, 1983.
8. P Nielsen, P Rauschenbach, A Bacher. Preparation, properties, and separation by high-performance liquid chromatography of riboflavin phosphates. Methods Enzymol 122:209–220, 1986.
9. S Ghisla, V Massey, JM Lhoste, SG Mayhew. Fluorescence and optical characteristics of reduced flavines and flavoproteins. Biochemistry 13:589–597, 1974.
10. C Lin, DE Robertson, M Ahmad, AA Raibekas, MS Jorns, PL Dutton, AR Cashmore. Association of flavin adenine dinucleotide with the Arabidopsis blue light receptor CRY1. Science 269:968–970, 1995.
11. A Bacher, W Eisenreich, K Kis, G Richter, J Scheuring, S Weinkauf. Recent advances in the biosynthesis of flavins and deazaflavins. Trends Org Chem 4:335–349, 1993.
12. B Lingens, TA Schild, B Vogler, P Renz. Biosynthesis of vitamin B_{12}. Transformation of riboflavin ^{2}H-labeled in the 1′R position or 1′S position into 5,6-dimethylbenzimidazole. Eur J Biochem 207:981–985, 1992.

13. RC Rose. Intestinal absorption of water-soluble vitamins. Proc Soc Exp Biol Med 212:191–198, 1996.

14. T Akiyama, J Selhub, IH Rosenberg. FMN phosphatase and FAD pyrophosphatase in rat intestinal brush borders: role in intestinal absorption of dietary riboflavin. J Nutr 112:263–268, 1982.

15. CP Chia, R Addison, DB McCormick. Absorption, metabolism, and excretion of 8α(amino acid)riboflavins in the rat. J Nutr 108:373–381, 1978.

16. H Ushijima, H Okamura, Y Nishina, K Shiga. Effects of pH and ionic strength on the binding of egg white riboflavin binding protein with flavins. J Biochem 105: 467–472, 1989.

17. CVR Murthy, PR Adiga. Isolation and characterisation of a riboflavin-carrier protein from human pregnancy serum. Biochem Int 5:289–296, 1982.

18. JL Chastain, DB McCormick. Flavin catabolites: identification and quantitation in human urine. Am J Clin Nutr 46:830–834, 1987.

19. HJ Amin, AK Shukla, F Snyder, E Fung, NM Anderson, HG Parsons. Significance of phototherapy-induced riboflavin deficiency in the full-term neonate. Biol Neonate 61:76–81, 1992.

20. National Research Council. Recommendation Dietary Allowances. 10th ed. Washington: National Academic Press, 1989, pp 132–137.

21. J Zempleni, JR Galloway, DB McCormick. Pharmacokinetics of orally and intravenously administered riboflavin in healthy humans. Am J Clin Nutr 63:54–66, 1996.

22. UA Walker, E Byrne. The therapy of respiratory chain encephalomyopathy: a critical review of the past and current perspective. Acta Neurol Scand 92:273–280, 1995.

23. HN Christensen. Riboflavin can protect tissues from oxidative injury. Nutr Rev 51: 149–150, 1995.

24. CP Mack, DE Hultquist, M Shlafer. Myocardial flavin reductase and riboflavin: a potential role in decreasing reoxygenation injury. Biochem Biophys Res Commun 212:35–40, 1995.

25. JM Perel. Compliance during tricyclic antidepressant therapy: pharmacokinetic and analytical issues. Clin Chem 34:881–887, 1988.

26. ZK Roughead, DB McCormick. Qualitative and quantitative assessment of flavins in cow's milk. J Nutr 120:382–388, 1990.

27. HA Hartman, DE Edmondson, DB McCormick. Riboflavin 5′-pyrophosphate: a contaminant of commercial FAD, a coenzyme for FAD-dependent oxidases, and an inhibitor of FAD synthethase. Anal Biochem 202:348–355, 1992.

28. K Matsui, S Kasai. Identification of nekoflavin as 7α-hydroxyflavin. J Biochem 119: 441–447, 1996.

29. NU Perisic-Janjic, MR Popovic, TL Djakovic. Quantitative determination of vitamin B complex constituents by fluorescence quenching after TLC separation. Acta Chromatogr 5:144–150, 1995.

30. AN Diaz, AG Paniagua, FG Sanchez. Thin-layer chromatography and fibre-optic fluorimetric quantitation of thiamine, riboflavin and niacin. J Chromatogr 655:39–43, 1993.

31. E Postaire, M Cisse, MD Le Hoang, D Pradeau. Simultaneous determination of water-soluble vitamins by over-pressure layer chromatography and photodensitometric detection. J Pharm Sci 80:368–370, 1991.

32. I Chibata, T Tosa, Y Matuo. Purification of flavin-adenine dinucleotide and coenzyme A on p-acetomercurianiline-agarose. Methods Enzymol 66:221–226, 1980.

33. SI Huang, MJ Caldwell, KL Simpson. Reverse phase open column chromatography determination of urinary riboflavin. Int J Vitam Nutr Res 63:217–222, 1993.

34. SI Huang, MJ Caldwell, KL Simpson. Urinary riboflavin determination by C18 reversed-phase open-column chromatography. Methods Enzymol 280:343–351, 1997.

35. A Kozik. Microtitre-plate enzyme-linked ligand-sorbent assay of riboflavin (vitamin B_2) in human plasma and urine. Analyst 121:333–337, 1996.

36. VM Kodentsova, OA Vrzhesinskaya, VB Spirichev. Fluorometric riboflavin titration in plasma by riboflavin-binding apoprotein as a method for vitamin B_2 status assessment. Ann Nutr Metab 39:355–360, 1995.

37. G Massolini, ED Lorenzi, MC Ponci, C Gandini, G Caccialanza, HL Monaco. Egg yolk riboflavin binding protein as a new chiral stationary phase in high-performance liquid chromatography. J Chromatogr A 704:55–65, 1995.

38. D Blanco, LA Sanchez, MD Gutierrez. Determination of water soluble vitamins by liquid chromatography with ordinary and narrow-bore columns. J Liq Chromatogr 17:1525–1539, 1994.

39. WG Chase, AM Soliman. Analysis of thiamin, riboflavin, pyridoxine and niacin in multivitamin premixes and supplements by high performance liquid chromatography. J Micronutr Anal 7:15–25, 1990.

40. Y Maeda, M Yamamoto, K Owada, S Sato, T Masui, H Nakazawa. Simultaneous liquid chromatographic determination of water-soluble vitamins, caffeine, and preservative in oral liquid tonics. J AOAC Int 72:244–247, 1989.

41. WE Lambert, PM Cammaert, AP DeLeenheer. Liquid-chromatographic measurement of riboflavin in serum and urine with isoriboflavin as internal standard. Clin Chem 31:1371–1373, 1985.

42. N Bilic, R Sieber. Determination of flavins in dairy products by high-performance liquid chromatography using sorboflavin as internal standard. J Chromatogr 511:359–366, 1990.

43. MW Dong, J Lepore, T Tarumoto. Factors affecting the ion-pair chromatography of water-soluble vitamins. J Chromatogr 442:81–95, 1988.

44. NR Raju, P Srilakshimi, GA Vinolia, GJ Bhounsule, VD Gaitonde. Simultaneous determination of water soluble vitamins in pharmaceutical oral liquids by gradient high performance liquid chromatography-sample preparation with reversed-phase Sep Pak cartridge. Indian Drugs 26:697–701, 1989.

45. S Akiyama, K Nakashima, K Yamada, N Shirakawa. High-performance liquid chromatographic determination of components in eye lotion using methoxy(3-morpholinopropyl)silanediyl modified silica gel column. Bull Chem Soc Jpn 64:3171–3172, 1991.

46. A Rizzolo, S Polesello. Chromatographic determination of vitamins in foods. J Chromatogr 624:103–152, 1992.

47. E Barna, E Dworschak. Determination of thiamine (vitamin B_1) and riboflavin (vitamin B_2) in meat and liver by high-performance liquid chromatography. J Chromatogr A 668:359–363, 1994.

48. C Kanno, K Shirahuji, T Hoshi. Simple method for separate determination of three

flavins in bovine milk by high performance liquid chromatography. J Food Sci 56: 678–681, 1991.

49. M Munari, M Miurin, G Goi. Didactic application to riboflavin HPLC analysis. A laboratory experiment. J Chem Educ 68:78–79, 1991.

50. R Gauch, U Leuenberger, U Müller. Determination of the water soluble vitamins B_1, B_2, B_6, and B_{12} in milk by HPLC. Z Lebensm Unters Forsch 195:312–315, 1992.

51. GM Greenway, N Kometa. On-line sample preparation for the determination of riboflavin and flavin mononucleotides in foodstuffs. Analyst 119:929–935, 1994.

52. S Albala-Hurtado, MT Veciana-Nogues, M Izquierdo-Pulido, A Marine-Font. Determination of water-soluble vitamins in infant milk by high-performance liquid chromatography. J Chromatogr 778:247–253, 1997.

53. GW Chase, WO Landen, RR Eitenmiller, AM Soliman. Liquid chromatographic determination of thiamine, riboflavin, and pyridoxine in infant formula. J AOAC Int 75:561–565, 1992.

54. L Li, W Chen, J Cui. Determination of vitamin B_1 and B_2 in fortified milk powder by high performance liquid chromatography. Chem Abstr CA118-146384. Sepu 11: 49–50, 1993.

55. AK Hewavitharana. Method for the extraction of riboflavin for high-performance liquid chromatography and application to casein. Analyst 121:1671–1676, 1996.

56. S Otles, Y Hisil. High pressure liquid chromatographic analysis of water-soluble vitamins in eggs. Ital J Food Sci 5:69–75, 1993.

57. M Hägg. Effect of various commercially available enzymes in the liquid chromatographic determination with external standardization of thiamine and riboflavin in foods. J AOAC Int 77:681–686, 1994.

58. A Sims, D Shoemaker. Simultaneous liquid chromatographic determination of thiamine and riboflavin in selected foods. J AOAC Int 76:1156–1160, 1993.

59. V Ollilainen, P Mattila, P Varo, P Koivistoinen, J Huttunen. The HPLC determination of total riboflavin in foods. J Micronutr Anal 8:199–207, 1991.

60. Y Han, Y Dai, X Song, J Zhao. Simultaneous determination of thiamin, riboflavin, niacin and pyridoxine in foods by HPLC. Chem Abstr CA121-007565. Yingyang Xuebao 15:448–453, 1993.

61. C Hasselmann, D Franck, P Grimm, PA Diop, C Soules. High-performance liquid chromatography analysis of thiamin and riboflavin in dietetic foods. J Micronutr Anal 5:269–279, 1989.

62. SM Fernando, PA Murphy. HPLC determination of thiamine and riboflavin in soybeans and tofu. J Agric Food Chem 38:163–167, 1990.

63. TS Agostini, HT Godoy. Simultaneous determination of nicotinamide, nicotinic acid, riboflavin, thiamin, and pyridoxine in enriched Brazilian foods by HPLC. J High Resol Chromatogr 20:245–248, 1997.

64. E Barna. Comparison of data obtained by HPLC and microbiological determination of riboflavin in ready-to-eat foods. Acta Alim 21:3–9, 1992.

65. G Ayranci, S Kya. Kinetic analysis of the loss of some B-vitamins during the cooking of macaroni. Nahrung 37:153–155, 1993.

66. PCH Hollman, JH Slangen, PJ Wagstaffe, U Faure, DAT Southgate, PM Finglas. Intercomparison of methods for the determination of vitamins in foods. Part 2. Water-soluble vitamins. Analyst 118:481–488, 1993.

67. JB Fox, DW Thayer, RK Jenkins, JG Phillips, SA Ackermann, GR Beecher, JM Holden, FD Morrow, DM Quirbach. Effect of gamma irradiation on the B vitamins of pork chops and chicken breasts. Int J Radiat Biol 55:689–703, 1989.

68. R Marsh, P Kajda, J Ryley. The effect of light on the vitamin B_2 and the vitamin A content of cheese. Die Nahrung 38:527–532, 1994.

69. CJ Bates. Human riboflavin requirements, and metabolic consequences of deficiency in man and animals. World Rev Nutr Diet 50:215–265, 1987.

70. T Seki, K Noguchi, Y Yanagihara. Determination of riboflavin in human urine by the use of a hydrophilic gel column. J Chromatogr 385:283–285, 1987.

71. Lopez-Ananya A, Mayersohn M. Quantification of riboflavin, riboflavin 5'-phosphate and flavin adenine dinucleotide in plasma and urine by high-performance liquid chromatography. J Chromatogr 423:105–113, 1987.

72. L Rokitzki, A Berg, J Keul. Serum-und Vollblutkonzentrationen von fett- und wasserlöslichen Vitaminen bei Normalpersonen und Sportlern. Z Gesamte Hyg 35:16–21, 1989.

73. J Zempleni. Determination of riboflavin and flavocoenzymes in human blood plasma by high-performance liquid chromatography. Ann Nutr Metab 39:224–226, 1995.

74. J Zempleni, G Link, W Kübler. The transport of thiamine, riboflavin and pyridoxal 5'-phosphate by human placenta. Int J Vitam Nutr Res 62:165–172, 1992.

75. J Zempleni, G Link, W Kübler. Intrauterine vitamin B_2 uptake of preterm and full-term infants. Pediatr Res 38:585–591, 1995.

76. DW Batey, CD Eckhert. Analysis of flavins in the ocular tissues of the rabbit. Invest Ophthalmol Vis Sci 32:1981–1985, 1991.

77. DW Batey, CD Eckhert. Identification of FAD, FMN, and riboflavin in the retina by microextraction and high-performance liquid chromatography. Anal Biochem 188:164–167, 1990.

78. DW Batey, KK Daneshgar, CD Eckhert. Flavin levels in the rat retina. Exp Eye Res 54:605–609, 1992.

79. T Ryll, R Wagner. Improved ion-pair high-performance liquid chromatographic method for the quantification of a wide variety of nucleotides and sugar-nucleotides in animal cells. J Chromatogr 570:77–88, 1991.

80. N Hohl, P Galland, H Senger. Altered flavin patterns in photobehavioral mutants of *Phycomyces blakesleeanus*. Photochem Photobiol 55:247–255, 1992.

81. S Susin, J Abian, F Sanchez-Baeza, ML Peleato, A Abadia, E Gelpi, J Abadia. Riboflavin 3'- and 5'-sulfate, two novel flavins accumulating in the roots of iron-deficient sugar beet (*Beta vulgaris*). J Biol Chem 268:20958–20965, 1993.

82. S Harbron, HJ Eggelte, BR Rabin. Amplified colorimetric assay of alkaline phosphatase using riboflavin 4'-phosphate: a simple method for measuring riboflavin and riboflavin 5'-phosphate. Anal Biochem 198:47–51, 1991.

83. P Nielsen, J Harksen, A Bacher. Hydrolysis and rearrangement reactions of riboflavin phosphates. An explicit kinetic study. Eur J Biochem 152:465–473, 1985.

84. P Pietta, A Calatroni, A Rava. Hydrolysis of riboflavin nucleotides monitored by high-performance liquid chromatography. J Chromatogr 229:445–449, 1982.

85. DR Light, C Walsh, MA Marletta. Analytical and preparative high-performance liquid chromatography separation of flavin and flavin analog coenzymes. Anal Biochem 109:87–93, 1980.

86. N Asakawa, H Ohe, M Tsuno, Y Nezu, Y Yoshida, T Sato. Liquid chromatography–mass spectrometry system using column-switching techniques. J Chromatogr 541: 231–241, 1991.

87. Z Zhao, JH Wahl, HR Udseth, SA Hofstadler, AF Fuciarelli, RD Smith. On-line capillary electrophoresis-electrospray ionisation mass spectrometry of nucleotides. Electrophoresis 16:389–395, 1995.

88. Packman LC, Mewies M, Scrutton NS. The flavinylation reaction of trimethylamine dehydrogenase. Analysis by directed mutagenesis and electrospray mass spectrometry. J Biol Chem 270:13186–13191, 1995.

89. E Kenndler, C Schwer, D Kaniansky. Purity control of riboflavin-5'phosphate (vitamin B_2 phosphate) by capillary zone electrophoresis. J Chromatogr 508:203–207, 1990.

90. R Huopalahti, J Sunell. Use of capillary zone electrophoresis in the determination of B vitamins in pharmaceutical products. J Chromatogr 636:133–135, 1993.

91. L Fotsing, M Fillet, I Bechet, P Hubert, J Crommen. Determination of six water-soluble vitamins in a pharmaceutical formulation by capillary electrophoresis. J Pharm Biomed Anal 15:1113–1123, 1997.

92. CW Wang, ES Yeung. Temperature programming in capillary zone electrophoresis. Anal Chem 64:502–506, 1992.

10
Vitamin B_6

Johan B. Ubbink
University of Pretoria, Pretoria, South Africa

I. INTRODUCTION

A. History

Vitamin B_6 was first identified as an essential nutritional component to prevent a florid dermatitis called acrodynia in rats (1). In 1938, a crystalline substance having vitamin B_6 activity was isolated and identified (2,3). This component, named pyridoxine, was required for growth of several lactic acid bacteria (4). However, Snell and coworkers found that extraordinarily large amounts of pyridoxine were required to support growth of *Lactobacillus casei* or *Streptococcus faecalis* (5). Heat sterilization of the pyridoxine-containing growth medium reduced the amount of vitamin required for growth of above-mentioned bacteria (5). These results indicated that heat treatment altered pyridoxine to form growth factors more potent than pyridoxine; subsequently, pyridoxal and pyridoxamine were discovered in 1944 (6).

B. Chemical Properties

The generic term vitamin B_6 refers to all 3-hydroxy-2-methylpyridine derivatives that exhibit the biological activity of pyridoxine in rats (7). Pyridoxine (3-hydroxy-4,5-bis-(hydroxymethyl)-2-methyl pyridine) should not be used as generic term synonymous with vitamin B_6. Two other forms of vitamin B_6, pyridoxal (PL) and pyridoxamine (PM), differ from pyridoxine (PN) in the respective location of an aldehyde and amine group at the 4 position of the pyridine ring structure (Fig. 1). The 5'-phosphoric ester of pyridoxal, pyridoxal-5'-phosphate (PLP), is the metabolically active form of vitamin B_6. Pyridoxamine-5'-phosphate

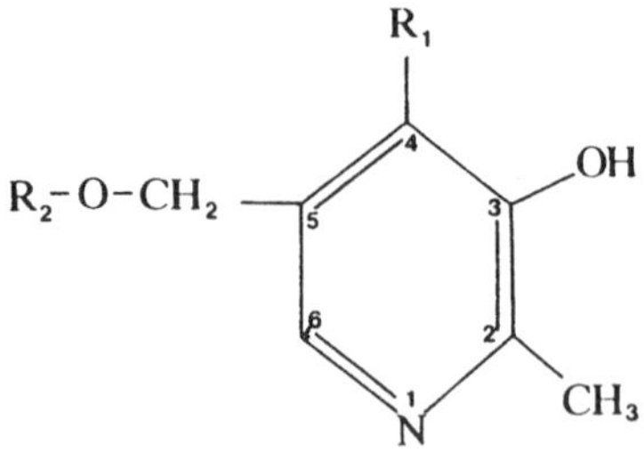

NAME	R_1	R_2
PYRIDOXINE	CH_2OH	H
PYRIDOXINE-5'-PHOSPHATE	CH_2OH	HO–P(=O)–OH
PYRIDOXAL	CHO	H
PYRIDOXAL-5'-PHOSPHATE	CHO	HO–P(=O)–OH
PYRIDOXAMINE	CH_2NH_2	H
PYRIDOXAMINE-5'-PHOSPHATE	CH_2NH_2	HO–P(=O)–OH
4-PYRIDOXIC ACID	COOH	H

Figure 1 Structural formulae of pyridoxine derivatives.

(PMP) and pyridoxine-5′-phosphate (PNP) are also widely distributed in animal and plant tissues. Glycosylated forms of pyridoxine are commonly found in plants (8). The major glycosylated form of vitamin B_6 in plants is 5′-O-β-D glucopyranosyl pyridoxine (PN-glucoside) (8). The oxidized metabolites of pyridoxal, 3-hydroxy-5-hydroxymethyl-2-methylpyridine-4-carboxylic acid, and the corresponding lactone are designated 4-pyridoxic acid (4-PA) and 4-pyridoxolactone (4-PA lactone), respectively.

C. Biochemistry

The different forms of vitamin B_6, as found in animal tissues, are interconvertible. Intracellular PLP is derived by phosphorylation of PL (9); the reaction requires

ATP (10,11) and is catalyzed by pyridoxal kinase (EC 2.7.1.35). Some tissues (liver, brain, erythrocytes) may utilize PN or PM as precursors for PLP; these vitamers are first phosphorylated by pyridoxal kinase and then oxidized (12,13) to PLP by PM(PN)-5'-phosphate oxidase (EC 1.4.3.5). Synthesis of PLP is balanced by dephosphorylation (14) to PL in a reaction catalyzed by PLP phosphatase. PL may then be oxidized to 4-PA, the end product of vitamin B$_6$ metabolism.

Vitamin B$_6$ is absorbed from the jejunum as PL, PM and PN; phosphorylated B$_6$ vitamers are first hydrolyzed before absorption can occur (15,16). The liver plays a major role in the metabolism of vitamin B$_6$ (17,18). Liver uptake of PN, PM, and PL occurs by diffusion followed by 5' phosphorylation catalyzed by PL kinase, which results in effective metabolic trapping of the B$_6$ vitamers (19). In humans, the liver releases PLP, PL and 4-PA into the circulation (17,18). Circulating PL presumably serves as transport form of vitamin B$_6$, which crosses into extrahepatic tissues to be phosphorylated to the active coenzyme PLP (20,21). Circulating PLP is predominantly albumin bound (22–24) and does not cross cellular membranes (20,22). It has been suggested that binding of PLP to serum albumin is important in regulating the availability of PLP to tissues as well as the catabolism of PLP (24). It should be recognized that PLP binding to proteins may cause an obstacle in the assessment of vitamin B$_6$ status. If the release of PLP from proteins during the analysis procedure is incomplete, it may result in falsely low plasma PLP concentrations (25).

D. Physiological Function

Intracellular PLP, which is derived from extracellular PL, functions as a very versatile coenzyme. Various reactions in amino acid metabolism, including transaminations, α decarboxylations, α,β eliminations, β,γ eliminations, aldolizations, and racemizations, are PLP dependent.

Decarboxylases function in the synthesis of major neurotransmitters (26); adequate PLP supply in brain tissue is therefore essential for normal brain function. Aminotransferases have a key function in both amino acid biosynthesis and catabolism. PLP has also a structural function (27) in glycogen phosphorylase (EC 2.4.1.1), and a role for PLP in lipid metabolism has been proposed (28,29).

E. Pharmacological Function

Genetic conditions of vitamin B$_6$ dependency which respond to high dose PN supplementation include certain idiopathic sideroblastic anaemias (30) and disturbed amino acid metabolism, e.g., cystathionine β-synthase deficiency (31).

Pyridoxine supplementation has been reported to benefit sufferers from premenstrual tension (32), nausea of pregnancy (33), carpal tunnel syndrome (34), and gestational diabetes (35). A previous report that PN supplementation is beneficial in the treatment of asthma (36) could not be substantiated in a placebo-controlled trial (37).

The role of vitamin B_6 supplementation in the treatment of hyperhomocyst(e)inemia, a recognized risk factor for premature cardiovascular disease (38–40), has been widely investigated (41–44). Supplemental PN attenuates the post-methionine load plasma homocyst(e)ine concentrations (45,46) and is therefore perceived to reduce cardiovascular disease risk.

F. Different Approaches to Vitamin B_6 Analysis

The measurement of vitamin B_6 in biological material is complicated by (1) the natural occurrence of vitamin B_6 in six different forms (Fig. 1); (2) the relatively low levels of vitamin B_6 in most biological samples; (3) the water solubility of vitamin B_6, which excludes the use of organic solvent extraction procedures for purification and enrichment prior to analysis; (4) the photosensitivity of vitamin B_6, which requires the availability of dark room facilities; and (5) the protein binding of PLP.

Various methods other than chromatography have been designed to satisfy different requirements in vitamin B_6 analysis, but none of these is suitable for the simultaneous analysis of all six natural forms of vitamin B_6. Enzymatic methods are only suitable for PLP quantification (47–49). Microbiological assay of different forms of vitamin B_6 is cumbersome, indirect and very time consuming. These methods employ different micro-organisms sensitive to the different forms of vitamin B_6 (50), but growth promoting and/or growth inhibitory substances in blood and plasma may render microbiological assays in these samples invalid (51,52).

Direct fluorimetric quantification of PLP and PL has been described (53–55), but these methods are either unreliable due to interfering fluorescent compounds, or tedious due to the prior requirement for sample cleanup procedures. Isopotential synchronous fluorescence spectrometry has recently been applied to human plasma to determine plasma PL concentrations with minimum background interference and without any prior separation procedures (56).

Immunoassays for the different B_6 vitamers have been reported (57,58). The need to develop different antibodies against the various B_6 vitamers and the possibility of cross reactivity are serious practical limitations that prohibit the use of immunoassay in vitamin B_6 analysis.

Compared to the methods mentioned above, chromatographic methods are superior since separation and quantification of the different B_6 vitamers can be achieved simultaneously. Especially HPLC, and to a lesser extent GLC, have been exploited to develop various methods suitable for different applications in vitamin B_6 analysis.

II. THIN-LAYER CHROMATOGRAPHY

Due to the advent of HPLC, thin-layer chromatography is today seldomly used as a quantitative technique in vitamin B$_6$ analysis. However, it is sometimes used as qualitative method to study vitamin B$_6$ metabolism. A recent example is provided by the study of Coburn and Mahuren, who investigated vitamin B$_6$ metabolism in cats (59). They found that although about 70% of the ingested [^{14}C]PN dose appeared in the urine, only 2% to 3% of the excreted dose was 4-pyridoxic acid. Cation exchange liquid chromatography revealed that two unknown radioactive components were excreted in cat urine. Using thin-layer chromatography, pyridoxine-3-sulfate and pyridoxal-3-sulfate were identified as the major urinary metabolites of ingested pyridoxine; this result was confirmed by infrared spectroscopy of the vitamin B$_6$ metabolites isolated from cat urine.

III. HIGH-PERFORMANCE LIQUID CHROMATOGRAPHY

A. Introduction

Before the introduction of HPLC, several different liquid column chromatographic procedures to separate mixtures of B$_6$ vitamers from each other were developed. Resolution of mixtures of pure vitamins was found to be quite simple. However, application to the analysis of food and biological material was difficult, mainly due to the low concentrations of the vitamers encountered and unsatisfactory means of detection. Quantification often involved laborious microbiological determination for the different B$_6$ vitamers in column eluate fractions (60).

Several chromatographic procedures published before 1980 relied on UV detection for B$_6$ vitamer quantification; these methods lacked sensitivity for B$_6$ analysis in food and biological fluids (61). Yet these methods form the foundation of modern vitamin B$_6$ analysis by HPLC. Today, a wide variety of HPLC methods, optimized for different applications, are available.

1. For nutrition survey studies, sensitive methods for determination of plasma/whole blood PLP have been developed. Many of these methods include PL, and sometimes even 4-PA.
2. Specific methods have been described to determine urinary 4-PA excretion, which is frequently used as index of vitamin B$_6$ nutritional status.
3. HPLC methods designed to separate and quantify the six major forms of vitamin B$_6$ and 4-PA in biological material, are termed comprehensive methods in this review. These methods were usually applied in studies of human vitamin B$_6$ metabolism or pharmacokinetics of pyridoxine supplementation.

4. Food analysis also requires comprehensive vitamin B_6 analysis, al-though differentiation between unphosphorylated vitamers and the 5' phosphoric acid esters is usually not required. Recent research has shown that PN-glucoside has a much lower bioavailability than pyri-doxine (8,62,63), and it is therefore sensible to assay PN-glucoside separately.

5. Analysis of pharmaceutical preparations is relatively simple when compared to analysis of biological material, because only PN is used as supplementary B_6 vitamer. Methods for PN determination are usually optimized to include determination of other vitamins of the B group.

It has already been pointed out that earlier attempts to use HPLC for B_6 vitamer quantification were hampered by either laborious (microbiological) or insensitive (UV) detection procedures. Modern HPLC analysis of the different B_6 vitamers now exploits the fluorescence characteristics of these vitamers, which allow quantification in the nanomolar range. Modern fluorescence detectors for HPLC are easy to use and are unrivaled as far as both sensitivity and selectivity for vitamin B_6 analysis are concerned. A thorough understanding of the fluores-cence characteristics of the different B_6 vitamers is therefore essential before different published HPLC methods can be evaluated.

B. Spectral Characteristics of B_6 Vitamers

1. Ultraviolet Absorption

The UV absorption spectra of PL, PN, PM, and their phosporylated derivatives dissolved in 0.1 M HCl are similar and the absorption maximum for each B_6 vitamer is at about 290 nm (67). At pH 7.0, PN, PM, PNP, and PMP have absorp-tion maxima at 253 and 325 nm. The spectra for PL and PLP at pH 7.0 differ from the other B_6 vitamers in displaying an absorption maximum at about 390 nm, which relates to the aldehyde group on the 4 position of the pyridine ring structure (67). Table 1 compares the molar extinction coefficients of the different B_6 vitamers at their respective absorption maxima as determined at pH 7.0.

2. Fluorescence

The fluorescence characteristics of hydroxypyridines, including the different B_6 vitamers, were thoroughly studied by Bridges and coworkers (68). Table 2 sum-marizes information on vitamin B_6 fluorescence which may be useful in the devel-opment or application of HPLC methods for vitamin B_6 analyses. PL, PN, and PM show strong fluorescence in both slightly acidic to neutral and strongly alkaline conditions. PNP and PMP show fluorescence characteristics similar to the respec-tive dephosphorylated vitamers.

Table 1 Molar Extinction Coefficients of B$_6$ Vitamers[a]

Vitamer	Absorbance maximum (nm)	Molar extinction coefficient
PN	253	3700
	325	7100
PNP	253	3700
	325	7400
PL	318	8200
	390	200
PLP	330	2500
	388	4900
PM	253	4600
	325	7700
PMP	253	4700
	325	8300

[a] Absorption spectra were obtained at pH 7.0 with 0.1 M sodium phosphate buffer as solvent.
Source: Ref. 67.

Table 2 Fluorescence Characteristics of B$_6$ Vitamers[a]

Vitamer	Excitation/emission wavelengths (nm)	Fluorescence intensity	pH range of maximum fluorescence
PN	332/400	238	6.5–7.5
	320/380	168	12.0–14.0
PL (hemiacetal)	330/382	207	6.0
	310/365	283	12.0
PLP	330/410	11	6.0
	315/370	17.5	12.0–14.0
PM	337/400	370	4.0–5.5
	320/370	410	14.0
4-PA	320/420	770	1.5–4.0
	315/425	650	6.1–9.4
4-Pyridoxolactone	365/423	32	2.5–4.8
	360/430	2150	8.7–13.0

[a] Fluorescence intensity is expressed relative to the fluorescence intensity of the 3-hydroxypyridine anion at pH = 11.0, which was arbitrarily chosen as 100.
Source: Ref. 68.

It should be noted that PL can exist in solution as a cyclic hemiacetal, aldehyde hydrate, or free aldehyde. The cyclic hemiacetal form is predominantly found in solution (69); the fluorescence characteristics of PL listed in Table 2 therefore refer to the cyclic hemiacetal form of the vitamer. If pyridoxal occurred in the aldehyde form, the fluorescence observed should be considerably less, since the aldehyde group tends to diminish the fluorescence of aromatic systems due to its tendency to withdraw electrons from the ring structure (68,70).

Unfortunately, PLP is only weakly fluorescent when compared to PL, PN, or PM, presumably due to a withdrawal of electrons from the aromatic ring structure as discussed above. For PLP the phosphoric acid ester on the 5 position of the pyridine ring structure prohibits hemiacetal formation. The natural fluorescence shown by PLP is therefore much lower and usually inadequate for accurate quantification of this vitamer in biological samples. Several methods have been described to enhance PLP fluorescence. Bonavita and Scardi reported in 1959 that PLP reacts with potassium cyanide in the presence of a phosphate buffer, resulting in characteristic changes in the absorption and fluorescence spectra of the coenzyme (71). Except for pyridoxal, none of the other B_6 vitamers was able to react with potassium cyanide. The cyanide reaction greatly enhanced the fluorescence of both PL and PLP. The spectral changes of PL and PLP treated with potassium cyanide were initially thought to be the result of cyanohydrin derivative formation of these vitamers (72). However, it was later discovered that cyanide acts as catalyst in the oxidation (by air) of PL and PLP to 4-pyridoxolactone and 4-pyridoxic acid 5'-phosphate respectively (73). The application of this method to biological samples usually involves extraction of the sample with an aliquot of 1 M $HClO_4$, adjustment of the supernatant to pH 3.5 with a concentrated KOH solution, addition of 0.1 M KCN and incubation at 50°C for 30 min. The pH of the solution is then again adjusted to 3.5 with 0.1 M HCl and the sample is left in the dark at room temperature until HPLC analysis the following day (74). Figure 2 shows the excitation and emission spectra of 4-pyridoxic acid 5'-phosphate.

Reagents that attack the aldehyde group usually enhance PLP fluorescence, presumably by elimination of the electron withdrawing effect of the aldehyde group. Aldehydes are known to react on sodium bisulfite to form hydroxysulphonic acid salts (75). The hydroxysulfonate derivative of PLP is much more fluorescent than native PLP; fluorescence shown by the hydroxysulfonate derivative at pH 7.5 is strong enough to allow PLP quantification in the nanomolar range (76). The reaction between PLP and bisulfite is rapid and requires no heating or incubation times as described for KCN catalyzed oxidation of PLP. Figure 3 shows the relative excitation and emission spectra of PLP and its bisulfite adduct as recently reported by Kimura et al. (77).

Ammonia derivatives may react with carbonyl groups resulting in products containing a carbon-nitrogen double bond with elimination of a molecule of wa-

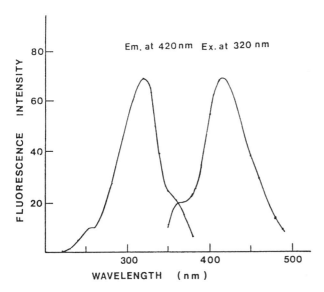

Figure 2 Excitation and emission spectra of 4-pyridoxic acid 5'-phosphate at pH 5.5. (From Ref. 74.)

ter. PLP has thus been shown to react with semicarbazide to form PLP-semicarbazone, which is not very fluorescent in slightly acid to neutral solutions (78) but shows strong fluorescence at pH 12 (79). Pyridoxal semicarbazone is also very fluorescent at alkaline pH (Fig. 4). The semicarbazone reaction has been shown by Conant et al. (80) to be a general acid-catalyzed reaction with a slightly acidic pH optimum. The correct acidity of the reaction medium is important, because addition involves nucleophilic attack by the basic semicarbazide on the carbonyl carbon (Fig. 5). When the carbonyl oxygen is protonated, the carbonyl carbon becomes more susceptible to nucleophilic attack, indicating that addition will be favored by high acidity. However, semicarbazide (: $NH_2NHCONH_2$) might also undergo protonation to form an ion ($^+NH_3NHCONH_2$) which lacks unshared electrons and is no longer nucleophilic. As far as semicarbazide is concerned, addition will be favored by lower acidity.

We studied PLP-semicarbazone formation under various pH conditions and found that semicarbazone formation was quantitative even in presence of a 3.3% solution of trichloroacetic acid (pH < 2.0) if semicarbazide was present in concentrations 4000-fold more than the PLP concentration (81). This implied that trichloroacetic acid protein precipitation and the semicarbazide reaction could be performed in a single step, a possibility that has been exploited in the development of a HPLC method for PLP and PL quantification in biological material

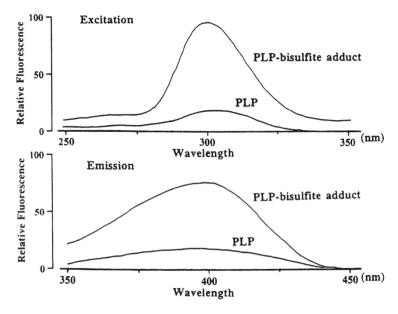

Figure 3 Excitation (λ_{max} 300 nm) and emission (λ_{max} 400 nm) spectra of pyridoxal 5′-phosphate (PLP) and the PLP-bisulfite adduct. (From Ref. 77.)

(81). An attractive, additional feature of the semicarbazide reaction is that semicarbazone derivatives of PLP and PL are stable under ordinary laboratory light conditions, thus eliminating the need for light-protective precautions (82).

Attaching of fluorescent probes to PLP has also been used to determine PLP. Chauhan and Dakshinamurti employed reductive amination of PLP with methyl anthranilate and sodium cyanoborohydride to form a highly fluorescent amine (83). Durko et al. (54) prepared diphenyl-isobenzofurane derivatives of B_6 vitamers to enable fluorometric measurement of vitamin B_6. However, the use of fluorescent probes are generally not suitable for routine analytical work, mainly because the unreacted fluorescent probe may interfere in B_6 vitamer quantification (83).

C. Pharmaceutical Preparations

Analysis of vitamins in multivitamin preparations is essential to ensure quality control and to establish the shelf life of the multivitamin product. The form of vitamin B_6 usually present in pharmaceutical preparations is PN; this vitamer is notably more stable than the aldehyde derivative, PL (84,85). Compared to

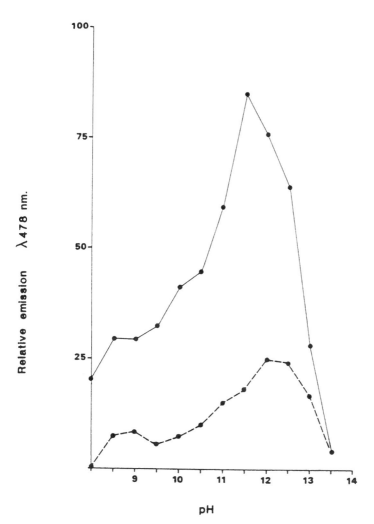

Figure 4 pH-dependent fluorescence of the semicarbazone derivatives of PLP (40 nmol/ L; dashed line) and PL (60 nmol/L; solid line). A stock solution of PLP (40 μmol/L) and of PL μmol/L) were separately made up in 0.025 mol/L semicarbazide. A blank solution containing only semicarbazide was also prepared. All three solutions were heated to 37°C in a water bath for 30 min to effect semicarbazone formation. Each solution was diluted 1000-fold in the following buffers: 0.1 M Na$_2$HPO$_4$/NaH$_2$PO$_4$ at pH 8.0, 8.5, and 9.0; 0.1 M NaHCO$_3$/Na$_2$CO$_3$ at pH 9.5, 10.0, and 10.5; 0.1 M Na$_2$HPO$_4$/NaOH at pH 11.0, 11.5, and 12.0; 0.1 M KCl/NaOH at pH 12.5, 13.0, and 13.5. Fluorescence emission intensity at 478 nm after excitation at 367 nm, was determined for PLP semicarbazone (PLPSC, dashed line) and PL semicarbazone (PLSC, solid line) at the above-mentioned pH values, using the different solutions as indicated.

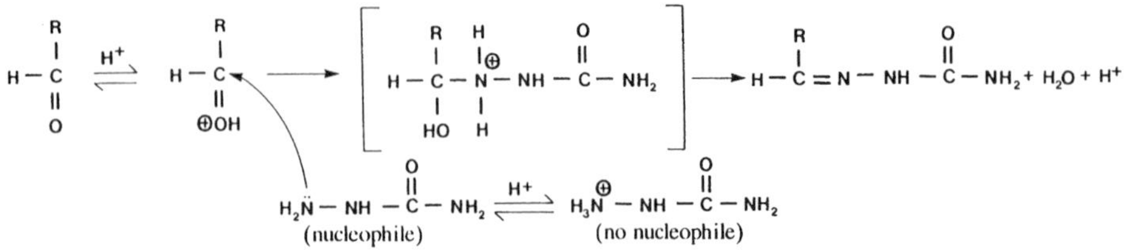

$$H-\underset{\underset{O}{\overset{R}{|}}}{C} \quad \overset{H^+}{\rightleftharpoons} \quad H-\underset{\underset{\oplus OH}{\overset{R}{|}}}{C} \quad \longrightarrow \quad \left[H-\underset{\underset{HO}{\overset{R}{|}}}{\overset{H}{\underset{|}{C}}} - \underset{\overset{|\oplus}{\underset{H}{N}}}{N} - NH - \overset{O}{\overset{||}{C}} - NH_2 \right] \quad \longrightarrow \quad H-\overset{R}{\overset{|}{C}} = N - NH - \overset{O}{\overset{||}{C}} - NH_2 + H_2O + H^+$$

$$H_2\overset{.}{N} - NH - \overset{O}{\overset{||}{C}} - NH_2 \quad \overset{H^+}{\rightleftharpoons} \quad H_3\overset{\oplus}{N} - NH - \overset{O}{\overset{||}{C}} - NH_2$$

(nucleophile)   (no nucleophile)

Figure 5 Proposed mechanism and pH dependency of semicarbazone derivatization of aldehyde containing compounds.

biological matrixes, vitamin B_6 analysis in pharmaceutical preparations is relatively simple, because PN is the only B_6 vitamer present and analysis of the complete B_6 vitamer profile is therefore not required. Moreover, PN concentrations in multivitamin products are usually high enough to allow UV detection. A summary of available methods is given in Table 3.

Table 3 HPLC Determination of Pyridoxine in Pharmaceutical Preparations

Vitamins determined	Chromatographic conditions[a]	Detection	Ref.
PN, thiamine, niacinamide, riboflavin, niacin	µBondapak C18; 5mM S6S in 1% acetic acid: methanol (75:25)	UV (270 nm)	86
PN, thiamine, riboflavin, niacinamide	µBondapak C18; 5mM S6S in 1% acetic acid: methanol (75:25)	UV (254/280 nm)	87
PN	Zorbax C8; 5mM NaClO₄, 5mM S6S, pH 2.5 with HClO₄, containing 10–20% methanol	Colorometric[c] (650 nm)	88
PN, thiamine, riboflavin, vitamin B_{12}	Lichrosorb RP-18; methanol:water (50:50)	UV, 254 nm	89
PN, thiamine, riboflavin, vitamin B_{12}, folate, nicotinamide	µBondapak C18; binary SP, 0.2 M ammoniumphosphate, pH 5.1 (solvent A); 30% methanol in water (solvent B)[b]	UV, 254 nm	90

[a] Only the most pertinent characteristics of the chromatographic separation are listed. Separations were achieved isocratically, unless indicated otherwise.
[b] Solvent program: 100% solvent A for 21 min, change to 100% solvent B. Flow rate: 2.0 mL/min.
[c] Pyridoxine is reacted with 2.6-dibromoquinone-4-chlorimide.
Abbreviations: S6S = sodium hexanesulphonate, SP = solvent program.

Kirchmeier and Upton (86) developed an ion-pair HPLC procedure with UV detection for the simultaneous determination of niacin, niacinamide, pyridoxine, thiamine and riboflavin extracted from multivitamin preparations (Table 3). Minimal interference from other vitamins, dyes, or preservatives was experienced. Use of sodium hexanesulfonate as ion pair agent allowed use of a mobile phase containing 25% of methanol and vitamins A, D, and E. Folic acid and calcium pantothenate were reported either to be insoluble in the aqueous portion of the mobile phase or to elute at the solvent front (86).

Walker et al. (87), using essentially the same chromatographic conditions as Kirchmeier and Upton (86), optimized pyridoxine detection by dual-wavelength UV detection. Pyridoxine elution was monitored at 280 nm, while other vitamins (thiamine, riboflavin, and niacinamide) were measured at 254 nm.

Kawamoto and coworkers used HPLC and post-column derivatization with 2,6-dibromoquinone-4-chlorimide to quantify PN in pharmaceutical products (88). The condensation product shows strong absorption in the visible spectrum (650 nm); detection has been reported to be almost seven times more sensitive than UV (250 nm) detection of PN. The main advantage of this method is selectivity; other B vitamins or caffeine (often present in vitamin preparations) were shown not to interfere in PN quantification. The disadvantage of the method is that two postcolumn reaction pumps with reaction coils are required to deliver reagents to the column eluate. Furthermore, only PN can be analyzed by this method, while most modern quality control programs tend to optimize vitamin analyses by simultaneous analysis of several vitamin compounds.

Amin and Reusch (89) described a rapid HPLC separation of pyridoxine, riboflavin, thiamine, and vitamin B$_6$. Separation has been reported to be complete within 3 min of injection. The work of Kothari and Taylor (90) provides yet another example of the use of HPLC in vitamin analysis for the phamaceutical industry. This method is suitable for simultaneous analysis of PN, thiamine, riboflavin, nicotinamide, and folic acid. This comprehensive vitamin analysis was possible by using a binary solvent program, which distinguishes this method from the above isocratic ones (Table 3).

HPLC analyses of multivitamin preparations are today widely accepted to ensure quality control. Colorimetric, fluorometric, and microbiological methods have often been replaced by HPLC analysis. The main advantage of HPLC analysis is its versatility and flexibility; published vitamin analysis methods are usually easily adapted to suit different requirements of the pharmaceutical industry.

D. Analysis of Food

1. Comprehensive Food Analysis

Vitamin B$_6$ analysis in foods is a difficult analytical problem, because all six vitamers occur in relatively low concentrations in complex, organic matrices.

Due to the complicated composition of food, it is clear that reliable vitamin B_6 determinations require high chromatographic efficiency and detection sensitivity, as well as excellent detection specificity. Fluorescence detection fulfills these requirements. This mode of detection is therefore virtually exclusively used in modern chromatographic vitamin B_6 analysis of food. Several methods suitable for food analysis, summarized in Table 4, have been described in the literature.

Using anion exchange HPLC, Vanderslice and coworkers pioneered the separation of the six nutritionally active B_6 vitamers from each other within 70 min (91). This early method required two fluorescence detectors in series and was not sensitive enough to measure PLP concentrations. In order to increase sensitivity, the method was later modified to include on-column semicarbazide derivarization of PLP and PL (92). Separation of PLP-semicarbazone from the internal standard could only be achieved by using a dual-column system, while the fluorescence detector wavelengths had to be changed within each run to allow optimum detection of the different B_6 vitamers.

The method of Vanderslice and coworkers was time consuming and the anion exchange resin used was not commercially available in prepacked HPLC columns. This prompted other investigators to develop simpler HPLC methods suitable for vitamin B_6 food analysis. Based on their reversed-phase HPLC method for the determination of PLP and PL in animal tissues as semicarbazone derivatives (78), Gregory and coworkers modified this technique to analyze selected foods from animal origin (93). In animal tissues, vitamin B_6 activity is almost entirely due to PL, PM, and their 5'-phosphorylated derivatives. Glyoxylate treatment of processed food samples before semicarbazone derivatization has been shown to deaminate PMP and PM to PLP and PL, respectively (93); analysis of food samples with and without glyoxylate treatment thus allowed calculation of the PMP and PM content of the sample. Although this method was not suitable for plant foods which usually contain PN, Gregory et al. pointed out that manganese dioxide could possibly be used to oxidize PN and PNP to PL and PLP, respectively (93). This would allow measurement of PN and PNP as semicarbazone derivatives of PL and PLP. However, this possibility was not further pursued, presumably due to later developments which allowed direct HPLC determination of the different B_6 vitamers (94–96).

Gregory and Feldstein (94) developed an ion-paired, reversed-phase HPLC method for individual B_6 vitamers extracted with sulphosalicylic acid from different foods. Using a ternary solvent program, elution of nutritionally active B_6 vitamers from the analytical column was complete within 30 min. PLP was determined as its hydroxysulfonate derivative, following postcolumn introduction of a buffered solution of sodium bisulfite. This method was found suitable for vitamin B_6 analysis in foods of both plant and animal origin. Recoveries for PLP and PL from pork loin were <90%; it was suggested that these vitamers were not completely released from muscle proteins, even in the presence of 5% sulfosalicylic acid.

Table 4 HPLC Determination of B₆ Vitamers in Food

Vitamers determined	Chromatographic conditions[a]	Detection[b]	PLP derivative	Foods analyzed	Reference
PL, PLP, PM, PMP	Ultrasphere IP; 0.033 M potassium phosphate (pH 2.2) with 2.5% acetonitrile	Fl (365/400)	SC pre-C	Cow's milk, raw and cooked beef liver	93
PL, PLP, PM, PMP, PN, PN-glucoside	Perkin Elmer 3 μM ODS; ternary SP with 0.033M potassium phosphate (pH 2.2) 8 mM S8S and propanol[c]	Fl (330/400)	Bisulfite post-C	Frozen broccoli, pork chops, cow's milk	8, 94
PL, PLP, PM, PMP, PN, 4-PA	LiChrospher RP-18, binary SP with: methanol (solvent A) and 0.03 M potassium phosphate (pH 2.2) + 4mM S8S (solvent B)[d]	Fl (330/400)	Bisulfite post-C	Pork liver, cow's milk	95
All[e]	TSK-gel, ODS-120A; 0.1 M NaHClO₄, 0.1 M potassium phosphate, 1% acetonitrile (pH 3.5)	Fl (305/390)	KCN pre-C	Fruit juice, wheat flour, asparagus, eggs, milk, cheese, rice bran	96
PL, PM, PN[f]	Spherisorb-ODS; 0.08 M sulfuric acid	Fl (290/395)	—	Pork, beef, eggs, milk, potatoes, frozen peas	97
All[e]	TSK-gel, ODS-80 T_m; 50 mM potassium phosphate + 120 mM NaHClO₄, pH 3.5 and 1% methanol	Fl (305/390)	Bisulfite post-C	Broccoli, cow's milk, orange juice	99
PN, PM, PL[g]	ODS hypersil, 0.1 M potassium phosphate, 1.25 mM S8S (pH 2.15) + 3% methanol	Fl (333/375)	—	Bananas, chicken breast, broccoli	98

[a] Columns and eluants used are tabled. Separations were achieved isocratically, unless indicated otherwise.

[b] Figures in parentheses indicate excitation and emission wavelengths used for fluorescence detection.

[c] Solvent program: Linear gradient from 100% solvent A to 100% solvent B over 12 min (1.8 mL/min flow rate), followed by a programmed switch to 100% solvent C 15 min after injection. Solvent A = 0.033 M potassium phosphate, 8 mM S8S (pH 2.2); solvent B = 0.033 potassium phosphate, 8 mM S8S, 2.5% propanol (pH 2.2); solvent C = 0.033 M potassium phosphate, 6.5% 2-propanol (pH 2.2).

[d] Solvent program: 90% B and 10% A from 0 to 2 min, linear gradient from 90% B at 2 min to 60% B at 12 min, 60% B and 40% A from 12 to 17 min, 60% B from 17 min to 90% B at 19 min. Flow rate: 1.5 mL/min.

[e] All refers to the six nutritionally active B₆ vitamers and 4-PA.

[f] Sulfuric acid treatment at 120°C was used to hydrolyze PLP, PMP, and PNP.

[g] Enzymatic hydrolysis of the phosphorylated B₆ vitamers was performed prior to HPLC analysis.

Abbreviations: SC, semicarbazide; SP, solvent program; on-C, on-column; post-C, postcolumn; pre-C, precolumn; Fl, fluorescence; S6S, sodium hexanesulphonate; S7S, sodium heptanesulphonate; S8S, sodium octanesulphonate.

Bitsch and Möller (95) modified the method of Gregory and Feldstein by changing the mode of elution to a binary gradient. A further modification included the use of perchloric acid for B_6 vitamer extraction from food and omission of the preparative anion exchange chromatography step for purification of food samples. Recovery for PLP and PL from liver was better than that obtained by Gregory and Feldstein on pork loin, presumably indicating that perchloric acid is a more efficient B_6 vitamer extraction agent from animal tissues than sulfosalicylic acid. This view is supported by Toukairin-Oda et al. (96), who found that extraction with 1 M perchloric acid gave more reliable data. Neither the method of Bitsch and Möller (95) nor the method of Gregory and Feldstein (94) was able to quantify PNP. However, the exclusion of PNP is of little importance, as this vitamer occurs only in minute amounts in food.

Toukairin-Oda reported an isocratic, reversed-phase HPLC method for determination of all six nutritionally active B_6 vitamers in food (96). PLP fluorescence was enhanced by precolumn potassium cyanide treatment to convert PLP to the highly fluorescent 4-pyridoxic acid-5′-phosphate. However, potassium cyanide causes oxidation of PL to 4-pyridoxic acid lactone, which shows little fluorescence at the acid pH (pH = 3.5) of the mobile phase used (Table 2). This problem is circumvented by duplicate analysis (1) without prior potassium cyanide treatment to determine all the B_6 vitamers except PLP, and (2) after potassium cyanide treatment to determine PLP as 4-pyridoxic acid-5-phosphate. This method has been applied to fruit juices, wheat flour, cream cheese, eggs, and baker's yeast.

A simple and popular approach to food vitamin B_6 analysis is to determine PL, PM, and PN after dephosphorylation of the 5′-phosphorylated vitamers (97,98). HPLC separation of PL, PN, and PM is then usually achieved with an isocratic solvent system. Bognar (97) used sulfuric acid treatment at 120°C to hydrolyze the phosphoric acid esters of PL, PN, and PM. Quantification of PL, PN, and PM with reversed-phase HPLC and fluorescence detection was subsequently easily accomplished. The advantage of this method is that derivatization steps to enhance PLP fluorescence become redundant. A possible disadvantage of this procedure is that hydrolysis of B_6 vitamer glycosides may also occur, resulting in an overestimation of the biologically available vitamin B_6 content of food, as discussed above. This problem may be circumvented by enzymatic hydrolysis of the phosphorylated B_6 vitamers in the food extract prior to HPLC analysis. This approach was followed by Van Schoonhoven and coworkers, who extracted B_6 vitamers from bananas, chicken breasts, or broccoli by homogenization in 5% TCA (98). Following centrifugation, the supernatant was filtered and the pH adjusted by a concentrated sodium acetate buffer (pH = 6.0). Enzymatic dephosphorylation was achieved by addition of Taka-Diastase, followed by incubation at 45°C for 30 min. The Taka-Diastase was subsequently removed by TCA precipitation and the supernatant was analyzed by reversed-phase HPLC and fluorescence detection. Van Schoonhoven and coworkers found that the B_6 content of food products measured by their HPLC method was on average 40% higher compared with the *Saccharomyces uvarum* microbiological assay. These differences were ascribed to systematic errors in the microbiological method.

PN-glucosides in plant foods may be determined by direct or indirect methods. The indirect method utilizes β-glucosidase treatment of homogenized food samples. The vitamin B_6 content of the sample is determined before and after enzyme treatment and the increase in pyridoxine concentration after β-glucosidase treatment reflects the PN glucoside concentration in the food sample (62). In the direct method, PN glucoside is quantitated by ion pair, reversed-phase HPLC separation and fluorescence detection (8,99). Gregory and Ink (8) used a ternary gradient programme to separate PN-glucoside and the other B_6 vitamers within 30 min (Fig. 6). Recently, Todera and Naka used a TSK gel ODS 80 T$_m$ reversed-phase column to achieve above mentioned separation isocratically (99). The analysis time was 40 min, and sodium bisulfite was used as postcolumn reagent to enhance PLP fluorescence.

In summary, analytical methods of food analysis which include the separate determination of PN glucoside are preferred. These methods offer a more accurate assessment of the B_6 content of food bioavailable to humans.

2. Determination of Pyridoxine in Fortified Food Products and Elemental Diets

The majority of breakfast cereals in the United States are fortified with PN, and additional PN is also added to infant formula products to ensure adequate vitamin B_6 supply to the infant. Gregory (100) reported an isocratic HPLC method for the determination of PN in breakfast cereals (Table 5). Other investigators attempted simultaneous determination of PN and other vitamins used in food fortification. Wehling and Wetzel used ion pair HPLC to separate pyridoxine, riboflavin and thiamine from each other after acid extraction of the vitamins from cereals (101). Using a dual fluorescence detector setup, pyridoxine and riboflavin were monitored by the first detector. After the column eluate had passed the first detector, an alkaline ferricyanide solution was introduced, resulting in the formation of a fluorescent thiochrome derivative of thiamine, which was detected by the second fluorescence detector. A similar method for simultaneous determination of pyridoxine and riboflavin in infant formula products has also been described (102).

Vitamin analyses in elemental diets are frequently required for process and quality control. Van der Horst et al. developed reversed phase methodology to determine the water-soluble vitamins, including PN, in total parenteral nutrition solutions (103). Iwase described a HPLC method to analyze the aqueous extract from an elemental pediatric diet for PN and nicotinamide (104). Chromatography involved a two-column, double-UV detector system to allow simultaneous determination of both PN and nicotinamide (104).

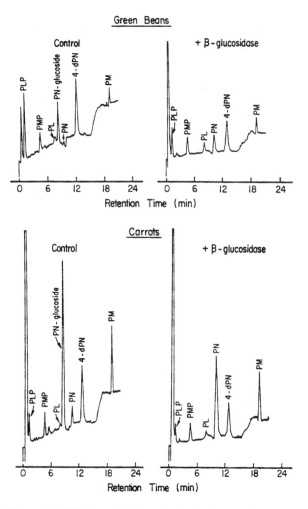

Figure 6 HPLC analysis for vitamin B$_6$ in green beans and carrots. Note the presence of PN-glucoside, which disappears after treatment with β-glucosidase (refer also to Table 4). (From Ref. 8.)

E. Biological Materials

1. Methods for the Determination of PLP and/or PL

Population studies to determine vitamin B$_6$ nutritional status preferably require rapid and reliable methods to measure vitamin B$_6$. PLP is the predominant B$_6$ vitamer in the circulation, and the plasma PLP level has been thought to be a

Table 5 Determination of Pyridoxine in Fortified Food Products or Elemental Diets

Vitamins determined	Chromatographic conditions[a]	Detection[b]	Foods analyzed	Reference
PN	μBondapak C18; 0.033 M potassium phosphate (pH 2.2)	Fl (295/405)	Breakfast cereal	100
PN, riboflavin, thiamine	μBondapak C18; methanol:water:acetic acid (30:69:1) containing 5 mM S7S	Fl (288/418)	Breakfast cereal	101
PN, PL, riboflavin	Spherisorb ODS-1; 7.5 mM S8S in 0.04 M triethyla-mine (pH to 3 with H$_3$PO$_4$):acetonitrile:methanol (85:5:10)	Fl (285/546)	Infant formula products	102
PN, nicotinamide	Lichrosorb RP8; 0.1% triethylamine in 0.067 M phosphate buffer (pH 6.7)	Fl (335/390) UV (260)	Parenteral nutrition	103
PN, nicotinamide	Capcellpak C18 dual-column system with column switching; 1.5% acetonitrile in water, containing 1.5 mM S7S	UV (290)	Elemental pediatric diet	104

[a] Methods cited used isocratic elution.
[b] Figures in parentheses indicate excitation and emission wavelengths used for fluorescence detection, or wavelength setting of UV detector.
Abbreviations: Refer to Table 4.

sensitive parameter for vitamin B_6 nutritional status (105). Several methods suitable for plasma/erythrocyte PLP determination have been described.

Some methods to measure PLP by HPLC are complementary to enzymatic determinations of PLP. An often used enzymatic method (47) for assaying plasma PLP is based on the coenzyme dependent decarboxylation of tyrosine catalyzed by L-tyrosine decarboxylase (EC 4.1.1.25). By using a well-resolved apo-enzyme preparation, it was shown that the reaction rate is directly proportional to the amount of PLP added to the reaction mixture. Conventionally, the reaction is monitored by $^{14}CO_2$ liberation from L-tyrosine-$^{14}C_1$; the liberated $^{14}CO_2$ is trapped in a potassium hydroxide solution and subsequently quantitated by liquid scintillation counting.

The enzymatic method described above has two disadvantages: (1) trapping of $^{14}CO_2$ is a cumbersome procedure, and (2) the use of a radioactive substrate requires special precautions for use and disposal of reagents. Measurement of the primary amine formed by decarboxylation of the amino acid can also be exploited to monitor the PLP-dependent, enzyme-catalyzed reaction. This principle has been applied by Allenmark et al. (106), who used L-3,4-dihydroxyphenyl-alanine (L-DOPA) as substrate for tyrosine decarboxylase; the dopamine produced by the decarboxylation reaction was determined by HPLC followed by amperometric detection. Both Hamfelt (107) and Lequeu et al. (108) utilized apo-tyrosine decarboxylase with tyrosine as substrate. The tyramine produced by the decarboxylation reaction was separated from the substrate (tyrosine) by HPLC and quantitated by either amperometric (108) or fluorometric (107) detection. The procedures discussed above are still subject to the main disadvantage of enzymatic methods: possible interference by other materials present in the PLP containing extract which could either inhibit reconstitution of the holoenzyme or alter the reaction rate of enzyme catalysis. Moreover, HPLC with amperometric detection can hardly be described as less cumbersome than $^{14}CO_2$ trapping; difficulties in baseline-stabilization encountered with these detectors are well known.

Several investigators developed HPLC methods for the direct quantification of PLP (Table 6). The detection of PLP as its semicarbazone derivative seemed a viable alternative to enzymatic methods. Gregory described an isocratic HPLC procedure for the determination of semicarbazone derivatives of PLP and PL after perchloric acid extraction from animal tissues (78). The low pH (2.2) of the mobile phase resulted in an excellent chromatographic efficiency, but also in a relatively low fluorescence intensity of the semicarbazone derivatives.

Schrijver et al. (79) described an isocratic HPLC separation for PLP and PL, followed by postcolumn semicarbazone derivatization. The pH of the column eluate was increased to 12 to 13 by introduction of an 8% NaOH solution to the column eluate before it passed through the fluorometer; in this way much higher fluorescence intensities of the semicarbazone derivatives of PLP and PL were obtained. The increased fluorescence obtained by postcolumn alkalinization re-

Table 6 Methods for the Determination of Plasma PLP and PL

Vitamers determined	Chromatographic conditions[a]	PLP derivative	Detection	Reference
PLP	Nucleosil SA; 0.05 M sodium acetate, 0.03 M citric acid, 0.06 M NaOH, 0.017 M acetic acid, pH 5.2	Enzymatic conversion[b]	Amperometric (+0.55 V)	106
PLP	MPLC RP-18; 0.07 M sodium acetate, pH 3.75	Enzymatic conversion[c]	Fl (280/353)	107
PLP	Lichrosorb RP-18; 0.1 M sodium phosphate, containing 2 mM S8S and 10% acetonitrile, pH 4.9	Enzymatic conversion[c,d]	Amperometric (+0.85 V)	108
PL, PLP	ODS-Hypersil; 0.05 M potassium phosphate, pH 2.9	SC; post-C	Fl (367/478)[e]	79
PL, PLP	Partisil 10 ODS-3; 0.05 M potassium phosphate, containing 7% acetonitrile, pH 2.9	SC; pre-C	Fl (367/478)[e]	81, 82
PL	Nucleosil 120 5 C18, 0.05 M HClO$_4$ and 0.02 M triethylamine	Enzymatic conversion to PL SC; post-C	Fl (365/480)	116, 117
PLP	Shimadzu STR ODS-H; 2 M potassium acetate containing 1 mM S7S, pH 2.9	KCN; pre-C	Fl (318/418)	126
PLP	μBondapak TM C-18; 0.033 M potassium phosphate containing 0.05 M semicarbazide and 3% acetonitrile, pH 2.5	KCN; pre-C	Fl (284/470)	127
PLP, PL	RP-18, 5 μm Nucleosil; 95% 0.033 M KH$_2$PO$_4$ buffer, pH 3.5, and 5% methanol	KCN; pre-C[f]	Fl (318/418)	128

[a] All the reported methods used isocratic elution. See also footnotes to Table 4. For abbreviations, refer to Table 4.
[b] Based on the measurement of PLP in deproteinized plasma by addition of L-3,4-dihydroxyphenylalanine and apo-tyrosine decarboxylase; the dopamine formed is measured by HPLC and the concentration found is directly proportional to the PLP level in the sample.
[c] Based on measurement of PLP in deproteinized plasma by addition of tyrosine (substrate) and apo-tyrosine decarboxylase; the tyramine formed is measured by HPLC, and the concentration found is directly proportional to the PLP level of the sample.
[d] Plasma samples were not deproteinized.
[e] For optimum detection, the pH of the column eluate was adjusted to 12.
[f] Each sample has to be analyzed twice. In the first analysis, the amount of 4-PA is determined. After KCN mediated oxidation, PLP is determined as pyridoxic acid by subtracting the amount of pyridoxic acid already existing before oxidation.

sulted in adequate sensitivity required for reliable PLP and PL quantification in human whole blood (79). The method could be easily automated, and has been used in several clinical studies (109–112).

Ubbink et al. demonstrated that extraction of human plasma with trichloroacetic acid and semicarbazone derivatization could be performed simultaneously (81). The semicarbazone derivatives of PLP and PL were then separated by reversed-phase HPLC, using an acidic (pH 2.9) mobile phase in order to achieve optimum chromatographic efficiency as described by Gregory (78). Postcolumn alkalinization before fluorometric detection resulted in excellent sensitivity, and the method has been applied in analysis of human plasma (45,81,113), whole blood (114), erythrocytes (115), and lymphocytes (Fig. 7).

Mascher (116) also used postcolumn semicarbazide derivatization to measure PL eluted from a Nucleosil 120 S C-18 column. Mascher opted to measure PLP and PL together as "total PL" (116,117). The method is based on the extraction of PLP and PL from plasma with 3M $HClO_4$, a 40-h-long acid phosphatase treatment of the pH-adjusted supernatant, precipitation of the enzyme by a second addition of a volume of 3 M $HClO_4$, followed by HPLC analysis of the supernatant. This analytical procedure described by Mascher is unduly long and it should be realized that the acid phosphatase step is superfluous. As described above, PLP and PL may be easily separated with reversed-phase HPLC. It is much simpler to determine "total PL" by measuring the semicarbazone derivatives of PLP and PL as separate entities and to calculate the sum of the two vitamers. Furthermore, the measurement of "total PL" may be misleading and is therefore not recommended. For example, Mascher measured "total PL" to determine the pharmacokinetic parameters of vitamin B_6 supplementation (117), thus ignoring the fact that PL and PLP display completely different pharmacokinetic properties in humans (118,119).

The simultaneous determination of PLP and PL proved to be an important advantage of methods based on semicarbazone derivatization. Determination of PL may sometimes offer an explanation for depressed PLP levels in various physiological or clinical conditions. Using semicarbazone derivatization and HPLC analysis, Barnard et al. found that declining plasma PLP levels during human pregnancy (120) or starvation of beagle dogs (121) could be explained by elevated plasma PL levels, thus indicating altered vitamin B_6 homeostasis. Similarly, a study on asthmatics revealed severely depressed plasma PLP levels, but normal PL levels, indicating impaired PL phosphorylation during asthma (122).

Quantification of PLP and PL as semicarbazone derivatives were also utilized for the analysis of PL kinase (EC 2.7.1.35), PM (PN) 5′-phosphate oxidase (EC 1.4.3.5) and PLP phosphatase activities (Fig. 8) in various tissues including erythrocytes (123), human placenta (124) and lymphocytes (Fig. 7). The detection of semicarbazone derivatives of both PLP and PL is very sensitive and the method is suitable for enzyme-kinetic studies (125).

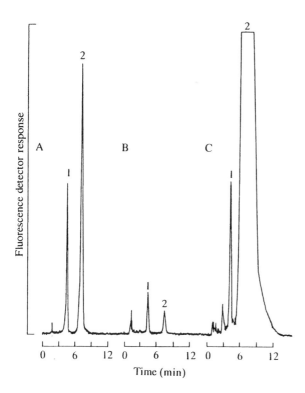

Figure 7 Application of HPLC to study vitamin B₆ metabolism in human lymphocytes. HPLC conditions: Column: Partisil 10 ODS-3 (Whatman, Clifton, NJ, USA). Mobile phase: Isocratic elution with 0.05 mol/L KH₂PO₄ buffer, pH = 2.9, containing 3% acetonitrile, at flow rate 1.1 mL/min. Detection: Fluorescence, excitation 367 nm, emission 478 nm. Postcolumn alkalinization of mobile phase to pH 12.0 was achieved by addition of 4% NaOH solution. Peaks: 1 = PLP semicarbazone; 2 = PL semicarbazone. PLP and/ or PL concentrations were determined using an external standard (A: 300 nmol/L of PLP and PL, respectively) and are expressed per mg of lymphocyte protein: (A) standards, (B) lymphocyte extract, (C) quantitation of PL kinase activity using added PL as a substrate.

The sensitivity of semicarbazone derivatization in PLP and PL analysis is illustrated by the application of HPLC to study vitamin B₆ metabolism in human lymphocytes. Lymphocytes were isolated from 10 mL of blood; 5 μL of packed lymphocytes was diluted to 100 μL using a 10-mM triethanolamine buffer (pH 7.4) containing 0.1% Triton X-100. After homogenization (MSE ultrasound homogenizer), 60 μL of the homogenized suspension was diluted to 400 μL in buffer A (10 mmol/L triethanolamine, 90 mmol/L K₂HPO₄, 2 mmol/L MgCl 2, 2.6 mmol/L ATP; pH 7.4). To a 100-μL aliquot, 50 μL of 10% trichloroacetic

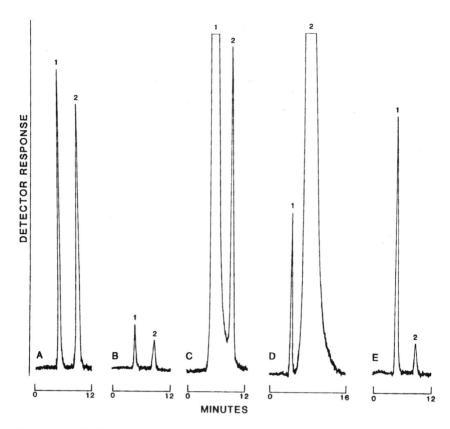

Figure 8 HPLC assay of enzymes involved in vitamin B metabolism. (A) Standard solution, containing PLP (226 nmol/L) and PL (147 nmol/L); (B) hemolysate; (C–E) the same hemolysate to which PLP (C), PL (D), and PMP (E) were added to measure PLP phosphatase, PL kinase and PM (PN) 5′-phosphate oxidase activities, respectively. Peaks: 1 = PLP; 2 = PL. (From Ref. 123.)

acid and 30 µL of 0.5 mol/L semicarbazide were added simultaneously; the mixture was incubated at 37°C for 30 min and the supernatant, obtained after low-speed centrifugation, was used for HPLC analysis of PLP semicarbazone and PL semicarbazone (Fig. 7B). Figure 7C depicts analyses of lymphocyte pyridoxal kinase activity. A 100-µL aliquot of homogenized cell suspension diluted in buffer A was incubated at 30°C for 10 min; subsequently, 10 µL of a 1.8 mmol/L pyridoxal (substrate) solution was added. The reaction was terminated after 1 hour by simultaneous addition of 50 µL of 10% trichloroacetic acid and 30 µL of 0.5 mol/L semicarbazide. HPLC was used to quantify PLP production from PL.

Potassium cyanide catalyzed oxidation of PLP with subsequent quantification of the highly fluorescent oxidation product, 4-pyridoxic acid-5′-phosphate, has been used to determine PLP levels in human plasma, brain tissue and cell cultures (74,126,127). The sensitivity of this method has been reported to be even better than methods based on PLP semicarbazone formation (127). PL is not determined by these methods; this vitamer is oxidized to 4-pyridoxic acid lactone, which shows little fluorescence at pH 2.5 to 3.8 of the mobile phases used (Table 2). Hess and Vuilleumier (128) overcame this problem by alkaline delactonization of pyridoxic acid lactone to yield the highly fluorescent 4-PA. The determination of PL as 4-PA suffers from the disadvantage that each sample has to be analyzed twice. In the first analysis, cyanide oxidation is omitted and the basal 4-PA concentration is measured. In the second analysis, PL is converted to 4-PA as described. The concentration of PL is determined from the difference between the two assays for 4-PA.

2. Comprehensive Vitamin B$_6$ Analysis

Table 7 summarizes several methods for comprehensive vitamin B$_6$ analysis in biological materials. In biological material, comprehensive vitamin B$_6$ analysis is even more complicated than in food, because (1) determination of 4-PA, the end product of vitamin B$_6$ metabolism, is now also required, and (2) levels of individual B$_6$ vitamers in certain biological matrices (e.g., human plasma) are very low. The first attempts to measure all the nutritionally active B$_6$ vitamers in a single analysis include the dual-column system described by Vanderslice et al. (92,129), or acid phosphatase hydrolysis of the extracted vitamers followed by reversed-phase HPLC to separate PN, PM, and PL (130). Coburn and Mahuren successfully developed a cation exchange HPLC method for comprehensive vitamin B$_6$ analysis (76,131). The cation exchange method was based on a ternary solvent program to separate the B$_6$ vitamers (Table 7), while a sodium bisulfite solution in 1M phosphate buffer (pH = 7.5) was introduced as postcolumn reagent. This served two functions: (1) the bisulfite ion reacts with PLP to form a hydroxysulfonate derivative with enhanced fluorescence, and (2) the pH of the column eluate is raised to about 7, which further enhances the fluorescence of the other vitamers.

The method of Coburn and Mahuren has been applied to plasma samples of several animal species (132) and to a variety of biological tissues (124,133). The method has been reproduced and established in other laboratories (133,134). Unfortunately, the Vydac 401TP packing used by Coburn and Mahuren has been discontinued by the manufacturer. Mahuren and Coburn have adapted their

Table 7 Comprehensive Measurement of B$_6$ Vitamers in Biological Tissues

Vitamers determined	Chromatographic conditions[a]	Detection	PLP derivative	Matrix	Reference
All[b]	Vydac 401 TP-B; ternary SP with 0.02 M HCl (solvent A), 0.1 M sodium phosphate, pH 3.3 (solvent B), 0.1 M sodium phosphate, pH 5.9 (solvent C)[c]	Fl (330/400)	Bisulfite post-C	Human plasma, plasma of domestic animals, organs	76, 124, 131–133
All	μBondapak ODS. Methanol: water, 85:15 (solvent A) and 0.005 M S7S, 0.005 M S8S in 1% acetic acid (solvent B)[d]	Fl (300/375)	Natural	Human plasma	136–138
All	Ultramex C18; 0.033 M H$_3$PO$_4$ and 0.01 M S8S, pH 2.2 (solvent A); 0.33 M H$_3$PO$_4$ in 10% propanol (solvent B)[e]	Fl (325/400)	Bisulfite post-C	Plasma, cerebrospinal fluid, rat brain	139
All	μBondapak C18; binary SP with 4 mM S7S in 0.09% acetic acid (solvent A), 20% acetonitrile in 0.09% acetic acid (solvent B)[f]	Fl (330/400)	Bisulfite post-C	Human blood and plasma	141, 142

PL, PLP, PM, PMP, PN	Biosphere ODS; 0.033 M potassium phosphate (pH 2.9) containing 3% methanol	Fl (not specified)	Bisulfite post-C	Human milk	143
All	TSK ODS-120 T (LKB); 0.075 M sodium phosphate containing 0.075 M NaClO$_4$, 0.85% acetonitrile, 0.05% triethanolamine (pH 3.4)	Fl (325/400)	Bisulfite post-C	Human plasma	144, 145
All	AQ-302 ODS; 0.1 M potassium phosphate buffer (pH 3.0) containing 0.1 M NaHClO$_4$ and 0.5 g/L sodium bisulfite	Fl (300/400)	Bisulfite on-C	Human and rat plasma	77
All	Tonen Carbonex column; 15% acetonitrile containing 1% HClO$_4$, and 0.05% sodium bisulfite	Fl (300/400 and 318/418)	Bisulfite on-C	Human plasma	146

a,b Refer to footnotes to Table 4.

c Solvent program: 100% A from 0 to 13 min, 100% B from 13 to 17 min, linear gradient from 100% B at 17 min to 88% B and 12% C at 25 min, linear gradient from 25 min to 100% C at 30 min, maintaining 100% C until 40 min. Flow rate: 1.5 mL/min. Method may also be applied with Nucleosil 5 SA column, with the following solvent program: 100% A from 0 to 10 min, linear gradient from 100% A to 100% B from 10 to 20 min, 100% B from 20 to 28 min, linear gradient from 100% B to 100% C from 28 to 30 min, 100% C from 30 to 35 min.

d Solvent program: 100% B to 75% B/25% A (convex curve) from 0 to 5 min, to 25% B/75% A (convex curve) from 5 to 8 min, to 60% B/40% A (concave curve) from 8 to 12 min, 100% B from 12 to 20 min.

e Solvent program: Linear gradient from 100% A to 100% B from 0 to 10 min, followed by 100% B for the next 15 min.

f Solvent program: Linear gradient from 100% A to 50% A and 50% B over 10 min, linear gradient from 10 min to 30% A and 70% B at 20 min, isocratic elution at 30% A and 70% B until 40 min. Flow rate: 1.0 mL/min.

Figures in parentheses indicate excitation and emission wavelengths for fluorescence detection. Abbreviations: Refer to Table 4.

method for a Nucleosil 5 SA (Phenomenex, Torrance, CA) column (Table 7), but found that the performance of the latter column was less than previously experienced with the Vydac column (135).

Chrisley et al., who used reversed-phase HPLC and a complex gradient elution program, also managed to separate the six biologically active B_6 vitamers from each other and from 4-PA (136). PLP was not derivatized to enhance fluorescence, implying that this method may be unsuitable for PLP measurements in vitamin B_6 deficiency states. Nevertheless, the method has been used in at least two population studies to characterize vitamin B_6 nutritional status (137,138). Sharma and Dakshinamurti (139) also used a binary solvent program and a reversed-phase column to separate the B_6 vitamers and 4-PA from each other, but used postcolumn sodium bisulfite addition to enhance PLP fluorescence. Tryfiates and Sattsangi described an ion-paired, reversed-phase HPLC separation of all the B_6 vitamers (including 4-PA) within 40 min (140). The separation is accomplished by using a binary solvent program, but since a UV detector was used, this method was not directly applicable to the analysis of biological samples. Using a fluorescence detector and postcolumn sodium bisulfite addition to enhance PLP fluorescence, Henderson and coworkers applied the method of Tryfiates and Sattsangi to the analysis of biological samples (141,142). Greater column stability renders reversed-phase, ion pair chromatography preferable to ion exchange chromatography. However, more laborious sample preparation is required to avoid interferences in ion pairing. The supernatant obtained after trichloroacetic acid protein precipitation had to be washed extensively with diethyl ether to remove residual trichloroacetic acid, which could interfere in ion pairing (141,142). Subsequently, the aqueous phase was dried under nitrogen; the extracts were then reconstituted in mobile phase prior to injection. The sample preparation is time-consuming, and only a limited number of samples could therefore be analyzed daily.

More recent developments focused on isocratic separation of the B_6 vitamers. Hamaker et al. used an isocratic method for the analysis of human milk (143), while Edwards and coworkers reported an isocratic method suitable for the determination of the six B_6 vitamers and 4-PA in human plasma (144,145). Edwards and coworkers used a 4.6×250 mm LKB analytical column packed with 5-μm particles of reversed-phase (TSK ODS-120T) material, and the separation was achieved within 23 min. The eluent was a 75-mM NaH_2PO_4 buffer containing 75 mM $NaClO_4$, 0.85% acetonitrile, and 0.05% triethanolamine with pH adjusted to 3.4 with concentrated $HClO_4$. Postcolumn sodium bisulfite addition and fluorescence detection were used to quantify the B_6 vitamers.

Kimura (77) used a mobile phase containing sodium metabisulfite to achieve on-column PLP derivatization. Using a AQ-302 ODS column, the separation of the B_6 vitamers and 4-PA was completed within 13 min. Only PLP, PL, and 4-PA were detected in plasma samples from unsupplemented individuals.

The method of Kimura et al. is fast and suitable for routine measurement of vitamin B_6 status, yet it also has flexibility to measure the other B_6 vitamers when required in, e.g., metabolic studies of PN supplementation. Figure 9 shows a chromatogram obtained with plasma analysis according to Kimura et al. (77). On-column derivatization with sodium bisulfite has also been used by Kurioka et al., who used a graphitic carbon column to separate the biologically active B_6 vitamers and 4-PA within 30 min (146).

3. Urine

Vitamin B_6 is predominantly excreted in urine as 4-pyridoxic acid (147). The relatively high urinary concentration and strong fluorescence of 4-PA enable the direct determination of 4-PA in urine after deproteinization (147,148). However, spontaneous lactonization of 4-PA in acid medium may create an analytical prob-

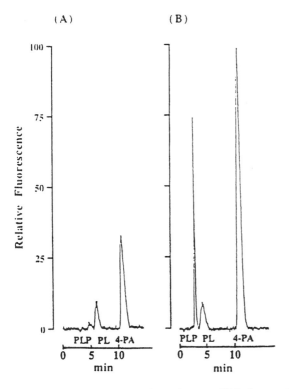

Figure 9 Determination of B_6 vitamers (A) before, and (B) after addition of 0.5 g/L sodium bisulfite to the mobile phase. (From Ref. 77.)

lem (147); a separate determination of recovery is therefore needed to account for 4-PA lactonization. This could be circumvented by precolumn lactonization of 4-PA with subsequent HPLC analysis of 4-PA lactone (149). Figure 10 illustrates the determination of urinary 4-PA levels as 4-PA lactone.

Analysis of urine samples for the other B_6 vitamers is difficult due to high concentrations of other interfering compounds present in urine. Ubbink et al. (150) demonstrated that an anion exchange column cleanup procedure was effective in removing interfering urinary compounds. The purified urine extract was then analyzed by cation exchange HPLC and fluorescence detection. Urinary B_6 vitamer excretion in a fasting person was below the sensitivity limit of the HPLC procedure used. However, the method was suitable to study urinary B_6 vitamer excretion in pharmacokinetic studies of pyridoxine supplementation (150).

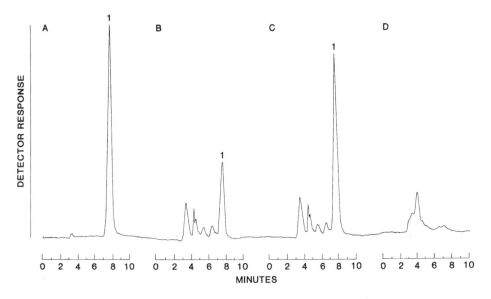

Figure 10 The determination of urinary 4-pyridoxic acid levels as 4-pyridoxic acid lactone: (A) Standard of 11 μmol/L 4-PA; (B) urine sample from a normal, fasting person; (C) the same urine sample spiked with 5.5 μmol/L 4-PA; (D) the same urine sample without lactonization. Each sample was injected at time = 0 min. Peaks: 1 = 4-pyridoxo-lactone, unmarked peaks were unidentified background components. HPLC conditions: Column: Partisil 10 SCX (Whatman). Mobile phase: isocratic elution with 0.025 M ammoniumdihydrogenphosphate (pH 2.8). Detection: fluorescence, excitation 360 nm, emission 430 nm. Postcolumn alkalinization of mobile phase to pH 12.0 was achieved by addition of a 4% NaOH solution. (From Ref. 143.)

IV. GAS-LIQUID CHROMATOGRAPHY

Although gas-liquid chromatography (GLC) is one of the most widely used separation techniques to date, its application in B$_6$ vitamer analysis has been hampered by the high polarity of these compounds (151). The presence of polar groups in vitamin B$_6$ compounds tends to keep them from being volatile enough for GLC. Nevertheless, GLC of a variety of B$_6$ vitamer derivatives, including acetates, trimethylsilyl ethers, benzyl ethers, and isopropylidene derivatives, have been reported (151,152). GLC of the fully acetylated derivatives of vitamin B$_6$ is suitable for the separation and quantification of the three unphosphorylated B$_6$ vitamers (152).

Williams showed that the heptafluorobutyryl derivatives of the three unphosphorylated forms of vitamin B$_6$ may be separated by GLC. The highly electronegative nature of the derivatives made electron capture detection suitable for the sensitive assay of the B$_6$ vitamers (152). Patzer and Hilker used N-methyl-bis-trifluoroacetamide (MBTFA) to form derivatives of B$_6$ vitamers which could be separated by GLC and quantitated by flame ionization detection (153). However, the sensitivity of the flame ionization detector was insufficient to allow quantitation of the B$_6$ vitamers in biological samples. Lim et al. showed that the MBTFA derivatives of B$_6$ vitamers could be measured at increased sensitivity with a GLC equipped with a ^{63}N electron capture detector (154). Lim and coworkers used this method to determine the vitamin B$_6$ content of bread, skimmed milk, and peas (154).

Compared with HPLC, the GLC techniques mentioned above suffer from lack of sensitivity, poor peak shapes, and excessive tailing as well as inconsistent derivatization of certain vitamers (i.e., pyridoxine), which often resulted in appearance of more than one peak per vitamer on the chromatogram. GLC is not often used in routine B$_6$ analysis; HPLC, which offers substantial advantages in simplicity and sensitivity, is generally preferred.

V. MASS SPECTROMETRY

Mass spectrometry is increasingly used in vitamin B$_6$ analysis. Hachey et al. (155) described analysis of B$_6$ vitamers in biological samples by isotope dilution mass spectrometry. Deuterated forms of the different B$_6$ vitamers were added in the early stages of the sample preparation procedure; these deuterated vitamers were used as internal standards to compensate for analytical losses during the isolation and derivatization steps. The B$_6$ vitamers in the homogenized tissue sample were separated by cation exchange HPLC as described previously (76). Acetylation was chosen as derivatization procedure for GLC. However, prior to acetylation, the 5'-phosphoric esters (PLP, PMP) were hydrolyzed in acid and PL was reduced to PN with sodium borohydride. Results obtained by GC/MS analysis of acetylated B_6 vitamers compared well with those of the cation exchange HPLC method previously reported by Coburn and Mahuren (76). Samples with a vitamin B_6 content as low as 0.02 nmol/mL could be analyzed with this method. However, this method is labor-intensive and requires expensive instrumentation, excluding its use in routine analysis of biological samples. Due to its high sensitivity and accurate compensation for analytical losses, the method might become a valuable reference method to verify results obtained by other analytical techniques.

GC-MS has been successfully utilized to study the kinetics of vitamin B_6 turnover in experimental animals. Beynon et al. used GC-MS to determine urinary 4-PA excretion in a metabolic study of the turnover of PLP bound to glycogen phosphorylase (156). Dideuterated PN ($[^2H_2]$PN) was added to the drinking water of vitamin B_6-deficient mice, and urine samples were collected over the following 50 days. 4-PA was isolated from the urine samples by HPLC, dried, and used for the preparation of the *tert*-butyldimethylsilyl (TBDMS) ether derivative of 4-PA lactone (4-PA-TBDMS). A Hewlett-Packard 5890 series II gas chromatograph interfaced with a 5971A mass-selective detector was operated in the selected ion monitoring (SIM) mode, acquiring data at m/z 222 (TBDMS-4-PA lactone derivative) and m/z 224 (TBDMS-$[^2H_2]$ 4-PA lactone derivative). The change in isotope abundance of urinary 4-PA following administration of $[^2H_2]$PN reflected the kinetics of labeling of the vitamin B_6 body pools and was used to estimate the rate of glycogen phosphorylase degradation (156).

Nakano et al. compared the bioavailability of PN and PN-glucoside by administration of the respective deuterium-labeled vitamers to healthy volunteers (157). Urine samples were collected for 48 h after administration of the dose, and 4-PA was isolated as described by Hachey et al. (155) and evaporated to dryness. Isolated 4-PA was converted to the 3-O-acetyl-4-PA lactone derivative, and GC-MS analysis was performed in the electron capture negative ionization mode with SIM at m/z 207 and 209 for the nonlabeled and $[^2H_2]$ 4-PA lactone derivatives, respectively. Using this technique, Nakano et al. found that the bioavailability of PN-glucoside was about 50% relative to PN.

The combination of GC with mass spectrometry has been used to study pyridoxine metabolism in tumor cells. Using radiolabeled PN and paired-ion, reversed-phase HPLC, Tryfiates and coworkers (158) demonstrated that about 30% of the radiolabel was associated with a product showing a retention time different from any of the known B_6 vitamers. Using mass spectrometry, with and without prior GLC separation, the product was eventually identified as adenosine-N^6-diethylthioether-N^1-pyridoximine-5'-phosphate (158,159). Recent studies from Tryfiates and Bishop indicate that the above-mentioned novel B_6 vitamer may be used as tumor marker which may be used for the detection of different malignancies in humans (159).

VI. RECENT TECHNIQUES

Recent progress in food vitamin B$_6$ analysis incorporates enzymatic hydrolysis of the phosphorylated vitamers prior to HPLC analysis (98). Vitamin B$_6$ is subsequently measured by HPLC as the sum of PM, PN, and PL. The advantage of enzymatic hydrolysis compared with previously utilized acid hydrolysis is that B$_6$ vitamer glycosides are not hydrolyzed by phosphatases.

Recent developments in comprehensive vitamin B$_6$ analysis focused on the isocratic separation of the B$_6$ vitamers. Various isocratic methods have recently been described for the determination of the B$_6$ vitamers in human milk (143) and plasma (77,144–146).

Semicarbazide derivatization, as described and applied by Gregory (78), Schrijver et al. (79), and Ubbink et al. (81), is still often used in the determination of PLP and/or PL. A recent modification of this work involves hydrolysis of PLP to PL, followed by the HPLC determination of PL as a semicarbazone derivative. However, this modification is unnecessary, because the semicarbazone derivatives of PL and PLP are easily separated by HPLC. In fact, the 40-h hydrolysis step of the modified procedure renders this method impractical for routine use.

The availability of dideuterated PN and the application of GC-MS to B$_6$ vitamer analysis is another major recent development which allows accurate assessment of vitamin B$_6$ metabolism and turnover in experimental animals (156), humans (157), and tissue cultures (159).

VII. FUTURE TRENDS

In the previous edition of this book (160) I predicted that future developments in column technology would allow simple, isocratic separation of B$_6$ vitamers encountered in biological material. In the past few years the ideal of simple, isocratic separation of the B$_6$ vitamers has come true. HPLC methods developed in the 1980s usually relied on binary or even ternary solvent programs to separate the B$_6$ vitamers in a single run (76,136,139,141). The HPLC methods of the 1990s are usually isocratic, and some of them allow the separation of the B$_6$ vitamers and 4-PA in a relatively short analysis time (77,144,146). The method developed by Kimura et al. (77) serves as an example. Kimura et al. achieved above-mentioned separation within 13 min. Furthermore, the addition of sodium bisulfite to the mobile phase allows on-column derivatization of PLP to its highly fluorescent hydroxysulfonate derivative. This step simplifies HPLC analysis of the B_6 vitamers because it circumvents postcolumn addition of the derivatizing agent.

New, comprehensive methods such as the one developed by Kimura et al. will eventually render HPLC methods for the determination of PLP redundant. The latter methods had the advantage of simplicity and a high turnover potential, but yielded information only on PLP and sometimes also PL. Now it has become possible to use isocratic HPLC and obtain information on all the B_6 vitamers without the inconvenience of unacceptably long analysis times. It has already been pointed out that the simultaneous analysis of PLP and PL sometimes yielded explanations for depressed PLP levels observed under certain circumstances. It is expected that a more comprehensive vitamin B_6 assay, which includes at least 4-PA together with PLP and PL, will be even more informative when applied in studies of vitamin B_6 metabolism in various clinical conditions. Perhaps it will be possible to explain certain observations of a low PLP status by increased formation of 4-PA? Perhaps the new HPLC methods will enable us to find an explanation for ethnic differences in PLP levels which are not dietary induced (161)?

New techniques to measure B_6 vitamers with capillary electrophoresis may be expected in the near future. Capillary electrophoresis has become a powerful analytical technique, and more and more applications are found for this separation method on a virtually daily basis.

Another major advance of the past decade is the study of vitamin B_6 metabolism by GC-MS. The availability of dideurated PN and the analyses of dideurated 4-PA lactone by GC-MS make it possible to gain new insights in the kinetics of vitamin B_6 metabolism. Fascinating results on the use of this technique to characterize the bioavailability of various B_6 vitamers have already been published (157). It is reasonable to expect further advances in this field, which may help us to understand many of the unresolved issues of vitamin B_6 metabolism.

It is anticipated that our understanding of vitamin B_6 metabolism will increase with the newer methods at our disposal to analyze biological tissues for the B_6 vitamers. This will eventually lead to a better understanding of the clinical consequences of vitamin B_6 deficiency states and the therapeutic potential of pyridoxine supplementation.

REFERENCES

1. P Gÿorgy. Vitamin B-2 and the pellagra like dermatitis in rats. Nature 133:498–499, 1934.
2. S Lepovsky. Crystalline factor I. Science 87:169, 1938.
3. P Gÿorgy. Vitamin B-6. J Am Chem Soc 60:1267–1268, 1938.

4. ET Müller. Vitamin B-6 (Adermin) als Wuchstoff für Milchsäurebakterien. Physiol Chem 254:285–286, 1938.

5. EE Snell. Effect of heat sterilization on growth promoting activity of pyridoxine for *Streptococcus faecalis*. Proc Soc Exp Biol Med 51:356–358, 1942.

6. EE Snell. The vitamin activities of "pyridoxal" and "pyridoxamine." J Biol Chem 154:313–314, 1944.

7. IUPAC-IUB Commission on Biochemical Nomenclature. Nomenclature of vitamins, coenzymes and related compounds. Eur J Biochem 2:1–8, 1967.

8. JF Gregory, SL Ink. Identification and quantification of pyridoxine-beta-glucoside as a major form of vitamin B$_6$ in plant-derived foods. J Agric Food Chem 35:76–82, 1987.

9. JE Leklem. Vitamin B-6 metabolism and function in humans. In: JE Leklem, RD Reynolds, eds. Current Topics in Nutrition and Disease, Vol. 19. Clinical and Physiological Applications of Vitamin B$_6$. New York: Alan R Liss, 1988, pp 3–28.

10. DB McCormick, ME Gregory, EE Snell. Pyridoxal phosphokinases. Assay, distribution, purification, and properties. J Biol Chem 236:2076–2084, 1961.

11. A Hamfelt. Pyridoxal kinase activity in blood cells. Clin Chim Acta 16:7–10, 1967.

12. H Wada, EE Snell. The enzymatic oxidation of pyridoxine and pyridoxamine phosphates. J Biol Chem 236:2089–2095, 1961.

13. DB McCormick, AH Merril. Pyridoxamine (pyridoxine) 5′-phosphate oxidase. In: GP Tryfiates, ed. Vitamin B$_6$ metabolism in growth. Westport, CT: Food & Nutrition Press, 1980, pp 1–26.

14. M Ebadi. Catabolic pathways of pyridoxal phosphate and derivatives. In: D Dolphin, R Poulson, O Avramovic, eds. Coenzymes and Cofactors, Vol. 1. Vitamin B$_6$—Pyridoxal Phosphate: Chemical, Biochemical and Medical Aspects, part B. New York: John Wiley & Sons, 1986, pp 449–476.

15. HA Serebro, HM Solomon, JH Johnson, TR Hendrix. The intestinal absorption of vitamin B$_6$ compounds by the rat and hamster. Bull Johns Hopkins Hosp 119:166–171, 1966.

16. HM Middleton. Uptake of pyridoxine by in vivo perfused segments of rat small intestine: a possible role for intracellular vitamin metabolism. J Nutr 115:1079–1088, 1985.

17. L Lumeng, A Lui, T-K Li. Plasma content of B$_6$ vitamers and its relationship to hepatic vitamin B$_6$ metabolism. J Clin Invest 66:688–695, 1980.

18. AH Merrill Jr, JM Henderson, E Wang, BW McDonald. Metabolism of vitamin B-6 by human liver. J Nutr 114:1664–1674, 1984.

19. H Mehansho, DD Buss, MW Hamm, LM Henderson. Transport and metabolism of pyridoxine in rat liver. Biochim Biophys Acta 631:112–123, 1980.

20. BB Anderson, CE Fulford-Jones, JA Child, MEJ Beard, CJT Bateman. Conversion of vitamin B$_6$ compounds to active forms in the red blood cell. J Clin Invest 50:1901–1909, 1971.

21. H Mehansho, LM Henderson. Transport and accumulation of pyridoxine and pyridoxal by erythrocytes. J Biol Chem 255:11901–11907, 1980.

22. L Lumeng, RE Brashear, T Li. Pyridoxal 5′-phosphate in plasma: source, protein-binding, and cellular transport. J Lab Clin Med 84:334–343, 1974.

23. BB Anderson, PA Newmark, M Rawlins, R Green. Plasma binding of vitamin B_6 compounds. Nature 250:502–504, 1974.

24. ML Fonda, C Trauss, UM Guempel. The binding of pyridoxal 5'-phosphate to human serum albumin. Arch Biochem Biophys 288:79–86, 1991.

25. RD Reynolds. Importance of deproteinized serum samples for pyridoxal 5'-phosphate determination. Am J Clin Nutr 60:148–149, 1994.

26. M Ebadi, P Govitrapong. Pyridoxal phosphate and neural transmitters in brain. In: GP Tryfiates, ed. Vitamin B_6 Metabolism in Growth. Westport, CT: Food & Nutrition Press, 1980, pp 223–256.

27. YC Chang, RD Scott, DJ Graves. Function of pyridoxal 5'-phosphate in glycogen phosphorylase: ^{19}F NMR and kinetic studies of phosphorylase reconstituted with 6-fluoropyridoxal and 6-fluoropyridoxal phosphate. Biochemistry 25:1932–1939, 1986.

28. SC Cunnane, MS Manku, DF Horrobin. Accumulation of linoleic and gamma-linolenic acids in tissue lipids of pyridoxine-deficient rats. J Nutr 114:1754–1761, 1984.

29. Y Cho, JE Leklem. In vivo evidence for a vitamin B-6 requirement in carnitine synthesis. J Nutr 120:258–265, 1990.

30. LR Solomon, RS Hillman. Vitamin B_6 metabolism in idiopathic sideroblastic anaemia and related disorders. Br J Haematol 42:239–253, 1979.

31. MH Lipson, J Kraus, LE Rosenberg. Affinity of cystathionine synthase for pyridoxal 5'-phosphate in cultured cells. J Clin Invest 66:188–193, 1980.

32. RM Salkeld, H Gerster. Premenstrual syndrome: a role for vitamin B-6. Gynaecol Pract 5:10–16, 1988.

33. V Sahakian, D Rouse, S Sipes, N Rose, J Niebyl. Vitamin B_6 is effective therapy for nausea and vomiting of pregnancy: a randomized, double-blind placebo-controlled study. Obstet Gynecol 78:33–36, 1991.

34. AL Bernstein, JS Dinesen. Brief communication: effects of pharmacologic doses of vitamin B_6 on carpal tunnel syndrome, electroencephalographic results, and pain. J Am Coll Nutr 12:73–76, 1993.

35. KS Rogers, C Mohan. Vitamin B-6 metabolism and diabetes. Biochem Med Metabol Biol 52:10–17, 1994.

36. RD Reynolds, CL Natta. Depressed plasma pyridoxal phosphate concentrations in adult asthmatics. Am J Clin Nutr 41:684–688, 1985.

37. S Sur, M Camara, A Buchmeier, S Morgan, HS Nelson. Double-blind trial of pyridoxine (vitamin B_6) in the treatment of steroid-dependent asthma. Ann Allergy 70: 147–152, 1993.

38. K Dalery, S Lussier-Cacan, J Selhub, J Davignon, Y Latour, J Genest. Homocysteine and coronary artery disease in French Canadian subjects: relation with vitamins B-12, B-6, pyridoxal phosphate and folate. Am J Cardiol 75:1107–1111, 1995.

39. D Murphy-Chutorian, EL Alderman. The case that hyperhomocysteinemia is a risk factor for coronary artery disease. Am J Cardiol 73:705–707, 1994.

40. IM Graham, LE Daly, HM Refsum, K Robinson, LE Brattstrom, PM Ueland, RJ Palma-Reis, GH Boers, RG Sheahan, B Israelsson, CS Uiterwaal, R Meleady, D McMaster, P Verhoef, J Witteman, P Rubba, H Bellet, JC Wautrecht, HW de Valk, AC Sales Luis, FM Parrot-Rouland, KS Tan, I Higgins, D Garcon, G Andria.

Plasma homocysteine as a risk factor for vascular disease. The European Concerted Action Project. JAMA 277:1775–1781, 1997.

41. JW Miller, JD Ribaya-Mercado, RM Russell, DC Shepard, FD Morrow, EF Cochary, JA Sadowski, SN Gershoff, J Selhub. Effect of vitamin B-6 deficiency on fasting plasma homocysteine concentrations. Am J Clin Nutr 55:1154–1160, 1992.

42. NPB Dudman, DEL Wilcken, J Wang, JF Lynch, D Macey, P Lundberg. Disordered methionine/homocysteine metabolism in premature vascular disease. Its occurrence, cofactor therapy, and enzymology. Arterioscler Thromb 13:1253–1260, 1993.

43. DG Franken, GHJ Boers, HJ Blom, FJM Trijbels, PWC Kloppenborg. Treatment of mild hyperhomocysteinemia in vascular disease patients. Arterioscler Thromb 14:465–470, 1994.

44. HJ Naurath, E Joosten, R Riezler, SP Stabler, RH Allen, J Lindenbaum. Effects of vitamin B-12, folate, and vitamin B-6 supplements in elderly people with normal serum vitamin concentrations. Lancet 346:85–89, 1995.

45. JB Ubbink, A Van der Merwe, R Delport, RH Allen, SP Stabler, R Riezler, WJH Vermaak. The effect of a subnormal vitamin B-6 status on homocysteine metabolism. J Clin Invest 98:177–184, 1996.

46. AG Bostom, RY Gohh, A Beaulieu, MR Nadeau, AL Hume, PF Jacques, J Selhub, IH Rosenberg. Treatment of hyperhomocyst(e)inemia in renal transplant patients. Ann Intern Med 127:1089–1092, 1997.

47. B Chabner, D Livingston. A simple enzymic assay for pyridoxal phosphate. Anal Biochem 34:413–423, 1970.

48. C Plese, W Fox, K Williams. Vitamin B$_6$ measured in plasma with a CO_2-selective electrode. Clin Chem 29:407, 1983.

49. BE Haskell, EE Snell. An improved apotryptonase assay for pyridoxal phosphate. Anal Biochem 45:567–576, 1972.

50. EM Benson, JM Peters, MA Edwards, MR Malinow, CA Storvick. Vitamin B$_6$ in blood, urine, and liver of monkeys. J Nutr 96:83–88, 1968.

51. JC Rabinowitz, EE Snell. Vitamin B$_6$ group. XV. Urinary excretion of pyridoxal, pyridoxamine, pyridoxine, and 4-pyridoxic acid in human subjects. Proc Soc Exp Biol Med 70:235–240, 1949.

52. BE Haskell, U Wallnofer. D-alanine interference in microbiological assays of vitamin B$_6$ in human blood. Anal Biochem 19:569–577, 1967.

53. I Durko, Y Vladovska-Yukhnovska, CI Ivanow. A new fluorometric method for the determination of vitamin B$_6$ in blood. Clin Chim Acta 49:407–414, 1973.

54. MS Chauhan, K Dakshinamurti. Fluorometric assay of pyridoxal and pyridoxal 5′-phosphate. Anal Biochem 96:426–432, 1979.

55. GP Smith, D Samson, TJ Peters. A fluorimetric method for the measurement of pyridoxal and pyridoxalphosphate in human plasma and leucocytes, and its application to patients with sideroblastic marrows. J Clin Pathol 36:701–706, 1983.

56. JJB Nevado, JAM Pulgarin, MAG Laguna. Determination of pyridoxal in human serum by matrix isopotential synchronous fluorescence spectrometry. J Pharm Biomed Anal 14:1487–1494, 1996.

57. DL Brandon, JW Corse, JJ Windle, LL Layton. Two homogeneous immunoassays for pyridoxamine. J Immunol Methods 78:87–94, 1985.

58. JW Thanassi, JA Cidlowski. A radioimmunoassay for phosphorylated forms of vitamin B_6. J Immunol Methods 33:261–266, 1980.

59. SP Coburn, JD Mahuren. Identification of pyridoxine 3-sulfate, pyridoxal 3-sulfate, and N-methylpyridoxine as major urinary metabolites of vitamin B_6 in domestic cats. J Biol Chem 262:2642–2644, 1987.

60. MM Polansky, EW Toepfer. Vitamin B-6 components in some meats, fish, dairy products, and commercial infant formulas. J Agric Food Chem 17:1394–1397, 1969.

61. SP Coburn. Chromatographic analysis of vitamin B_6 and derivatives. In: D Dolphin, R Poulson, O Avramovic, eds. Coenzymes and Cofactors, Vol. 1, Part A. Vitamin B_6—Pyridoxal Phosphate: Chemical, Biochemical and Medical Aspects. New York: John Wiley & Sons, 1986, pp 497–544.

62. H Kabir, J Leklem, LT Miller. Measurement of glycosylated vitamin B-6 in foods. J Food Sci 48:1422–1425, 1983.

63. K Tadera, T Kaneko, F Yagi. Isolation and structural elucidation of three new pyridoxine-glycosides in rice bran. J Nutr Sci Vitaminol 34:167–177, 1988.

64. H Kabir, JE Leklem, LT Miller. Relationship of the glycosylated vitamin B_6 content of foods to vitamin B_6 bioavailability in humans. Nutr Rep Int 28:709–716, 1988.

65. Trumbo PR, Gregory JF, Sartain DB. Incomplete utilization of pyridoxine-beta-glucoside as vitamin B_6 in the rat. J Nutr 118:170–175, 1988.

66. H Nakano, LG McMahon, JF Gregory. Pyridoxine-5′-β-D-glucoside exhibits incomplete bioavailability as a source of vitamin B-6 and partially inhibits the utilization of co-ingested pyridoxine in humans. J Nutr 127:1508–1513, 1997.

67. EA Peterson, HA Sober. Preparation of crystalline phosphorylated derivatives of vitamin B_6. J Am Chem Soc 76:169–175, 1954.

68. JW Bridges, DS Davies, RT Williams. Fluorescence studies on some hydroxypyridines including compounds of the vitamin B_6 group. Biochem J 98:451–467, 1966.

69. DE Metzler, EE Snell. Spectra and ionization constants of the vitamin B-6 group and related 3-hydroxypyridine derivatives. J Am Chem Soc 77:2431–2437, 1955.

70. JW Bridges, RT Williams. Fluorescence of some substituted benzenes. Nature 196: 59–61, 1962.

71. V Bonavita, V Scardi. Spectrophotometric determination of pyridoxal-5-phosphate. Anal Chim Acta 20:47–50, 1959.

72. V Bonavita. The reaction of pyridoxal-5′-phosphate with cyanide and its analytical use. Arch Biochem Biophys 88:366–372, 1960.

73. N Ohishi, S Fukui. Further study on the reaction products of pyridoxal and pyridoxal 5′-phosphate with cyanide. Arch Biochem Biophys 128:606–610, 1968.

74. N Hirose, N Kubo, H Tsuge. Highly sensitive determination of PLP in human plasma with HPLC method. J Nutr Sci Vitaminol 36:521–529, 1990.

75. IL Finar. Organic Chemistry, Vol. 1, 5th ed. London: Longmans, Green, and Co Ltd, 1967, pp 176–177.

76. SP Coburn, JD Mahuren. A versatile cation-exchange procedure for measuring the seven major forms of vitamin B_6 in biological samples. Anal Biochem 129:310–317, 1983.

77. M Kimura, K Kanehira, K Yokoi. Highly sensitive and simple liquid chromatographic determination in plasma of B$_6$ vitamers, especially pyridoxal 5'-phosphate. J Chromatogr 722:295–301, 1996.
78. JF Gregory III. Determination of pyridoxal 5'-phosphate as the semicarbazone derivative using HPLC. Anal Biochem 102:374–379, 1980.
79. J Schrijver, AJ Speek, WHP Schreurs. Semi-automated fluorometric determination of pyridoxal-5'-phosphate (vitamin B$_6$) in whole blood by HPLC. Int J Vitam Nutr Res 51:216–222, 1981.
80. B Conant, PD Barlett. A quantitative study of semicarbazone formation. J Am Chem Soc 54:2881–2899, 1932.
81. JB Ubbink, WJ Serfontein, LS de Villiers. Analytical recovery of protein-bound pyridoxal-5'-phosphate in plasma analysis. J Chromatogr 375:399–404, 1986.
82. JB Ubbink, WJ Serfontein, LS de Villiers. Stability of pyridoxal-5-phosphate semicarbazone: application in plasma vitamin B-6 analysis and population surveys of vitamin B-6 nutritional status. J Chromatogr 342:277–284, 1985.
83. MS Chauhan, K Dakshinamurti. Fluorometric assay of B$_6$ vitamers in biological material. Clin Chim Acta 109:159–167, 1981.
84. CYW Ang. Stability of three forms of vitamin B$_6$ to laboratory light conditions. J Assoc Off Anal Chem 62:1170–1173, 1979.
85. B Saidi, JJ Warthesen. Influence of pH and light on the kinetics of vitamin B$_6$ degradation. J Agric Food Chem 31:876–880, 1983.
86. RL Kirchmeier, RP Upton. Simultaneous determination of niacin, niacinamide, pyridoxine, thiamine, and riboflavin in multivitamin blends by ion-pair HPLC. J Pharm Sci 67:1444–1446, 1979.
87. MC Walker, BE Carpenter, EL Cooper. Simultaneous determination of niacinamide, pyridoxine, riboflavin and thiamine in multivitamin products by HPLC. J Pharm Sci 70:99–100, 1981.
88. T Kawamoto, E Okada, T Fujita. Post-column derivatization of vitamin B-6 using 2,6-dibromoquinone-4-chlorimide. J Chromatogr 267:413–419, 1983.
89. M Amin, J Reusch. HPLC of water-soluble vitamins. II. Simultaneous determination of vitamins B-1, B-2, B-6 and B-12 in pharmaceutical preparations. J Chromatogr 390:448–453, 1987.
90. RM Kothari, MW Taylor. Simultaneous separation of water-soluble vitamins and coenzymes by reversed-phase HPLC. J Chromatogr 247:187–192, 1982.
91. JT Vanderslice, KK Stewart, MM Yarmas. Liquid chromatographic separation and quantification of B-6 vitamers and their metabolite, pyridoxic acid. J Chromatogr 176:281–285, 1979.
92. JT Vanderslice, CE Maire, RF Doherty, GR Beecher. Sulfosalicylic acid as an extraction agent for vitamin B$_6$ in food. J Agric Food Chem 28:1145–1149, 1980.
93. JF Gregory, DB Manley, JR Kirk. Determination of vitamin B-6 in animal tissues by reverse-phase HPLC. J Agric Food Chem 29:921–927, 1981.
94. JF Gregory, D Feldstein. Determination of vitamin B-6 in foods and other biological materials by paired-ion HPLC. J Agric Food Chem 33:359–363, 1985.
95. R Bitsch, J Möller. Analysis of B-6 vitamers in foods using a modified HPLC method. J Chromatogr 463:207–211, 1989.
96. T Toukairin-Oda, E Sakamoto, N Hirose, M Mori, T Itoh, H Tsuge. Determination

of vitamin B-6 derivatives in foods and biological materials by reversed-phase HPLC. J Nutr Sci Vitaminol 35:171–180, 1989.

97. A Bognar. Bestimming von vitamin B-6 in lebensmitteln mit hilfe der HPLC. Z Lebensm Unters Forsch 181:200–205, 1985.

98. J van Schoonhoven, J Schrijver, H van den Berg, GRMM Haenen. Reliable and sensitive high-performance liquid chromatographic method with fluorometric detection for the analysis of vitamin B_6 in foods and feeds. J Agric Food Chem 42: 1475–1480, 1994.

99. K Tadera, Y Naka. Isocratic paired-ion HPLC method to determine B6 vitamers and pyridoxine glucoside in foods. Agric Biol Chem 55:563–564, 1991.

100. JF Gregory. Comparison of HPLC and Saccharomyces uvarum methods for the determination of vitamin B-6 in fortified breakfast cereals. J Agric Food Chem 28: 486–489, 1980.

101. RL Wehling, DL Wetzel. Simultaneous determination of pyridoxine, riboflavin, and thiamine in fortified cereal products by HPLC. J Agric Food Chem 32:1326–1331, 1984.

102. BK Ayi, DA Yuhas, NJ Deangelis. Simultaneous determination of vitamins B-2 (riboflavin) and B-6 (pyridoxine) in infant formula products by reverse phase liquid chromatography. J Assoc Off Anal Chem 69:56–59, 1986.

103. A van der Horst, HJM Martens, PNFC de Goede. Analysis of water-soluble vitamins in total parenteral nutrition solution by high pressure liquid chromatography. Pharm Week Sci 11:169–174, 1989.

104. H Iwase. Determination of nicotinamide and pyridoxine in an elemental diet by column-switching high-performance liquid chromatography with UV detection. J Chromatogr 625:377–381, 1992.

105. JE Leklem, RD Reynolds. Challenges and directions in the search for clinical applications of vitamin B_6. In: JE Leklem, RD Reynolds, eds. Current Topics in Nutrition and Disease, Vol. 19. Clinical and Physiological Applications of Vitamin B_6. New York: Alan R. Liss, 1988, pp 437–454.

106. S Allenmark, E Hjelm, U Larsson-Cohn. New method for quantitative analysis of pyridoxal-5'-phosphate in biological material. J Chromatogr 146:485–489, 1978.

107. A Hamfelt. A simplified method for determination of pyridoxal phosphate in biological samples. Uppsala J Med Sci 91:105–109, 1986.

108. B Lequeu, J Guilland, J Klepping. Measurement of plasma pyridoxal 5'-phosphate by combination of an enzymatic assay with HPLC/electrochemistry. Anal Biochem 149:296–300, 1985.

109. H van den Berg, ES Louwerse, HW Bruinse, JTNM Thissen, J Schrijver. Vitamin B_6 status of women suffering from premenstrual syndrome. Hum Nutr Clin Nutr 40:441–450, 1986.

110. J Schrijver, J Alexieva-Figusch, N van Breederode, HA van Gilse. Investigations on the nutritional status of advanced breast cancer patients. The influence of long-term treatment with megestrol acetate or tamoxifen. Nutr Cancer 10:231–245, 1987.

111. EJ van der Beek, W van Dokkum, J Schrijver, M Wedel, AWK Gaillard, A Wesstra, H van de Weerd, RJJ Hermus. Thiamin, riboflavin, and vitamins B_6 and C: impact of combined restricted intake on functional performance in man. Am J Clin Nutr 48:1451–1462, 1988.

112. FJ Kok, J Schrijver, A Hofman, JCM Witteman, DACM Kruyssen, WJ Remme, HA Valkenburg. Low vitamin B_6 status in patients with acute myocardial infarction. Am J Cardiol 63:513–516, 1989.
113. JB Ubbink, WJH Vermaak, A van der Merwe, PJ Becker, R Delport, HC Potgieter. Vitamin requirements for the treatment of hyperhomocysteinemia in humans. J Nutr 124:1927–1933, 1994.
114. WJH Vermaak, HC Barnard, EMSP van Dalen, GM Potgieter. Correlation between pyridoxal-5'-phosphate levels and the percentage activation of aspartate aminotransferase enzyme in haemolysate and plasma during in vitro incubation studies with different vitamin B_6 vitamers. Enzyme 35:215–224, 1986.
115. JB Ubbink, R Delport, PJ Becker, S Bissbort. Evidence of a theophylline-induced vitamin B_6 deficiency caused by noncompetitive inhibition of pyridoxal kinase. J Lab Clin Med 113:15–22, 1989.
116. H Mascher. Determination of total pyridoxal in human plasma following oral administration of vitamin-B_6 by high-performance liquid chromatography with postcolumn derivatization. J Pharm Sci 82:972–974, 1993.
117. HJ Mascher. High-performance liquid chromatography determination of total pyridoxal in human plasma. Methods Enzymol 280:12–21, 1997.
118. JB Ubbink, WJ Serfontein. The response of the plasma B_6 vitamers to a single, oral pyridoxine supplement. In: JE Leklem, RD Reynolds, eds. Current Topics in Nutrition and Disease, Vol. 19. Clinical and Physiological Applications of Vitamin B_6. New York: Alan R. Liss, 1988, pp 29–34.
119. J Zempleni, W Kubler. The utilization of intravenously infused pyridoxine in humans. Clin Chim Acta 229:27–36, 1994.
120. HC Barnard, JJ de Kock, WJH Vermaak, GM Potgieter. A new perspective in the assessment of vitamin B_6 nutritional status during pregnancy in humans. J Nutr 117:1303–1306, 1987.
121. HC Barnard, WJH Vermaak, GM Potgieter. The effect of acute, prolonged starvation on the concentrations of vitamin B_6 aldehyde derivatives in whole blood. Int J Vitam Nutr Res 56:351–354, 1986.
122. R Delport, JB Ubbink, WJ Serfontein, PJ Becker, L Walters. Vitamin B-6 nutritional status in asthma: the effect of theophylline therapy on plasma pyridoxal-5'-phosphate and pyridoxal levels. Int J Vitam Nutr Res 58:67–72, 1988.
123. JB Ubbink, AM Schnell. High-performance liquid chromatographic assay of erythrocyte enzyme activity levels involved in vitamin B_6 metabolism. J Chromatogr 431:406–412, 1988.
124. S Schenker, RF Johnson, JD Mahuren, GI Henderson, SP Coburn. Human placental vitamin B-6 (pyridoxal) transport: normal characteristics and effects of ethanol. Am J Physiol 262:R966–974, 1992.
125. JB Ubbink, S Bissbort, WJH Vermaak. Inhibition of pyridoxal kinase by methylxanthines. Enzyme 43:72–79, 1990.
126. M Naoi, H Ichinose, T Takahashi, T Nagatsu. Sensitive assay for determination of pyridoxal-5-phosphate in enzymes using HPLC after derivatization with cyanide. J Chromatogr 434:209–214, 1988.
127. H Millart, D Lamiable. Determination of pyridoxal 5'-phosphate in human serum by reversed-phase HPLC combined with spectrofluorimetric detection of 4-pyridoxic acid 5'-phosphate as a derivative. Analyst 114:1225–1228, 1989.

128. D Hess, JP Vuillemier. Assay of pyridoxal-5′-phosphate, pyridoxal and pyridoxic acid in biological material. Int J Vitam Nutr Res 59:338–343, 1989.

129. JT Vanderslice, CE Maire, GR Beecher. B_6 vitamer analysis in human plasma by HPLC: a preliminary report. Am J Clin Nutr 34:947–950, 1981.

130. JA Pierotti, AG Dickinson, JK Palmer, JA Driskell. Liquid chromatographic separation and quantitation of B_6 vitamers in select rat tissues. J Chromatogr 306:377–382, 1984.

131. SP Coburn, JD Mahuren. Cation-exchange HPLC analysis of vitamin B_6. Methods Enzymol 122:102–111, 1986.

132. SP Coburn, JD Mahuren, TR Guilarte. Vitamin B_6 content of plasma of domestic animals determined by HPLC, enzymatic and radiometric microbiological methods. J Nutr 114:2269–2273, 1984.

133. SP Coburn, PJ Ziegler, DL Costill, JD Mahuren, WE Fink, WJ Schalten. Response of vitamin B-6 content of muscle to changes in vitamin B-6 intake in men. Am J Clin Nutr 53:1436–1442, 1991.

134. A Lui, L Lumeng, T Li. The measurement of plasma vitamin B_6 compounds: comparison of a cation-exchange HPLC method with the open-column chromatographic method and the L-tyrosine apodecarboxylase assay. Am J Clin Nutr 41:1236–1243, 1985.

135. JD Mahuren, SP Coburn. Determination of 5-pyridoxic acid lactone, and other vitamin B_6 compounds by cation-exchange high-performance liquid chromatography. Methods Enzymol 280:22–29, 1997.

136. B McChrisley, FW Thye, HM McNair, JA Driskell. Plasma B_6 vitamer and 4-pyridoxic acid concentrations of men fed controlled diets. J Chromatogr 428:35–42, 1988.

137. B McChrisley, HM McNair, JA Driskell. Separation and quantification of the B_6 vitamers in plasma and 4-pyridoxic acid in urine of adolescent girls by reversed phase high-performance liquid chromatography. J Chromatogr 563:369–378, 1991.

138. JA Driskell, B McChrisley. Plasma B-6 vitamer and plasma and urinary 4-pyridoxic acid concentrations in young women as determined using high performance liquid chromatography. Biomed Chromatogr 5:198–201, 1991.

139. SK Sharma, K Dakshinamurti. Determination of vitamin B_6 vitamers and pyridoxic acid in biological samples. J Chromatogr 578:45–51, 1992.

140. GP Tryfiates, S Sattsangi. Separation of vitamin B_6 compounds by paired-ion HPLC. J Chromatogr 227:181–186, 1982.

141. B Hollins, JM Henderson. Analysis of B-6 vitamers in plasma by reversed-phase column liquid chromatography. J Chromatogr 380:67–75, 1986.

142. JM Henderson, MA Codner, B Hollins, MH Kutner, AH Merrill. The fasting B_6 vitamer profile and response to a pyridoxine load in normal and cirrhotic subjects. Hepatology 6:464–471, 1986.

143. B Hamaker, A Kirksey, A Ekanayake, M Borschel. Analysis of B-6 vitamers in human milk by reverse-phase liquid chromatography. Am J Clin Nutr 42:650–655, 1985.

144. P Edwards, PKS Liu, GA Rose. A simple liquid-chromatographic method for measuring vitamin B_6 compounds in plasma. Clin Chem 35:241–245, 1989.

145. P Edwards, PKS Liu, GA Rose. Liquid chromatographic studies of vitamin B$_6$ metabolism in man. Clin Chim Acta 190:67–80, 1990.

146. S Kurioka, N Ishioka, J Sato, J Nakamura, T Ohkubo, M Matsuda. Assay of vitamin-B$_6$ in human plasma with graphitic carbon column. Biomed Chromatogr 7: 162–165, 1993.

147. JF Gregory III, JR Kirk. Determination of urinary 4-pyridoxic acid using HPLC. Am J Clin Nutr 32:879–883, 1979.

148. K Schuster, LB Bailey, JJ Cerda, JF Gregory. Urinary 4-pyridoxic acid excretion in 24-hour versus random urine samples as a measurement of vitamin B$_6$ status in humans. Am J Clin Nutr 39:466–470, 1984.

149. JB Ubbink, WJ Serfontein, PJ Becker, L de Villiers. Determination of urinary 4-pyridoxic acid levels as 4-pyridoxic acid lactone using HPLC. Am J Clin Nutr 44: 698–703, 1986.

150. JB Ubbink, WJ Serfontein, PJ Becker, L de Villiers. Effects of different levels of oral pyridoxine supplementation on plasma pyridoxal-5'-phosphate and pyridoxal levels and urinary vitamin B-6 excretion. Am J Clin Nutr 46:78–85, 1987.

151. W Korytnyk. Gas chromatography of vitamin B$_6$. Methods Enzymol 18:500–504, 1970.

152. AK Williams. Vitamin B$_6$: gas-liquid chromatography of pyridoxol, pyridoxal, and pyridoxamine. J Agric Food Chem 22:107–109, 1974.

153. EM Patzer, DM Hilker. New reagent for vitamin B$_6$ derivative formation in gas chromatography. J Chromatogr 135:489–492, 1977.

154. K Lim, RW Young, JK Palmer, JA Driskell. Quantitative separation of B-6 vitamers in selected foods by a gas-liquid chromatographic system equipped with an electron-capture detector. J Chromatogr 250:86–89, 1982.

155. DL Hachey, SP Coburn, LT Brown, W Erbelding, B DeMark, PD Klein. Quantitation of vitamin B$_6$ in biological samples by isotope dilution mass spectrometry. Anal Biochem 151:159–168, 1985.

156. J Beynon, DM Leyland, RP Evershed, RHT Edwards, SP Coburns. Measurement of the turnover of glycogen phosphorylase by GC/MS using stable isotope derivatives of pyridoxine (vitamin B$_6$). Biochem J 317:613–619, 1996.

157. H Nakano, LG McMahon, JF Gregory. Pyridoxine 5'-β-D-glucoside exhibits incomplete bioavailability as a source of vitamin B-6 and partially inhibits the utilization of co-ingested pyridoxine in humans. J Nutr 127:1508–1513, 1997.

158. GP Tryfiates, RE Bishop. Tentative structure of a novel vitamin B$_6$ tumor product. Prog Clin Biol Res 259:295–305, 1988.

159. GP Tryfiates, RE Bishop. Vitamin B$_6$ and cancer: adenosine-N6-diethylthioether N^1-pyridoximine 5'-PO$_4$, a circulating human tumor marker. Anticancer Res 15: 379–384, 1995.

160. JB Ubbink. Vitamin B$_6$. In: AP de Leenheer, WE Lambert, HJ Nelis, eds. Modern Chromatographic Analyses of the Vitamins. Chromatographic Science Series Vol. 60. New York: Marcel Dekker, 1992, pp 399–439.

161. WJH Vermaak, HC Barnard, GM Potgieter, H Theron. Vitamin B$_6$ and coronary artery disease. Epidemiological observations and case studies. Atherosclerosis 63: 235–238, 1987.

11
Biotin

Olivier Ploux
University of Pierre and Marie Curie, Paris, France

I. INTRODUCTION

A. History

Biotin, hexahydro-2-oxo-1H-thieno(3,4-d)imidazole-4-pentanoic acid (Fig. 1), has been isolated from egg yolk. The structure of that vitamin, also called bios II, factor X, coenzyme R, anti–egg white injury, vitamin B_8, or vitamin H, was established in 1942 and its synthesis was achieved a year later (1). The absolute configuration of biotin was elucidated by x-ray crystallographic analysis on biotin and on a N_1-carboxy derivative (reviewed in 2). NMR studies (3,4) demonstrated that the conformation of the thiophene ring is the same in the solid state and in solution: the sulfur atom lies above the thiophene ring and the valeric chain adopts a quasi-equatorial position (Fig. 1). The same conformation is encountered in the biotin sulfoxides (Fig. 2) (3,4). Numerous total syntheses of biotin have been published and this topic has been thoroughly reviewed recently (5), but the bulk of biotin is still produced according to the original Hoffmann–La Roche process (5).

B. Chemical Properties

1. Stability

In contrast to the reputation of lability of vitamins, biotin is very stable and can be autoclaved without being affected. This molecule can even resist autoclaving in concentrated sulfuric acid (4 M, 120°C, 2 h), conditions used to extract total biotin from biological samples (6). However, biotin can be easily oxidized into biotin sulfoxides and biotin sulfone (Fig. 2). This process, which is negligible

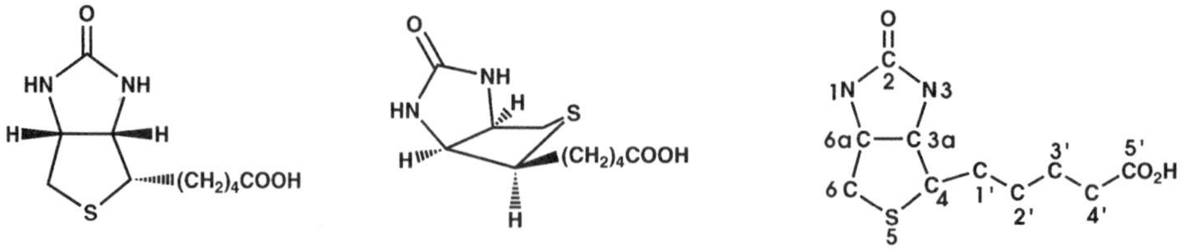

Figure 1 Structure, conformation, and atom numbering of biotin.

in concentrated solutions, can become predominant in very dilute solutions, for instance, during liquid or thin-layer chromatography of biological samples or even during purification of radiolabeled biotin. This oxidation can be prevented by using carefully degassed solvents (Gaudry, personal communication, 1990). In addition to this oxidation process, microbial growth can drastically lower the concentration of nonsterile dilute solutions of the vitamin.

2. Chemical Reactivity

Biotin exhibits normal reactivity at the potentially reactive centers of the molecule. The carboxylate group of biotin has a pK_a around 4.5 and can be esterified without any problem by classical methods (treatment by diazocompounds for example). Likewise, amides are easily formed by activation of the carboxylate (as its *N*-succinimidyl derivative for instance). Numerous esters and amides have been prepared as biotinylation reagents in relation to the avidin-biotin technology (7). The sulfur atom can be easily oxidized with various reagents (hydrogen peroxide, sodium periodate . . .) yielding a mixture of diastereoisomeric (+)- and (−)-biotin sulfoxides (4), which can further react, leading to biotin sulfone (Fig. 2). Biotin can readily be reduced to dethiobiotin (Fig. 3) by Raney nickel desulfurization (8). Reactions of the ureido ring occur almost exclusively at the N_1

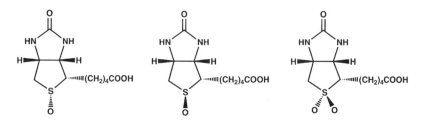

Figure 2 Structure of (−)biotin sulfoxide, (+)biotin sulfoxide, and biotin sulfone.

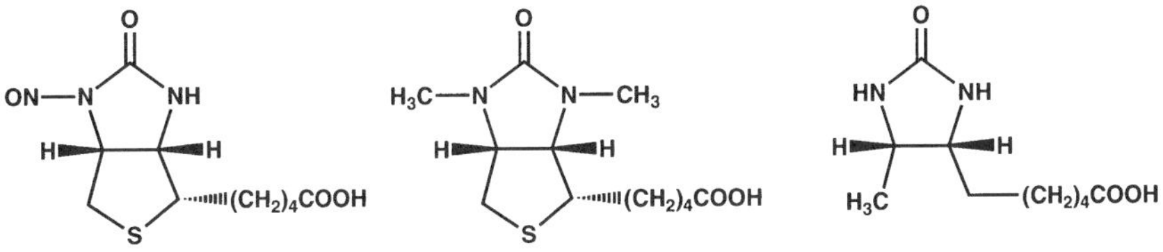

Figure 3 Structure of N_1-nitrosobiotin, N_1N_2-dimethylbiotin, and dethiobiotin.

atom upon nitrosation or methoxycarboxylation, whereas both N_1 and N_3 atoms are methylated by a mixture of formaldehyde and formic acid (9) (Fig. 3).

C. Biochemical Properties

Biotin is required by all living cells but only biosynthesized by plants, fungi, and the majority of micro-organisms. Sources of exogenous biotin for animals are for instance found in yeast extracts, liver, kidneys, egg yolks, milk, and cereals.

1. Biotin Biosynthesis and Degradation

Biotin biosynthesis has been mainly studied in bacteria such as *Escherichia coli* and *Bacillus sphaericus*, in fungi (10), and more recently in higher plants (11) (Fig. 4). However, the recent cloning and sequencing of the *bio* genes from various organisms (bacteria, yeast, and plants) showed that the biosynthetic enzymes catalyzing step 2 to 5 in Figure 4 are very homologous from species to species (12). Such data strongly suggest that this pathway has been conserved throughout evolution. However, the origin of pimeloyl-CoA varies among biotin producing organisms; for instance, while it derives from pimelate in *B. sphaericus* (13), it seems that it derives from fatty acids in *E. coli* (14).

The *bio* genes so far identified in *E. coli* are represented in Figure 5. In addition to the biosynthetic genes *bioABFCD*, which are clustered on a divergent operon that is negatively controlled by the BirA-biotinyl-5′-AMP complex (15,16), there are three other *bio* genes: *bioR*, which codes for the repressor (BirA protein) that also supports the holocarboxylase synthetase activity (see below); *bioH*, probably involved in early steps of the biosynthesis (17); and *bioP*, which codes for the specific carrier for biotin (15,18). The biotin repressor, for which a three-dimensional structure is now available, has both been studied as a model for bacterial DNA repressors and as a model for holocarboxylase synthetases (19,20).

The potential microbiological production of biotin as an economical substitute to chemical synthesis has promoted intensive studies on the enzymology of

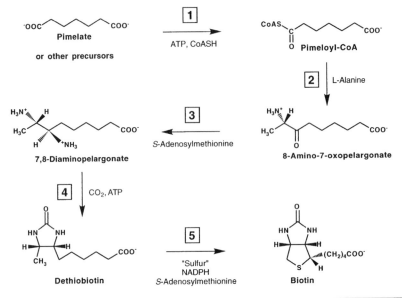

Figure 4 Biosynthesis of biotin.

Step	Enzyme	E. coli genes	B. sphaericus genes
1	Pimeloyl-CoA synthase	bioC, bioH ?	bioW
2	8-Amino-7-oxopelargonate synthase	bioF	bioF
3	7,8-Diaminopelargonate synthase	bioA	bioA
4	Dethiobiotin synthase	bioD	bioD
5	Biotin synthase	bioB	bioB

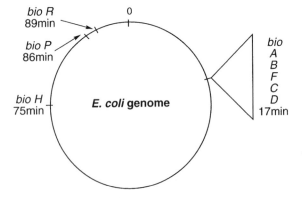

Figure 5 Biotin genes in *Escherichia coli*.

biotin biosynthesis. Almost all the enzymes involved in this pathway have been overproduced, purified, and carefully studied: pimeloyl-CoA synthase (13), 8-amino-7-oxopelargonate synthase (21), dethiobiotin synthase (22,23), and biotin synthase (24). The three-dimensional structure of dethiobiotin synthase has been solved and implications for reaction mechanism have been discussed (22,23), while the determination of the 8-amino-7-oxopelargonate synthase structure is under way (25). We are now starting to decipher the reaction mechanism of biotin synthase, which catalyzes an unprecedented reaction in biochemistry, that is very likely a radical type mechanism. It is clear that all these efforts will be helpful in the near future for the development of microbiological production of biotin and in the design of interesting inhibitors of biotin biosynthesis, as antibiotics or herbicides.

Little is known about biotin degradation (26), but the main pathway so far identified in micro-organisms involves a β-oxidation of the side chain via biotinyl-5′-coenzyme A leading to bisnorbiotin and teranorbiotin (Fig. 6). Interestingly, bisnorbiotin has been identified as the main biotin metabolite in human, pig, and rat urine (27–29).

2. Biotin Transport

Biotin uptake has been extensively studied in micro-organisms such as *Lactobacillus plantarum, Saccharomyces cerevisiae*, and *Escherichia coli* (30 and references cited). The micro-organisms are able to recover biotin from the medium and to concentrate it intracellularly by an active process, that is, against a concentration gradient, mediated by a specialized protein and dependent on an energy source. The biosynthesis of the transport system is regulated by the level of external biotin (30). Since biotin is not biosynthesized by mammalian cells, it must be obtained from exogenous sources by absorption. This uptake of free biotin has been studied, in relation to biotin deficiency, in the small intestine, liver, kidney, and placenta (reviewed in 31). Different models have been utilized (whole animal studies, everted sacs, brush border membrane vesicles . . .), but the preferred model is cultured cell lines (31). It seems that biotin uptake in

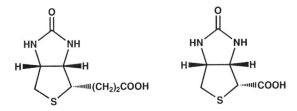

Figure 6 Structures of biotin metabolites: bisnorbiotin and tetranorbiotin.

mammalian cells is mediated by a specialized carrier and that this process is energy dependent and coupled to sodium transport (31). The carrier cDNA from rat placenta has recently been cloned and the functional protein expressed in HeLa cells (32).

3. Attachment of Biotin to the Apoenzymes and Hydrolysis of Biocytin

In biotin-dependent enzymes the cofactor is covalently linked to the protein through an amide bound to a lysine residue (Fig. 7). The attachment of free biotin to apo-enzymes is catalyzed by holocarboxylase synthetases (HCS). This reaction occurs in two steps, an activation step yielding biotinyl-5′-AMP (Eq. 1), which then reacts in a second step with the lysine ε-amino group of the apoenzyme, yielding the active holoenzyme (Eq. 2):

$$ATP + biotin \rightarrow biotinyl\text{-}5'\text{-}AMP + PP_i \qquad (1)$$

$$biotinyl\text{-}5'\text{-}AMP + apoenzyme \rightarrow holoenzyme + AMP \qquad (2)$$

The sequence surrounding the lysine to which biotin is linked is highly conserved among the different carboxylases (33) with a common Ala-Met-Lys-Met sequence. HCSs are not very specific for the holocarboxylases and for instance in humans, one HCS is responsible for the biotinylation of the four biotin-dependent carboxylases found in human cells, although HCS is distributed in the cytosol and the mitochondria (34). In *E. coli* the HCS activity is supported by the BirA protein, the *bioR* gene product, and there is only one protein substrate, the biotin carboxylase carrier protein (BCCP). The BirA protein has been well studied (thermodynamics, kinetics . . .) and is the prototype for HCS (20). Interestingly, sequence homologies were found between human and bacterial HCS, and the human HCS gene was able to complement an *E. coli bioR* mutant (35). HCS from bovine liver has been recently purified and characterized (36).

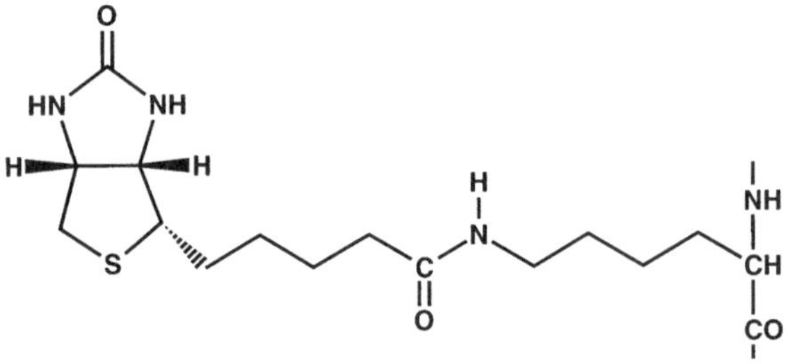

Figure 7 Structure of biotin linked to a lysine residue in biotin-dependent enzymes.

Bound biotin is recovered as free biotin in animals after proteolysis of biotin-dependent enzymes (from endogeneous holocarboxylases and during digestion), leading to biocytin (biotinyl-5′-lysine), which in turn is hydrolyzed to lysine and biotin by biotinidase. Biotinidase (37) has been extensively studied (cloned and purified) because it is related to biotin deficiency in human and numerous assays for biotinidase activity have been designed (38,39). It has been shown that biotinidase also supports biotinyl transferase activity and that this may actually be the main catalytic function for this enzyme, particularly in neuronal cells (40).

4. Biotin-Dependent Enzymes

The biotin-dependent enzyme group includes three classes of enzymes: carboxylases, transcarboxylase, and decarboxylases (41). In all the reactions catalyzed by these enzymes, biotin acts as a CO_2 carrier via N_1-carboxybiotin (Fig. 8). The reaction mechanism of biotin-dependent enzymes has been reviewed (42), as has their evolutionary relatedness (43).

Carboxylases. The carbon dioxide ligase group comprises six enzymes, and four of them are of prime importance since they are found in animals. Pyruvate carboxylase is an important enzyme since it converts pyruvate into oxaloacetate and is involved in gluconeogenesis (Eq. 3). The structure and mechanism of pyruvate carboxylase has been reviewed recently (44).

$$CH_3COCOOH + ATP + HCO_3^- \rightarrow$$
$$^-O_2CCH_2COCOOH + ADP + P_i \quad (3)$$

Acetyl-CoA carboxylase catalyzes the formation of malonyl-CoA from acetyl-CoA and is involved in fatty acid synthesis (Eq. 4). The mechanism, the different isoforms, and the regulation of the activity of this enzyme from animal tissues have been reviewed (45).

$$CH_3COSCoA + ATP + HCO_3^- \rightarrow$$
$$^-O_2CCH_2COSCoA + ADP + P_i \quad (4)$$

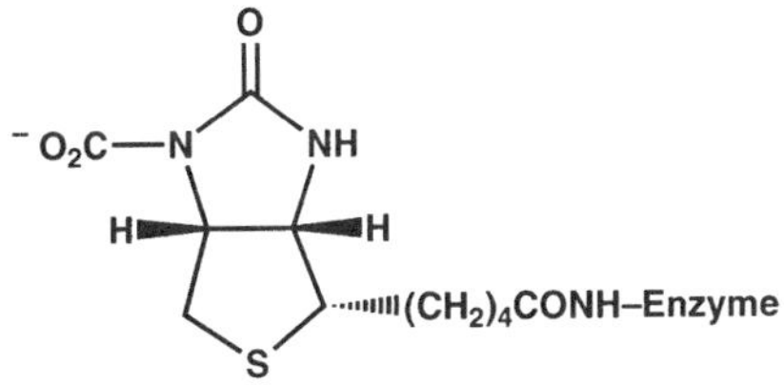

Figure 8 Structure of N_1-carboxybiotinyl-5′-enzyme, the CO_2 carrier in biotin-dependent enzymes.

Propionyl-CoA carboxylase converts propionyl-CoA into methylmalonyl-CoA (Eq. 5) (46). This enzyme is essential in isoleucine degradation.

$$CH_3CH_2COSCoA + ATP + HCO_3^- \rightarrow$$
$$CH_3(CO_2^-)\,CHCOSCoA + ADP + P_i \quad (5)$$

β-Methylcrotonyl-CoA carboxylase, which transforms β-methylcrotonyl-CoA into β-methylglutaconyl-CoA (Eq. 6), is a key enzyme in leucine degradation (47):

$$(CH_3)_2C\!=\!CHCOSCoA + ATP + HCO_3^- \rightarrow$$
$$^-O_2CCH_2(CH_3)C\!=\!CHCOSCoA + ADP + P_i \quad (6)$$

Other carboxylases are urea carboxylase and geranyl-CoA carboxylase.

Transcarboxylase. Methylmalonyl-CoA transcarboxylase, which is induced in *Propionibacterium shermanii*, catalyzes the transcarboxylation between methylmalonyl-CoA and pyruvate (48). This ATP-independent enzyme is important in the metabolism of propionate.

Decarboxylases. Four decarboxylases, methylmalonyl-CoA decarboxylase, oxaloacetate decarboxylase, glutaconyl-CoA decarboxylase, and malonate decarboxylase, are encountered in anaerobic procaryotes. These biotin-dependent enzymes do not require ATP, are membrane bound, and are coupled to sodium transport across the membrane (49).

5. Biotin Deficiency

Because daily requirements are very low (around 100 μg for humans) and because biotin is present in many foodstuffs, acquired deficiency is rare but can be induced in animals by feeding them with raw egg whites, which contain avidin, a glycoprotein that has a very high affinity for biotin (see below). Cases of spontaneous biotin deficiency in humans were not detected before 1976, but since then several cases have been diagnosed. Two inherited disorders have been described: holocarboxylase synthase deficiency and biotinidase deficiency (50,51). Both lead to multiple carboxylase deficiency that is deficiency in the four biotin-dependent carboxylases found in animals. The clinical symptoms (cutaneous and neurological) can be eliminated by biotin absorption (up to 10 mg/day). In HCS deficiency, a lower affinity of the altered HCS for biotin impairs the formation of holocarboxylases. In biotinidase deficiency the patient is unable to recover biotin from biocytin, which is then excreted.

Cloning and sequencing of HCS and biotinidase cDNAs from patients with biotin-deficiency has allowed the mutational analysis of these disorders. Deletions, insertions, and missense mutations were found (34,52).

6. Biotin Binding Proteins

Avidin-biotin technology, which takes advantage of the extremely tight binding between biotin and avidin, has been extensively used in research and technology (assays, molecular biology, diagnostics . . .). An entire monograph, published in 1990 (53), is devoted to this technology and the reader is referred to this excellent book, but since then novel applications have been published every year. We shall only briefly review the biotin binding proteins because they are used in several biotin determination protocols.

Avidin and Streptavidin. Avidin is a glycoprotein isolated from egg white. It is composed of four identical subunits of Mw = 15,600 Da, as deduced from the sequence, and it binds 4 moles of biotin per mole of tetramer with a very high affinity ($K_D = 0.6 \ 10^{-15}$ M). The nonglycosylated form exhibits the same binding properties toward biotin. A very similar protein, streptavidin, found in *Streptomyces avidinii*, has homologous molecular and binding properties: it binds 4 moles of biotin per mole of protein. Contrary to previous reports it has been shown that the binding is noncooperative (54). Both proteins have been cloned, sequenced, and overproduced in *E. coli*. Their x-ray crystal structures (free or complexed with biotin or 2-(4′-hydroxyazobenzene)benzoic acid, HABA) have been solved (55,56). The protein-biotin complexes are stabilized by numerous noncovalent interactions (van der Waals, aromatic side chain contacts, multiple hydrogen bonds . . .) and individual amino acid contributions are now studied by site-directed mutagenesis (57). Other biotin binding proteins related to streptavidin have been isolated from *Streptomyces lavendulae* (58).

Egg Yolk and Chicken Plasma Biotin Binding Proteins. Apart from avidin, streptavidin, and biotin-dependent enzymes, a few proteins bind biotin: biotinidase, HCS (already discussed), and two proteins found in the egg yolk of chicken (59). The role of these proteins is to transport and store biotin for the developing embryo.

Antibodies. The first antibodies to biotin were developed in the late 1970s by Berger (60). Several monoclonal antibodies have now been produced and characterized (61) and might be useful as substitutes in avidin-biotin technology.

II. BIOTIN DETERMINATIONS

The usual methods for biotin determination are based on the biochemical properties of the vitamin and do not involve chromatographic separation. Depending on whether avidin is used, they can be classified into two categories.

A. Methods Without Avidin

1. Reaction with Paradimethylaminocinnamaldehyde

Upon reaction with this compound (62) in acidic medium, a reddish color develops, which is characteristic of the ureido ring and can be used to determine biotin concentrations spectrophotometrically at 533 nm. This method is easy to use but its sensitivity is low, in the microgram range, and is not specific for biotin, since dethiobiotin and other analogs also respond to this reagent.

2. Enzymatic Determination

This assay is based on the formation of holopyruvate carboxylase from apopyruvate carboxylase and biotin, followed by the determination of the pyruvate carboxylase activity. The activity is proportional to the amount of biotin available to form the active holoenzyme (63). This method is not very easy to use and involves unpurified yeast enzymes, but its sensitivity is good (as low as 0.5 ng).

3. Microbiological Determinations

These methods have been used for a long time and are among the most sensitive assays for biotin (0.1 ng). Several micro-organisms have been tested (64), but the most commonly used are *Saccharomyces cerevisiae* (Fleischmann strain 139 ATCC 9896), whose growth allows the determination of biotin and its "vitamers," i.e., biotin, biotin-sulfoxides, biocytine, dethiobiotin, 8-amino-7-oxopelargonate, and 7,8-diaminopelargonate (see Fig. 4) (65), and *Lactobacillus plantarum* 17.5 (ATCC 8014), which responds to "true biotin," i.e., biotin and biotin-sulfoxides (66). Other useful strains are *E. coli bio* mutants (11). The bioassay using *L. plantarum* is very sensitive and specific, and it can be used with complex samples. Interferences are rare although it has been reported that large amounts of dethiobiotin enhanced the growth response (67). This problem was circumvented, in that case, by coupling with the yeast assay of Snell et al. (65).

Two different techniques have been designed: measurements of growth response in liquid media or on agar plates (68). Both techniques are very sensitive (0.1 ng per assay) and easy to set up for routine analysis. Recently, an improved agar plate method has been designed and used for the determination of biotin in urine and serum (69). The following is an experimental part describing the agar plate method (Ploux, unpublished data, 1990):

> *L. plantarum* is grown in *Lactobacillus* MRS broth (from DIFCO) at 37°C (shaking is not required) and kept at −80°C in 25% glycerol, 50% reconstituted milk (12% milk powder) in MRS. Plates (245 × 245 mm² Petri dishes) are prepared daily (2% agar in Biotin Assay Medium from DIFCO) autoclaved at 0.8 atm for 15 min and inoculated with 1% of a fresh overnight culture of *L. plantarum* previously washed twice with sterile water. The plate

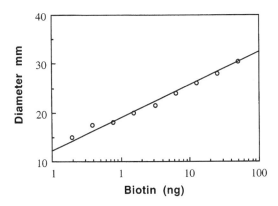

Figure 9 Typical standard curve for the determination of biotin by the agar diffusion plate technique.

is then dried and 5 μL of biotin standards (0.01 to 1 μg/mL) and 5 μL of unknowns are applied to 6-mm paper disks. After an overnight incubation at 37°C, the diameters of the circles corresponding to growth are measured and plotted against the logarithm of the biotin concentration. The standard curves are linear and determination of unknowns is straightforward (Fig. 9).

B. Methods Involving Avidin or Streptavidin

1. 2-(4′-Hydroxyazobenzene)benzoic Acid Method (HABA)

This method (70) takes advantage of the formation of a complex between avidin and HABA (Fig. 10), for which a three-dimensional structure is now available. Biotin, which has a higher affinity than HABA for avidin, quantitatively displaces the dye from the complex. The titration is generally monitored spectrophotometrically at 500 nm. This method is easy to use, but its sensitivity is low (around 1 μg).

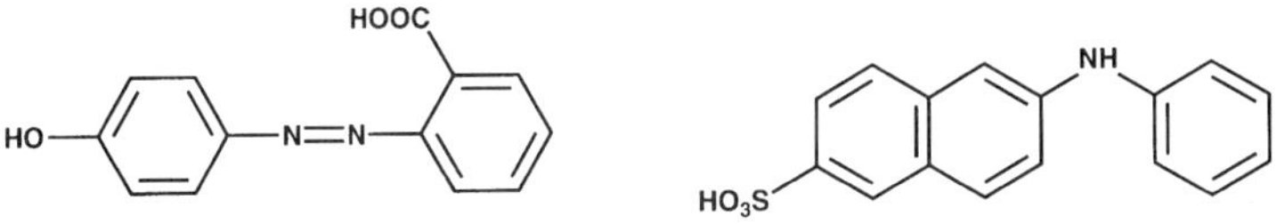

Figure 10 Structures of HABA and 2,6-ANS.

2. Spectral Shift Method

Upon complexation with biotin, the avidin absorbance at 233 nm is modified (70). The sensitivity of this method is similar to that of the HABA method.

3. Fluorescence Determinations

A few methods based on fluorescence have been described for biotin and avidin determinations. A first one is based on the quenching of the avidin tryptophan fluorescence by biotin upon binding (71). A second one involves the increase of the fluorescence of avidin labeled with fluorescein isothiocyanate upon binding of biotin (72). The latter technique has been applied to HPLC postcolumn detection of biotin (see below). The sensitivities are, respectively, 20 and 0.5 ng. Another method is based on the variation of the fluorescence polarization of a biotin-fluorescein derivative upon interaction with avidin (73). Minimal detectable concentrations reported were 5 ng for avidin and 0.1 ng for biotin (73). Mock et al. reported another technique relying on the displacement by biotin of the fluorescent probe 2-anilinonaphthalene-6-sulfonic acid (2,6-ANS) (Fig. 10) when bound to avidin (74). The advantage of this method is obviously the large increase of fluorescence of 2,6-ANS when bound to avidin as compared to the unbound form in water solution. The detection limit was around 1 ng. This technique has also been applied to postcolumn detection of biotin (see below).

4. Chemiluminescence and Bioluminescence

The chemiluminescence method (75) is based on the competition between biotin and a biotin luminol derivative for avidin. The titration is monitored using the chemiluminescence of the luminol derivative (Fig. 11). Williams and Campbell have improved the sensitivity of this kind of assay by quenching the chemiluminescence of the biotin luminol with fluorescein-labeled avidin (76). The detection limit reported seems to be <10 μg per sample.

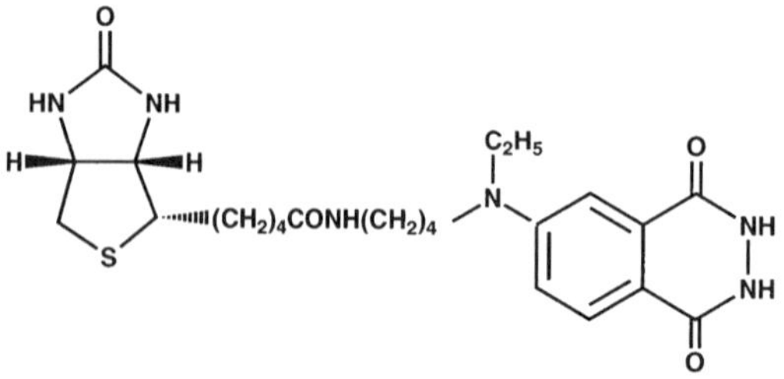

Figure 11   Structure of a chemiluminescent derivative of biotin.

A competitive binding assay based on bioluminescence has been described (77). An aequorin-biotin conjugate competes with free biotin for binding to immobilized avidin. After washing out unbound ligands, bound aequorin is detected by bioluminescence. The signal is related to the concentration of free biotin in the sample. An alternative homogeneous assay has also been described in which the quenching of the bioluminescence is measured upon binding of the aequorin-biotin conjugate (77). Both assays are extremely sensitive with a limit of detection around 10^{-14} M for biotin, but require specific instrumentation.

5. Methods Based on Radioactivity Detection

These methods have been widely used. Isotopic dilution methods are based on the dilution of the biotin to be estimated by a known amount of labeled biotin, followed by the determination of the specific activity of the avidin complexed or the free biotin after thorough separation of both species. [^{14}C]Biotin (78,79), [^{3}H]biotin (80), and ^{125}I-labeled derivatives (Fig. 12) of biotin (81) have been used. The separation of free and avidin-complexed biotin has been achieved with bentonite (80), zinc hydroxide (78), dextran-coated charcoal (79), polyethyleneglycol (82), antibodies (83) or by the use of covalently linked avidin on cellulose (84). The sensitivity depends on the specific activity of the labeled biotin and can range from 20 ng with [^{14}C]biotin, to 1 ng with [^{3}H]biotin, down to 1 pg with ^{125}I-labeled derivatives.

Another radioactive assay based on ^{125}I-labeled avidin has been reported (85). The principle of this solid phase assay is the following: a fixed amount of

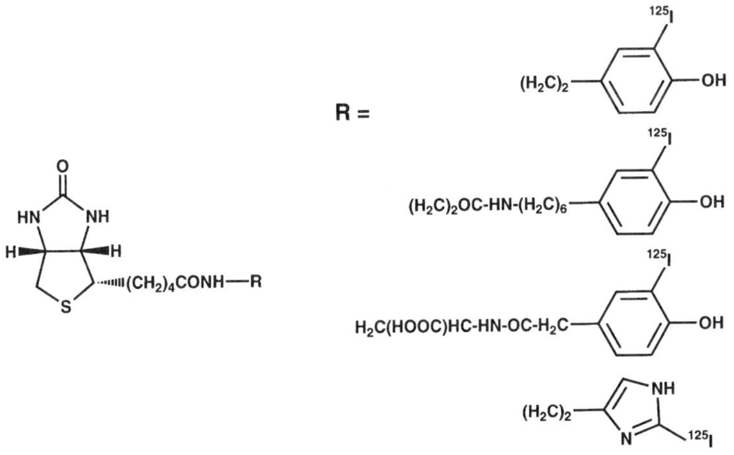

Figure 12 [^{125}I]-Derivatives of biotin used for radioactive assays.

[^{125}I]avidin is incubated with biotin (standard or unknown sample) and the avidin with unoccupied sites is then bound to a plate precoated with biotinylated bovine serum albumin. After washing, the plate wells are counted, giving a measure of the biotin concentration. This method has also been used for postcolumn detection of biotin and its metabolites separated by HPLC (27). The sensitivity is 2 pg per well.

6. Enzyme-Linked Assays

These methods are based on the detection of an enzyme activity, usually using visible spectrophotometry, and are therefore now preferred over radioactive methods. The enzyme can be linked to either avidin, streptavidin, or biotin. Various enzymes have been used: alkaline phosphatase (86), glucose-6-phosphate dehydrogenase (87,88), adenosine deaminase (89), and pyruvate carboxylase (90), but the preferred enzyme is horseradish peroxidase (91–94). The principle of these assays is a competition between free biotin (standards or unknowns) and biotinylated bovine serum albumin or biotinylated immunoglobulin, immobilized on a multiwell plate, for binding to streptavidin linked to the enzyme (Fig. 13). After washing, the bound enzyme is detected colorimetrically. The sensitivity is very good (pg range) and this method has been applied to the determination of biotin in biological fluids.

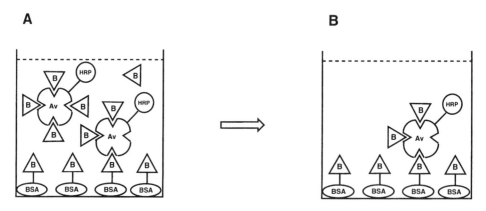

Figure 13 Illustration of the principle of the competitive assay using an avidin-horseradish peroxidase conjugate. (A) Incubation of the components. (B) After washing, only bound avidin-HRP is detected. The more free biotin in the incubation mixture, the less bound avidin-HRP after the washing step. B, biotin; Av, avidin; HRP, horseradish peroxidase.

The methods involving horseradish peroxidase and biotinylated bovine serum albumin are very easy to use since all reagents are commercially available. Other advantages are that they can be used with 96-well plates and that detection is based on visible spectroscopy. Automation is therefore possible. Coupled to HPLC as postcolumn detection they were invaluable in determining biotin and its metabolites in human serum and other biological fluids (see below).

7. Bioaffinity Sensors

This technique has been developed by Aizawa and coworkers. The goal is to build a convenient and specific detector using an enzymatic activity as signal. In the case of biotin determination, avidin coupled with catalase is bound to a membrane bearing covalently linked HABA residues. Addition of biotin destroys quantitatively the HABA-avidin-catalase complex. Washing the membrane and measurement of the remaining catalase activity afforded a sensitive (0.5 ng) and convenient titration of biotin (95).

8. Miscellaneous

An electrochemical assay for biotin has been described (96). This assay involves an electroactive biotinyl-5'-daunomycin derivative. Upon binding to avidin the biotinyldaunomycin loses its electrochemical properties. Therefore free biotin will compete for binding to avidin and will increase the electrode response. This method has the advantage that separation of bound and free biotin is not necessary. The sensitivity was good (0.1 μg range) for biotin, but the response was not linear.

A competitive agglutination assay of biotin has been proposed (97). The method is based on the agglutination of biotin-coated latex particles caused by avidin. Free biotin competes for binding to avidin and thus disrupts the agglutination, which yields a turbidity that is detected by spectrophotometry. Although this assay could be automated it lacks sensitivity and selectivity.

III. ION EXCHANGE, PAPER, AND TLC CHROMATOGRAPHY

A. Ion Exchange Chromatography

This technique has been widely used in early studies to isolate and purify biotin and biotin vitamers and their metabolites from the culture filtrate of many microorganisms. Ogata (6) used a Dowex 1X2 formate column. After washing with deionized water, the biotin vitamers (biotin, biotin sulfoxides, and dethiobiotin) were eluted with formic acid (0.012 M). The same Dowex 1X2 chromatography technique proved to be very good to monitor and to titrate biotin and vitamers in biological samples during investigations on biotin biosynthesis (98). The elution pattern in Figure 14 illustrates, for instance, the chromatographic separation of a mixture of [³H]dethiobiotin and [¹⁴C]biotin. This technique is still very useful when treating large amounts of biological sample.

B. Paper Chromatography

Paper chromatography has only been used in early studies (99). Ascending paper chromatography has been carried out with n-butanol-1 M HCl (6:1), n-butanol-

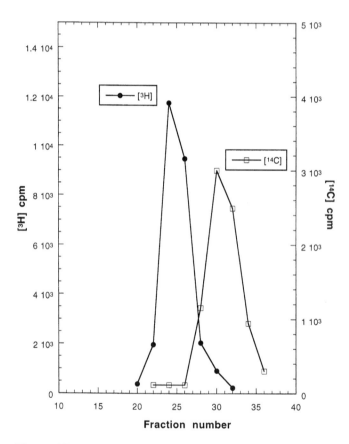

Figure 14 Elution pattern from column chromatography of [³H]dethiobiotin (50 mg, SA = 7400 MBq/mmol) and [¹⁴C]biotin (5 mg, SA = 1110 MBq/mmol). The mixture was poured over a Dowex 1X2 (formate) column 1.5 × 16.5 cm, and eluted with a linear gradient of formic acid (0 to 1 M, 600 mL). Fractions (12 mL) were collected. (From F. Frappier, personal communication, 1985.)

acetic acid-water $(60:15:25)$, n-butanol-acetic acid water $(2:1:1)$, or n-butanol-formic acid-water $(4:1:1)$ as developing solvents. Biotin and selenobiotin have been separated by descending paper chromatography using Whatman 1 filter paper with the solvent system ethanol-t-butanol-formic acid-water $(24:8:2:6)$ (100).

C. Thin-Layer Chromatography

Thin-layer chromatography (TLC) has been applied to analyze biotin in biotin vitamers or multivitamin mixtures. This technique is more convenient and gives better results than paper chromatography and is still routinely used for the identification of biotin and its metabolites. Silicagel plates have generally been used as support, but cellulose powder has also been proposed. Some developing solvents are listed in Table 1. Several detection reagents have been proposed to detect biotin (101): 1% potassium permanganate, 1% dimethylaminobenzaldehyde in HCl (1 M), or iodine vapor. The detection limits obtained with such agents are relatively high, about 5 µg. Better results have been obtained by spraying with mixtures of o-toluidine-potassium iodide (102) or o-toluidine-chlorine (103); biotin is revealed as a deep blue spot. The detection limit was found to be 0.3 µg. However, the best detection agent is p-dimethylaminocinnamaldehyde (p-DACA). In sulfuric acid-methanol mixtures, biotin yielded an intense reddish color with maximal absorbance at 533 nm (104). This reagent is quite specific for biotin and its analogs, since the ureido ring is a prerequisite for suitable color response. The reaction proceeds by formation of the conjugated imine (Schiff's base) with cyclic compounds (104). A spray reagent composed of equal volumes of 2% sulfuric acid in ethanol and 0.2% p-DACA has been described (104).

Another, very convenient visualization technique for TLC plates is the bioautography: the TLC plate is inverted on the surface of an agar plate seeded with *L. plantarum* or *S. cerevisiae* allowing the diffusion of the chromatographed

Table 1   TLC Developing Solvents

Solvents	Proportions (v/v)	References
Acetone-acetic acid-benzene-methanol	1:1:14:4	101–103
Butanol-water-acetic acid	4:5:1	101
Butanol-water-acetic acid	4:1:1	114
Chloroform-methanol-4% formic acid	10:1:1	101
Chloroform-methanol-formic acid	17:3:0.2	117
Ethyl acetate-heptane-acetic acid-water	45:45:1:9	11
Butanol-water-benzene-methanol	2:1:1:1	98

compounds (biotin or its vitamers, for instance). The selectivity and sensitivity have already been discussed.

Nuttall and Bush (102) described a TLC chromatographic method for the analysis of multivitamin preparations. After extraction of fat-soluble vitamins, water-soluble vitamins and water-soluble materials were separated in three TLC systems. Biotin was resolved with acetone-acetic acid-benzene-methanol (1:1: 14:4) as solvent and visualized by spraying o-toluidine-potassium iodide. Standards can be included if quantitative results are required. However, the reproducibility of the technique has not been tested. Groningsson and Jansson (105) worked out a TLC method for the determination of biotin in the presence of other water-soluble vitamins. After dissolution of the lyophilized preparation and addition of the internal standard (2-imidazolidone), the sample was applied on a silicagel plate and eluted with chloroform-methanol-formic acid (70:40:2). Biotin was visualized by spraying with p-DACA and determined in situ by reflectance measurements. The sensitivity of the method could be increased by spraying with paraffin after the coloring procedure. Under these conditions the detection limit was 10 ng.

IV. LIQUID CHROMATOGRAPHY

HPLC is probably the best method for separating biotin and its analogs. The problem here lies in the detection procedure since biotin itself absorbs UV light poorly. However, when large amounts of material are separated, UV detection is applicable. Early work described the separation of biotin, dethiobiotin, biotin sulfoxides, biotin sulfone, bisnorbiotin, and other analogs using a reversed-phase C18 column (linear gradient of acetonitrile, from 0% to 30%, in 0.05% aqueous trifluoroacetic acid pH 2.5) or an anion exchange column (linear gradient of 50 mM Tris-HCl from pH 4.5 to pH 6.5) and UV detection at 220 nm (106). Similarly, biotin methyl ester has been identified and separated from its metabolic precursor dethiobiotin methyl ester by C18 reversed-phase chromatography using a methanol-water (4:6) mixture in the isocratic mode (Azoulay and Frappier, personal communication, 1985).

The use of radiolabeled biotin or analogs, as tracers or in biosynthetic experiments, provides a way to increase the sensitivity of the detection after HPLC separation. For instance, Birch et al. have used a reversed-phase column (C18 column, isocratic elution, 10% acetonitrile in 50 mM sodium phosphate buffer pH 3.5 containing 20 mM triethylamine) with on-line radioactive detection to monitor the transformation of dethiobiotin to biotin catalyzed by biotin synthase (107). Using reversed-phase HPLC with radiometric flow detection, Mock et al. have detected biotin metabolites (biotin sulfoxides and bisnorbiotin) in pig and rat urine, after injection of the animal with radiolabeled biotin (28,29).

Electrochemical detection of biotin after separation by HPLC has also been described (108). A C18 reversed-phase column was used, with acetonitrile-0.05 M potassium phosphate buffer pH 2.0 (15:85 v/v) as eluent, to separate different vitamins. The detection limit for biotin was 10 ng.

Precolumn and postcolumn derivatization of biotin and analogs have also been described when small quantities were analyzed by HPLC.

A. Precolumn Derivatization

Conversion of biotin and dethiobiotin to UV-absorbing BAP esters by reaction with 4-dibromoacetophenone (DBAP) and to fluorescent Mmc esters by reaction with 4-bromomethoxycoumarin (Br-Mmc) (Fig. 15) has been described (109). These derivatives are easily separated using a reversed-phase C18 column with methanol-water or tetrahydrofuran (THF)-water as mobile phase. A linear THF-water gradient is the best system to achieve a good chromatographic separation of the BAP or Mmc esters of dethiobiotin, biotin, biotin sulfone and sulfoxides (Fig. 16). UV ($\lambda = 254$ nm) detection limit was approximately 50 ng, whereas it fell to 5 ng with fluorescence detection. In each case the regressions between the detection response and the injected samples were linear for the concentration range usually seen in biological samples (from 0.15 to 4 nmol for BAP esters and 0.015 to 1 nmol for Mmc esters).

Similar precolumn derivatizations have been described by the use of 9-anthryl (110) or 1-pyrenyl esters (Fig. 16) (111). Both techniques were sensitive enough (0.1 ng) for the determination of biotin in serum and plasma. Panacyl esters of biotin (Fig. 16) have also been used for fluorometric detection (112,113). After extraction of biotin from biological samples (extraction by 5% trichloroacetic acid followed by centrifugation and prepurification on C18 Sep-Pak cartridge, DOWEX anion exchange column, and TLC on silica) the samples were esterified in the presence of a crown ether. Reversed-phase or normal-phase HPLC columns were used for separation although normal phase gave the best results. This technique has been applied in the determination of the biotin content of rat intestinal

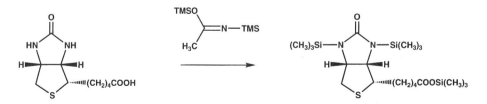

Figure 15 Trimethylsilylation of biotin.

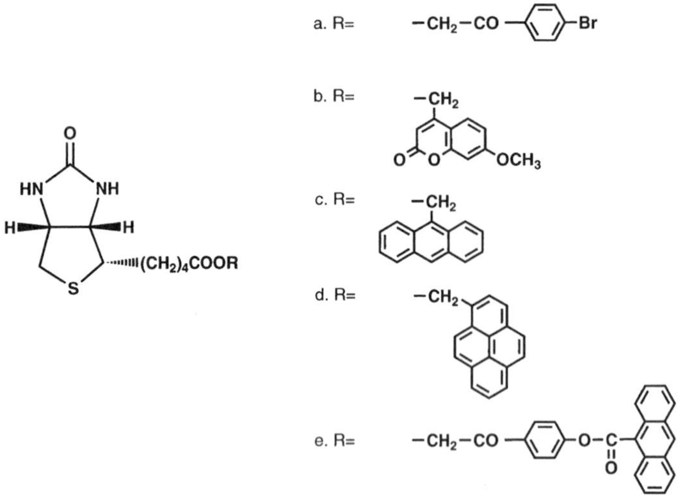

a. R= $-CH_2-CO-\langle\bigcirc\rangle-Br$

b. R= $-CH_2$

c. R= $-CH_2$

d. R= $-CH_2$

e. R= $-CH_2-CO-\langle\bigcirc\rangle-O-\underset{O}{\overset{\;}{C}}-$

Figure 16   Precolumn derivatization of biotin for HPLC. Structure of BAP (a), Mmc (b), anthryl (c), pyrenyl (d), and panacyl (e) esters of biotin.

tissues using dethiobiotin as internal standard. The detection limit is about 2 ng per injection.

B. Postcolumn Derivatization

Mock has described a very sensitive method (the limit of detection is around 5 fmol) for the detection of biotin and its metabolites, involving C18 reversed-phase HPLC separation followed by detection using an avidin-horseradish peroxidase conjugate (92). The principle of this method has already been discussed. This methodology has allowed the identification and quantification of biotin, bisnorbiotin, biotin sulfoxides, and biotin sulfone in human and animal serum and urine. Furthermore, new biotin metabolites (tetranorbiotin sulfoxide and bisnorbiotin methyl ketone) were detected in human urine (114). This technique should potentiate more accurate studies on human biotin metabolism (see, e.g., 115).

In another study Bachas and coworkers have described two different postcolumn detection techniques for biotin and biocytin after HPLC separation based on fluorescence (116). One method involves avidin using 2,6-ANS (Fig. 10) as the fluorophore, while the other one relies on fluoresceine covalently linked to streptavidin. The principle of these fluorescence detections has already been discussed. The HPLC separation was achieved on a C18 column using 0.1 M potassium phosphate buffer pH 7.0 with methanol (8:2) as mobile phase. The postcolumn reagent was directly pumped into the HPLC effluent. Using 2,6-ANS, a negative peak (a decrease in fluorescence) was observed for biotin and biocytin. The detection limit was around 2 ng for biotin. Using the streptavidin-fluoresceine conjugate, a positive peak was observed with a detection limit of 0.1 ng for biotin.

V. GAS-LIQUID CHROMATOGRAPHY

Biotin is not volatile enough for direct GLC analysis but it is possible to use this technique after derivatizing the carboxylic group. Using bis-trimethylsilyl acetamide as silylating agent, Viswanathan et al. (117) described the quantitative determination of biotin after its conversion into its silyl ester. The column was a 2% OV 17 on diatomite (chromosorb G, AW/HMDS treated). n-Octosane was added as internal standard, and detection was achieved with a flame ionization detector. The detection limit for biotin trimethylsilyl (TMS) ester was approxi-

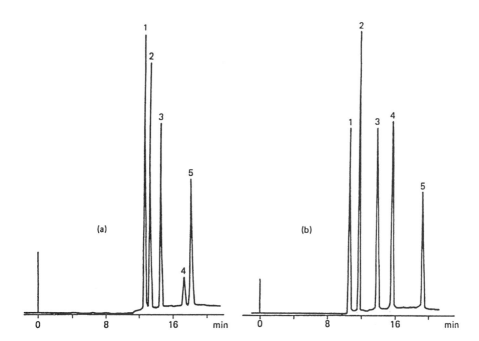

Figure 17 HPLC separation of BAP esters (a) and Mmc esters (b) of biotin and some analogs. Peaks: (1) (−) biotin sulfoxide; (2) (+) biotin sulfoxide; (3) biotin sulfone; (4) biotin; (5) dethiobiotin. Conditions: two 330 × 4 mm C18 bonded columns (Waters) directly connected were used; gradient: THF-water from 30:70 to 60:40; flow rate: 1 mL/ min. (From Ref. 109.)

mately 3 µg. Viswanathan et al. observed that bis-trimethylsilyl acetamide was suitable for complete silylation, whereas the hexamethyldisilazane-trimethyl-chlorosilane (2:1) mixture was not. Furthermore, they assumed that the reaction occurred exclusively at the carboxyl group and that no reaction took place on the two amido NH groups. It was later (118) proved that the proposed structure was incorrect. In fact, the silyl derivative prepared according to Bowers-Komro et al. (106) was not the silyl ester but the trisilyl derivative (Fig. 17). This trisilyl derivative can be formed upon heating biotin with a solution of bis-trimethylsilyl acetamide and pyridine (1:1). The same silylating procedure can be applied to the biotin methyl ester to prepare the *N*-silylated methyl ester.

Janecke and Voege (119) attempted to determine biotin and its oxidation products in commercial multivitamin preparations using the silylation technique. However, the separation of biotin, biotin sulfone, and sulfoxides from a multivitamin mixture was not complete, and the two isomeric sulfoxides could not be separated.

In conclusion, the silylation derivatization seems to give the best results for analysis of biotin by GLC. However, this reaction is hardly reproducible and no direct application to the identification or estimation of biotin in biological mixture has been described in the literature so far.

VI. MASS SPECTROMETRY

A. GC-MS

Johnson and Boyden (118) analyzed biotin with gas chromatography–mass spectrometry (GC-MS) using a 3% OV 17 column. They characterized the trisilylated biotin derivative (Fig. 17) and the corresponding methyl ester. The mass spectra consisted of the molecular ion and a base peak at m/z = 73, probably corresponding to the loss of one trimethylsilyl group.

B. HPLC-MS

Interesting results were obtained by using the combined HPLC-MS technique (120). Mass spectrometry is very specific and appeared as sensitive as UV detection (Table 2). After treatment with diazomethane, biotin has been identified in biological samples as its methyl ester, using a C8 reversed-phase column and an isocratic system methanol-water (4:6) as mobile phase. The mass spectrometer was operated in positive-ion chemical ionization mode, and the limit of detection was <10 ng. The regression between the detector response and the injected sample was found to be linear up to 300 ng.

Except for the methods described above, mass spectrometry has only been used in a few cases mainly during biosynthetic experiments involving heavy iso-

Table 2 Comparison of the Detection Limits of the
Main Biotin Determination Methods

Method	Detection limit (ng)	References
p-DACA	10^3	62
Microbiological	0.1	64–69
Avidin, HABA	10^3	70
Avidin fluorescence	20	71
Avidin-fluoresceine	0.5	72
Avidin, 2,6-ANS	1	74
Enzyme-coupled detection	10^{-3}	86–94
Radioactive detection		
^{14}C	20	78, 79
^3H	1	80
^{125}I	10^{-3}	81, 85
TLC	1–500	102–105
HPLC		
UV detection	50	106, 109
Radioactive detection	10^{-3}	115
Fluorescence detection	1 to 5	116
Avidin-HRP	10^{-3}	92
GLC	10^3	117
LC-MS	10	120

tope labeling precursors of biotin (see, e.g., 121). It has not been applied to the
detection of biotin in biological fluids or tissues.

VII. RECENT TECHNIQUES

It appears that new separation techniques have not yet been applied to biotin
analysis, and only one example was found in the literature. Capillary zone electrophoresis separation of biotin and multivitamin mixtures was achieved on a 600 ×
0.05 mm fused silica capillary under 30 kV using 20 mM phosphate buffer pH
8.0 as eluent (122). Although the separation was good, the sensitivity was limited
by the spectrophotometric detection.

VIII. FUTURE TRENDS AND CONCLUSIONS

The determination methods for biotin that have been designed since the discovery
of that vitamin have always relied on the biochemical properties of biotin. There

are two major methodologies: the microbiological methods based on the fact that some micro-organisms depend on biotin for growth, and the methods involving avidin or streptavidin, two proteins that bind biotin with unsurpassed affinities (K_D in the picomolar range). Because Nature provided these extraordinary tools, chromatographic separations were not necessary and therefore were not thoroughly developed. However, because avidin and streptavidin cannot differentiate between biotin and close analogs, separation techniques are now required for precise analysis. It seems that nowadays the state of the art in biotin determination relies on methods based on HPLC separations with fluorimetric or avidin-horseradish peroxidase based detections. The use of his type of technique has been, and will be, invaluable in assessing biotin status and biotin metabolism in humans and animals, particularly in clinical studies related to biotin deficiency.

Yet, for routine laboratory practice the choice of the detection method is highly dependent on the biotin concentration range to be determined (Table 2). At high levels, the HABA methods remain the more convenient ones, since they require no radioactive materials or sophisticated detection techniques. The microbiological methods, in spite of their lack of accuracy, are still very useful for routine analysis. Perhaps the best methods rely on the use of HPLC separation coupled to a sensitive detection method (avidin-based or fluorimetric). The author's guess is that the enzyme-linked methods based on avidin-horseradish peroxydase conjugate will be the future standard because they are compatible with the 96-well plate format and depend on visible spectrophotometry, being therefore easily automated.

ACKNOWLEDGMENT

This review is dedicated to Michel Gaudry (1942–1998), who contributed to many aspects in the research field of biotin biochemistry (biotin metabolism, biotin transport, biotin deficiency, and so on). His outstanding contribution to the previous editions of this book chapter is gratefully acknowledged.

REFERENCES

1. S Budavari. The Merck Index. 11th ed. Rahway, NJ: Merck and Co., 1989, p 192.
2. R Bentley. The configuration of biotin and related compounds. Trends Biochem Sci 11:51–56, 1985.
3. R Lett, A Marquet. Analyse conformationnelle de dérivés du thiophane, des sulfoxydes et sulfones correspondants. Tetrahedron Lett 30:3365–3377, 1974.
4. R Lett, A Marquet. Determination de configurations de sulfoxydes cycliques par RMN. Tetrahedron Lett 30:3379–3392, 1974.

5. PJ De Clercq. Biotin: a timeless challenge for total synthesis. Chem Rev 97:1755–1792, 1997.

6. K Ogata. Microbial synthesis of dethiobiotin and biotin. Methods Enzymol 18A: 390–394, 1970.

7. M Wilcheck, EA Bayer. Biotin-containing reagents. Methods Enzymol 184:123–138, 1990.

8. G Guillerm, F Frappier, JC Tabet, A Marquet. Deuterium or tritium labeling by ionic hydrogenation. A convenient route to specifically labeled dethiobiotin. J Org Chem 42:3776–3778, 1977.

9. ABA Jansen, PJ Stokes. A search for biotin antagonists and the isolation of γ-biotin. J Chem Soc 4909–4914, 1962.

10. E De Moll. Biosynthesis of biotin and lipoic acid. In: FC Neidhart, ed. *Escherichia coli* and *Salmonella typhimurium*: Cellular and Molecular Biology. 2nd ed. Washington, DC: American Society for Microbiology, Vol. 1, 1996, pp 704–709.

11. P Baldet, H Gerbling, S Axiotis, R Douce. Biotin biosynthesis in higher plant cells. Eur J Biochem 217:479–485, 1993.

12. K Hatakeyama, M Kobayashi, H Yukawa. Analysis of biotin biosynthesis pathway in coryneform bacteria: *Brevibacterium flavum*. Methods Enzymol 279:339–348, 1997.

13. O Ploux, P Soularue, A Marquet, R Gloecker, Y Lemoine. Investigation of the first step of biotin biosynthesis in *Bacillus sphaericus*. Biochem J 287:685–690, 1992.

14. I Sanyal, SL Lee, DH Flint. Biosynthesis of pimeloyl-CoA, a biotin precursor in *E. coli*, follows a modified fatty acid synthesis pathway: ^{13}C-labeling studies. J Am Chem Soc 116:2637–2638, 1994.

15. MA Eisenberg. Regulation of the biotin operon in *E. coli*. Ann Rev NY Acad Sci 447:335–349, 1985.

16. AJ Otsuka, MR Buoncristiani, PK Howard, J Flamm, C Johnson, R Yamamoto, K Uchida, C Cook, J Ruppert, J Matsuzaki. The *E. coli* biotin biosynthetic enzyme sequences predicted from the nucleotide sequence of the bio operon. J Biol Chem 263:19577–19585, 1988.

17. M O'Regan, R Gloeckler, S Bernard, C Ledoux, I Ohsawa, Y Lemoine. Nucleotide sequence of the bioH gene of *E. coli*. Nucleic Acids Res 17:8004, 1989.

18. DL Daniels, G Plunkett III, V Burland, FR Blattner. Analysis of the *E. coli* genome: DNA sequence of the region from 84.5 to 86.5 minutes. Science 257:771–778, 1992.

19. D Beckett, BW Matthews. *E. coli* repressor of biotin biosynthesis. Methods Enzymol 279:362–376, 1997.

20. Y Xu, D Beckett. Biotinyl-5'-adenylate synthesis catalyzed by *E. coli* repressor of biotin biosynthesis. Methods Enzymol 279:405–421, 1997.

21. O Ploux, A Marquet. Mechanistic studies on the 8-amino-7-oxopelargonate synthase, a PLP-dependent enzyme involved in biotin biosynthesis. Eur J Biochem 236:301–308, 1996.

22. D Alexeev, RL Baxter, L Sawyer. Mechanistic implications and family relationships from the structure of dethiobiotin synthetase. Structure 2:1061–1072, 1994.

23. H Kack, KJ Gibson, Y Lindqvist, G Schneider. Snapshot of a phosphorylated substrate intermediate by kinetic crystallography. Proc Natl Acad Sci USA 95:5495–5500, 1998.

24. A Méjean, B Tse, D Florentin, O Ploux, A Marquet. Highly purified biotin synthase can transform dethiobiotin into biotin in the absence of any other protein, in the presence of photoreduced deazaflavin. Biochem Biophys Res Commun 217:1231–1237, 1995.

25. S Spinelli, O Ploux, A Marquet, C Anguille, C Jelsch, C Cambillau, C Martinez. Crystallisation and preliminary x-ray study of the 8-amino-7-oxopelargonate synthase from *Bacillus sphaericus*. Acta Cryst D 52:866–868, 1996.

26. M Osakai, Y Izumi, K Nakamura, H Yamada. Bacterial degradation of biotin and dethiobiotin. Agric Biol Chem 50:311–316, 1986.

27. DM Mock, GL Lankford, JJ Cazin. Biotin and biotin analogs in human urine: biotin accounts for only half of the total. J Nutr 123:1844–1851, 1993.

28. DM Mock, KS Wang, GL Kearns. The pig is an appropriate model for human biotin catabolism as judged by the urinary metabolite profile of radioisotope-labeled biotin. J Nutr 127:365–369, 1997.

29. KS Wang, A Patel, DM Mock. The metabolite profile of radioisotope-labeled biotin in rats indicates that rat biotin metabolism is similar to that of humans. J Nutr 126: 1852–1857, 1996.

30. A Piffeteau, M Gaudry. Biotin uptake: influx, efflux and countertransport in *E. coli* K12. Biochem Biophys Acta 816:77–82, 1985.

31. DL Dyer, HM Said. Biotin uptake in cultured cell lines. Methods Enzymol 279: 393–405, 1997.

32. PD Prasad, H Wang, R Kekuda, T Fujita, YJ Fei, LD Devoe, FH Leibach, V Ganapathy. Cloning and functional expression of a cDNA encoding a mammalian sodium-dependent vitamin transporter mediating the uptake of pantothenate, biotin and lipoate. J Biol Chem 273:7501–7506, 1998.

33. HG Wood, RE Barden. Biotin enzymes. Annu Rev Biochem 46:385–413, 1977.

34. Y Suzuki, Y Aoki, Y Ishida, Y Chiba, A Iwamatsu, T Kishino, N Niikawa, Y Matsubara, K Narisawa. Isolation and characterization of mutations in the human holocarboxylase synthetase cDNA. Nat Genet 8:122–128, 1994.

35. A Leon-Del-Rio, D Leclerc, B Akerman, N Wakamatsu, RA Gravel. Isolation of a cDNA encoding human holocarboxylase synthetase by functional complementation of a biotin auxotroph of *E. coli*. Proc Natl Acad Sci USA 92:4626–4630, 1995.

36. Y Suzuki, K Narisawa. Purification and properties of bovine and human holocarboxylase synthetases. Methods Enzymol 279:386–393, 1997.

37. J Hymes, K Fleischhauer, B Wolf. Biotinidase in serum and tissue. Methods Enzymol 279:422–434, 1997.

38. K Hayakawa, K Yoshikawa, J Ozumi, K Yamauchi. Determination of biotinidase activity with biotinyl-6-aminoquinoline as substrate. Methods Enzymol 279:434–442, 1997.

39. E Livaniou, SE Kakavakos, SA Evangelatos, GP Evangelatos, DS Ithakissios. Determination of serum biotinidase activity with radioiodinated biotinylamide analogs. Methods Enzymol 279:442–451, 1997.

40. J Hymes, B Wolf. Biotinidase and its role in biotin metabolism. Clin Chim Acta 255:1–11, 1995.

41. J Moss, MD Lane. The biotin dependent enzymes. Adv Enzymol 46:321–442, 1977.

42. JR Knowles. The mechanism of biotin dependent enzymes. Annu Rev Biochem 58:195–221, 1989.
43. D Samols, CG Thornton, VL Murtif, GK Kumar, FC Haase, HG Wood. Evolutionary conservation among biotin enzymes. J Biol Chem 263:6461–6464, 1988.
44. PA Atwood. The structure and the mechanism of action of pyruvate carboxylase. Int J Biochem Cell Biol 27:231–249, 1995.
45. RW Brownsey, R Zhande, AN Boone. Isoforms of acetyl-CoA carboxylase: structures, regulatory properties and metabolic functions. Biochem Soc Trans 25:1231–1238, 1997.
46. F Kalousek, MD Darigo, LE Rosenberg. Isolation and characterization of propionyl-CoA carboxylase from normal human liver. J Biol Chem 255:60–65, 1980.
47. U Schiele, F Lynen. 3-Methylcrotonyl-CoA carboxylase from *Achromobacter*. Methods Enzymol 71:781–791, 1981.
48. CG Thornton, GK Kumar, BC Shenoy, FC Haase, NF Phillips, VM Park, WJ Magner, DP Hejlik, HG Wood, D Samols. Primary structure of the 5S subunit of transcarboxylase as deduced from the genomic DNA sequence. FEBS Lett 330:191–196, 1993.
49. P Dimroth. Primary sodium ion translocating enzymes. Biochim Biophys Acta 1318:11–51, 1997.
50. ER Baumgartner, T Suormala. Multiple carboxylase deficiency: inherited and acquired disorders of biotin metabolism. Int J Vitam Nutr Res 67:377–384, 1997.
51. WL Nyhan. Multiple carboxylase deficiency. Int J Biochem 20:363–370, 1988.
52. HC Knight, TR Reynolds, GA Meyers, RJ Pomponio, GA Buck, B Wolf. Structure of the human biotinidase gene. Mamm Genome 9:327–330, 1998.
53. M Wilcheck, EA Bayer. Avidin-biotin technology. Methods Ezymol 184, 1970.
54. ML Jones, GP Kurzban. Noncooperativity of biotin binding to tetrameric streptavidin. Biochemistry 34:11750–11756, 1995.
55. O Livnah, EA Bayer, M Wilcheck, JL Susman. The structure of the complex between avidin and the dye, HABA. FEBS Lett 328:165–168, 1993.
56. L Puglese, M Malcovati, A Coda, M Bolognesi. Crystal structure of apo-avidin from hen egg-white. J Mol Biol 235:42–46, 1994.
57. LA Klumb, V Chu, PS Stayton. Energetic roles of hydrogen bonds at the ureido oxygen binding pocket in the streptavidin-biotin complex. Biochemistry 37:7657–7663, 1998.
58. EA Bayer, T Kulik, R Adar, M Wilchek. Close similarity among streptavidin-like, biotin-binding proteins from *Streptomyces*. Biochim Biophys Acta 1263:60–66, 1995.
59. N Subramanian, PR Adiga. Simultaneous purification of biotin-binding proteins I and II from chicken egg yolk and their characterization. Biochem J 308:573–577, 1995.
60. M Berger. Antibodies that bind biotin and inhibit biotin-containing enzymes. Methods Enzymol 62:319–326, 1979.
61. F Kohen, H Bagci, G Barnard, EA Bayer, B Gayer, DG Schindler, E Ainbinder, M Wilchek. Preparation and properties of anti-biotin antibodies. Methods Enzymol 279:451–466, 1997.

62. DB McCormick, JA Roth. Colorimetric determination of biotin and analogs. Methods Enzymol 18A:383–385, 1970.

63. S Haarasilta. Enzymatic determination of biotin. Anal Biochem 87:306–315, 1978.

64. RB Ferguson, HC Lichstein. Comparison of microorganisms for the assay of bound biotin. J Bacteriol 75:366, 1958.

65. EE Snell, RE Eakin, RJ Williams. A quantitative test for biotin and observations regarding its occurrence and properties. J Am Chem Soc 62:175–178, 1940.

66. HR Skeggs. In: F Kavanagh, ed. Analytical Microbiology. New York: Academic Press, 1963, pp 421–430.

67. E DeMoll, W Shive. Assay for biotin in the presence of dethiobiotin with *Lactobacillus plantarum*. Anal Biochem 158:55–58, 1986.

68. DS Genghof, CWH Partridge, FH Carpenter. An agar plate assay for biotin. Arch Biochem 17:413–420, 1948.

69. T Fukuii, K Iinuma, J Oizumi, Y Izumi. Agar plate method using *Lactobacillus plantarum* for biotin determination in serum and urine. J Nutr Sci Vitaminol 40: 491–498, 1994.

70. NM Green. Spectrophotometric determination of avidin and biotin. Methods Enzymol 18A:418–424, 1970.

71. HJ Lin, JC Kirsch. A rapid, sensitive fluorometric assay for avidin and biotin. Methods Enzymol 62:287–289, 1979.

72. MHH Al-Hakeim, J Landen, DS Smith, RD Nargessi. Fluorometric assays for avidin and biotin based on biotin-induced fluorescence of fluorescein labeled avidin. Anal Biochem 116:264–267, 1981.

73. KJ Schray, PG Artz, RC Hevey. Determination of avidin and biotin by fluorescence polarization. Anal Chem 60:853–855, 1988.

74. DM Mock, P Horowitz. Fluorometric assay for avidin-biotin interaction. Methods Enzymol 184:234–240, 1990.

75. HR Schroeder, PO Vogelhut, RJ Carris, RC Boguslaski, RT Buckler. Competitive protein binding assay for biotin monitored by chemiluminescence. Anal Chem 48: 1933–1937, 1976.

76. EJ Williams, AK Campbell. A homogeneous assay for biotin based on chemiluminescence. Anal Biochem 155:249–255, 1986.

77. S Lizano, S Ramanathan, A Feltus, A Witkowski, S Daunert. Bioluminescence competitive assays for biotin based on photoprotein aequorin. Methods Enzymol 279:296–303, 1997.

78. RL Hood. Isotopic dilution assay for biotin: use of [^{14}C]biotin. Methods Enzymol 62:279–283, 1979.

79. R Rettenmaier. Radioligand assay for biotin in liver tissue. Anal Chim Acta 113: 107–112, 1980.

80. K Dakshinamurti, R Allan. Isotopic dilution assay for biotin: use of [^{3}H]biotin. Methods Enzymol 62:284–287, 1979.

81. EV Gromann, JM Rothenberg, EA Bayer, M Wilchek. Enzymatic and radioactive assays for biotin, avidin, and streptavidin. Methods Enzymol 184:208–217, 1990.

82. SA Evangelatos, SE Kakabakos, E Livianou, GP Evangelatos, DS Ithakissios. Biotin radioligand assay with polyethylene glycol as separation reagent. Clin Chem 37:1306–1307, 1991.

83. LP Thuy, L Sweetman, WL Nyhan. A new immunochemical assay for biotin. Clin Chim Acta 202:191–197, 1991.

84. L Goldstein, SA Yankovsky, G Cohen. Solid-phase assay for biotin and avidin on cellulose disks. Methods Enzymol 122:72–82, 1986.

85. DM Mock. Sequential solid-phase assay for biotin based on ^{125}I-labeled avidin. Methods Enzymol 184:224–233, 1990.

86. EA Bayer, H Ben-Hur, M Wilchek. A sensitive enzyme assay for biotin, avidin, and streptavidin. Anal Biochem 154:367–370, 1986.

87. RS Niedbala, F Gertis, KJ Schray. A spectrophotometric assay for nanogram quantities of biotin and avidin. J Biochem Biophys Methods 13:205–210, 1986.

88. B Terouanne, M Bencheick, P Balaguer, AM Boussioux, J Nicolas. Bioluminescent assays using glucose-6-phosphate dehydrogenase: application to biotin and streptavidin detection. Anal Biochem 180:43–49, 1989.

89. TL Kjellström, L Bachas. Potentiometric homogeneous enzyme-linked competitive binding assays using adenosine deaminase as the label. Anal Chem 61:1728–1732, 1989.

90. S Daunert, BR Payne, L Bachas. Pyruvate carboxylase as a model for oligosubstituted enzyme-ligand conjugates in homogeneous enzyme immunoassays. Anal Chem 61:2160–2164, 1989.

91. JO Nyalala, E Livianou, L Leondiadis, GP Evangelatos, DS Ithakissios. Indirect enzyme-linked method for determining biotin in human serum. J Immunoassay 18: 1–19, 1997.

92. DM Mock. Determination of biotin in biological fluids. Methods Enzymol 279: 265–275, 1997.

93. EZ Huang, YH Rogers. Competitive enzymatic assay of biotin. Methods Enzymol 279:304–308, 1997.

94. D Shiuan, CH Wu, YS Chang, RJ Chang. Competitive enzyme-linked immunosorbent assay for biotin. Methods Enzymol 279:321–326, 1997.

95. Y Ikariyama, M Aizawa. Bioaffinity sensors. Methods Enzymol 137:111–124, 1988.

96. K Sugawara, S Tanaka, H Nakamura. Electrochemical assay of avidin and biotin using a biotin derivative labeled with an electroactive compound. Anal Chem 67: 299–302, 1995.

97. EZ Huang. Competitive agglutination assay of biotin. Methods Enzymol 279:308–320, 1997.

98. AG Salib, F Frappier, G Guillerm, A Marquet. On the mechanism of conversion of dethiobiotin to biotin in *E. coli*. Biochem Biophys Res Commun 88:312–319, 1979.

99. S Iwahara, DB McCormick, LD Wright. Isolation and characterization of bisnorbiotin, dehydrobisnorbiotin, and tetranorbiotin from catabolism of biotin by *Pseudomonas* sp. Methods Enzymol 18A:404–409, 1970.

100. C Lindblow-Kull, FJ Kull, A Schrift. Evidence for the biosynthesis of selenobiotin. Biochem Biophys Res Commun 93:572–576, 1980.

101. VM Svetlaeva, DS Danilova, MT Yanotovskii. Determination of biotin in the presence of dethiobiotin by a thin-layer chromatographic method. Khim Farm Zh 12: 140–142, 1978. Chem Abstr 90:12365u, 1979.

102. RT Nuttall, B Bush. The detection of ten components of a multivitamin preparation by chromatographic methods. Analyst 96:875–878, 1971.

103. H Thieleman. Trennung nachweisgrenzen einiger vitamine an fertigfolien uv 254 die dunnschichtchromatographie. Sci Pharm 42:221–227, 1974.

104. DB McCormick, JA Roth. Specificity, stereochemistry, and mechanism of the color reaction between p-dimethylaminocinnamaldehyde and biotin analogs. Anal Biochem 34:226–236, 1970.

105. K Groningsson, L Jansson. TLC determination of biotin in a lyophilized multivitamin preparation. J Pharm Sci 68:364–366, 1979.

106. DM Bowers-Komro, JL Chastain, DB McCormick. Separation of biotin and analogs by HPLC. Methods Enzymol 122:63–67, 1986.

107. OM Birch, M Fuhrmann, NM Shaw. Biotin synthase from *Escherichia coli*, an investigation of the low molecular weight and protein components required for activity in vitro. J Biol Chem 270:19158–19165, 1995.

108. K Kamata, T Hagiwara, M Takahashi, S Uehara, K Nakayama, K Akiyama. Determination of biotin in multivitamin pharmaceutical preparations by high-performance liquid chromatography with electrochemical detection. J Chromatogr 356:326–330, 1986.

109. PL Desbene, S Coustal, F Frappier. Separation of biotin and its analogs by high-performance liquid chromatography: convenient labeling for ultraviolet or fluorimetric detection. Anal Biochem 128:359–362, 1983.

110. K Hayakawa, J Oizumi. Determination of free biotin in plasma by liquid chromatography with fluorimetric detection. J Chromatogr 413:247–250, 1987.

111. T Yoshida, A Uetake, C Nakai. Liquid chromatographic determination of biotin by using 1-pyrenyldiazomethane as a pre-column fluorescent labelling reagent. J Chromatogr 456:421–426, 1988.

112. J Stein, A Hahn, B Lembcke, G Rehner. High-performance liquid chromatographic determination of biotin in biological materials after crown ether-catalyzed fluorescence derivatization with panacyl bromide. Anal Biochem 200:89–94, 1992.

113. GI Rehner, J Stein. High-performance liquid chromatographic determination of biotin in biological materials after crown ether-catalyzed fluorescence derivatization with panacyl bromide. Methods Enzymol 279:286–295, 1997.

114. J Zempleni, DB McCormick, DM Mock. Identification of biotin sulfone, bisnorbiotin methyl ketone, and tetranorbiotin-1-sulfoxide in human urine. Am J Clin Nutr 65:508–511, 1997.

115. DM Mock, NI Mock, SL Stratton. Concentrations of biotin metabolites in human milk. J Pediatr 131:456–458, 1997.

116. NG Hentz, LG Bachas. Fluorophore-linked assays for high-performance liquid chromatography postcolumn reaction detection of biotin and biocytin. Methods Enzymol 279:275–286, 1997.

117. V Viswanathan, FP Mahn, VS Venturella, BZ Senkowski. Gas-liquid chromatography of biotin. J Pharm Sci 59:400–402, 1970.

118. RN Johnson, GR Boyden. Characterization of biotin trimethylsilyl derivative. J Pharm Sci 66:1212–1213, 1977.

119. H Janecke, H Voege. On the gas-chromatographic determination of silylated biotin,

its oxidation products and other silylated vitamins. Z Anal Chem 254:355–359, 1971.

120. M Azoulay, PL Desbene, F Frappier, Y Georges. Use of liquid chromatography-mass spectrometry for the quantitation of dethiobiotin and biotin in biological samples. J Chromatogr 303:272–276, 1984.

121. F Frappier, M Jouany, A Marquet, A Olesker, JC Tabet. On the mechanism of the conversion of dethiobiotin to biotin in *E. coli*. Studies with deuterated precursors using tandem mass spectroscopic (MS-MS) techniques. J Org Chem 47:2257–2261, 1982.

122. J Schiewe, S Gobel, M Schwarz, R Neubert. Application of capillary zone electrophoresis for analyzing biotin in pharmaceutical formulations—a comparative study. J Pharm Biomed Anal 14:435–439, 1996.

12
Cobalamins

Jan Lindemans
University Hospital Rotterdam, Rotterdam, The Netherlands

I. INTRODUCTION

A. History

In 1926 Whipple et al. (1) described the successful treatment of dogs with an experimental form of anemia by feeding them with large amounts of raw liver. In the same year Minot and Murphy (2) showed that this treatment was equally effective in patients with pernicious anemia. However, it took another 20 years of research to isolate and identify the active substance: vitamin B_{12}, or cyanocobalamin. The red-colored material was isolated from liver in crystalline form almost simultaneously by Folkers and coworkers (3) in the United States and by Smith in Great Britain (4). Its corrinoid structure was defined by chemical analysis (5) and by x-ray crystallography (6) in 1956. Once it was recognized that the biologically active forms of vitamin B_{12} were highly photosensitive, Barker et al. (7–9) were able to isolate coenzyme B_{12} (adenosylcobalamin) from *Clostridium tetanomorphum* and from liver. A second biologically active form of vitamin B_{12} was detected in extracts from human plasma by Lindstrand and Ståhlberg and proved to be identical with already synthetically prepared methylcobalamin (10,11). From the very beginning, chromatographic techniques have enabled researchers to separate and isolate specific cobalamin compounds from natural substances and preparative mixtures. These techniques keep on contributing to the expansion of our detailed knowledge of cobalamin biosynthesis in microorganisms and the reaction mechanisms in cobalamin-dependent metabolic pathways.

It is the aim of this chapter to summarize established and modern techniques for chromatographic separation of cobalamins and to determine their value in present biochemical, pharmaceutical, and clinical investigations.

B. Chemical Properties

1. Structure

Vitamin B_{12} or cyanocobalamin belongs to the corrinoids. This is a group of compounds having in common a corrin nucleus, that is, a partially hydrogenated tetrapyrrole of which two pyrroles are joined directly rather than through methene bridges, and with a central cobalt atom bound by coordinate linkages to the nitrogen atoms of the four pyrroles. Vitamin B_{12} is further characterized by a specific number of methyl, propionamide, and acetamide side chains attached to the pyrroles, and one side chain on ring D, in which the propionic acid is amidated with 1-amino-2-propanol. The latter, in its turn, is esterified with α-D-ribofuranosyl-(5,6-dimethylbenzimidazole)-3′-phosphate. The second N atom of the 5,6-dimethylbenzimidazole forms the fifth coordinate linkage with the cobalt atom in the α-position (Fig. 1). The β position in the naturally occurring cobalamins is occupied by a CN- (cyanocobalamin), OH- (hydroxycobalamin), H_2O (aquoco-

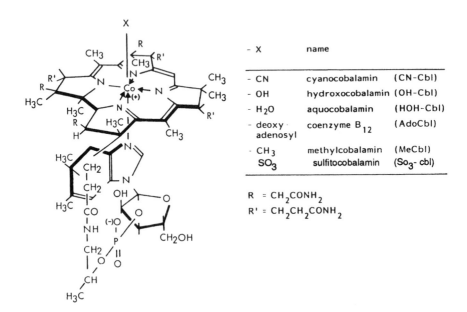

- X	name	
- CN	cyanocobalamin	(CN-Cbl)
- OH	hydroxocobalamin	(OH-Cbl)
- H_2O	aquocobalamin	(HOH-Cbl)
- deoxy adenosyl	coenzyme B_{12}	(AdoCbl)
- CH_3	methylcobalamin	(MeCbl)
SO_3	sulfitocobalamin	(SO_3- cbl)

R = CH_2CONH_2
R′ = $CH_2CH_2CONH_2$

Figure 1 Structural formula of cobalamin.

balamin), CH_3 (methylcobalamin), or desoxyadenosyl group (adenosylcobalamin or coenzyme B_{12}). Other members of the corrinoid group, described in the section on biosynthesis, are depicted in Figure 2. For an extensive treatise on corrinoid nomenclature the reader is referred to the IUPAC-IUB reports on this subject (12,13).

2. Stability

At room temperature cyanocobalamin is fairly soluble in water (12 g/L), lower alcohols, and phenol but almost insoluble in acetone, ether, and chloroform. The molecular weight is 1355.4 ($C_{63}H_{88}O_{14}N_{14}PCo$) and the molecule is neutral in water. Between pH 4 and 7 cyanocobalamin is stable in aqueous solution and can be heated at 120°C without significant loss.

Alkaline hydrolysis (0.1 M) at 100°C induces the formation of "dehydrovitamin B_{12}," a biologically inactive form with a lactam ring fused to ring B of the cobamide nucleus (14). Cyanide can be split off on exposure to light with the production of hydroxocobalamin. This process is stimulated by the addition of acid (15). Hydroxocobalamin occurs in neutral or acid solution as aquocobalamin. Hydroxocobalamin can be converted to cyano-, sulfito-, chloro-, cyanato-, nitrito-, bromo-, thiocyanato-, and azidocobalamin by the addition of the respective reactants under the appropriate conditions (16). At alkaline pH and excess of cyanide, the stable dicyanocobalamin is formed. Mild acid hydrolysis of corrinoids causes the loss of the peripheral amides from the propionamide side chains forming mono-, di-, and tricarboxylic acids (17). The most vulnerable is the propionamide in the e position. Deamidation of the more stable acetamide side chains requires more drastic conditions, in which the nucleotide moiety is also released together with the isopropylamine group at position f. The susceptibility to acid hydrolysis is influenced by the ligands in α and β position. Cyanocobinamide, 1-

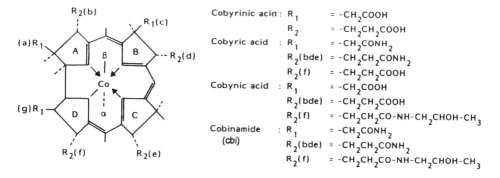

Figure 2 Structural formulas of cobalamin precursors.

α-D-ribofuranosyl-5,6-dimethylbenzimidazole, and phosphate are formed during alkaline hydrolysis of cyanocobalamin in a suspension of cerous hydroxyde, pH 8–9, at 100°C for 2 hr (18).

In cyano- and hydroxocobalamin the cobalt atom is in the 3^+ state. Reduction to the Co^{2+}, and even the Co^{1+}, state can be brought about by electrochemical and purely chemical means (19). The color then changes from red (Co^{3+}) to brown (Co^{2+}) and gray-green (Co^{1+}). Co^{2+}-cobalamin is also formed when aquocobalamin is treated with monothiols (20,21).

The naturally occurring organocorrinoids, AdoCbl and MeCbl, share extreme photosensitivity with most other artificially synthesized organocorrinoids. This makes it necessary to perform all analytical work with these compounds in the dark or under dim red light.

In the absence of oxygen the photodecomposition of adenosylcobalamin leads to the formation of Co^{3+}-cobalamin (22) and a 5′-deoxyadenosyl that cyclizes to 8,5-cyclic-adenosine (23). In the presence of oxygen, aquocobalamin and adenosine-5′-carboxaldehyde are formed (24). Photolysis of methylcobalamin occurs very rapidly in aqueous solution with formation of formaldehyde and aquocobalamin as the major products. In the absence of oxygen the reaction is rather slow and gives rise to the formation of Co^{2+}-cobalamin and methane (25,26). Remarkably, photolysis of methylcobalamin in the presence of homocysteine yields methionine, a methylation reaction that under aerobic, intracellular conditions occurs only in an enzyme-catalyzed reaction with reductive activity (27).

The carbon-cobalt bond in methyl- and adenosylcobalamin is stable in neutral aqueous solution in the dark and withstands even heating for 20 min at 100°C (28). By heating in 0.1 M HCl the carbon-cobalt bond of adenosylcobalamin is cleaved; the alkylcobalamins are relatively stable in dilute acid or alkali. Cyanide forms another threat to the adenosyl-cobalt bond, in particular at alkaline pH and at elevated temperature, by inducing dicyanocobalamin. The alkylcorrinoids are more resistant to the effect of cyanide, but are spliced through electrophilic attacks on the methyl group, for instance by ionic mercury or thallium, into aquocobalamin and the methylmercury cation, or the methylthallic dication (29).

Treatment of Co^{3+}-corrinoids with sodium bisulfite yields sulfitocorrinoids, which behave very similarly to the alkylcorrinoids with respect to photodecomposition and reactions with various compounds (30,31).

3. Spectroscopy

All corrinoids have distinct colors, varying from yellow, red, and purple to brown-green and blue (28), which can be utilized advantageously to identify the various compounds. The most reliable standard for the determination of the extinction coefficient of a corrinoid is the γ band of the compounds dicyanocobalamin or dicyanocobinamide, which are readily formed from all other forms of cobalamin and cobinamide by cyanide, alkaline pH, and light. The molar extinction coefficient of the sharp γ band at 367 to 368 nm amounts to $30.8 \times 10^3\ \mathrm{M^{-1}}$ $\mathrm{cm^{-1}}$ (28,32).

The absorption spectra of some representative corrinoids are presented in Figure 3.

C. Biochemistry

1. Biosynthesis of Cobalamin

The biosynthesis of cobalamin (Fig. 4) is restricted to specific microorganisms and most of our knowledge on the biosynthetic pathways stems from studies with cultures of *Streptomyces griseus, Propionibacterium shermanii*, and *Clostridium tetanomorphum*. The process has been extensively reviewed by Friedman (33,34) and Battersby and McDonald (35). The first steps in cobalamin biosynthesis fol-

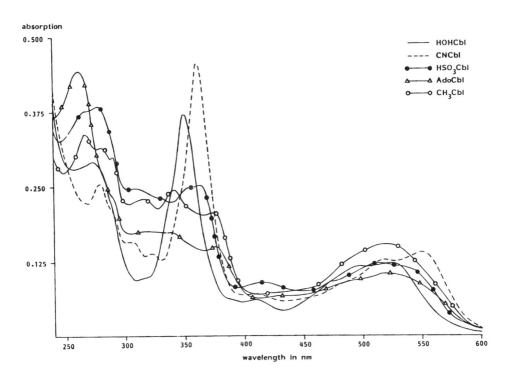

Figure 3 Absorption spectra of the naturally occurring cobalamins.

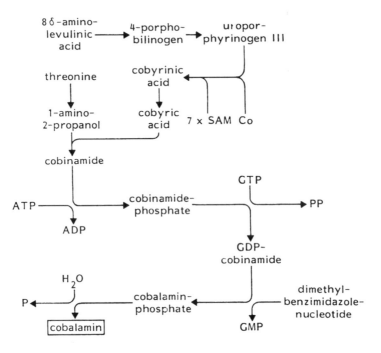

Figure 4 Biosynthesis of cobalamin.

low those of porphyrin biosynthesis. At the stage of uroporphyrinogen III, seven methyl groups, derived from S-adenosylmethionine, are attached and a hydrolytic ring contraction takes place by which the methene bridge between rings A and D is expelled. Subsequently, the cobalt atom is inserted and several side chain modifications take place. The acetate side chain of ring C is decarboxylated and becomes a methyl group. At this stage the intermediate is called cobyrinic acid. By amidation of the remaining carboxylic groups, with exception of the propionic acid side chain of ring D, cobyric acid is formed, which is converted to cobinamide by the attachment of 1-amino-2-propanol to the last propionic acid side chain. Cobinamide phosphate is formed at the expense of ATP through a cobinamide kinase activity.

The next step in the biosynthesis is the formation of an anhydride linkage between the N-β-glycosidic ribonucleoside-5'-phosphate of GTP and the phosphate of cobinamide phosphate to form GDP-cobinamide. The GMP moiety is subsequently exchanged for 5,6-dimethylbenzimidazole-N-α-ribose-5'-phosphate. The latter compound, with its rather unique N-α-glycosidic bond, is synthesized by the action of trans-N-glycosidase or nicotinate nucleotide phosphoribosyltransferase (E.C. 2.4.2.21). This enzyme has very little base specificity, and it is possible to induce the formation of different corrinoids by feeding the growing bacteria all kinds of different bases. The final product of the base exchange reaction, cobalamin phosphate, is ultimately dephosphorylated to cobalamin, preferentially in the "base-off" condition.

2. Cobalamin-Dependent Reactions

The natural, metabolically active forms of vitamin B_{12} are adenosylcobalamin (AdoCbl) and methylcobalamin(MeCbl), both of which can be formed from hydroxocobalamin. The conversion of hydroxo- and cyanocobalamin into the active coenzyme forms has mainly been studied in bacterial systems (36,37), but there is evidence that it is representative for the mammalian system as well (15,38,39). The first step in the conversion of hydroxocobalamin is the two-step reduction of the Co^{3+} atom to the Co^{1+} state by two different but cooperative reductases at the expense of NADH. After this reduction an adenosylating enzyme catalyzes the transfer of the adenosyl moiety from ATP to the Co^{1+}-cobalamin, yielding deoxyadenosylcobalamin and tripolyphosphate. The formation of methylcobalamin probably occurs in the process of the complex methyltransferase reaction catalyzed by N^5-methyltetrahydrofolate homocysteine methyltransferase (E.C. 2.1.1.13) (40), which is described later. Adenosylcobalamin-dependent reactions always involve the exchange of a hydrogen atom of one carbon atom with another group (R) from an adjacent carbon atom. The only exception is the AdoCl-dependent ribonucleotide reductase reaction in bacteria and the eukaryotic flagellate *Euglena gracilis*. An overview of AdoCl-dependent reactions is presented in Table 1 (41).

Of these only the methylmalonyl-CoA mutase (E.C. 5.4.99.2) reaction takes place in mammalian metabolism and forms the link between the metabolism of odd-chain fatty acids, cholesterol, isoleucine, valine, threonine, methionine, and propionic acid on the one hand, and the tricarboxylic acid cycle through succinyl-CoA on the other (Fig. 5). The reaction is of special importance for ruminants in which propionate, arising from cellulose degradation, forms a major source of energy. The enzyme involved is located in the mitochondria and is composed of two nonidentical subunits, one of which binds adenosylcobalamin. A deficiency of adenosylcobalamin or the apoenzyme results in the accumulation of methylmalonic acid, which is secreted by the kidneys and gives rise to methylmalonic aciduria (42).

Methylcobalamin-dependent reactions, involved in the synthesis of methionine in animals and micro-organisms, and in the formation of acetate and methane in bacteria, have been reviewed by Poston and Stadtman (43) and more recently by Taylor (40). As stated before, methylcobalamin is formed from reduced Co^{2+}-cobalamin in the course of the methyltetrahydrofolate homocysteine methyltransferase reaction (Fig. 6). Presumably, the Co^{2+}-cobalamin is bound by the apoen-

Table 1 Adenosylcobalamin-Dependent Reactions

Enzyme	Substrate	Transferred group (R)	Product
Glutamate mutase	L-Glutamate	Glycine	L-*threo*-β-Methyl aspartate
Methylmalonyl CoA mutase	L-Methylmalonyl CoA	—COSCoA	Succinyl CoA
α-Methyleneglutarate mutase	α-Methyleneglutarate	—C(=CH$_2$)COOH	Methylitaconate
Dioldehydrase	Ethylene glycol	OH	Acetaldehyde
	1,2 Propanediol		Propionaldehyde
Glycerol dehydrase	Glycerol	OH	β-Hydroxypropionaldehyde
Ethanolamine ammonia lyase	Ethanolamine	NH$_2$	Acetaldehyde + NH$_3$
β-Lysine mutase	L-β-Lysine	NH$_2$	L-Erythro-3,5-diaminohexanoic acid
α-Lysine mutase	D-Lysine	NH$_2$	D-2,5-Diaminohexanoic acid
α-Leucine 2,3-aminomutase	α-Leucine	NH$_2$	β-Leucine
Ornithine mutase	D-Ornithine	NH$_2$	D-*threo*-2,4-Diaminovaleric acid
Ribonucleotide reductase	Ribonucleotide	—	2'-Deoxyribonucleotide

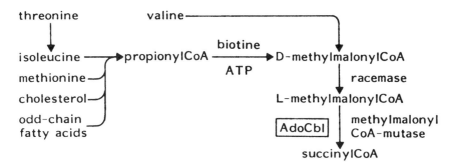

Figure 5   Propionic acid pathway.

zyme, further reduced to Co^{1+}-cobalamin in a $FADH_2$-requiring process, and methylated by 5'-methyltetrahydrofolate in the presence of catalytic amounts of S-adenosylmethionine. Further transfer of methyl groups from methyltetrahydrofolate to homocysteine occurs independently of S-adenosylmethionine. Through this reaction, cobalamin is involved in the regeneration of tetrahydrofolate, an indispensable folate intermediate in the de novo biosynthesis of purines and thymidilate, and in the regeneration of methionine.

Various bacteria from sewage sludge use methanol and acetate as major sources of energy. The methyl group is first transferred to cobalamin and then released as methane as a result of reductive cleavage (43). In *Clostridium thermoaceticum* acetate is synthesized from CO_2 in a folate- and MeCbl-dependent pathway. CO_2 is first reduced to formate which is used to form 10-formyltetrahydrofolate. After further reduction to 5-methyltetrahydrofolate, the methyl group is

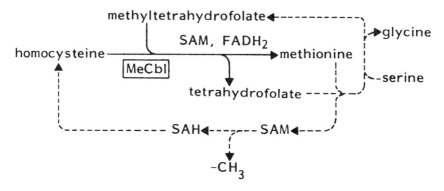

Figure 6   MeCbl-dependent regeneration of methionine and tetrahydrofolate.

transferred to cobalamin thus forming methylcobalamin. Subsequently, two molecules of acetate arise from a reaction between methylcobalamin and pyruvate, with carboxylmethyl-Cbl as a likely intermediate (44).

D. Physiology

The ultimate source of cobalamin for mammalian metabolism is microbial synthesis. The cobalamins from the microbial flora in the gastrointestinal system of herbivorous animals are absorbed by the host and stored by their tissues. From there cobalamin is passed on to other animals in the food chain. A normal daily adult human diet contains about 5 μg of cobalamin, which is about three times the minimum daily requirement to maintain cobalamin homeostasis. The total amount of cobalamin in the human body is ~3 mg.

The hydrophilic cobalamin molecule has to be transported from the intestine to the blood by an elaborate transfer system. Cobalamin is first released from binding substances in the food by peptic activity at low pH in the stomach and becomes bound to so-called R-binders present in saliva and gastric juice. In the ileum pancreatic proteolytic enzymes dissociate the cobalamins from the R-binders and in this way facilitate their binding to Intrinsic Factor, a glycoprotein that in man is synthesized and released by the gastric parietal cells. The cobalamin–Intrinsic Factor complex becomes attached to specific receptors on the mucosal surface of the distal part of the ileum (45). Through an incompletely defined mechanism cobalamin enters the mucosal cell and is passed on to the plasma transport protein transcobalamin II (46). This 38,000-MW polypeptide carries cobalamin through the portal circulation first to the liver and distributes it from there to the other tissues.

Transcobalamin II–bound cobalamin is taken up by those tissues through receptor-mediated endocytosis in which the protein moiety is degraded in the lysosomes (47). The daily turnover of transcobalamin II–bound cobalamin is 5 to 10 μg/day. The transport of cobalamin from peripheral tissues to the central storage organ, the liver, is taken care of by R-type cobalamin-binding proteins, which are a group of glycoproteins occurring in almost all body fluids and produced mainly by granulocytes.

The most frequent cause of cobalamin deficiency is insufficient absorption due to a lack of Intrinsic Factor. This condition is more generally known as pernicious anemia and is caused by the occurrence of autoantibodies against parietal cells and Intrinsic Factor, blocking its capacity to bind cobalamin and preventing its absorption. Less frequent causes are abnormal intestinal flora, partial or total gastrectomy, tropical sprue, fish tapeworm infestation, and the congenital Intrinsic Factor abnormality and Intrinsic Factor receptor dysfunction (Imerslund-Gräsbeck disease).

E. Pharmacology of Cobalamin

Pure nutritional cobalamin deficiency being extremely rare, most other causes of cobalamin deficiency require lifelong treatment with parenteral, mainly intramuscular, administration of cobalamin. Both cyano- and hydroxocobalamin are being used for this purpose. Unbound cobalamin in the circulation cannot be retained by the kidney, and is excreted. As hydroxocobalamin binds more strongly to plasma proteins other than the specific cobalamin binders, it is better retained by the body and thus more effective. After replenishment of the cobalamin stores by about five doses of at least 250 µg hydroxocobalamin on alternate days, cobalamin homeostasis can be maintained by bimonthly injections of 1000 µg cobalamin.

A particular application of hydroxocobalamin is its use in the treatment of cyanide-poisoning. Large and frequent doses can be given without significant side effects to scavenge the cyanide from the circulation. Allergic and anaphylactic reactions to cobalamin are extremely rare. Several reports have appeared on patients having developed circulating IgG antibodies against cobalamin or against the cobalamin–transcobalamin II complex (48,49). This leads to an accumulation of cobalamin in the circulation but seems of no clinical consequence.

Oral treatment of cobalamin deficiency due to malabsorption can only be successful when a dose of at least 500 to 1000 µg is given allowing the passive absorption of about 5 to 10 µg/day.

Specific treatment schedules are devised for patients with one of the various forms of congenital metabolic disturbance of cobalamin metabolism. In general these conditions require daily injections, sometimes in combination with folic acid, carnitine, betaine, or choline (50).

II. THIN-LAYER CHROMATOGRAPHY

A. Extraction Procedures

In living organisms cobalamin is for the greater part bound with high affinity to the enzymes for which it functions as cofactor. Chromatographic analysis has therefore to be preceded by dissociation and extraction from these complexes. An overview of established extraction methods has been given by Pawelkiewicz (51) and Friedrich (52). In summary, cell material is homogenized in either dilute buffer, 1% acetic acid, or 75%, 80%, or 90% ethanol and heated for 20 to 30 min at 80 or 100°C. The extraction efficiency improves by adding 0.1% cyanide, but then only cyano forms of the different corrinoids are recovered. For extraction of organocorrinoids, such as AdoCbl and MeCbl, the whole procedure must be carried out under dim red illumination to avoid photodecomposition.

After the heating, particulate matter is removed by filtration or centrifugation. Absorption of the corrinoids from the solution can be achieved by activated carbon, Montmorillonite, or Amberlite IRC-50-H cation exchanger. Elution from activated carbon is performed with alcohol-water mixtures; elution from Montmorillonite requires slightly alkaline buffers or a phenol solution. Cobalamin is released from the Amberlite by elution with water after neutralization of the ion exchanger with ammonia or after washing with dilute acid, by elution with 0.1 M HCl in 75% acetone (53) or 0.1% H_2SO_4 in 50% acetone (54). Further purification of the cobalamins is generally performed with phenol extraction (55,56). Similar extraction procedures have been found applicable for the isolation of the rhodium analogs of methylcobalamin and coenzyme B_{12} (57). Ford and Friedman have extracted various forms of cobyric acid, cobinamide, cobinamide phosphate, and GDP-cobinamide from *P. shermanii* using aqueous 75% acetone at room temperature in the dark. Further purification was obtained by phenol extraction (58). The use of phenol has been largely replaced by the nonpolar adsorbent Amberlite XAD or silanized silica gel 60 H (59). Columns are prepared in Pasteur pipettes plugged with glass wool or in 5 to 10 mL syringes filled with a few milliliters of Amberlite XAD in methanol.

The mesh size of the Amberlite must be reduced by crushing in a mortar to obtain a suitable flow rate. The silica gel 60 H is ready for use. The column has to be pretreated by subsequent elution of degased methanol, methanol/water/acetic acid (50:50:5, v/v), and water/acetic acid (100:5). An evaporated ethanol extract is dissolved in a small volume of 1% acetic acid and applied to the adsorbent column. Salts and other polar substances are washed from the column by 1% acetic acid, whereas corrinoids are eluted with degased water/methanol/acetic acid (50:50:5). Recoveries of 90% to 98%, as reported for OH-Cbl, SO_3-Cbl, and AdoCbl, are about 10% higher than achieved by phenol extraction. Elution of cobalamin from Amberlite XAD can also be carried out with alkaline solvents such as methanolic 1% ammonium hydroxide. Under these conditions OH-Cbl is strongly retained by silanized silica gel (60).

Amberlite XAD adsorption has been applied by Koppenhagen et al. (57) for the isolation of rhodium analogs of cobalamin cofactors. Gimsing et al. (61) used it for the desalting of chemically prepared radioactive cobalamin forms, and Stupperich et al. (62) extracted cobamides from various bacterial species with this material.

The reversed-phase cartridge Sep-Pak C18 (Waters) has been found applicable by Jacobson et al. (63) for the desalting of an extract from L1210 lymphoblasts. Elution of cobalamins from this type of column is brought about by 50% acetonitrile.

In all presented studies little attention has been paid to the recovery of individual cobalamin forms in extracts. Frenkel et al. (64) reported a variably decreased recovery of OH-Cbl in comparison with the other cobalamin forms and proposed a tentative correction for the apparent loss of OH-Cbl on the basis of the difference between total tissue cobalamin content and the sum of the collected cobalamin fractions after separation.

In many studies on plasma cobalamin forms, the method initially described by Lindstrand and Ståhlberg in 1963 (10) is still being used. One volume of plasma is poured into four volumes of absolute alcohol and heated for 20 min at 80°C. After removal of precipitate by filtration the alcohol in the filtrate is evaporated under vacuum at 30°C. The residual water phase is extracted three to six times with ether to remove lipid material. In the next step cobalamins are extracted from the water phase into phenol by shaking in four volumes of phenol with 15% water. The phenol phase is shaken with one part of acetone, three parts of ether, and a small amount of water. Phenol in the resulting water phase is further removed with ether. The water phase containing the cobalamin is evaporated to remove residual ether. Linnell et al. (65,66) used essentially the same procedure with smaller volumes of blood plasma.

Unsatisfactory extraction recoveries of cobalamin from plasma were first mentioned by Gimsing et al. (67). By adding cyano(^{57}Co)cobalamin as an internal standard they showed that a loss of 30% to 50% of the original amount of cobalamin occurred during the ethanol and phenol extractions. At first they reported equally low extraction recoveries for all cobalamin forms present in plasma; later a selective underestimation of OH-Cbl, due to inadequate extraction, was reported (68). Similar observations had been published by Mahoney and Rosenberg (69) and by Frenkel et al. (64). Inadequate extraction of one component obviously invalidates the values of the substance concentrations of all other individual forms when calculated on the basis of their relative amounts in a chromatographed extract and the total amount of cobalamin in the original sample.

Gimsing et al. (70) attributed the loss of hydroxo- and sulfitocobalamin during ethanol extraction to binding onto histidine residues of plasma proteins, a phenomenon already described by Bauriedel et al. (71), Taylor and Hanna (72), and Lien and Wood (73). This binding process is optimal at pH 7.3, has specificity for aquocobalamin but not sulfitocobalamin, increases in the presence of 5 M urea, and is competitively inhibited by Cd^{2+} ions. In the histidine-bound form, Co^{3+}-cobalamine should be more resistant to reductive agents than in the free form.

By utilizing the inhibiting effect of cadmium, Gimsing et al. (70) were able to improve the extraction recovery of OH-Cbl as well as sulfito-Cbl to levels comparable to the extraction of CN-, Ado-, and MeCbl, as calculated from recoveries of added radioactive cobalamin forms.

Studies by Van Kapel et al. (74) confirmed the selective loss of OH-Cbl during plasma extraction in hot ethanol and in 1% acetic acid by showing an extraction recovery of 83% to 93% for added CN-, Ado-, and MeCbl, but only 30% to 40% for added OH-Cbl and SO_3-Cbl. In this study silanized silica gel

60H, according to the method of Fenton and Rosenberg (60), was used to adsorb cobalamins from the extraction medium. The extraction efficiency could be improved by selective conversion of OH-Cbl into SO_3-, NO_2-, or N_3-cobalamin by adding $Na_2S_2O_5$, $NaNO_2$, or NaN_3, but this made it impossible to separate the individual cobalamin forms thereafter. In the course of these studies it was noticed that during the heating step in ethanol as well as in dilute acid a color change from red to brown occurred in plasma spiked with about 100 pmol/L OH-Cbl. Since thiols reduce Co^{3+}-corrinoids to brown Co^{2+}-corrinoid-thiol complexes (20,21), it was assumed that this reduction might originate from thiol groups of proteins that had become more accessible by denaturation during heating (75). Protein-thiol in plasma occurs in a concentration of 0.4 to 0.6 mmol/L (76), which is largely in excess to OH-Cbl. Various thiol-blocking agents were therefore investigated with respect to their ability to prevent the loss of OH- and SO_3-Cbl. In this respect chloroacetophenone, β-hydroxomercuribenzoate, ethylacrylate, N-ethylmaleimide, and iodoacetamide indeed improved the OH-Cbl recovery to the level of the other cobalamins, but only ethylacrylate and N-ethylmaleimide did not alter the distribution of the individual cobalamin forms during subsequent chromatography.

The extraction efficiencies of added cobalamins under the influence of N-ethylmaleimide, using silanized silica 60 H for desalting, are presented in Table 2.

Comparing the two proposed mechanisms for OH-Cbl binding to plasma proteins (70,74), various arguments favor the concept of SH binding. Incomplete

Table 2 Influence of N-Ethylmaleimide on the Extraction Efficiency of Exogenously Added Cobalamins

Cobalamin[a]	Control[b]	NEM[b]
OH-Cbl	19.5	77.2
SO_3-Cbl	23.1	73.6
CN-Cbl	78.4	86.8
AdoCbl	78.6	86.8
MeCbl	79.6	82.6

[a] Plasma enriched with about 800 pmol/L of each individual cobalamin.

[b] Extraction with four volumes of 1% acetic acid (+12.5 mmol/L N-ethylmaleimide) followed by desalting on silanized silica gel 60 H. Recovery is expressed as a percentage of the amount added, determined by radioisotope dilution assay.

OH-Cbl extraction has been observed both at pH 4.3 in acetate buffer and at neutral pH during hot ethanol extraction, whereas histidine binding apparently occurs only above pH 5.0 (72).

Moreover, OH-Cbl binding to plasma proteins can be prevented by N-ethyl-maleimide, p-hydroxomercuribenzoate, or iodoacetamide, which are all SH-directed agents. In addition, Cd^{2+} ions not only bind to histidine in proteins, but can also block SH groups (77). This observation has been confirmed in our laboratory in 1% acetic acid and hot ethanol extracts of human plasma using protein-SH group detection with 5,5'-dithio-bis-(2-nitrobenzoic acid, DTNB) according to Sedlak and Lindsay (78).

B. Paper Chromatography

Paper chromatography is one of the earlier techniques for the separation of cobalamins and corrinoids. Whatman nr.2 paper with *sec*-butanol/glacial acetic acid/water (100:3:50) (10) or *n*-butanol/isopropanol/water (10:7:10) (11) is most frequently used. Whatman 3MM paper has been found useful for the separation of acid hydrolysis products of cyanocobalamin using 2-butanol/glacial acetic acid/water (440:4.1:150) with 0.07% KCN as the ascending solvent (79). A similar system was applied by Kolhouse and Allen (80) for the preparation of various cobalamin analogs and by Kondo et al (81) for the analytical separation of analogs from animal tissues.

C. Thin-Layer Chromatography

The separation of a large number of different organocobalamins on cellulose thin-layer chromatography (TLC) with four different solvent systems has been described by Firth et al. (82) and is presented in Table 3. These systems have been found applicable also for the analysis of fluoroalkylcobalamins (56) and formylmethyl derivatives of cobalamin (83). Two-dimensional TLC separation of cobalamins from blood plasma has been introduced by Linnell et al. (65,84). The cobalamins in the chromatogram were made visible by a bioautography technique. This technique was first described by Lindstrand and Stålberg (10). They placed a paper chromatogram on an agar dish which was inoculated with the cobalamin-dependent *E. coli* stem 113-3 (ATCC 10586). After 16 hours at 37°C the chromatograms were removed and the dishes were further incubated. The presence of bacterial growth indicated the location of cobalamin subfractions. Linnell et al. (65) improved the sensitivity by incorporating the growth indicator 2,3,5-triphenyltetrazoliumchloride in the medium. The color intensity of the growth spots could be quantified by light reflection measurement (85). The growth response of the bacteria was found to be equal for OH-Cbl, CN-Cbl, and AdoCbl, but a little lower for MeCbl. Gimsing et al. (67,70) estimated the growth

Table 3 Thin-Layer Chromatography of Vitamin B$_{12}$ Derivatives on Cellulose[a]

Axial ligands	Solvent I	Solvent II	Solvent III	Solvent IV
		Cobalamins		
H$_2$O	0.15–0.75[b]	0.25–0.75[b]	0.30	0.55–0.75[b]
N≡C—	1.00[(0.25)]	1.00[(0.25)]	1.00[(0.31)]	1.00[(0.37)]
CH≡C—	1.50	1.45	1.40	1.15
CH$_2$=	1.75	1.85	1.60	1.20
CHBrCH=	1.80	1.85	1.60	1.20
CH$_2$OHCH$_2$	0.90	1.00	0.90	1.05
CF$_3$CH$_2^-$	1.95	2.10	1.60	1.20
—SO$_3$CH$_2^-$	0.25	0.30	0.25	0.55
HOOCCH$_2$	1.10	1.15	0.65[d]	1.10
CH$_3^-$	1.80	1.80	1.50	1.20
CH$_3$CH$_2^-$	1.90	1.80	1.60	1.20
CH$_3$CH$_2$CH$_2^-$	2.05	2.15	1.70	1.20
C5′-deoxyadenosyl	0.65	0.70	0.65	0.95
		Cobinamides		
H$_2$O	0.20–0.50[b]	0.35–0.50[b]	0.20[b]	0.80–1.00[b]
N≡C—	0.60, 110[c]	0.80, 1.20[c]	0.50, 0.80[c]	0.95, 1.05[c]
CH$_2$=CH—	1.45	1.65	1.35	1.15
CH$_3^-$	1.35	1.55	1.25	1.15
CH$_3$CH$_2^-$	1.40	1.55	1.30	1.15
CH$_3$CH$_2$CH$_2$	1.60	1.75	1.50	1.20
SO$_3^{2-}$	0.75	0.74	0.59	0.83

[a] Solvent I, sec-butanol/water, 9.5:4. Solvent II, sec-butanol/glacial acetic acid/water, 100:1:50. Solvent III, sec-butanol/0.88 ammonia/water, 9.5:0.675:4. Solvent IV, n-butanol/glacial acetic acid/water, 4:1:5. Retention values relative to the R$_f$ value of cyanocobalamin (in parenthesis). From Ref. 82.
[b] Elongated spots due to equilibration between aquo and hydroxo form on the plate.
[c] Double spots due to the existence of stereoisomers involving the axial ligands.
[d] Ionized carboxylic acid.

response by fotometric measurement of the indicator in an ethanol extract of the culture plate and have found that in the range between 10 and 80 fmol an almost identical and linear response could be achieved for OH-, CN-, Ado-, and MeCbl if the chromatograms were exposed to light and a bisulfite-soaked filter paper was placed between culture plate and chromatogram.

The ascending solvent for silica gel TLC was butanol/ammonia/water (75:2:25) in the first dimension and water-saturated benzyl alcohol in the second. A representative separation is shown in Figure 7. A one-dimensional separation on

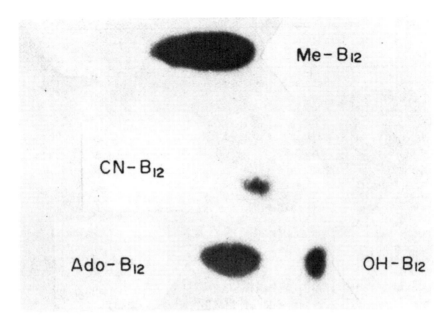

Figure 7 Two-dimensional chromatography and bioautography of plasma cobalamins form a normal subject, the origin is marked at the left-hand corner. (From Ref. 85.)

silica gel TLC with *sec*-butanol/isopropanol/water/ammonia (30:45:25:2) was developed by Nexo and Anderson (86). In this solvent R_f values for the different cobalamins were as follows: MeCbl, 0.40; AdoCbl, 0.23; CN-Cbl, 0.35; OH-Cbl, 0.00; sulfito-Cbl, 0.58; (Ado)CN-Cba, 0.22; and $(CN)_2Cbi$, 0.03.

Fenton and Rosenberg (60) introduced the principle of nonpolar adsorption chromatography to the field of TLC separation of cobalamins. The R_f values for various cobalamins on silanized silica gel TLC with different solvents are presented in Table 4. It shows that complete separation of all cobalamins found in human plasma, and in particular Ado- and CN-Cbl, still requires two-dimensional chromatography.

1. Cobalamins in Tissues

Linnell et al. (85) and Dillon et al. (87) have analyzed the cobalamin distribution in the tissues of normal humans, pernicious anemia patients, and patients with various congenital disorders of cobalamin metabolism, using TLC with bioautographic detection. The results for normal tissues, summarized in Table 5, show that AdoCbl is the most abundant cobalamin form, the concentration of MeCbl is quite variable, and CN-Cbl is hardly present except in blood cells. In pernicious

Table 4 R_f Values for Cobalamins on Reversed- and Normal-Phase TLC

	R_f				
	OH-Cbl	MeCbl	AdoCbl	SO$_3$Cbl	CN-Cbl
Reversed phase[a]					
1:60:140	0.10	0.16	0.40	0.71	0.42
1:80:120	0.11	0.40	0.63	0.83	0.66
1:100:100	0.25–0.45	0.64	0.80	0.89	0.80
5:30:65	0.37	0.21	0.19	0.65	0.46
10:30:60	0.46	0.35	0.40	0.63	0.56
Normal phase[b]	0.02	0.52	0.27	0.61	0.45

[a] Prepoured silanized silica gel plates developed for 1–2 h with acetic acid/methanol/water in the indicated proportions.
[b] Prepoured silica gel plates (IB 2) developed for 10 h with ammonium hydroxide/2-propanol'2-butanol/water (1:50:50:50).
Source: Ref. 60.

anemia the distribution of cobalamins remains quite normal in liver and kidney, but in leukocytes and bone marrow a remarkable rise in CN-Cbl and a decrease in Ado- and MeCbl is found. Similar distribution patterns were found in various animal species, although the relative amount of MeCbl is generally lower than in humans (88). In 50% of rat tissues and 60% of guinea pig tissues unidentified corrinoids were detected. In erythrocytes of rats, cats, and guinea pigs the levels of OH- and MeCbl were higher than in plasma, in contrast to human red cells. Cobalamin deprivation led to lower cobalamin levels in the tissues of rats without changing the cobalamin distribution (68). Choline-deficient food caused a relative

Table 5 Tissue Cobalamins in Normal Human Subjects

Tissue	Total cobalamin (nmol/g)	OH-Cbl (%)	CN-Cbl (%)	AdoCbl (%)	MeCbl (%)
Liver	615 ± 122	44	2.0	45	9.0
Kidney	170 ± 30	28	0.6	53	18
Spleen	24 ± 3	24	0.8	50	26
Brain	38 ± 7	24	0.5	63	13
Erythrocytes	0.15 ± 0.02	26	6	53	15
Leukocytes	2.90 ± 0.32	28	4	48	20
Bone marrow	9.6 ± 1.1	30	2	55	13

Source: Refs. 85 and 87.

decrease of MeCbl in the liver, whereas a methionine-deficient diet exerted the opposite effect (89).

2. Cobalamins in Plasma

The concentrations of the different cobalamin forms found with TLC/bioautography in human plasma are presented in Table 6. The difference in the results of the two quoted papers may be the result of the differences in extraction techniques and/or the difference in quantitation of the growth response. Gimsing, recently, reported on the distribution of cobalamins in plasma and myeloid cells from patients with chronic myeloid leukemia on the basis of his methodology (90).

D. Nonpolar Adsorption Chromatography

As already mentioned in the section on corrinoid extraction, the nonpolar adsorbent Amberlite XAD-2 has proven valuable for batch separation of cobalamins from polar materials (60). This technique is based on the work of Vogelmann and Wagner (91), who collected a vast amount of data on the chromatographic behavior of 50 different corrinoids, including rhodibamides and hydrogenobamides, and on the influence of C8, C10, and C13 substitutions, the axial ligands, and the nucleotide on retention times. Elution was generally performed with 3 to 20 vol% *tert*-butanol at various pH and cyanide concentrations.

The method was used successfully in the isolation of descobaltocobalamin from a crude extract of Chromatium by Koppenhagen et al. (92). An aqueous ethanol/acetic acid extract from 500 g wet cells was applied on an Amberlite XAD-2 column (4.0 × 7.0 cm) and sequentially eluted with 0%, 2%, 10%, 20%, and 50% *tert*-butanol (v/v) in water. Hydrogenobyric acid eluted at 10%, descobaltocobalamin at 20%, and uroporphyrins at 50% butanol. After insertion of the

Table 6 Distribution of Cobalamins in Human Plasma[a]

| | | Gimsing et al. (61,67)[b] | |
Cobalamin	Linnell et al. (85)[c]	−CdAc	+CdAc
OH− + SO₃Cbl	47 (35–145)	(0–30)	38 (10–104)
CN-Cbl	2 (0–40)	(0–70)	13 (2–48)
AdoCbl	60 (22–185)	(6–90)	20 (2–77)
MeCbl	276 (117–458)	(212–512)	250 (135–427)

[a] Mean concentrations in pmol/L, ranges in parentheses.
[b] Extraction with hot ethanol and XAD-2, TLC, and bioautography.
[c] Extraction with hot ethanol and phenol, TLC, and bioautography.

rhodium atom in the descobaltocobalamin, the reaction mixture was subjected to desalting on an XAD-2 column, followed by further purification on CM- and DEAE-cellulose and once more on a XAD-2 column to obtain the pure mono- and dicyanorhodibalamins, which eluted at 8% and 10% *tert*-butanol respectively.

E. Ion Exchange Chromatography

Ion exchange chromatography has been used mainly in preparative separations of cobalamin forms and their precursors. The use of Amberlite resins and cellulose derivatives has been reviewed by Pawelkiewicz (38). With these materials it was impossible to separate methyl-, cyano-, ado-, and hydroxocobalamin in one step. Tortolani et al. (93) first described a combination of CM-cellulose and Dowex 50W-X2, but later a one-step separation on SP-Sephadex (94). Subsequently, Gams et al. (95,96) and Gimsing and Hippe (97) used this procedure to separate the various coenzyme forms in L1210 cells, mitochondria, and plasma. Begley and Hall (98) measured the conversion of $CN(^{57}Co)Cbl$ into coenzyme forms in cells in tissue culture with this technique. They found that a fifth form of cobalamin, later identified as sulfitocobalamin, almost coeluted with the CN-Cbl peak, but by a small modification of the fraction collection procedure the SO_3-Cbl peak was eluted separately from the CN-Cbl peak (Fig. 8). The method is still in use as is apparent from the publication of Anes et al. (99) on the separation of nitrito- and nitrosocobalamin.

Separation of adenosyl forms of cobyric acid and various cobinamide and corrinoid intermediates in the biosynthesis of cobalamin in *P. shermanii* on

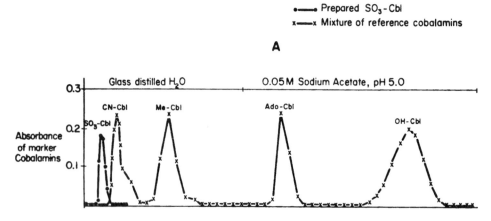

Figure 8 SP-Sephadex C-25 column chromatography of marker cobalamins at approximately 100 μg each. (From Ref. 98.)

Dowex 50W-X8 (200 to 400 mesh) by stepwise increase in pH of the eluting 0.05 M acetate buffer was carried out by Ford and Friedman (59). Ion exchange paper chromatography on Whatman CM-82 paper was applied by Lee (100) for the separation of biosynthetic intermediates of cobyrinic acid, isolated from *P. shermanii*. Ion exchange chromatography seems in particular suitable for the separation of the different carboxylic acid derivatives obtained by mild acid hydrolysis of cyanocobalamin. Each additional hydrolysis of an amide group on the corrin side chain results in longer retention on a QAE-Sephadex column eluted with an acetic acid gradient in 0.4 M pyridine (79). Katada et al. (101) described a novel isomeric form of vitamin B_{12} and its derivatives that could be separated from native cobalamin by DEAE-Sephadex chromatography, on which the isomeric form is retained, unlike the native cobalamin, or by SP-Sephadex chromatography, in which the isomeric form always moves just a little faster than its normal counterpart.

III. HIGH-PRESSURE LIQUID CHROMATOGRAPHY

A. Introduction

An overview of methods and applications of high pressure liquid chromatography is presented in Table 7. They can be divided in methods for separation of multivitamin mixtures, preparative separation of cobalamin coenzymes and analogs, and the analytical separation of biosynthetic intermediates, food cobalamins, and serum and tissue cobalamins.

B. Multivitamin Mixtures

For the separation of vitamin B_{12} from the vitamins B_2, K_3, C_1, D_2, E, and A in a multivitamin preparation, Schmit et al. (102) proposed a gradient elution on a Zibax ODS Permaphase column with a gradient changing from 0% to 100% methanol in water at 3%/min and a flow rate of 0.9 mL/min. The separation of vitamin B_{12} from vitamin B_2 on the one side and vitamin K_3 on the other was not impressive, but as the authors stated, chromatographic conditions can be adjusted to meet the specific requirements of a particular vitamin. Optimal conditions for the separation of a mixture of seven water-soluble vitamins on µBondapak C18 and µBondapak NH_2 have been more thoroughly investigated by Wills et al. (103), as demonstrated in Figures 9 and 10. The best eluent combination for each of the individual vitamins in the preparation can be chosen from these data. Maeda et al. (104) reported the simultaneous determination of nicotinamide, thiamine, riboflavin (phosphate), pyridoxine, caffeine, benzoate, cyanocobalamin, and folic acid using a reversed-phase C18-column eluted with acetonitrile/

Table 7 HPLC Separation of Cobalamins and Related Corrinoids

Column	Column dimensions	Mobile phase	Flow rate (mL/min)	Separated compounds	Reference
Partisil ODS 10 μm	250 × 4.5 mm	Methanol/0.2% ammonia (90:10 v/v)	2.6	OH-Cbl and methyl-/CN-Cbl	105
Lichrosorb RP8 10 μm	250 × 4.6 mm	Acetonitrile/water (30:70 v/v); or 10 min linear gradient from 10 to 25% acetonitrile in 0.083 M phosphoric acid/triethanolamine, pH 3.3	2.0	OH-, CN-, Ado-, and MeCbl	108
Lichrosorb RP8 10 μm	150 × 4.7 mm	6 min linear gradient from 15 to 45% methanol in 0.05% sulfuric acid	1.33	OH-, CN-, Ado-, and MeCbl from other water-soluble vitamins	106
Lichrosorb RP8 10 μm	250 × 4.6 mm	Acetonitrile/1 mM ammonium acetate, pH 4.4 (30:70 v/v); or 10 min linear gradient from 5 to 30% acetonitrile in 0.05 M sodium phosphate, pH 3.0	2.0	OH-, CN-, Ado-, and MeCbl	106
			2.0	Cbl and Cbi analogs	63
Lichrosorb RP8 10 μm	250 × 4.6 mm	17 min linear gradient from 0 to 25% acetonitrile in 0.085 M phosphoric acid/triethanol-amin, pH 3.3	2.0	OH-, CN-, Ado-, and MeCbl from human plasma	74
Lichrosorb Si-60 5 μm	250 × 4.6 mm	Methanol/0.01% sodium cyanide (40:60% v/v)	0.8	diCN-Cbl in human plasma	117
Zibax ODS "Permaphase"	100 × 2.1 mm	33 min linear gradient from 100% water to 100% methanol (50°C)	0.9	Multivitamin preparation (B_2, B_{12}, K_3, C, D_2, E, A)	102

Column	Dimensions	Mobile phase	Flow	Compounds	Ref.
μBondapak C18	300 × 3.9 mm	Methanol/1% KCl (20:80% v/v) Methanol/1.5% tetrabutylammonium phosphate (30:70% v/v)	1.0 1.0	Multivitamin preparation (B$_1$, B$_2$, B$_6$, B$_{12}$, C, folic acid, niacine, niacinamide)	103
μBondapak NH$_2$	300 × 3.9 mm	Methanol/0.125% citrate, pH 2.4 (80:20 v/v)	1.0	Multivitamin preparation (B$_1$, B$_2$, B$_6$, B$_{12}$, C, folic acid niacine, niacinamide)	103
μBondapak C18 Zorbax ODS 6 μm	300 × 3.9 mm 250 × 4.6 mm	10 min concave gradient from 23 to 70% methanol in 0.05 M NaH$_2$PO$_4$ (40°C)	1.8	OH-, CN-, Ado-, MeCbl, cobalamin analogs, SO$_3$Cbl, monobasic acids of CNCbl; alkanolamine analogs, amilide derivative of monobasic acids	107
μBondapak C18	300 × 3.9 mm	H$_2$O/acetic acid/isopropanol (90:1:9 v/v); or 10 min linear gradient from 3 to 21% isopropanol in 0.05 M 1-heptane sulfonic acid	0.3	CN-Cbl and analogs of CN-Cbl and CNCba	109
			1.0	SO$_3^-$, CN-, OH-, Ado-, MeCbl	109
μBondapak NH$_2$	300 × 3.9 mm	58 mM Pyridine acetate pH 4.4/tetrahydrofuran (96:4 v/v)	0.3	Cbl and Cbl monocarboxylic acids	109
μBondapak RP18	300 × 3.9 mm	Methanol/0.1% acetic acid (24:76), or methanol/0.1% acetic acid (5 min 25:75 followed by a gradient of 25 min to 65:35)	1.0	Bacterial cobamides	62

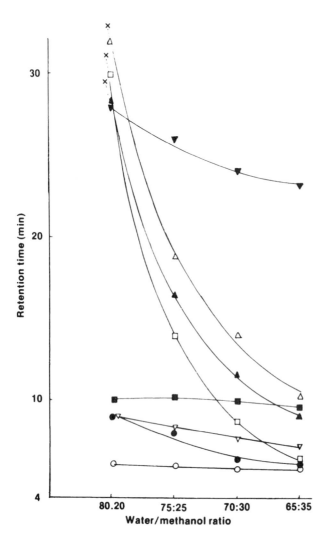

Figure 9 Effect of water/methanol ratio with tetrabutylammonium phosphate reagent on retention times of water-soluble vitamins on μBondapak C18: ascorbic acid (○), niacin (●), folic acid (□), pyridoxin (□), riboflavin (△), vitamin B$_{12}$ (▲), niacinamide (▽), thiamine (▽), x: not eluted from column. (From Ref. 103.)

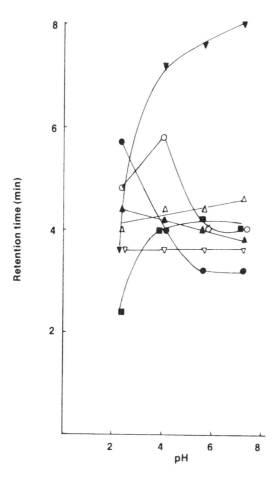

Figure 10   Effect of the pH of 1% citrate/methanol (20:80) on retention times of water soluble vitamins on µBondapak NH2. See Fig. 9 for explanation of symbols. (From Ref. 103.)

0.01 M potassium phosphate/triethylamine (8:91.5:0.5), containing 5 mM sodium octane–sulfonate, adjusted to pH 2.8.

C.   Preparative Separation of Cobalamin Coenzymes and Analogs

Partial separation of OH- from CN- and MeCbl on a Partisil ODS 10 µm column was described by Pellerin et al. (105) using methanol/0.2% ammonia (90:10) as a solvent at a flow rate of 2.6 mL/min. Complete separation of OH-, CN-, Ado-, and MeCbl was obtained by Mourot et al. (106) by the introduction of gradient elution from 15% to 45% methanol in 0.05% sulfuric acid with an increase of 5%/min at 80 mL/h on a Lichrosorb RP8 10-µm column. For the quantitative analysis of cyanocobalamin in veterinary drugs, these authors proposed isocratic elution with 0.05% sulfuric acid/acetonitrile (85:15) at 1 mL/min on a 10-cm Lichrosorb RP8 column. With UV monitoring at 350 nm, a lower detection limit of about 100 ng was reached.

A similar gradient elution for the separation of different cobalamin forms and various synthetic cobalamin analogs was described by Frenkel et al. (63,107). The detection was linear between 100 ng and 50 µg of injected cobalamin. Jacobsen and Green (108) achieved complete separation of the four cobalamin forms by isocratic elution on a C8 reversed-phase column using acetonitrile/water (30:70) at 2 mL/min, but separation was improved by elution with a 10-min gradient from 10% to 25% acetonitrile in 0.083 mol/L phosphoric acid titrated to pH 3.3 with triethanolamine. In a later report, Jacobsen et al. (63) have compared the separation of the naturally occurring cobalamins, coenzyme analogs, aminoalkylcobalamins, and of various cobinamide forms by gradient elution and isocratic elution (Table 8). In both elution systems a 10-µm Lichrosorb C8 column was used. Isocratic elution was carried out with 30% acetonitrile in 1 mM ammonium acetate pH 4.4; the gradient system consisted of 0.05 M phosphoric acid titrated to pH 3.0 with concentrated ammonia and acetonitrile from 5% to 30% or 50% (as required) at a rate of 2.5%/min and a flow rate of 2.0 mL/min (Fig. 11). In general, separation in the isocratic system was satisfactory despite minor disadvantages of small pH-dependent mobility changes of some cobalamin forms (Ado- and H_2O-Cbl) and prolonged retention of some analogs. The elution methods also proved excellently suitable for the analysis of the conversion of radioactive cyanocobalamin into other cobalamin forms after uptake by either *Lactobacillus leichmanii* or L1210 lymphoblasts.

An even larger number of cobalamin analogs were investigated for their retention times on either µBondapak C18 columns or µBondapak NH_2 columns by Binder et al. (109) (Table 9). The elution solvent on C18 columns was H_2O/acetic acid/isopropanol (90:1:9), and the flow rate was 0.3 mL/min. NH_2 columns were eluted with 58 mM pyridine acetate (pH 4.4)/tetrahydrofuran (96:4) at a flow rate of 0.3 mL/min. The latter column was used to separate different monocarboxylic derivatives of cyanocobalamin (Fig. 12). For the separation of the five naturally occurring cobalamins the authors used elution of a C18 column with a 10-min linear gradient from 3% to 21% isopropanol in 5 mM 1-heptane sulfonic acid. Cobalamins were monitored at 254 nm (Fig. 13). Stupperich et al. (62) published a list of retention times for still another series of cobamides from bacterial origin, obtained with a µBondapak RP18 column, 3.9 × 300 mm, eluted with a gradient of methanol in 0.1% acetic acid (5 min 25% methanol, followed

Table 8 Retention Times of Cobalamin Coenzymes and
Related Corrinoid Analogs on a Lichrosorb C8 Column

Corrinoid	Gradient elution (min)	Isocratic elution (min)
Naturally occurring		
AqCbl	9.0	4.6–4.8[a]
HSO$_3$Cbl	9.7	1.1
CN-Cbl	10.2	1.6
AdoCbl	11.4	2.8–3.0[a]
MeCbl	12.8	2.2
Coenzyme analogs		
CN-Cbl	10.9	5.2
TubCbl	11.1	9.6
AraACbl	11.2	3.3
ForCbl	11.1	4.3
ε-AdoCbl	11.7	3.5
PrCbl	14.5	9.1[b]
DapCbl	16.8	27.1[b]
Aminoalkylcobalamins		
AC$_2$Cbl	9.8	5.9
AC$_3$Cbl	10.5	10.4[b]
AC$_5$Cbl	11.2	nd
AC$_8$Cbl	12.8	nd
AC$_{11}$Cbl	14.9	nd
Cobinamides		
(Aq)$_2$Cbi	7.8	33.8
(CN)$_2$Cbi	—[c]	1.6
CN-Cbi	9.0,10.5[d]	4.9,5.7[d]
AC$_3$Cbi	9.2	nd
AdoCbi	10.2	6.1
MeCbi	12.7	9.6
DapCbi	17.0	23.9

[a] Sharp peaks but positions somewhat variable.
[b] Spreading.
[c] Compounds unstable at pH 3.0; monocyano isomers observed.
[d] Isomeric forms.
Source: Ref. 63.

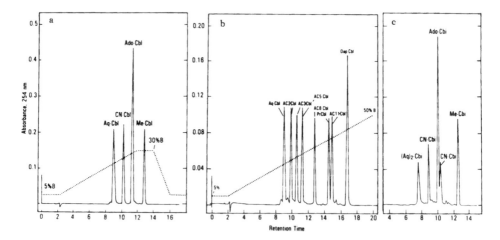

Figure 11 Gradient separation of cobalamin and cobinamide standards on a Lichrosorb C8 column. The gradient profile is indicated in (a) and (b) by the dotted line. (a) Separation of the naturally occurring cobalamins, 5 nmol each. (b) Separation of a mixture of cobalamins containing different upper-axional ligands. (c) Separation of cobinamide analogs, 5 nmol each, with a gradient identical to that in (a). (From Ref. 63.)

by a linear gradient to 65% methanol in 20 min) at 1 mL/min (Table 9). The system has been used in several other studies (110,111).

HPLC has proven to be very useful in the preparation of p-tri-n-butylstannyl hippurate conjugates to cobalamin-diaminododecane adducts, which themselves were prepared after linking diaminododecane to the isolated free b-, d-, and e-carboxylate products of mild acid hydrolysis of cobalamin (112). The final purpose of this study was the preparation of a radioiodinated cobalamin analog, capable of binding to transcobalamin II and traceable in mice or rats. With a similar aim in mind Pathare et al. (113) studied the possibilities to prepare bioactive biotinylated cobalamin analogs.

D. Analytical Separation of Plasma and Tissue Cobalamin

Chromatography of only 25 pg CN(^{57}Co)Cbl on a Lichrosorb RP8 column, according to Jacobsen and Green (108), demonstrated that this small amount could be recovered in a sharp peak for >99% (74). Consequently, it also proved possible to separate cobalamins in extracts from human plasma by HPLC with a sensitive radioisotope dilution assay for the detection of cobalamins in the column fractions. In the search for a solution of the specific problems of cobalamin extraction, van Kapel et al. (74) investigated the effects of NaN$_3$, Na$_2$S$_2$O$_5$, NaNO$_2$,

Table 9 Separation of Cobalamin Analogs by HPLC[a]

	Retention time	
Cobalamin	Min 30.0	Relative to CN-Cbl 1.00
Alterations adjacent to corrin ring		
CN-Cbl[C_8-OH, c-OH]	22.0	0.73
CN-Cbl[c-lactam]	24.0	0.80
CN-Cbl[c-lactone]	35.0	1.17
CN-Cbl[10-Cl,c-lactone]	68.3	2.28
CN-Cbl[13-epi]	21.1	0.72
CN-Cbl[b-OH]	38.5	1.30
CN-Cbl[d-OH]	41.8	1.38
CN-Cbl[e-OH]	45.8	1.53
CN-Cbl[C_8-OH,dc-OH]	23.7	0.79
CN-Cbl[C_8-OH,ec-OH]	27.7	0.92
CN-Cbl[C_8-OH,bc-OH]	26.7	0.89
CN-Cbl[de-OH]	61.4	2.04
CN-Cbl[bd-OH]	52.3	1.74
CN-Cbl[be-OH]	53.2	1.77
CN-Cbl[bde-OH]	90.2	3.01
CN-Cbl[13-epi.d-OH]	29.4	0.97
CN-Cbl[13-epi.e-OH]	42.3	1.38
CN-Cbl[13-epi.h-OH]	27.2	0.92
CN-Cbl[c-lactam.d-OH]	27.8	0.93
CN-Cbl[c-lactam.e-OH]	24.2	0.81
CN-Cbl[c-lactam.b-OH]	20.9	0.70
CN-Cbl[c-lactone.d-OH]	52.3	1.74
CN-Cbl[c-lactone.e-OH]	46.6	1.55
CN-Cbl[c-lactone.b-OH]	38.4	1.28
Alterations in nucleotide		
[5,6-Cl_2BZA] CN-Cba	43.1	1.44
[NZA]CN-Cba	34.8	1.16
[5,(6)-MeBZA]CN-Cba	25.7	0.86
[5,(6)-ClBZA]CN-Cba	28.8	0.96
[5,(6)-NO_2BZA]CN-Cba	19.5	0.65
[5,(6)-OCH_3BZA]CN-Cba	23.4	0.78
[5,(6)-COOH BZA]CN-Cba	19.5	0.65
[5,(6)-OH BZA]CN-Cba	21.1	0.70
[BZA]CN-Cba	23.6	0.79
[3,5,6-Me_3BZA](CN,OH)Cba	36.4;121.7[b]	1.21;4.06
[CN,OH]Cbi	22.7;14.0[b]	0.76;0.47
[2-MeAde]CN-Cba	20.4	0.68
[Ade]CN-Cba	21.0	0.70

[a] Separation was performed on a Waters μBondapak C18 reversed-phase column (10 μm particle size, 300 × 3.9 mm ID) using H_2O-acetic acid-isopropanol (90:1:9) at a flow rate of 0.3 mL/min. Samples of 0.1–1.0 nmol were injected in a volume of 10 μL and detected based on absorbance at 365 nm. From Ref. 109.

[b] Two peaks are seen with these analogs in which the base is absent or cannot coordinate as the lower axial ligand; thus, the CN can be present as either the upper or the lower axial ligand.

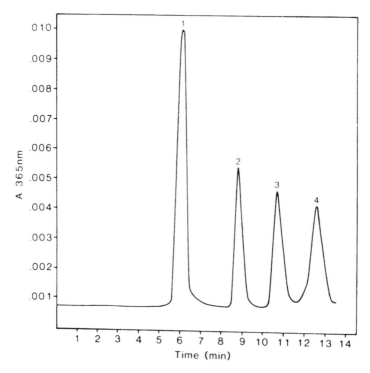

Figure 12   HPLC separation of Cbl (20 pmol) and Cbl-monocarboxylic acids (10 pmol each) on μBondapak NH₂. (1) CN-Cbl, (2) CN-Cbl {d-OH}, (3) CN-Cbl {b-OH}, and (4) CN-Cbl {e-OH}. (From Ref. 109.)

chloroacetophenone, p-hydroxomercuribenzoate, iodoacetamide, and N-ethylmaleimide on the separation of a mixture of reference cobalamins (Figs. 14, 15). Only N-ethylmaleimide left the separation pattern unchanged. A typical distribution of cobalamin and cobalamin analogs extracted from human plasma in the presence of SH-blocking N-ethylmaleimide is presented in Figure 16. Recovery of injected cobalamins (about 1 pmol) in the eluate fractions varied from 97% to 100%. Day-to-day variations of retention time for individual cobalamins were less than 10 sec.

Table 10 contains the data for the normal distribution of cobalamin forms in human plasma as found with this method. This procedure has proven its value in the analysis of two patients with an atypical form of congenital homocystinuria. In the plasma we found a severe deficiency of methylcobalamin in one patient and a moderate deficiency in another patient with a compensatory rise in adeno-

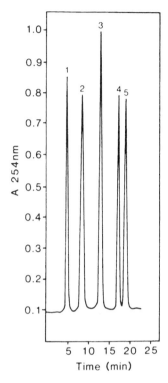

Figure 13 HPLC separation of the naturally occurring cobalamins on μBondapak C18 with an isopropanol gradient from 3 to 21% in 5 mmol/L 1-heptane sulfonic acid: (1) SO₃Cbl, (2) CN-Cbl, (3) OH-Cbl, (4) MeCbl, and (5) AdoCbl. (From Ref. 109.)

sylcobalamin (114,115). The same separation technique was used to measure specifically the distribution of transcobalamin II–bound cobalamins in human plasma. For that purpose, 10 mL plasma was incubated with 15 mg CM-Sephadex C50 in 5 mL 0.4 M glycine buffer, pH 3.0. After an overnight rotation at room temperature the Sephadex was collected by centrifugation and washed twice with 15 mL 25 mM phosphate/175 mM NaCl, pH 5.6. The supernatant was kept for later analysis of R-binder-bound cobalamin coenzymes; 50 μL 30% acetic acid, and 10 μL N-ethylmaleimide (final concentration 10 mM) were added to the pellet and the mixture was heated in a boiling water bath for 15 min. After cooling, 150 mg sodium acetate was added and the Sephadex was removed by centrifugation; 300 μl of the supernatant was used for injection on the HPLC column.

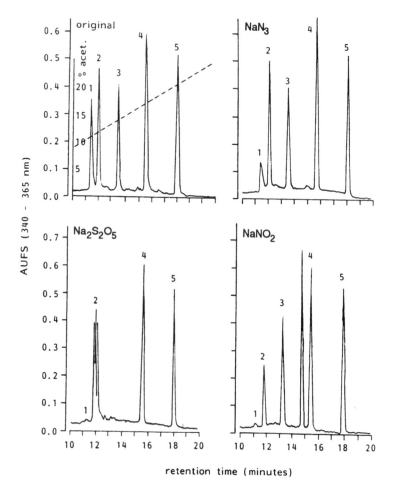

Figure 14 HPLC separation of a mixture of reference cobalamins (1.4 nmol each) on Lichrosorb RP8 with a 8–25% acetonitrile gradient in 85 mmol/l phosphate/triethanolamine buffer, pH 3.3, after extraction from plasma without (original) or with the additives NaN$_3$ (10 mmol/l), Na$_2$S$_2$O$_5$ (5 mmol/l), or NaNO$_2$ (5 mmol/l). (1) OHCbl, (2) SO$_3$Cbl, (3) CNCbl, (4) AdoCbl, (5) MeCbl. (From Ref. 74.)

The remainder was used to measure the total amount of cobalamin in the transcobalamin II extract.

The transcobalamin II–free plasma was mixed with four volumes of 1% acetic acid and N-ethylmaleimide to a concentration of 10 mM and treated further as a normal plasma sample. The results of the analyses are presented in Table

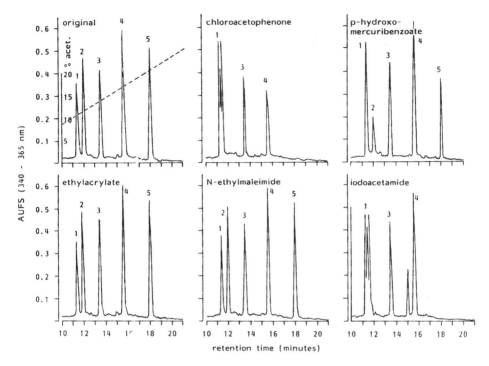

Figure 15 HPLC separation of a mixture of reference cobalamins (1.4 nmol each) on Lichrosorb RP8 with a 8–25% acetonitrile gradient in 85 mmol/l phosphate/triethanol-amine buffer pH 3.3, after extraction from plasma without (original) or with thiol-blocking agents (5 mmol/l each). (1)OHCbl. (2) SO₃Cbl, (3) CNCbl, (4) AdoCbl, (5) MeCbl. (From Ref. 74.)

11 and confirm earlier semiquantitative observations by Nexo (116) that transco-balamin II carries relatively less MeCbl and more AdoCbl. The differences, though, were considerably less impressive than in the study of Nexo. The interest-ing point of this methodology is that now the more rapidly metabolized pool of transcobalamin II ($T_{1/2} \pm 1$ h) can be analyzed separately from the slowly ($T_{1/2} \pm$ 10 d) exchanging pool of R-binder-Cbl.

Butte et al. (117) used HPLC for the quantitative separation of CN-Cbl as a tool in the estimation of the glomerular filtration rate. CN-Cbl (2 mg) is injected intravenously to saturate binding sites in blood and cells. Then 5 mg CN-Cbl is administered intravenously 8 to 14 hr later, and blood samples are taken at time intervals up to 90 min. The plasma is extracted with trichloroacetic acid (TCA), and bromocresyl green is added as an internal standard. Desalting is carried out on Sep-pak C18. After elution from the Sep-pak with methanol, a sample is in-

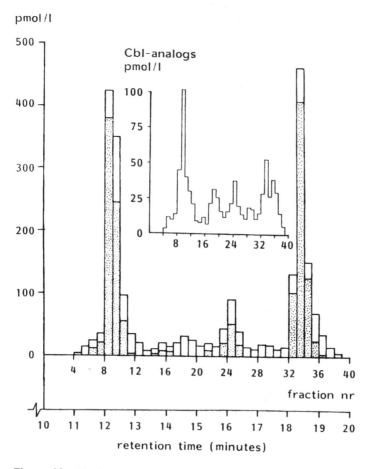

Figure 16 Distribution pattern of cobalamins and cobalamin analogs (insert) in normal human plasma, measured in the HPLC-eluate by radio-isotope dilution assay using R-binder (□) or R-binder-blocked Intrinsic Factor (□): OH/SO₃Cbl (fractions 7–12). CNCbl (fractions 15–17). AdoCbl (fractions 22–26), MeCbl (fractions 33–36), fraction volume 0.5 ml. (From Ref. 74.)

jected on a Lichrosorb Si60 column and eluted with methanol–sodium cyanide, 0.1 g/L (40:60) at a flow rate of 0.8 mL/min. The relative peak area is taken as a measure for the amount of cobalamin in the sample in comparison with a series of standard CN-Cbl concentrations in plasma. The analytical recovery of the procedure is about 65%. Precision varies from 4% at 500 µg/L to 7% at 200 µg/L cobalamin in plasma.

Table 10 Distribution of Cobalamins in Human
Plasma Measured with HPLC and RIDA[a]

Cobalamin	−NEM	+NEM
OH- + SO$_3$Cbl	47 (0–95)	161 (45–280)
CN-Cbl	35 (0–90)	15 (0–30)
AdoCbl	55 (25–85)	32 (10–52)
MeCbl	252 (80– 428)	182 (80–282)

[a] Mean concentrations in pmol/L ranges in parentheses. Extraction with 1% acetic acid and silanized silica 60 H.

Table 11 Distribution of Human Plasma Cobalamin Coenzymes

Cbl-coenzyme	Transcobalamin II-bound					
	Percentage (%)			Concentration (pmol/L)		
	Range	Mean	SD	Range	Mean	SD
OH-Cbl	30–62	49	9	13–74	30	19
CN-Cbl	0.7–10.5	4.6	3.2	1–4	2.4	1.3
Ado-Cbl	4.3–17.1	12.2	4.1	1–13	7.6	4.2
MeCbl	20–53	34	8	6–41	20	11
	R-binder-bound					
OH-Cbl	28–56	38	9	14–196	86	56
CN-Cbl	2–18	6.1	4.7	1–30	11	7
Ado-Cbl	2–12	6.6	2.9	2–37	20	15
MeCbl	31–61	49	10	17–319	147	105

The earlier-mentioned application of intravenous OH-Cbl as an antidote for severe cyanide poisoning has for the first time been explored using HPLC separation in two papers from Astier and Baud in which they demonstrate that not only plasmatic OH-Cbl absorbs cyanide but also that intracellular cyanide is scavenged by the administered OH-Cbl (118,119).

IV. GAS CHROMATOGRAPHY

Corrinoids lend themselves very poorly to gas chromatographic separations due to their low volatility and thermal instability. As a result, it has found no real application in the analysis of these compounds.

V. MASS SPECTROMETRY

Low volatility and thermal instability have also been an obstacle for mass spectrometric analysis. Still, considerable progress has been made in finding new, mild
methods for ionization of such complex molecules. Siebel and Schulten (120)
have written an excellent review on the subject of soft ionization of biomolecules,
in which they describe the results of ten different methods with corrins and vitamin B_{12}. In fact, vitamin B_{12} has become a model compound for the development
of new mass-spectrometric methods. Mass spectrometry has contributed and no
doubt will contribute in future to our knowledge of the biosynthesis and biological
activity of corrins, vitamin B_{12}, and their synthetically prepared analogs. The first
systematic mass spectrometric investigations on corrins were carried out by Seibl
(121) on methyl-substituted corrin complexes and were published in 1968. He
used electron impact ionization, but this technique led to massive thermal and
ionization-induced degradation of the molecule. Less destructive but still unsuitable was electron capture ionization as reported by Von Ardenne et al. (122). The
experiences with chemical ionization, heavy-ion-induced desorption, and field
ionization have not been much better, and it was the technique of field desorption
that turned out to be a breakthrough in that it yielded abundant molecular ions
of all types of corrins and even of intact vitamin B_{12} (123). The essential difference from the earlier-mentioned technique was that the ionization process in field
desorption takes place when the substance is absorbed on the emitter surface,
reducing the thermal stress on the molecule before ionization (hence soft ionization). The much higher sensitivity of the method makes it possible to detect substance amounts even in the nanogram range and allows experiments using stable-
isotope-labeled corrins.

  Field desorption has more recently been extended with several specific ionization techniques, such as secondary ion mass spectrometry or SIMS. When
SIMS is applied to samples taken up in a liquid matrix such as glycerol, it is
known as fast atom bombardment or FABMS. An advantage of FABMS in corrin
research is the high amount of structural information that arises from the particular fragment patterns. Stupperich et al. (110) used FABMS for the identification
of various corrinoids occurring in acetogenic bacteria. Kräutler et al. (111) did
the same for the various forms of cobamides, differing in their nucleotide bases,
in some sulfate-reducing and sulfur-metabolizing bacteria. FABMS has also been
applied for the identification of a high-performance liquid chromatography–purified product from the reaction catalyzed by the enzyme nicotinate-mononucleotide: 5,6-dimethylbenzimidazole phophoribosyltransferase (CobT) from *Salmonella typhimurium* (124). Still another development in mass spectrometry of
corrins is the online coupling with high-performance liquid chromatography using a thermospray interface (125).

VI. FUTURE TRENDS

In the last 10 years HPLC has become the method of choice for the separation of cobalamin metabolites and coenzyme forms from all sources. Thin-layer chromatography, in combination with bioautography, still has the advantage of a very high sensitivity and thus requiring only small quantities of tissue and plasma for a full analysis. Nevertheless, more detailed knowledge of cobalamin biosynthesis and reaction mechanisms of cobalamin-mediated enzymatic reactions has been gained thanks to the application of HPLC. This is in particular true for the investigations concerning the corrins from acetogenic, methanogenic, and sulfur-metabolizing bacteria, which are of special significance in view of their role in the environment.

Still, little progress has been made in defining the nature of the cobalamin analogs in human plasma, but this is not so much the result of insufficient separation power as the inability to obtain enough patient sample for the molecular analysis. Following this trail, it might be worthwhile to develop a reference method for the measurement of cobalamins in human plasma, based on HPLC separation. Such a development might end the discussion on the true value of the cobalamin concentration in human plasma and might provide us with established serum standards with assayed values to be used in less specific competitive protein-binding assays.

The latest developments in mass spectrometry with respect to soft ionization of nonvolatile and thermolabile compounds have brought this technique to the level where it may open, and indeed has opened, new ways in cobalamin research, especially when integrated with HPLC.

REFERENCES

1. GH Whipple, FS Robscheit, CN Hooper. Blood regeneration following simple anemia. IV. Influence of meat, liver and various extractives, alone or combined with standard diets. Am J Physiol 3:236, 1920.
2. GR Minot, WP Murphy. Treatment of pernicious anemia by a special diet. JAMA 87:470, 1926.
3. EL Rickes, NG Brink, FR Koniuszy, TR Wood, K Folkers. Cristalline vitamin B_{12}. Science 107:396–397, 1948.
4. EL Smith. Purification of antipernicious anemia factors from liver. Nature 161: 638–639, 1948.
5. R Bonnett, JR Cannon, VM Clark, AW Johnson, LFJ Parker, EL Smith, AR Todd. Chemistry of the vitamin B_{12} group. Part V. The structure of the chromophoric grouping. J Chem Soc 1:1158, 1957.

6. DC Hodgkin, J Kamper, M Mackay, JW Pickworth, KN Trueblood, JG White. Structure of vitamin B_{12}. Nature 178:64, 1956.

7. HA Barker, H Weissbach, RD Smyth. A coenzyme containing pseudo-vitamin B_{12}. Proc Natl Acad Sci USA 44:1093–1097, 1958.

8. H Weissbach, JI Toohey, HA Barker. Proc Natl Acad Sci USA 45:521, 1959.

9. JI Toohey, HA Barker. Isolation of coenzyme B_{12} from liver. J Biol Chem 236: 560–563, 1961.

10. K Lindstrand, K-G Ståhlberg. On vitamin B_{12} forms in human plasma. Acta Med Scand 174:665–669, 1963.

11. K-G Ståhlberg. Forms of plasma vitamin B_{12} in health and in pernicious anemia, chronic myeloid leukemia and acute hepatitis. Scand J Haematol 1:220–222, 1964.

12. IUPAC-IUB Commission on Biochemical Nomenclature. Biochemistry 13:1555, 1974.

13. IUPAC-IUB Commission on Biochemical Nomenclature. The nomenclature of corrinoids. Biochem J 147:1–10, 1975.

14. R Bonnett, JR Cannon, AW Johnson, AR Todd. J Chem Soc 1148–1168, 1957.

15. MJ Mahoney, LE Rosenberg. Synthesis of cobalamin coenzymes by human cells in tissue culture. J Lab Clin Med 78:302–308, 1971.

16. RA Firth, HAO Hill, JM Pratt, RG Thorp, RJP Williams. The chemistry of vitamin B_{12}. Some further formation constants. J Chem Soc A:381–386, 1969.

17. JB Armitage, JR Cannon, AW Johnson, LFJ Parker, EL Smith, WH Stafford, AR Todd. Chemistry of the vitamin B_{12} group. Part III. The course of hydrolytic degradations. J Chem Soc 3849–3864, 1953.

18. W Friedrich, K Bernhauer. Z Naturforsch 9b:685, 1954.

19. D Dolphin. Preparation of the reduced forms of vitamin B_{12} and of some analogues of the vitamin B_{12} coenzyme containing a cobalt-carbon bond. Methods Enzymol 18c:34–52, 1971.

20. JL Peel. The catalysis of the auto-oxidation of 2-mercaptoethanol and other thiols by vitamin B_{12} derivatives. Biochem J 88:296–308, 1963.

21. GN Schrauzer, JW Sibert. Electron transfer reactions catalyzed by vitamin B_{12} and related compounds: the reduction of dyes and of riboflavin by thiols. Arch Biochem Biophys 130:257–266, 1969.

22. RO Brady, HA Barker. Biochem Biophys Res Commun 4:373, 1961.

23. HPC Hogenkamp. J Biol Chem 238:477–480, 1963.

24. HPC Hogenkamp, JN Ladd, HA Barker. J Biol Chem 237:1950, 1962.

25. JN Pratt. Chemistry of vitamin B_{12}. Part II. Photochemical reactions. J Chem Soc 5154–5160, 1964.

26. HPC Hogenkamp. The photolysis of methylcobalamin. Biochemistry 5:417–422, 1966.

27. AW Johnson, N Shaw, F Wagner. A chemical synthesis of S-adenosyl-homocysteine on the vitamin B_{12} coenzyme. Biochim Biophys Acta 72:107–110, 1963.

28. HPC Hogenkamp. The chemistry of cobalamins and related compounds. In: BM Babior, ed. Cobalamin. New York: Wiley Interscience, 1975, pp 21–74.

29. JM Wood, FS Kennedy, CG Rosen. Synthesis of methyl-mercury compounds by extracts of a methanogenic bacterium. Nature 220:173–175, 1968.

30. K Bernhauer, P Renz, F Wagner. Synthesen auf dem vitamin B_{12}-gebiet, über das Diaquocobinamid-ion. Biochem Z 335:443–452, 1962.

31. D Dolphin, AW Johnson, N Shaw. Sulphitocobalamin. Nature 199:170–171, 1963.

32. JA Hill, JM Pratt, RJP Williams. The chemistry of vitamin B_{12}. Part I. The valency and spectrum of the coenzyme. J Chem Soc :5149–5153, 1964.

33. HC Friedman. Biosynthesis of corrinoids. In: BM Babior, ed. Cobalamin, New York: Wiley Interscience, 1975, pp 75–110.

34. HC Friedman. In: B Zagalak, W Friedrich, eds. Vitamin B_{12}: Berlin: Walter de Gruyter, 1979, p 331.

35. AR Battersby, E McDonald. In: D Dolphin, ed. B_{12}, Vol. 1. New York: Wiley Interscience, 1982, p 107.

36. RO Brady, EG Castanera, HA Barker. The enzymatic synthesis of cobamide coenzymes. J Biol Chem 237:2325–2332, 1962.

37. GA Walker, S Murphy, FM Huennekens. Enzymatic conversion of vitamin B_{12a} to adenosyl-B_{12}: evidence for the existence of two separate reducing systems. Arch Biochem Biophys 134:95–102, 1969.

38. J Pawelkiewicz, M Gorna, W Fenrych, S Magar. Ann NY Acad Sci 112:641, 1964.

39. SS Kerwar, C Spears, B McAuslan, H Weissbach. Studies on vitamin B_{12} metabolism in Hela cells. Arch Biochem Biophys 142:231–237, 1971.

40. RT Taylor. In: D Dolphin, ed. B_{12}, Vol. 2. New York: Wiley Interscience, 1982, p 307.

41. BM Babior. Cobamides and cofactors: adenosylcobamide-dependent reactions. In: BM Babior, ed. Cobalamin. New York: Wiley Interscience: 1975, pp 141–214.

42. G Morrow, LA Barness, GJ Cardinale, RH Abeles, JG Flaks. Congenital methylmalonic acedemia: enzymatic evidence for two forms of the disease. Proc Natl Acad Sci USA 63:191–197, 1969.

43. JM Poston, TC Stadtman. Cobamides as cofactors: methylcobamides and the synthesis of methionine, methane and acetate. In: BM Babior ed. Cobalamin. New York: Wiley Interscience, 1975, pp 111–140.

44. LGJ Ljungdahl, HG Wood. In: D. Dolphin, ed. B_{12}, Vol. 2. New York: Wiley Interscience, 1982, p 165.

45. JS Levine, RH Allen, DH Alpers, B Seetharam. Immunocytochemical localization of the intrinsic factor–cobalamin receptor in dog ileum: distribution of intracellular receptor during cell maturation. J Cell Biol 98:1111–1118, 1984.

46. I Chanarin, H Muir, A Hughes, AV Hoffbrand. Evidence for intestinal origin of transcobalamin II during vitamin B_{12} absorption. Br Med J i:143–145, 1978.

47. J Lindemans. Transcobalamin II–mediated uptake of vitamin B_{12} by isolated rat liver cells. Thesis, Erasmus University, Rotterdam, 1979.

48. H Olesen, BL Hom, M Schwartz. Turnover of [57]Co-labelled vitamin B_{12}-TC II and autologous ^{131}I-labeled IgG in a patient with antibody to transcobalamin II. Scand J Haematol 5:107–115, 1968.

49. AP Skouby, E Hippe, H Olesen. Antibody to transcobalamin II and B_{12}-binding capacity in patients treated with hydroxocobalamin. Blood 38:769–774, 1971.

50. DW Bartholomew, ML Batshaw, RH Allen, CR Rae, D Rosenblatt, DL Valle, CA Francomano. Therapeutic approaches to cobalamin C methylmalonic acidemia and homocystinuria. J Paediatr 112:32–39, 1988.

51. J Pawelkiewicz. The methods of the preparation of corrinoids including the analytical methods. In: HC Heinrich, ed. Vitamin B_{12} und Intrinsic Factor. Stuttgart: F. Enke Verlag, 1962, pp 280–294.

52. W Friedrich. Vitamin B_{12} und verwandte corrinoide. In: R Ammon, W Dirscherl, eds. Fermente, Hormone und Vitamine. Stuttgart: Georg Thieme Verlag, 1965, pp 1–4.

53. G Matsuda. J Vitaminol 1:221, 1955.

54. H Kubo, S Fujita, H Ogino. Chem Abstr 54:13153, 1960.

55. W Friedrich, K Bernhauer. Z Naturforsch 9b:755, 1954.

56. MW Penley, DG Brown, JM Wood. Chemical and biological studies with fluoroalkylcobalamins. Biochemistry 22:4302–4310, 1970.

57. VB Koppenhagen, B Elsenhaus, F Wagner, JJ Pfiffner. Methylrhodibalamin and 5'-deoxyadenosylrhodibalamin, the rhodium analogues of methylcobalamin and cobalamin-coenzyme. J Biol Chem 249:6532–6540, 1974.

58. EV Quadros, A Hamilton, DM Matthews, JC Linnell. Isolation of 57-cobalamin coenzymes at high specific activity from *Streptomyces griseus*. J Chromatogr 160: 101–108, 1978.

59. SH Ford, HC Friedman. Vitamin B_{12} biosynthesis: in vitro formation of cobinamide from cobyric acid and L-threonine. Arch Biochem Biophys 175:121–130, 1975.

60. WA Fenton, LE Rosenberg. Improved techniques for the extraction and chromatography of cobalamins. Anal Biochem 90:119–125, 1978.

61. P Gimsing, E Nexo, E Hippe. Determination of cobalamin in biological material. II. The cobalamins in human plasma and erythrocytes after desalting on non-polar adsorbent material and separation by one-dimensional thin-layer chromatography. Anal Biochem 129:296–304, 1983.

62. E Stupperich, I Steiner, M Rühlemann. Isolation and analysis of bacterial cobamides by high-performance liquid chromatography. Anal Biochem 155:365–370, 1986.

63. DW Jacobsen, R Green, EV Quadros, YD Montejano. Rapid analysis of cobalamin coenzymes and related corrinoid analogues by high-performance liquid chromatography. Anal Biochem 120:394–403, 1982.

64. EP Frenkel, R Prough, RL Kitchens. Measurement of tissue vitamin B_{12} by radioisotopic competitive inhibition assay and quantitation of tissue cobalamin fractions. Methods Enzymol 67:31–40, 1980.

65. JC Linnell, HM Mackenzie, J Wilson, DM Matthews. Patterns of plasma cobalamins in control subjects and in cases with vitamin B_{12} deficiency. J Clin Pathol 22: 545–550, 1969.

66. JC Linnell, AV Hoffbrand, HA-A Hussein, IJ Wise, DM Matthews. Tissue distribution of coenzyme and other forms of vitamin B_{12} in control subjects and patients with pernicious anemia. Clin Sci Mol Med 46:163–172, 1974.

67. P Gimsing, E Hippe, E Nexo. Determination of the plasma cobalamins by one-dimensional thin-layer chromatography. In: W Friedrich and B Zagalak, eds. Vitamin B_{12}. Berlin: Walter de Gruyter, 1979, pp 665–669.

68. P Gimsing, E Hippe, I Elleberg-Rasmusen, M Moergaard, J Lanng Nielson, P Bastrup-Madsen, R Berlin, T Hansen. Cobalamin forms in plasma and tissue during treatment of vitamin B_{12} deficiency. Scand J Haematol 29:311–318, 1982.

69. MC Mahoney, LE Rosenberg. Inherited defects of B_{12} metabolism. Am J Med 48: 584–593, 1970.

70. P Gimsing, E Nexo, E Hippe. Determination of cobalamins in biological material. I. Improvement of the unequal recovery of cobalamins by preincubation with cadmium acetate. Anal Biochem 129:288–295, 1983.

71. WR Bauriedel, JC Picken Jr, LA Underkofler. Proc Soc Exp Biol Med 91:377, 1956.

72. RT Taylor, ML Hanna. Binding of aquocobalamin to the histidine residues in bovine serum albumin. Arch Biochem Biophys 141:247–257, 1970.

73. EL Lien, JM Wood. The specificity of aquocobalamin binding to bovine serum albumin. Biochim Biophys Acta 264:530–537, 1972.

74. J van Kapel, LJM Spijkers, J Lindemans, J Abels. Improved distribution analysis of cobalamins and cobalamin analogues in human plasma in which the use of thiol-blocking agents is a prerequisite. Clin Chim Acta 131:211–224, 1983.

75. G Markus, F Karush. The disulfide bonds of human serum albumin and bovine gamma-globulin. J Am Chem Soc 79:134–139, 1957.

76. A Lorber, CC Chang, D Masuoka, I Meacham. Effects of thiols in biological systems on protein sulfhydryl content. Biochem Pharmacol 19:1551–1556, 1970.

77. RB Martin, JT Edsall. J Am Chem Soc 81:4044, 1959.

78. J Sedlak, RH Lindsay. Estimation of total, protein-bound and nonprotein sulfhydryl groups in tissues with Ellman's reagent. Anal Biochem 25:192–205, 1968.

79. RH Allen, PW Majerus. Isolation of vitamin B_{12}–binding proteins using affinity chromatography. I. Preparation and properties of vitamin B_{12}–Sepharose. J Biol Chem 247:7695–7701, 1972.

80. JF Kolhouse, RH Allen. Absorption, plasma transport, and cellular retention of cobalamin analogues in the rabbit. Evidence for the existence of multiple mechanisms that prevent the absorption and tissue dissemination of naturally occurring cobalamin analogues. J Clin Invest 60:1381–1392, 1977.

81. H Kondo, JF Kolhouse, RH Allen. Presence of cobalamin analogues in animal tissues. Proc Natl Acad Sci USA 77:817–821, 1980.

82. RA Firth, HAO Hill, JM Pratt, RG Thorp. Separation and identification of organo-cobalt derivatives of vitamin B_{12} on thin-layer cellulose. Anal Biochem 23:429–432, 1968.

83. TM Vickrey, RN Katz, GN Schrauzer. Synthesis and reactions of formylmethylcobalamin and of related compounds. J Am Chem Soc 97:7248–7253, 1975.

84. JC Linnell, HA-A Hussein, DM Matthews. A two-dimensional chromato-bioautographic method for complete separation of individual plasma cobalamins. J Clin Pathol 23:820–821, 1970.

85. JC Linnell, AV Hoffbrand, TJ Peters, DM Matthews. Chromatographic and bioautographic estimation of plasma cobalamins in various disturbances of vitamin B_{12} metabolism. Clin Sci 40:1–16, 1971.

86. E Nexo, J Andersen. Unsaturated and cobalamin-saturated transcobalamin I and II in normal human plasma. Scand J Clin Lab Invest 37:723–728, 1977.

87. MJ Dillon, JM England, D Gompertz, PA Goodey, DB Grant, HA-A Hussein, JC Linnell, DM Matthews, SH Mudd, GH Newns, JWT Seakins, BW Uhlendorf, IJ Wisse. Clin Sci Mol Med 47:43, 1974.

88. EV Quadros, DM Matthews, IJ Wisse, JC Linnell. Tissue distribution of endogenous cobalamins and other corrins in the rat, cat, and guinea pig. Biochim Biophys Acta 421:141–152, 1976.

89. JC Linnell, MJ Wilson, YB Mikol, LA Poirier. J Nutr 113:124, 1983.

90. P Gimsing. Cobalamin forms and analogues in plasma and myeloid cells during chronic myelogenous leukaemia related to clinical condition. Br J Haematol 89: 812–819, 1995.

91. H Vogelmann, F Wagner. Separation of corrinoids by elution-chromatography on columns of the non-polar adsorbent Amberlite XAD-2. J Chromatogr 76:359–379, 1973.

92. VB Koppenhagen, F Wagner, JJ Pfiffner. α-(5,6-dimethylbenzimidazolyl)rhodibamide and rhodibinamide, the Rhodium analogues of vitamin B_{12} and cobinamide. J Biol Chem 248:7999–8002, 1973.

93. G Tortolani, P Bianchini, V Mantovani. Separation of various cobalamins by column chromatography. Farm Ed Prat 25:772–775, 1970.

94. G Tortolani, P Bianchini, V Mantovani. Separation and determination of cobalamins on an SP-Sephadex column. J Chromatogr 53:577–579, 1970.

95. RA Gams, EM Ryel, LM Meyer. Proc Soc Exp Biol Med 149:384, 1975.

96. RA Gams, EM Ryel, F Ostroy. Protein mediated uptake of vitamin B_{12} by isolated mitochondria. Blood 47:923–930, 1976.

97. P Gimsing, E Hippe. Increased concentration of transcobalamin I in a patient with metastatic carcinoma of the breast. Scand J Haematol 24:243–249, 1978.

98. JA Begley, CA Hall. Overestimation of cyanocobalamin due to coelution of sulfitocobalamin on SP-Sephadex C-25. J Chromatogr 177:360–362, 1979.

99. JM Anes, RA Beck, JJ Brink, RJ Goldberg. Nitritocobalamin and nitrosocobalamin may be confused with sulfitocobalamin using cation-exchange chromatography. J Chromatogr B Biomed Appl 660:180–185, 1994.

100. SL Lee. Separation of cobyrinic acid and its biosynthetic precursors by ion-exchange paper chromatography. Methods Enzymol 67:3–5, 1980.

101. M Katada, S Tyagi, A Nath, RL Petersen, RK Gupta. A novel form of vitamin B_{12} and its derivatives. Biochim Biophys Acta 584:149–163, 1979.

102. JA Schmit, RA Henry, RC Williams, JF Dieckman. J Chromatogr Sci 9:645, 1971.

103. RBH Wills, CG Shaw, WR Day. Analysis of water soluble vitamins by high pressure liquid chromatography. J Chromatogr Sci 15:262–266, 1977.

104. Y Maeda, M Yamamoto, K Owada, S Sato, T Masui, H Nakazowa. Simultaneous liquid chromatographic determination of water-soluble vitamins, caffeine and preservatives in oral liquid tonics. J Assoc Off Anal Chem 72:244–247, 1989.

105. E Pellerin, J-F Letavernier, N Chanon. Identification des cobalamines par chromatographie liquide haute pression. Ann Pharm Fr 9–10:413–415, 1977.

106. D Mourot, B Delepine, J Boisseau, G Gayot. Séparation totale des cobalamines et du coenzyme B_{12} par chromatographie liquide haute pression. Ann Pharm Fr 5–6: 235–238, 1979.

107. EP Frenkel, RL Kitchens, R Prough. High performance liquid chromatographic separation of cobalamins. J Chromatogr 174:393–400, 1979.

108. DW Jacobsen, R Green. Rapid determination of corrinoids by high performance liquid chromatography. In: W Friedrich, B Zagalak, eds. Vitamin B_{12}. Berlin: Walter de Gruyter, 1979, pp 663–664.

109. M Binder, JF Kolhouse, KC van Horne, RH Allen. High pressure liquid chromatography of cobalamins and cobalamin analogs. Anal Biochem 125:253–258, 1982.

110. E Stupperich, HJ Eisinger, B Kräutler. Diversity of corrinoids in acetogenic bacteria. Eur J Biochem 172:459–464, 1988.

111. B Kräutler, H-PE Kohler, E Stupperich. 5'-Methylbenzimidazolyl-cobamides are the corrinoids for some sulfate-reducing and sulfur metabolizing bacteria. Eur J Biochem 176:461–469, 1988.

112. DS Wilbur, DK Hamlin, PM Pathare, S Heusserm, RL Vessella, KR Buhler, JE Stray, J Daniel, EV Quadros, P McLoughlin, AC Morgan. Synthesis and nca-radio-iodination of arylstannyl-cobalamin conjugates. Evidence of aryliodo-cobalamin conjugate binding to transcobalamin II and biodistribution in mice. Bioconj Chem 7:461–474, 1996.

113. PM Pathare, DS Wilbur, S Heusser, EV Quadros, P McLoughlin, AC Morgan. Synthesis of cobalamin-biotin conjugates that vary in the position of cobalamin coupling. Evaluation of cobalamin derivative binding to transcobalamin II. Bioconj Chem 7:217–232, 1996.

114. B Fowler, RBH Schutgens, DS Rosenblatt, GPA Smit, J Lindemans. Folate-responsive homocystinuria and megaloblastic anaemia in female patient with functional methionine synthase deficiency (CblE disease). J Inher Metab Dis 10:731–741, 1997.

115. EA Kvittingen, S Spangen, J Lindemans, B Fowler. Methionine synthase deficiency without megaloblastic anemia. Eur J Pediatr 156:925–930, 1997.

116. E Nexo. Characterization of the cobalamins attached to transcobalamin I and transcobalamin II in human plasma. Scand J Haematol 18:358–360, 1977.

117. W Butte, H-H Riemann, AJ Walle. Liquid chromatographic measurement of cyanocobalamin in plasma, a potential tool for estimation of glomerular filtration rate. Clin Chem 28:1778–1781, 1982.

118. A Astier, FJ Baud. Simultaneous determination of hydroxocobalamin and its complex cyanocobalamin in human plasma by high-performance liquid chromatography. Application to pharmacokinetic studies after high-dose hydroxocobalamin as an antidote for severe cyanide poisoning. J Chromatogr B Biomed Appl 667:129–135, 1995.

119. A Astier, FJ Baud. Complexation of intracellular cyanide by hydroxocobalamin using a human cellular model. Hum Exp Toxicol 15:19–25, 1996.

120. HM Siebel, H-R Schulten. Soft ionization of biomolecules: a comparison of ten ionization methods for corrins and vitamin B_{12}. Mass Spectrom Rev 5:249–311, 1986.

121. J Seibl. Adv Mass Spectrom 4:317, 1968.

122. M Von Ardenne, K Steinfelder, R Tümmler. Elektronenanlagerungs-Massenspektrographie organischer substanzen. Berlin: Springer, 1971.

123. H-R Schulten, HM Schiebel. Naturwissenschaften 65:223, 1978.

124. JR Trzebiatowski, JC Escalante-Semerena. Purification and characterization of CobT, the nicotinate-mononucleotide: 5,6-dimethylbenzimidazole phosphoribosyl-transferase enzyme from *Salmonella typhimurium* LT2. J Biol Chem 272:17662–17667, 1997.

125. ML Vestal. High-performance liquid chromatography-mass spectrometry. Science 226:275–281, 1984.

13
Pantothenic Acid

Jan Velíšek and Jiří Davídek
Institute of Chemical Technology, Prague, Czech Republic

I. INTRODUCTION

The principal biologically active natural forms of pantothenic acid (vitamin B_5), one of the B vitamin complex, are coenzyme A (CoA or CoASH) and acyl-carrier protein (ACP). In solid pharmaceuticals, foods, and feeds, because of simple handling and increased stability, the pantothenic acid sodium and calcium salt are often used as additives. Panthenol is usually used in liquid pharmaceutical preparations and in cosmetics.

Some review articles and books on pantothenic acid and its natural and synthetic derivatives—their chemical, physical, and pharmaceutical properties, biological function, biosynthesis, nutritional aspects, and use—have appeared during the last decade (1–8).

A. History

The existence of pantothenic acid as a B-group vitamin, a dietary factor, whose absence caused a form of dermatitis in chicks, was anticipated around 1930 and discovered in 1931 by Williams (2,9). The name pantothenic acid, given to the active substance by Williams and his coworkers in 1933 (10), indicated its widespread occurrence in nature. The structure of pantothenic acid (Fig. 1) was established by Williams in 1938 (11) and the total synthesis was first achieved by the US Merck group in 1940 (12).

The first active form of pantothenic acid, coenzyme A (CoA, CoASH), was isolated and identified as the acyl transfer agent in two-carbon unit metabolism

556   Velíšek and Davídek

HO-CH₂-C(CH₃)(CH₃)-CH(OH)-CO-NH-CH₂-CH₂-COOH

Figure 1 (R)-Pantothenic acid structure.

in 1946 (13,14). Its structure (Fig. 2) was elucidated in 1950 (15) and confirmed synthetically in 1961 (16,17).

The second active form of pantothenic acid is the acyl-carrier protein (ACP) involving the 4′-phosphopantetheine moiety linked with a phosphoserine residue in the respective molecule (Fig. 3). It was described by Pugh and Wakil in 1965 (18) and found to be a coenzyme of several enzymes (19).

B. Chemistry

Pantothenic acid, still designated as D-(+)-pantothenic acid, D-(+)-α,γ-dihydroxy-β,β-dimethyl butyrylalanine, or N-(2,4-dihydroxy-3,3-dimethylbutyryl)-β-alanine, is composed of D-(+)-pantoic acid (2,4-dihydroxy-3,3-dimethylbutyric acid) linked by an amide bond to β-alanine (3-aminopropionic acid) (Fig.

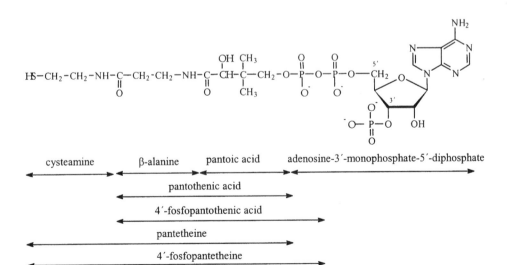

Figure 2 Coenzyme A structure.

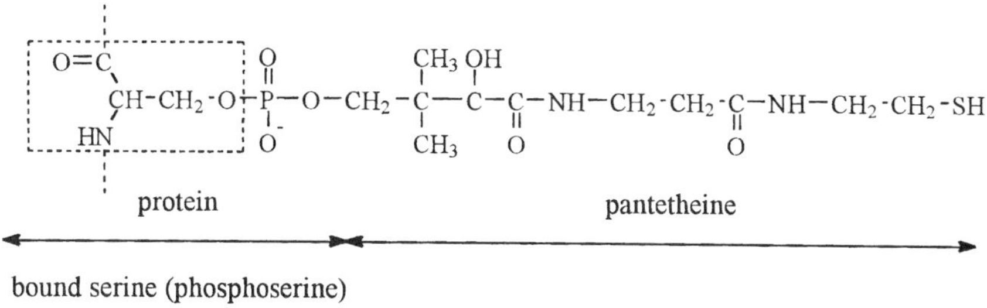

Figure 3   Acyl-carrier protein structure.

1). Pantothenic acid occurs in nature only as the D-(+)- or the (R)-enantiomer; the L-(−)- or (S)-form has no vitamin activity, as biological effects of this group of compounds are specific and connected solely with the optically active dextrorotatory forms.

Pantothenic acid is a yellowish viscous oily liquid, which is readily soluble in water, alcohols, and dioxane, slightly soluble in diethyl ether and acetone, and virtually insoluble in benzene and chloroform. The stability of pantothenic acid in aqueous solutions is very pH dependent. It is most stable in slightly acidic medium (pH 4 to 5). Both in acidic and alkaline media it is hydrolytically cleaved to yield pantoic acid and its salts, respectively, and β-alanine. In acidic solutions, pantoic acid spontaneously eliminates one molecule of water, forming (R)-2-hydroxy-3,3-dimethyl-4-butanolide (α-hydroxy-β,β-dimethyl-γ-butyrolactone), referred to as pantoyl lactone or pantolactone (Fig. 4).

Pantothenic acid itself is not used as it is hygroscopic and unstable. Its more stable sodium salt and, especially, the calcium salt are synthesized chemically and used pharmaceutically, mainly in solid multivitamin preparations and as an additive compound for some foods (such as infant milk formulas) and domestic animal feeds. Calcium pantothenate is most stable in almost neutral media (pH 6 to 7). Salts of pantothenic acid are colorless crystals and less hygroscopic (especially the calcium salt) than pantothenic acid. The solubility of calcium pantothenate in water at 25°C is 0.356 g/mL, whereas the sodium salt also is very water soluble.

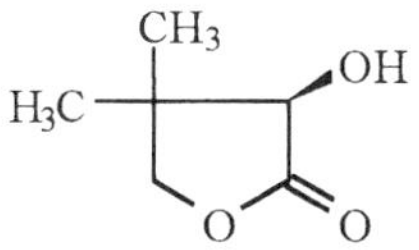

Figure 4   (R)-Pantolactone structure.

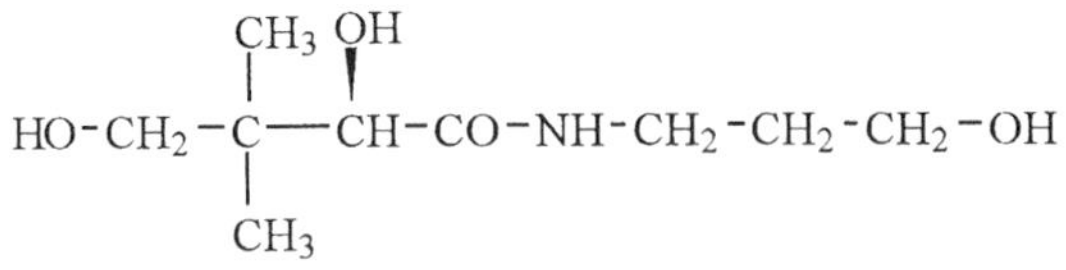

$$\text{HO-CH}_2-\overset{\overset{\displaystyle CH_3}{|}}{\underset{\underset{\displaystyle CH_3}{|}}{C}}-\overset{\overset{\displaystyle OH}{|}}{CH}-CO-NH-CH_2-CH_2-CH_2-OH$$

Figure 5   (*R*)-Panthenol structure.

An alcohol related to pantothenic acid and referred to as D-panthenol or D-pantothenyl alcohol also possesses vitamin activity (Fig. 5). It is commonly used in liquid pharmaceutical and cosmetic preparations. Panthenol is a hygroscopic viscous oil that can be crystallized. Panthenol is very slightly soluble in water but very soluble in alcohol. The products of hydrolysis of panthenol are pantoic acid and 3-amino-1-propanol (β-alanol).

Pantetheine and coenzyme A are amorphous, colorless, water-soluble powders. Solutions of coenzyme A are relatively stable at pH 2 to 6. Both compounds form oxidation products containing disulfide bonds, that is, pantethine and the coenzyme A disulfide form (2,5,20).

Some of the chemical and physical constants of pantothenic acid and the related compounds are summarized in Table 1.

C. Biochemistry

Pantothenic acid is synthesized in plants and some microorganisms from pantoic acid and β-alanine. Pantoic acid is formed from 2-oxopantoic acid (4-hydroxy-3,3-dimethyl-2-oxobutyric acid) and 2-oxoisovaleric acid (3-methyl-2-oxobutyric acid), a precursor of valine. β-Alanine is formed by decarboxylation of L-aspartic acid. Enzymes involved include pantothenate synthetase (EC 6.3.2.1), oxopantoate reductase (EC 1.1.1.169), oxopantoate hydroxymethyltransferase (EC 4.1.2.12), and aspartate 1-decarboxylase (EC 4.1.1.12).

A degradative enzyme of some bacteria, pantothenase (EC 3.5.1.22) specifically splits pantothenic acid to pantoic acid and β-alanine.

Animals do not synthesize pantothenic acid. However, they, like yeasts and bacteria, convert the exogenous vitamin derived from the diet to coenzyme A (CoA) and acyl-carrier protein (ACP), the two metabolically active forms. The reaction pathway is shown in Figure 6.

The biosynthesis of CoA in animals requires five enzymes. The first three are found only in the cytosol, and the other two are also found in mitochondria. Reaction IV is reversible. Four moles of ATP are required for the biosynthesis of one mole CoA from one mole of pantothenic acid. Two additional enzymes

Table 1 Some Physicochemical Constants of Pantothenic Acid and Related Compounds

Compound	Molecular weight	Bruto formula	λ max (nm)	Optical rotation $[\alpha]_D^t$	Melting point (°C)	pK$_a$ at 25°C
Pantothenic acid	219.2	$C_9H_{17}O_5N$		+37.5°		4.41
Calcium pantothenate	476.5	$C_{18}H_{32}O_{10}N_2Ca$		+24.3° (+26 to +27.5°)	200 (dec.)	
Sodium pantothenate	241.2	$C_9H_{16}O_5NNa$		+26.5 to +28.5°	160–165 (122–124)	
Panthenol	205.3	$C_9H_{19}O_4N$		+29.5 to +31.5°		
Pantetheine	278.4	$C_{11}H_{23}O_4N_2S$				
4'-phosphopantetheine	358.4	$C_{11}H_{23}O_7N_2SP$		+10.8°		
Pantethine	556.7	$C_{22}H_{44}O_8N_4S_2$		+13.5 to +17.7°		
Coenzyme A	767.6	$C_{21}H_{36}O_{16}N_7SP_3$	257			10.35
Pantolactone	130.1	$C_6H_{10}O_3$		+49.8°	92–93	

Source: Refs. 2, 20.

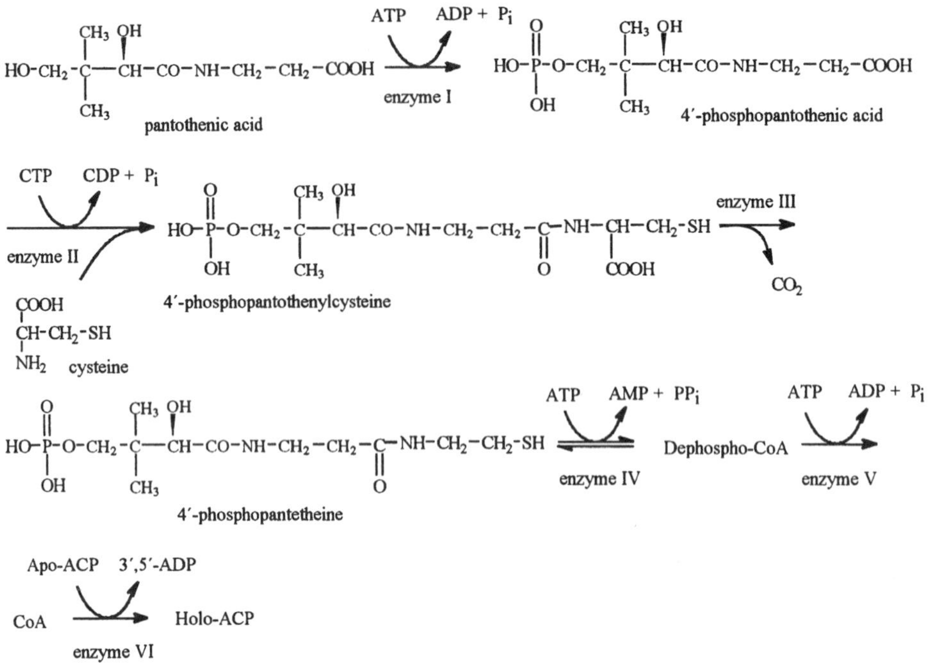

Figure 6 Biosynthesis of coenzyme A and acyl-carrier protein. Enzyme I = pantothenate kinase (EC 2.7.1.33); enzyme II = phosphopantothenylcystein synthetase (EC 6.3.2.5); enzyme III = phosphopantothenylcystein decarboxylase (EC 4.1.1.33); enzyme IV = pantetheine phosphate adenyltrasferase (EC 2.7.7.3); enzyme V = dephospho-CoA kinase (EC 2.7.1.24); enzyme VI = transferase.

in animal tissues are involved in the biosynthesis of holo-acyl-carrier protein (holo-ACP) and its hydrolysis to 4′-phosphopantetheine, respectively (2,21,22).

Ingested pantothenic acid is transported by the blood to the organs. The previously described five steps are required to re-form CoA from the pantothenic acid. The ingested CoA is first hydrolyzed in the intestinal lumen to pantothenic acid and pantetheine via 4′-phosphopantetheine. Panthenol is more easily absorbed than pantothenic acid and is converted to the latter (21). The excretion of pantothenic acid or its derivatives occurs mainly via the urine, in which the free form of the vitamin predominates.

Coenzyme A can form high-energy bonds with acetic acid via its sulfhydryl group to yield acetyl-coenzyme A (acetyl-CoA), also referred to as activated acetate. Through a variety of biochemical sequences, CoA can also be converted to a number of other acyl-CoA derivatives, such as malonyl-CoA, methylmalonyl-CoA, succinyl-CoA, etc. The energy required for the transformation of CoA to acetyl-CoA is derived either from the cleavage of ATP or from a thioclastic cleavage or oxidative decarboxylation. Acetyl-CoA is formed, for example, from fatty acids, amino acids, and carbohydrate metabolism or comes from its own metabolic pool. Acetyl-CoA and short-chain acyl-CoA esters subsequently enter various biochemical reactions that are of particular importance in the metabolism of carbohydrates, fats, and nitrogen-containing compounds such as citrate cycle, fatty acid synthesis, and synthesis of phospholipids, steroids, heme, and some secondary plant metabolites. Acetyl-CoA facilitates the interchange of two-carbon units between donors such as pyruvate, acetoacetate, and acetyl phosphate, and acceptors such as acetoacetate, oxaloacetate, and choline. It is capable of "head condensation," carboncarbonyl condensation, and "tail condensation" such as the acetyl group of acetyl-CoA condensing through its methyl group with oxaloacetic acid–forming citric acid; it is capable of acetylation of amines such as choline, aldol condensation of aldehydes, etc. The biosynthesis of the long-chain fatty acids takes place on the fatty acid–synthase complex. The growing chain of the fatty acid remains covalently bound to the enzyme, where the panto-thenic acid serves as the binding agent. It is bound to a serine residue on the protein via a phosphate group forming the above-mentioned complex (Fig. 3) of pantetheine and protein called ACP (2,5,21,22).

The natural higher homologue of pantothenic acid called homopantothenic or hopantenic acid (Fig. 7) occurs in biological fluids of animals (23,24).

D. Physiology

About 6 to 8 mg/day has been established as the desirable pantothenic acid intake for adult human beings. Adolescent individuals (over 12 years) and pregnant and nursing women have a particularly high requirement for pantothenic acid (the recommended daily allowance is 10 to 15 mg). The normal diet contains 6 to 12 mg pantothenic acid per day, so that severe deficiency symptoms are rarely seen and are difficult to produce experimentally (2,4,8).

However, with prolonged maintenance on diets deficient in pantothenic acid and/or administration of its antagonists, it is possible to produce an extensive

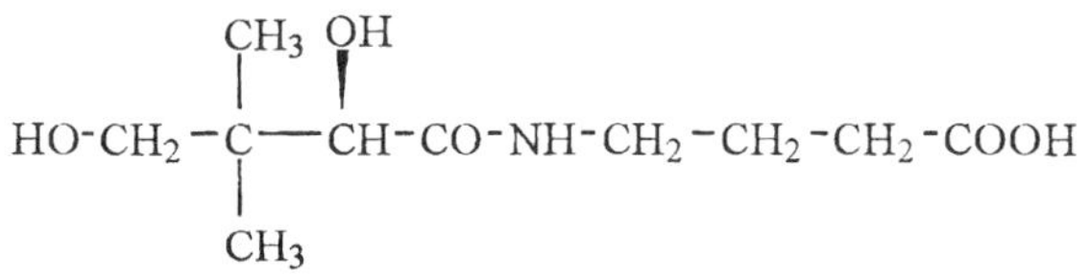

$$HO\text{-}CH_2\text{-}\underset{\underset{CH_3}{|}}{\overset{\overset{CH_3}{|}}{C}}\text{---}\underset{}{\overset{\overset{OH}{|}}{CH}}\text{-}CO\text{-}NH\text{-}CH_2\text{-}CH_2\text{-}CH_2\text{-}COOH$$

Figure 7 (*R*)-Hopantenic acid structure.

array of metabolic abnormalities and accompanying symptoms (dermatitis, burning sensations, intestinal disorders, vasomotor instability, psychoses, depression, increased susceptibility to infection, etc.).

Although deficiency conditions are known, and although addition of pantothenic acid salts to food is known to improve the nitrogen balance, few efforts have been made so far to supplement foods with this vitamin. Recent studies indicate that the increased use of processed foods has reduced the pantothenic acid levels in the diet of industrial countries, which suggests the future need for pantothenic acid supplements (25).

Severe deficiency in animals, such as chick, produces retardation of growth, depigmentation, dermatitis, impaired reproductivity, and neurological abnormalities. Therefore, the feed of domestic animals is currently supplemented to give an optimal level. For example, estimated minimum requirements for pantothenic acid for growing chicks and breeding hens are 10 mg/kg diet and for breeding turkeys 16 mg/kg diet (9).

Even small changes in (R)-pantothenic acid molecular structure reduce its biological activity very strongly or completely. Thus, (S)- or L-(−)-pantothenic acid has no biological activity and is an antimetabolite of (R)-pantothenic acid. Another effective antimetabolite of (R)-pantothenic acid is the so-called pantoyltaurine (sulfopantothenic acid) having a sulfhydryl group instead of carboxyl group in the β-alanine moiety. In animals, pantoyltaurine has no biological effect (2). The only universally effective antimetabolite of pantothenic acid is ω-methylpantothenic acid (with methyl group instead of hydroxymethyl group at the end of the molecule), which evokes symptoms of pantothenic acid deficiency even in human beings.

The calcium salt of hopantenic acid (Fig. 7) has been used for improving blood circulation and metabolism in the brain (26).

E. Occurrence and Stability in Foods

Pantothenic acid is found in practically all foods of plant and animal origin, but usually only in small quantities. It occurs rarely in free form as it is usually a component of coenzyme A, acyl-coenzymes A, and the acyl-carrier protein. The total levels of pantothenic acid in individual foods vary greatly (6). Relatively large amounts are present in meats (especially in organs such as liver), fish, yeasts, eggs, several kinds of cheese, whole-grain products, and legumes. The lowest amounts have been found in milk, fruits, and vegetables (Table 2).

Pantothenic acid in foods has proven to be relatively labile during storage and, especially, during thermal processing. The stability, even at elevated temperatures, is highest at pH 4 to 5. The losses upon dissolution in water during operations such as washing and leaching into the cooking water during blanching and boiling often exceed losses due to the degradation of the vitamin by hydrolysis (27).

Table 2 Pantothenic Acid Content of Some Foods

Food	Content in mg/kg (edible portion)	Food	Content in mg/kg (edible portion)
Pork	3.0–30	Cabbage	1.0–3.0
Beef	3.0–20	Spinach	1.8–27
Chicken	5.3–9.6	Tomatoes	3.0–4.0
Liver (pork)	4.0–200	Carrots	3.0
Fish	1.2–25	Potatoes	3.0
Milk	0.4–4.0	Apples	1.0
Cheese	2.9–4.0	Citrus fruits	2.0
Eggs	16–55	Bananas	2.0
Wheat flour	8.0–13	Nuts	1.0
Bread	4.0–5.0	Bakers' yeasts	50–200
Legumes	9.4–14	Mushrooms	20

Source: Refs. 2, 5, and 27.

Losses of 12% to 50% were reported in various cooked meats and fish, mainly caused by transfer to the drip or broth, depending on type of heat treatment, volume of water used, and other factors (28,29).

Pasteurization of milk apparently does not influence the vitamin content (30). During sterilization of milk (heating to 112°C for 10 min) the vitamin content reportedly decreases by 14%. The mean retention in UHT (ultra-high temperature) sterilized milk was approximately 96% (31). After 6 weeks of storage at room temperature the losses increase to 30%, and the total pantothenic acid losses caused by UHT heat treatment and storage amount to 20% to 35% (32). In dried milk after storage for 8 weeks at 60°C the loss was 18%. The natural vitamin content in fermented milk products was affected only slightly by the fermentation (33,34).

In grains, the highest levels of pantothenic acid are present in the outer layer and the vitamin is mostly lost during milling. Relatively large amounts are present in whole-grain products including certain types of bread (5). The vitamin retention in bread is relatively high (being about 90%), while cooking of pasta products and of legumes results in low retention (55% to 75% and 44% to 75%, respectively) of the vitamin (35).

The losses of pantothenic acid in canned fruits and fruit juices were about 50%; in canned vegetables the losses ranged from 46% to 78% (5).

F. Analysis

Biological, chemical, and physical methods used for the analysis of pantothenic acid and its derivatives have been reviewed in several excellent books and review articles (2,36–38).

The formerly frequently used animal tests have been gradually replaced by simpler, more rapid, and cheaper microbiological methods based on measurements of the growth of yeasts and bacteria. The microbiological methods are especially well suited for the determination of the vitamin in samples such as foods and feeds, where the vitamin is usually present in bound forms. The detection limit of these methods is about 0.03 to 0.05 µg pantothenic acid/mL (5,39). The most commonly employed analytical assay is currently the assay procedure using *Lactobacillus plantarum*. This procedure can be extremely time-consuming, needs careful control of assay conditions, requires regular maintenance of cultures, and can be susceptible to interferences from unidentified compounds causing inhibition or stimulation of the micro-organisms. Therefore numerous alternative biochemical, chemical, and physicochemical methods have been developed.

Pantothenic acid and pantetheine may be assayed using several enzymatic tests with pantothenase (40). Immunoassays and, especially, radioimmunoassays have been used in the analysis of pantothenic acid in tissue fluids, because of their sensitivity, specificity, and high rates of sample throughput. The radioimmunoassay (RIA) methods have about the same sensitivity as microbiological tests (approximately 0.05 µg/mL). As an alternative to the radioimmunoassay, the enzyme-linked immunosorbent assay (ELISA) is particularly suitable for the routine analysis of foods (41–45).

Chemical methods are mainly based on analysis of pantothenic acid hydrolysis products by spectrophotometry or fluorometry. Although relatively rapid to perform, these methods lack the specificity and sensitivity needed to determine vitamin in foods or differentiate between D and L forms. Several spectrophotometric determinations of pantothenic acid and its salts have been developed. They are based on the reaction of pantolactone with hydroxylamine and 2,7-naphthalenediol. Most published methods use the determination of β-alanine. Ninhydrin, 1,2-naphthoquinone, *o*-phthalaldehyde, acetylacetone, and some other reagents have been introduced for the spectrophotometric determination of pantothenates via β-alanine (5,46,47).

Some other chemical and physical methods, such as thermal analysis and nuclear magnetic resonance analysis, have been occasionally used for the analysis of pantothenates in pharmaceutical preparations. Pure compounds can also be determined chelatometrically or using alkalimetric titration in nonaqueous solvents or can be analyzed after decomposition to ammonia (5,48,49).

Panthenol can be determined polarographically. A sensitive fluorometric determination in multivitamin preparations (hydrolysis, reaction with ninhydrin) has been described (5).

Because of the importance of coenzyme A and its acyl thioesters in metabolic control, several enzymatic assays in conjunction with spectrometric or fluorometric monitoring have been developed to determine their total content in tissues (5).

II. THIN-LAYER CHROMATOGRAPHY

Thin-layer chromatography (TLC) and paper chromatography (PC) were widely used for the separation of pantothenic acid, its salts, and panthenol about two decades ago. With the development of GLC and HPLC methods, TLC lost its leading position in both the qualitative and quantitative analysis of these compounds. TLC is a simple, rapid, and inexpensive method, so it is still of some importance for controlling the purity of pharmaceutical products.

The method described recently by Nag and Das (47) is an example of modern TLC procedures useful for the determination of pantothenic acid and/or panthenol in pharmaceutical preparations containing other vitamins, amino acids, saccharides, and enzymes. The procedure is based on extraction of the vitamin with ethanol (tablets and capsules) or benzyl alcohol (liquid preparations). The extract is then separated on a silica gel layer with isopropyl alcohol-water (85:15, v/v) as the developing solvent. β-alanine and/or β-alanol is liberated by heating the plate for 20 min at 160°C and the liberated amines are visualized by ninhydrin and determined in situ densitometrically at 490 nm. Recoveries for panthenol and pantothenic acid were $99.8 \pm 2.3\%$ and $100.2 \pm 1.7\%$, respectively.

Details about TLC (and PC) can be found in the previous edition of this book (5).

III. HIGH-PERFORMANCE LIQUID CHROMATOGRAPHY

A. General Considerations

In the last two decades high-performance liquid chromatography (HPLC) has become a valuable method for the separation and quantitation of pantothenic acid and some of its derivatives. Several methods for the determination of pantothenic acid and its salts (mainly calcium pantothenate), or panthenol, in pharmaceutical preparations have been developed and are applied on a routine basis in a few research and control laboratories (50–52).

An HPLC method was also applied for the determination of calcium pantothenate in infant milk formulas fortified with this salt (53). However, none of HPLC methods are yet useful for the determination of the total content of pantothenic acid in complex matrices such as foods and feeds.

In the last few years, chiral stationary phases have been developed and used in HPLC for the determination of optical purity of synthetic pantothenic acid and related compounds (54–56).

Of particular interest are the HPLC methods developed for the separation and quantitation of coenzyme A and its short-chain acyl analogues in biological materials (57–59).

HPLC techniques present several potential advantages over earlier spectro-photometric, fluorometric, and official microbiological assay methods, which are very tedious and often difficult to perform. These advantages include simple sample preparation, direct analysis of compounds without any derivatization, and potential simultaneous determination of several compounds in a single run, in addition to a reduction of analysis time, excellent precision, and relatively good sensitivity.

Almost entirely reversed-phase materials have been used, and most separations have been performed on octadecyl columns. Mixtures of polar organic solvents such as methanol or acetonitrile with water or appropriate buffers are the most common eluents (60,61). The ion pair technique of a reversed-phase HPLC has been recently used (51).

Only spectrophotometric and exceptionally refractometric detection are suitable for direct routine analysis. To facilitate the analysis and increase the sensitivity, suitable derivatives allowing fluorometric detection have been prepared.

Recently, HPLC-MS (electron impact, chemical ionization in positive-ion and negative-ion modes) has been used for the analysis of pantothenic acid in an artificial mixture with other water-soluble vitamins (62).

Overviews of recent methods and applications are given in Tables 3 and 4. Older methods have been reviewed in the previous edition of this book (5).

B. Pantothenic Acid and Pantothenates

1.   Multivitamin Pharmaceuticals

Pantothenic acid and its salts as well as its degradation products such as pantoic acid and β-alanine do not exhibit significant absorption above 220 nm. As a result, the limitation of the direct HPLC assays lies in the lack of a selective detection wavelength. Detection is mainly performed by UV absorption at low wavelength. Analysis using UV detection below 220 nm has inherent problems because of the limited number of common mobile-phase solvents that have appropriate cutoff and because of dissolved oxygen that has to be removed via sonication under vacuum.

Pantothenic acid and pantothenates may also be analyzed following derivatization to extend the chromophore and hence allow UV detection at higher wavelengths or fluorometric detection. Hudson et al. (63) have attempted to analyze the vitamin as a β-alanine-fluorescamine complex. The derivatization procedure was lengthy and required extensive sample cleanup before the hydrolysis step due to the interference of riboflavin, niacinamide, and some minerals such as zinc, copper, manganese, and molybdenum. Although these interferences were eliminated, the method did not yield reproducible results.

Table 3 HPLC Conditions for the Analysis of Pantothenic Acid and Its Derivatives

Compound separated	Sample preparation	Stationary phase	Mobile phase	Detection	Reference
Pantothenic acid	Dissolved in water	Spherisorb ODS C18	Octylamine o-phosphate or octylamine salicylate, pH 6.4	UV 210 nm	51
	Dissolved in 0.005 M KH$_2$PO$_4$, pH 5	LiChrosorb NH$_2$	Acetonitrile-0.005 M KH$_2$PO$_4$, pH 5 (87:13, v/v)	UV 210 nm	63, 64
	Dissolved in water	Hypersil ODS C18	Acetonitrile-0.25 M NaH$_2$PO$_4$, pH 2.5 (3:97, v/v)	UV 205 nm	50
Panthenol	Dissolved in 0.5 M HCl, hydrolysis, derivatization with fluorescamine	Chromegabond C18	Methanol-sodium borate buffer, pH 8 (30:70, v/v)	Fluorescence: ex 390 nm, em 475–490 nm or UV 390 nm	67
	Dissolved in 0.5 M HCl, hydrolysis, derivatization with fluorescamine	μBondapak C18	Acetonitrile-0.1 M ammonium acetate (18:82, v/v)	Fluorescence: ex 390 nm, em 475–490 nm	63, 64
D,L-pantothenic acid, D,L-panthenol, and D,L-pantolactone	Hydrolyzed to pantoic acid with 0.5 M NaOH	MCI gel CRS 10W	0.002 M CuSO$_4^-$ acetonitrile (90:10, v/v)	UV 254	54
D,L-pantothenic acid	Derivatization with methanol/HCl and 3,5-dinitrophenyl isocyanate	(1) YMC A-KO3; (2) Sumipax OA-4000; (3) TSK gel Enantio P1	(1) hexane-dichloromethane-ethanol (70:30:8); (2) and (3) hexane-dichloromethane-ethanol (70:30:5)	UV 254	54
D,L-panthenol	Derivatization with 3,5-dinitrobenzoyl chloride	(1) YMC A-KO3; (2) Sumipax OA-4000; (3) TSK gel Enantio P1	(1) hexane-dichloromethane-ethanol (70:30:10), (2) hexane-dichloromethane-ethanol (70:30:1); (3) hexane-dichloromethane-ethanol (70:30:5)	UV 254	54
D,L-pantolactone	Derivatization with 3,5-dinitrophenyl isocyanate	(1) YMC A-KO3; (2) Sumipax OA-4000; (3) TSK gel Enantio P1	(1) hexane-dichloromethane-ethanol (70:30:8); (2) and (3) hexane-dichloromethane-ethanol (70:30:5)	UV 254	54

Table 4 HPLC Conditions for the Analysis of CoA and Short-Chain Acyl-CoA Esters

Sample preparation	Stationary phase	Mobile phase	Detection	Reference
Neutralized $HClO_4$ extract	ODS Ultrasphere C18	NaH_2PO_4 buffer-acetonitrile gradient	UV 254 nm	57
Neutralized $HClO_4$ extract, concentrated on a Sep-Pak C18 cartridge	Develosil ODS	Water-acetonitrile gradient	UV 260 nm	58
Extract with sulfosalicylic acid	Hypersil ODS with Pelliguard LC-18 guard column	$0.1\ M$ NaH_2PO_4/$0.075\ M$ sodium acetate buffer, pH 4.6–70% buffer in methanol, gradient	UV 254 nm	60
Extract with KH_2PO_4 buffer, pH 7.5 (further purified)	Hypersil ODS with LiChrosorb RP18 guard column	$0.2\ M$ NaH_2PO_4 buffer, pH 5.0-methanol (100:17, v/v)	UV 254 nm	61
Neutralized $HClO_4$ extract	Spherisorb ODS II C18	(1) $0.22\ M$ KH_2PO_4 buffer, pH 4; (2) chloroform-methanol (2:98, v/v)	UV 254 nm	69

The first successful separation of calcium pantothenate in multivitamin preparations that allowed the quantitative determination in <5 min was reported by Jonvel et al. (65). They used a 20-cm Nucleosil 7 C_{18} reversed-phase column, an acetic acid–water (5:95, v/v) mobile phase at a flow rate of 2 mL/min, and a refractometer as a detector. Samples were simply dissolved in the mobile phase and analyzed without any internal standard. Mean recoveries of the assays of commercially available calcium pantothenate tablets containing also thiamine, riboflavin, niacinamide, and pyridoxine ranged from 96.8% to 104.4%. The detection limit was 50 ng injected.

Franks and Stodola (66) analyzed calcium panthothenate employing a 250×4.6 mm Zorbax C-8 column equipped with a guard MPLC 30-mm cartridge column packed with RP-18 Spheri-5 sorbent. Calcium pantothenate was extracted from tablets with a mixture of methanol-water (25:75, v/v) containing adipic acid as the internal standard. The mobile phase was a mixture of methanol-NaH_2PO_4 buffer (pH 3.5), and the effluent was monitored at 214 nm. A highly retained excipient peak eluted approximately 2 h after calcium pantothenate peak (retention time 8 min) and hence interfered with subsequent chromatograms. A column-switching arrangement was employed to shorten the chromatography run time. Average recovery was 99.7%, and relative standard deviation ranged from 0.83% to 2.32%.

Pantothenic acid/calcium pantothenate in pharmaceutical products and vitamin premixes was also analyzed using low-wavelength ultraviolet (UV) detection (64,66). The vitamin was extracted from tablets or powdered premixes with 0.005 M NaH_2PO_4 buffer (pH 4.5) and separated from other water-soluble vitamins on an aminopropyl-bonded silica column (LiChrosorb NH_2) eluted with an acetonitrile–0.005 M NaH_2PO_4 buffer (pH 4.5) (87:13, v/v) and detected at 210 nm. Quantitative recoveries (>95%) and relative standard deviations 0.79% to 2.2% were obtained for multivitamin tablets, vitamin premixes, fortified yeasts, and raw materials. The limit of sensitivity was approximately 1 mg/g sample. The results were compared with those obtained by the standard microbiological procedure. Low levels of calcium pantothenate (<3 mg per tablet) were more precisely analyzed by the HPLC procedure than by the microbiological method.

A fast, simple, and sensitive method was developed for the determination of calcium pantothenate in commercial multivitamin tablet formulations and raw materials (50). The chromatographic system included a 5-μm reversed-phase C_{18} column (150 $\times$ 4.6 mm) and a mobile phase consisting of acetonitrile and 0.25 M NaH_2PO_4 buffer of pH 2.5 (97:3, v/v). The column effluent was monitored by UV detection at 205 nm. The sample preparation involved only extraction in water followed by filtration of the extract. The method had a detection limit of approximately 50 ng/mL sample, with mean recovery ranging from 98.7% to 99.8%. The relative standard deviation ranged from 0.3% to 2.0% (Fig. 8).

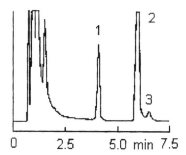

Figure 8 Determination of calcium pantothenate in chewable multivitamin tablets. Chromatographic conditions: see Table 3. Peak identification: (1) pantothenate; (2) saccharin; (3) 2-sulfamoylbenzoic acid. (From Ref. 50.)

Gennaro (51) proposed a method for the separation of water-soluble vitamins by means of the ion interaction reagent using octylamine o-phosphate or octylamine salicylate buffer (at pH 6.4) as the interaction reagent and the mobile phase at a flow rate of 1 mL/min, and a 2.5-μm Spherisorb ODS C18 column (250 × 4.6 mm) as the stationary phase. The column effluent was monitored at 210 nm. Retention times of pantothenic acid obtained with octylamine o-phosphate and octylamine salicylate were 64.0 and 9.8 min. The method was used for the determination of pantothenic acid in a model mixture of water-soluble vitamins and also in a commercial multivitamin isotonic salt dietetic drink (Fig. 9).

2. Foodstuffs

A reversed-phase method was adopted for the assay of pantothenic acid added to infant milk formulas as a vitamin premix containing calcium pantothenate (53). Sample preparation consisted of deproteination of the formulas with acetic acid/ sodium acetate solutions, followed by centrifugation and filtration. The chromatographic system included a Supersphere C-18 column (5 μm, 250 × 4.6 mm) and a mobile phase consisting of a 0.25 M NaH_2PO_4 buffer (pH 2.5)–acetonitrile (97:3, v/v). The column effluent was monitored at 197 nm. The chromatogram obtained analyzing the starting infant formula containing 55 mg pantothenic acid per kilogram is given in Figure 10. The recoveries of pantothenic acid ranged from 89% to 98%, and the coefficients of variation ranged from 1.2% to 3.2%. The results obtained with the HPLC method and the microbiological standard official assay with *Lactobacillus plantarum* were highly correlated.

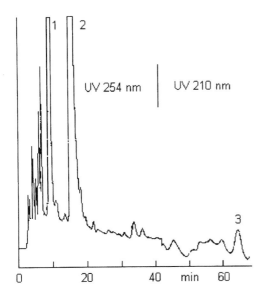

Figure 9 Determination of pantothenic acid in a multivitamin isotonic salt dietetic drink. Chromatographic conditions: see Table 3. Peak identification: (1) L-ascorbic acid; (2) niacin; (3) pantothenic acid. (From Ref. 51.)

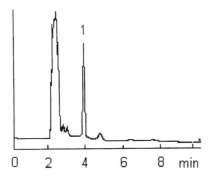

Figure 10 Determination of pantothenic acid in starting infant formula. Column: Supersphere C18 (5 μm, 250 × 4.6 mm); mobile phase: 0.25 M NaH_2PO_4 buffer (pH 2.5)-acetonitrile (97:3, v/v); flow rate: 1 mL/min; detection: UV 197 nm; peak identification: (1) pantothenic acid. (From Ref. 53.)

C. Panthenol

Panthenol can be determined using the procedure developed by Umgat and Tscherne (67), which is useful for the analysis of panthenol in premixes and multivitamin preparations. The method is based on the acid hydrolysis of panthenol with 0.5 M HCl at 85°C and derivatization of the released 3-amino-1-propanol (β-alanol) with fluorescamine, a reagent that reacts specifically with primary amines. The condensation product of aminopropanol and the corresponding derivative of 6-aminohexanoic acid (ε-aminocaproic acid) used as the internal standard were separated on a 300 × 4.6 mm Chromegabond C_{18} column and detected by using either a spectrofluorometer with the excitation wavelength at 390 nm and the emission wavelength of 475 to 490 nm or by measuring the absorbance of 390 nm (Fig. 11). The fluorometric measurement had to be used for trace analysis. The mobile phase was prepared by mixing methanol with 0.1 M borate buffer (adjusted to pH 8 with 2 M NaOH) in a ratio of 30:70, v/v (Table 3). In comparison to the earlier methods, the HPLC procedure was faster, more specific, and also suitable to monitor natural trace levels of aminopropanol in panthenol. The results obtained by the HPLC method were in close agreement with those obtained using nonaqueous titration with perchloric acid, spectrophotometric determination with 1,2-naphthoquinone 4-sulfonate, or microbiological assay. Recoveries found for pure D- and DL-panthenol were 99.6% and 100.5%, respectively, whereas for pharmaceutical preparations containing D-panthenol

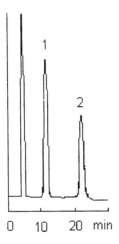

Figure 11   Determination of panthenol in multivitamin tablets. Chromatographic conditions: see Table 3. Peak identification: (1) 6-aminohexanoic acid (internal standard)-fluorescamine derivative; (2) 3-amino-1-propanol fluorescamine derivative. (From Ref. 67.)

they ranged from 97.9% to 103.1%. In multivitamin preparations where large amounts of riboflavin were present, the peak of panthenol was not fully resolved from that of the internal standard, which elutes immediately after riboflavin. In that case, 2-aminoethanol had to be used as the internal standard.

A modification of this procedure was described by Hudson et al. (63) and used for the determination of panthenol in liquid multivitamin products. A sample aliquot containing approximately 1 mg panthenol was hydrolyzed for 30 min at 85°C in 0.5 M HCl and centrifuged, and the supernate was mixed with fluorescamine reagent and diluted with 1% (w/w) borate buffer. The aminopropanol-fluorescamine complex was chromatographed on a μBondapak C_{18} column with acetonitrile-0.1 M ammonium acetate (18:82, v/v) as mobile phase and using fluorescence detection (the excitation and emission wavelengths were set as mentioned above at 390 and between 475 and 490 nm, respectively). This procedure was limited to samples containing at least 100 μg/g panthenol. The average recovery was 99% and precision was 2.9% (relative standard deviation) at 1.03 mg/g panthenol. Excellent agreement was found when the HPLC method was compared with the microbiological assay procedure.

D. Optical Isomers of Pantothenic Acid and Related Compounds

In the last few years, numerous chiral stationary phases have been developed for optical resolution by HPLC. Two enantiospecific HPLC methods have been described for the separation of pantothenic acid, panthenol, and pantolactone enantiomers (54).

The first method is based on the conversion of racemic pantothenic acid and its derivatives (panthenol and pantolactone) to DL-pantoic acid. The hydrolysis is carried out in 0.5 M NaOH at 70°C for either 30 min (pantolactone) or 60 min (pantothenic acid and panthenol). Pantoic acid is directly resolved on a ligand-exchange chiral stationary phase MCI gel CRS 10W column (Fig. 12; Table 3).

The second method needs derivatization with either 3,5-dinitrophenyl isocyanate or 3,5-dinitrobenzoyl chloride which is followed by separation on a chiral acrylic polymer YMC A-KO3 (a conjugated D-naphthylethyl amine on a silica gel surface), Sumipax OA-4000, and/or TSK gel Enantio P1.

Pantothenic acid (10 mg) is first esterified with 5 mL of 1.5 M HCl in methanol at 50°C for 30 min and then derivatized with 3,5-dinitrophenyl isocyanate in toluene (2 mL) containing pyridine (0.05 mL). The arised 3,5-dinitrophenyl carbamate (Fig. 13) dissolved in chloroform is used for HPLC separation (Fig. 14).

Samples of DL-panthenol and DL-pantolactone are converted to 3,5-dinitrophenyl carbamates in the same way as pantothenic acid, but without esterification. 3,5-Dinitrobenzoyl esters of these substances are prepared analogously as the

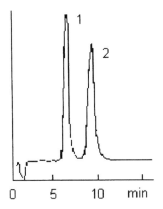

Figure 12 Separation of racemic pantoic acid converted from panthenol. Column: MCI gel CRS 10W (3 μm, 50 × 4.6 mm); mobile phase: 0.002 *M* CuSO₄-acetonitrile (90:10, v/v); flow rate: 0.8 mL/min; temperature: 35°C; detection: UV 254 nm; peak identification: (1) D-pantoic acid, (2) L-pantoic acid. (From Ref. 54.)

corresponding carbamates using 3,5-dinitrobenzoyl chloride in tetrahydrofuran and pyridine. Typical chromatograms of enantiomeric 3,5-dinitrophenyl carbamates of pantolactone (Fig. 15) and 3,5-dinitrobenzoyl esters of panthenol (Fig. 16) can be seen in Figures 17 and 18, respectively.

E. Coenzyme A and Related Compounds

Several methods have been developed to separate and determine coenzyme A and short-chain coenzyme A esters in tissue extracts by HPLC and were recently reviewed (5,68). The older methods provided chromatographic separation of a

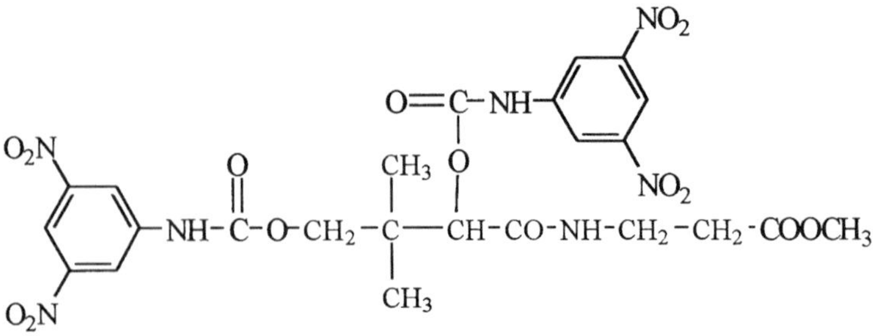

Figure 13 Structure of 3,5-dinitrophenyl carbamate of pantothenic acid methylester.

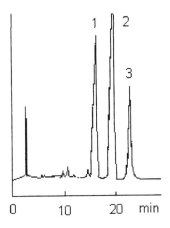

Figure 14 Separation of racemic pantothenic acid. Column: YMC A-KO3 (250 × 4.6 mm); mobile phase: hexane-dichloromethane-ethanol (70:30:8, v/v/v); flow rate: 0.8 mL/ min; temperature: 35°C; detection: UV 254 nm; peak identification: (1) 3,5-dinitrophenyl carbamate of D-pantothenic acid methylester; (2) unknown; (3) 3,5-dinitrophenyl carbamate of L-pantothenic acid methylester. (From Ref. 54.)

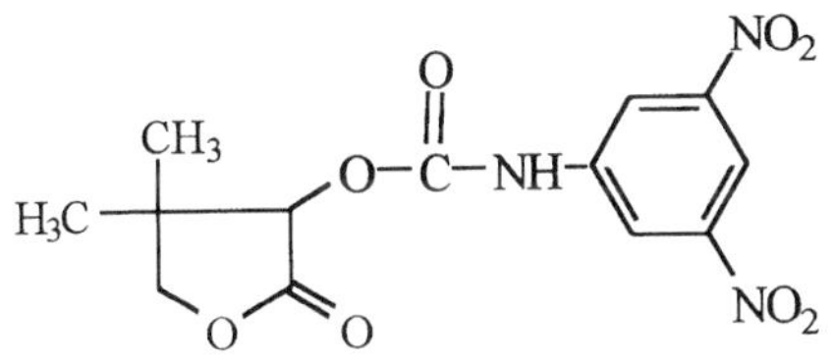

Figure 15 Structure of 3,5-dinitrophenyl carbamate of pantolactone.

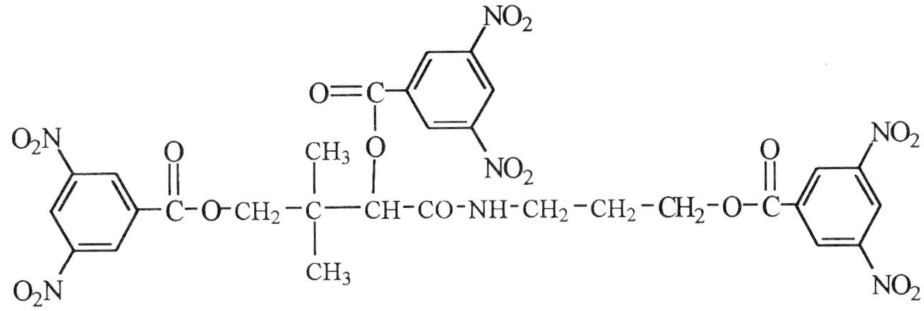

Figure 16 Structure of 3,5-dinitrobenzoyl ester of panthenol.

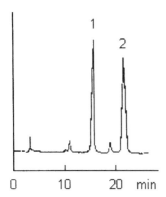

Figure 17   Separation of racemic pantolactone. Column: YMC A-KO3 (250 × 4.6 mm); mobile phase: hexane-dichloromethane-ethanol (70:30:8, v/v/v); flow rate: 0.8 mL/min; temperature: 35°C; detection: UV 254 nm; peak identification: (1) 3,5-dinitrophenyl carbamate of D-pantolactone; (2) 3,5-dinitrophenyl carbamate of L-pantolactone. (From Ref. 54.)

few CoA esters only, employing corrosive ion-pairing reagents or a combination of three consecutive chromatographic systems. More recently, a reversed-phase chromatographic separation system suitable for the determination of CoA, dephospho-CoA, and acetyl-CoA has been developed, modified, and employed for the separation of CoA and various short-chain acyl CoA esters in rat tissues (69). Separation of the various compounds was accomplished by using a 250 × 4.6

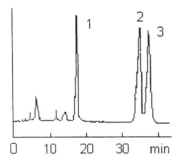

Figure 18   Separation of racemic panthenol. Column: YMC A-KO3 (250 × 4.6 mm); mobile phase: hexane-dichloromethane-ethanol (70:30:10, v/v/v); flow rate: 0.8 mL/min; temperature: 35°C; detection: UV 254 nm; peak identification: (1) unknown; (2) 3,5-dinitrobenzoyl ester of L-panthenol; (3) 3,5-dinitrobenzoyl ester of D-panthenol. (From Ref. 54.)

mm Spherisorb ODS II, 5-μm C_{18} column fitted with a 45×4 mm guard column filled with reversed-phase Bio-Sil P1B-ODS beads. The mobile phase solvents were (1) 0.22 M NaH_2PO_4 buffer of pH 4, containing 0.05% (v/v) thiodiglycol to prevent oxidation of CoA, and (2) chloroform-methanol (2:98, v/v). The effluent was monitored at 254 nm (at about the λ_{max} of adenosine), and nearly baseline separation was achieved for a standard mixture of free CoA, methylmalonyl-CoA, β-hydroxy-β-methylglutaryl-CoA, succinyl-CoA, acetoacetyl-CoA, acetyl-CoA, propionyl-CoA, β-methylcrotonyl-CoA, and isovaleryl-CoA. Profiles of metabolically accumulated CoA derivatives were determined in rat heart and liver and in isolated liver mitochondria. However, the method did not appear to be applicable to the direct determination of small amounts of these derivatives in tissues. The HPLC assay procedure gave results similar to the enzymatic assay of CoA (Table 4).

The first quantitative determinations of HPLC of CoA compounds have been presented by King and Reiss (57). The authors developed a reversed-phase HPLC procedure to measure CoA and short-chain CoA compounds in rat liver tissue. Seventeen CoA-related standards in model experiments were separated and quantified in a 37-min run employing a 75×4.6 mm ODS Ultrasphere octadecylsilica column (Fig. 19). In rat liver CoA, acetyl-CoA and six minor CoA-related metabolites could be quantitated (Fig. 20). The separation was achieved using an Na_2HPO_4-acetonitrile gradient. The eluate was monitored at 254 nm. Recovery of CoA standards added in tissue extracts ranged from 83% to 107%. The procedure allowed measurement of as little as 12 pmol of CoA compounds injected.

More sensitive HPLC reversed-phase procedures for the determination of CoA and short-chain acyl-CoA esters in normal rat liver extracts have recently been developed by Hosokawa et al. (58). The acyl-CoA esters present in the neutralized $HClO_4$ extract were concentrated on a Sep-Pak C_{18} cartridge. The cartridge was washed with water and a series of organic solvents and subsequently the acyl-CoA esters were eluted with ethanol-water (65:35, v/v) containing 0.1 M ammonium acetate. The eluate was analyzed using 150×4.6 mm and 250×4.6 mm Develosil ODS columns. The separation of eight CoA compounds was conducted with a linear gradient (from 1.75% to 10%) of acetonitrile in water (v/v). The elution was monitored at 260 nm. Isobutyryl-CoA was used as the internal standard. The lower detection limit of the individual acyl-CoA esters was about 50 pmol injected. Recoveries of authentic short-chain acyl-CoA esters ranged from 56% to 77%.

A reversed-phase method was used for the detection and quantification of [14C]-coenzyme A esters released from rat liver mitochondrial proteins (by treatment with thiols) modified with [1-[14C]pantothenic acid (59). For quantitative analysis, a 119×4 mm Supersphere 100 CH-18 column was used. It was equilibrated for 10 min at a flow rate of 1 mL/min in solvent A (0.02 M KH_2PO_4

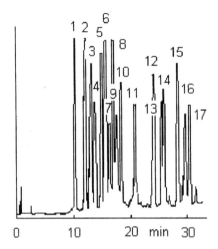

Figure 19 Separation of coenzyme A and its analogs (standards). Column: 3 µm, C_{18} (ODS Ultrasphere); 75 × 4.6 mm; temperature: 30°C; mobile phase: gradient elution from 0.6% to 18% acetonitrile with constant Na_2HPO_4 concentration of 0.2 M; flow rate: 1 mL/ min; detection: UV at 254 nm; peak identification: (1) malonyl-CoA; (2) glutathione-CoA; (3) CoA (thiol); (4) methylmalonyl-CoA; (5) succinyl-CoA; (6) 3-hydroxy-3-methylglu-taryl-CoA; (7) dephospho-CoA; (8) acetyl-CoA; (9) acetoacetyl-CoA; (10) CoA(disul-fide); (11) propionyl-CoA; (12) crotonyl-CoA; (13) isobutyryl-CoA; (14) butyryl-CoA; (15) 3-methylcrotonyl-CoA; (16) isovaleryl-CoA; (17) valeryl-CoA. (From Ref. 57.)

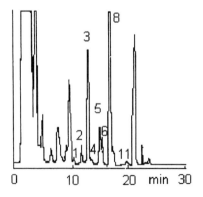

Figure 20 Determination of coenzyme A and its acyl derivatives in liver from a starved rat. Chromatographic conditions and peak identification: see Fig. 19. (From Ref. 57.)

buffer, pH 5.0). Following sample injection, the column was run for 10 min with solvent A and then the CoA esters were eluted with a linear gradient of 0% to 100% solvent B (80% 0.02 M KH$_2$PO$_4$ buffer, pH 5.0–20% methanol, v/v) for 30 min. The efluent was monitored at 254 nm.

A simple and rapid reversed-phase method suitable for the analysis of short-chain coenzymes A, a modification of the method published by Corkey and Deeney (70) and by Bartlett and Causey (71) has recently been described by Demoz et al. (60). Samples of liver, heart, and kidney tissues were homogenized in 5% sulfosalicylic acid containing 50 μM dithiothreitol in 1:9 w/v proportion. Following centrifugation, 20 μL of the supernatant were directly injected onto a 3-μm Hypersil ODS (C$_{18}$) column (100 × 4.6 mm). A gradient elution with sodium phosphate, sodium acetate, and methanol was employed at a constant flow rate of 1.5 mL/min. Compared to earlier methods, the total elution time of short-chain CoA compounds (acetyl-CoA, malonyl-CoA, methylmalonyl-CoA, succinyl-CoA, propionyl-CoA, and free CoASH) was reduced to <20 min. Furthermore, the separation of these short-chain coenzymes was achieved at a constant flow rate, rather than with a changing flow rate as described earlier.

An isocratic HPLC procedure with an RP-18 column Hypersil ODS (250 × 4.6 mm, 5 μm) was recently developed for the separation of short-chain

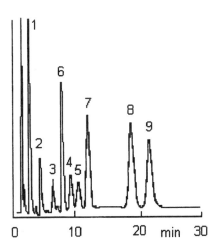

Figure 21 Separation of a model mixture of short-chain coenzyme A esters. Column: Hypersil ODS; 250 × 4.6 mm, 5 μm; mobile phase: 0.2 M Na$_2$HPO$_4$ buffer (pH 5.0)-methanol (100:17, v/v); flow rate 1.5 mL/min; detection: UV at 254 nm; peak identification: (1) guanosine (internal standard); (2) CoASH; (3) HMG-CoA; (4) acetoacetyl-CoA; (5) acetyl-CoA; (6) dephospho-CoA; (7) HMG-dephospho-CoA; (8) acetoacetyl-dephospho-CoA; (9) acetyl-dephospho-CoA. (From Ref. 61.)

CoA esters in *Catharanthus roseus* plant cell cultures by Hermans-Lokkerbol et al. (61). The method allowed baseline separation of CoASH, acetyl-CoA, 3-hydroxy-3-methylglutaryl-CoA (HMG-CoA), 3-methylglutaconyl-CoA (MG-CoA), and their respective 3′-dephospho derivatives (Fig. 21). The elution was done with 0.2 M Na_2HPO_4 buffer (pH 5.0)–methanol (100:17, v/v) at 1.5 mL/min and the detection was conducted at 254 nm. Detection limits were found to be between 2.5 and 6 pmol.

IV. GAS CHROMATOGRAPHY

A. General Considerations

Pantothenic acid, its salts, and panthenol as such are not volatile enough for direct gas-liquid chromatography (GLC). However, it is possible to use this chromatographic technique after derivatization of the polar hydroxyl and carboxyl groups of the vitamin (23,38,72,73,75–79). The majority of the developed methods are, however, applicable only to relatively pure and simple samples such as multivitamin preparations, and certain biological samples, such as urine (75–77). Only a few methods are suitable for the determination of the vitamin in complex matrices such as foods. An overview of methods was given by Velíšek et al. (5).

Two approaches are currently used for the preparation of volatile derivatives of the vitamin. The first approach is mostly based on the conversion of pantothenic acid and/or panthenol to acetyl (72) or trimethylsilyl derivatives (73). A simpler and more convenient approach for most applications seems to be the procedure based on the hydrolysis of the vitamin in acidic medium and analysis of the hydrolysis products. Pantothenic acid, its salts, and coenzyme A as well as its analogs undergo acid hydrolysis with formation of β-alanine and pantolactone. Panthenol breaks down to 3-amino-1-propanol (β-alanol) and pantolactone. For instance, β-alanine can be analyzed as the corresponding *N*-trifluoroacetyl methyl ester or *N*-trifluoroacetyl butyl ester. Pantolactone is sufficiently volatile to be amenable to direct GLC (23,76,77), but it can also be analyzed as the corresponding trimethylsilyl ether, trifluoroacetyl, or isopropylurethane derivative (5,78).

B. Pantothenic Acid and Pantothenates

The nature of the sample is indicative for the suitability for the determination of pantothenic acid or its salts. Most of the described methods are applicable to multivitamin and other pharmaceutical preparations, but only a few of them are also applicable to biological materials.

1. Multivitamin Pharmaceuticals

Calcium and sodium pantothenate can be chromatographed following derivatization, degradation, or a combination of the two. Derivatization is mostly based on the determination of the corresponding acetates or trimethylsilyl ethers. The determination of pantothenic acid and/or pantothenates as acetates (72) requires esterification of the carboxyl group prior to acetylation of the hydroxyl function.

Prosser and Sheppard (72) analyzed the vitamin as ethyl pantothenate diacetate. The pantothenate was first converted to ethyl pantothenate by using ethanolic HCl (2.5%) reaction (at room temperature, for 2 h) and then to pantothenate diacetate using a mixture of acetic acid anhydride-pyridine (1:1, v/v; at room temperature, for 1 h). The product was chromatographed on a polar stationary phase (Table 5; and Fig. 22). Sensitivity of the determination was 5 to 8 ng of injected compound.

Trifluoroacetyl derivatives of pantothenic acid and its salts have been prepared by Prosser and Sheppard (72).

A number of pharmaceutical preparations containing pantothenate have been analyzed after trimethylsilylation of the compound (73). The latter can be carried out using a 2:1:1 (v/v/v) mixture of bis(trimethylsilyl)acetamide (BSA), N-trimethylsilyl imidazole (TMSIM), or trimethylchlorosilane (TMCS) in dimethyl sulfoxide (at room temperature, for 10 min). The resulting product is analyzed on a nonpolar stationary phase (Fig. 23). The detection limit is 2 to 4 ng. Bis(trimethylsilyl)trifluoroacetamide (BSTFA) can be used instead of BSA. TMCS is required only in the derivatization of pantothenates, but pantothenic acid can be also derivatized using a simpler 4:1 (v/v) mixture of BSTFA with TMSIM.

A second group of analytical methods requires preliminary acid hydrolysis. This is conducted in 7 M HCl at 95 to 97°C for 4 h (73) or in 1 M HCl at 80°C for 3 h (23). The hydrolysis product β-alanine has been analyzed after its conversion to the corresponding N-trifluoroacetyl methyl ester or N-trifluoroacetyl butyl ester. Derivatization of β-alanine is a two-step procedure. The compound is converted first to its methyl or butyl ester using alcoholic HCl and then to its N-trifluoroacetate by the use of trifluoroacetic acid anhydride. The resulting derivative is analyzed using a polar stationary phase (23).

More convenient are the methods based on degradation to pantolactone. This compound can be simply extracted in a suitable organic solvent, such as dichloromethane, chloroform, or ethyl acetate and chromatographed on polar stationary phases (Fig. 24) (23,75).

2. Biological Materials

Neither acetylation nor trimethylsilylation is suitable as a derivatization technique for the determination of pantothenic acid in biological materials. Only methods

Table 5   GLC Conditions for the Analysis of Pantothenic Acid, Pantothenates, and Panthenol with the Use of FID

Compound analyzed	Derivative	Column (length × ID in mm)	Stationary phase	Solid support (mesh)	Column temperature (°C)	Reference
Pantothenates, panthenol	Acetyl	Glass, 2440 × 4	2% Neopentylgly-colsebacate	Anakrom ABS (110/120)	230	72
Pantothenates, panthenol	Trifluoroacetyl	Glass, 1830 × 4	3% OV-17	Diatomaceous earth (100/120)	135	72
Pantothenates, panthenol	Trimethylsilyl	Glass, 2440 × 4	5% SE-30	Gas Chrom Q (100/120)	185	73
Pantothenates, panthenol	Trimethylsilyl	Glass, 2440 × 4	3% OV-1	Anakrom ABS (110/120)	200	73
Pantothenates, panthenol	Pantolactone	Glass, 2100 × 4	1% Carbowax 20M	Gas-Chrom P (100/120)	115	23
D,L-Pantolactone	β-alanine-trifluoro-acetylalkylester/pantolactone	Glass, 2400 × 2.1	10% Carbowax 20M	Chromaton N-AW-DMCS (0.125–0.16 mm)	120–220 (5°C/min)	77
D,L-Pantothenic acid	Methylbis(trifluoro-acetyl)	Glass, 15,000-mm-long capillary	XE-60-L-Val-(R)-α-phenylethylamide		From 130 (2°C/min)	78
D,L-Panthenol	Trifluoroacetyl	Fused silica, 15,000-mm-long capillary	XE-60-L-Val-(S)-α-phenylethylamide		From 130 (1.5°C/min)	78
D,L-Pantolactone	None	Fused silica, 50,000-mm-long capillary	XE-60-L-Val-(S)-α-phenylethylamide		160	78
D,L-Pantolactone	Trifluoroacetyl	Glass, 40,000-mm-long capillary	XE-60-L-Val-(S)-α,α'-naphthylethylamide		80	78
D,L-Pantolactone	Isopropylurethane	Glass, 15,000-mm-long capillary	XE-60-L-Val-(R)-α-phenylethylamide		From 145 (2°C/min)	78

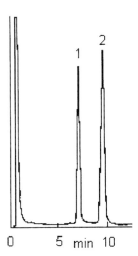

Figure 22 Determination of calcium pantothenate and panthenol as acetyl derivatives. Detector: FID; column: glass, 2440 × 4 mm; stationary phase: 2% NPGS (neopentylglycolsebacate); solid support: Anakrom ABS (110/120 mesh); temperatures: injector 280°C, column 230°C, detector 280°C; peak identification: (1) ethyl pantothenate diacetate; (2) panthenol triacetate. (From Ref. 72.)

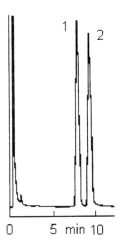

Figure 23 Determination of sodium pantothenate and panthenol as trimethylsilyl derivatives. Detector: FID; column: glass, 2440 × 4 mm; stationary phase 5% SE-30; solid support: Gas Chrom Q (100/120 mesh); temperatures: injector 270°C, column 185°C, detector 270°C; peak identification: (1) trimethylsilyl derivative of panthenol; (2) trimethylsilyl derivative of pantothenic acid (sodium pantothenate). (From Ref. 73.)

0   10   20   min

Figure 24 Determination of calcium pantothenate and panthenol as pantolactone. Detector: FID; column: glass, 2100 × 4 mm; stationary phase: 1% Carbowax 20M; solid support: Gas Chrom P (100/120 mesh); temperatures: injector (not stated), column 115°C, detector (not stated); peak identification: (1) *o*-toluidine (internal standard); (2) pantolactone. (From Ref. 23.)

based on the analysis of the hydrolysis products of pantothenic acid have been successfully employed. The determination of β-alanine is not applicable for the estimation of the pantothenic acid content, as β-alanine can arise not only from the pantothenic acid itself but also from some other substances such as aspartic acid or β-alanylhistidine dipeptides occurring in animal tissues (carnosine, anserine, balenine, etc.).

Schulze et al. (74) described a method useful for the analysis of pantothenic acid in urine. Samples were first purified by ion exchange chromatography and then hydrolyzed with HCl solutions; the generated pantolactone was extracted with dichloromethane and analyzed with methyl myristate as an internal standard on a polar stationary phase. The relative standard deviation of the method was 6% to 7%.

Determination of pantothenic acid in foods was described in detail by Davídek and Velíšek (38). This method is also based upon the acid hydrolysis of the vitamin and the separation of pantolactone. The hydrolysis is performed in 25% (w/w) HCl for 4 to 5 h at 95 to 100°C. Pantolactone can be extracted either directly from the hydrolysate or from the neutralized hydrolysate (at pH 5) into dichloromethane. The extract may be further purified by chromatography on a silica-gel column (75,76) or directly analyzed by GLC (77) using polar stationary phases such as polyethylene glycols (Fig. 25). Methyl myristate and ethyl laurate can be used as internal standards.

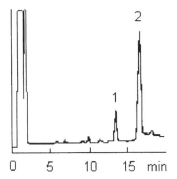

Figure 25 Gas chromatogram of fresh beef liver hydrolysate extract. Detector: FID; column: glass, 2400 × 2.1 mm; stationary phase: 10% Carbowax 20M; solid support: Chromaton N-AW-DMCS (0.125/0.160 mm); temperatures: injector 220°C, column 120–220°C (5°C/min), detector 250°C; peak identification: (1) ethyl laurate (internal standard); (2) pantolactone. (From Ref. 77.)

All of the described methods are suitable for the determination of the total content of the vitamin, that is, free pantothenic acid as well as its bound forms such as coenzyme A or pantetheine.

C. Panthenol

In pharmaceutical preparations, panthenol can be determined as a volatile derivative or as derivatives of its degradation products. Most commonly employed are acetyl and trimethylsilyl derivatives. The same derivatization procedures used for the acetylation of pantothenic acid can be employed for the conversion of panthenol into its triacetate, which can then be analyzed on polar stationary phases. Similarly, a tris(trimethylsilyl) ether of panthenol may be prepared using the derivatization procedures previously described. In addition, a 3:6:2 (v/v/v) mixture of TMCS-HMDS-dioxane (35°C, 100 min) has been used. Nonpolar polymethylsiloxanes are the stationary phases of choice.

The hydrolysis procedure leading to the formation of pantolactone, which can be analyzed as such, has been described above. The amount of residual pantolactone in panthenol can be estimated after conversion of both compounds to trimethylsilyl derivatives, which are chromatographed on nonpolar stationary phases with the trimethylsilyl ether of 2,6-dimethylphenol as the internal standard. For the derivatization, a 9:3:1 (v/v/v) pyridine-HMDS-TMCS mixture was used (room temperature, 30 sec). The procedures have been described by Velíšek et al. (5).

D. Optical Isomers of Pantothenic Acid and Related Compounds

GLC was the first chromatographic method to enable the resolution of enantiomers of pantothenic acid and hence the specific determination of the biologically active form of pantothenic acid.

Similarly to analysis of the other optically active compounds, the enantiomers of pantothenic acid, panthenol, and pantolactone can be separated either on optically active stationary phases (for instance, in the form of trifluoroacetates) or on common stationary phases (in the form of the corresponding diastereoisomers).

König and Sturm (78) separated racemic pantothenic acid in the form of methyl bis(trifluoroacetyl)pantothenate on a capillary column with a chiral polysiloxane phase XE-60-L-Val (R)-α-phenylethylamide (Fig. 26). The volatile derivative was prepared by esterification of pantothenic acid with methanolic HCl (20°C, for 14 h) and by acylation of methyl pantothenate formed with trifluoroacetanhydride in dichloromethane (20°C, for 20 min).

Enantiomers of panthenol can also be separated on chiral stationary phases. This separation can be performed either after conversion of panthenol isomers into their trifluoroacetates with a 4:1 (v/v) mixture of dichloromethane-trifluoroacetanhydride (20°C, 15 min) (78) or by analyzing free pantolactone formed in acidic medium.

Optical purity of D-pantothenic acid can also be determined indirectly, for example, by analysis of pantolactone, as no racemization of pantothenic acid/pantolactone occurs in acidic or in alkaline aqueous media (79). The D- and L-pantolactone were analyzed on a chiral stationary phase, either as such or as trifluoroacetyl ester or isopropylurethane derivative (Fig. 27) (78).

The trifluoroacetates were prepared using a 4:1 (v/v) mixture of dichloromethane-trifluoroacetanhydride (at 20°C, for 20 min), and isopropylurethane derivatives using a 1:1 (v/v) mixture of dichloromethane-isopropyl isocyanate (at 100°C, for 30 min) (78).

Not only chiral phases but also common stationary phases have been used for the separation of DL-pantolactone isomers. Takasu and Ohya (79) described the separation of MTPA derivatives (Fig. 27) of pantolactone isomers employing a packed column containing 2% OV-17 (Fig. 28). MTPA derivatives were prepared using a 5:5:2 (v/v/v) mixture of 1,2-dichloroethane-(−)-α-methoxy-α-(trifluoromethyl)-phenylacetyl chloride (MTPAC)-pyridine (at 80°C, for 30 min). N-Trifluoroacetyl-L-prolyl and some other derivatives were also tested but did not give adequate separation.

1
2
0
5
min
10

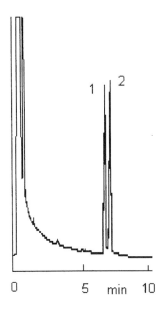

Figure 26 Separation of D- and L-pantothenic acid. Detector: FID; column: glass capillary (15 m long); stationary phase: XE-60-L-Val-(*R*)-α-phenylethylamide (film thickness not stated); temperatures: injector (not stated), column from 130°C (2°C/min), detector (not stated): peak identification: (1) D-pantothenic acid methylbis(trifluoroacetyl) derivative; (2) L-pantothenic acid methylbis(trifluoroacetyl) derivative. (From Ref. 78.)

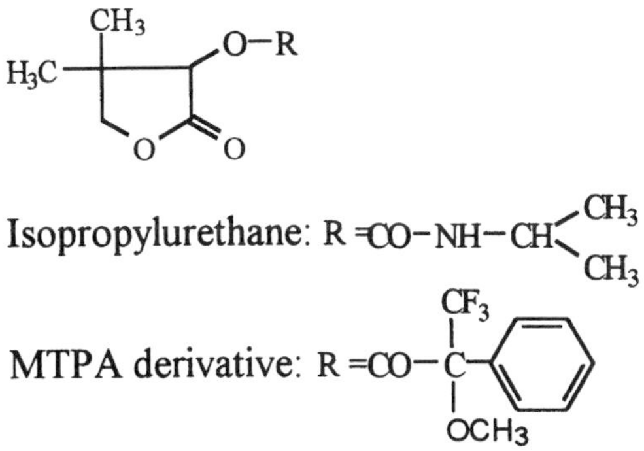

CH3
H3C
O−R
O
O
Isopropylurethane: R =CO−NH−CH
CH3
CH3
CF3
MTPA derivative: R =CO−C
OCH3

Figure 27 Some pantolactone derivative structures.

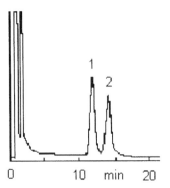

Figure 28   Separation of the MTPA derivatives of D- and L-pantolactone. Detector: FID; column: glass, 2000×3 mm; stationary phase: 2% OV-17; solid support: Gas Chrom Q (80/100 mesh); temperatures: injector 190°C, column 170°C, detector (not stated); peak identification: (1) MTPA derivative of L-pantolactone; (2) MTPA derivative of D-pantolactone. (From Ref. 79.)

E.   Coenzyme A

Coenzyme A and its analogs have not yet been subjected to GLC separation. Only the indirect method based on the analysis of pantolactone has been described and used for the evaluation of the degree of hydrolysis of CoA in HCl solutions (75). Pantolactone was extracted into dichloromethane and chromatographed on a polar stationary phase with methyl myristate as the internal standard. Recovery of the determination was 91% and the relative standard deviation was 3.6%.

Similarly to CoA, none of the other acyl-CoA analogues has been chromatographed using GLC. HPLC procedures and ion exchange chromatographic methods are the only methods of choice.

Reported levels of pantothenic acid in biological samples obtained by the pantolactone GLC approach pertain not only to the free acid but always to all other active forms, such as CoA, acetyl-CoA, etc.

V.   MASS SPECTROMETRY

A.   Liquid Chromatography–Mass Spectrometry

1.   Pantothenic Acid

In liquid chromatographic analysis of complex matrices, liquid chromatography–mass spectrometry (LC-MS) is a useful and powerful technique. In the past, LC-MS using direct liquid introduction, thermospray, frit-FAB interface, and other techniques has been proposed for the analysis of vitamins and has found certain application for some of the water-soluble vitamins such as thiamine. Less attention has been paid to pantothenic acid and related compounds.

Particle beam LC-MS was investigated by Careri et al. (62) for the analysis of pantothenic acid and 10 other water-soluble vitamins. A reversed-phase HPLC method making use of volatile buffers was set up for the simultaneous separation of this mixture of vitamins using narrow-bore columns.

At an MS source temperature of 200°C and in electron impact (EI), a relatively poor EI spectrum was obtained for pantothenic acid. Prominent peaks appear at m/z 57 ($[CO-NH-CH_2]^+$) and m/z 71 ($[CO-NH-(CH_2)_2]^+$), whereas the molecular ion at m/z 219 is of low intensity (relative abundance 2%).

The base peak at m/z 131 (its abundance was 10 times higher than that obtained in EI) in the corresponding positive-ion chemical ionization (PCI) is due to the cleavage of the peptide bond. The subsequent loss of 18 a.m.u. from this fragment results in the formation of an m/z 113 ion. In addition, an intense $[M]^+$ ion at the m/z 219 (20%) is present. The detection limit was relatively high, being 0.1 μg injected, and the relative standard deviations were 4.58% (0.2 μg) and 0.92% (0.5 μg).

A different ionization pattern was observed under negative-ion chemical ionization (NCI) conditions. The $[M-H]^-$ ion was detected with relatively high intensity (59%), the ion $[M-H-CH_2OH]^-$ at m/z 187 being the main fragment. The release of a CH_2CH_2COOH moiety from the deprotonated molecular ion produced the base peak at m/z 145.

An HPLC particle beam PCI-MS chromatogram of a water-soluble vitamin standard mixture is shown in Figure 29.

B. Gas Chromatography–Mass Spectrometry

1. Pantothenic Acid and Hopantenic Acid

The determination of trimethylsilylated pantoyllactone by gas chromatography–mass fragmentography (GC-MF) with a packed column has been developed for the assay of pantothenic acid by Tarli et al. (23). Umeno et al. (24) used the same method to investigate hopantenic acid levels in serum and urine after administration of calcium hopantenate.

Simultaneous rapid microanalysis of pantothenic acid and hopantenic acid in biological samples (plasma and brain samples) and foodstuffs (rice, green tea, and dried yeasts) by GC-MF was recently described by Banno et al. (55). Plasma samples were purified without deproteinization on an ion exchange resin MCI GEL CK08P (170 × 10 mm, H$^+$), and the eluate was extracted with ethyl acetate under acidic conditions. The organic layer was evaporated and the residue silylated with bis(trimethylsilyl)trifluoroacetamide. The brain samples were homoge-

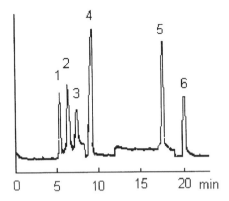

Figure 29   Total ion current HPLC-PB-PCI-MS chromatogram of a mixture of water-soluble vitamins. Column: 5 μm, Ultracarb ODS; 250 × 2 mm; mobile phase: methanol (solvent A), 0.02 M ammonium formiate buffer, pH 3.75 (solvent B); a binary gradient as follows: B-A (98:2, v/v) from 0 to 5 min, linear gradient from 98% B at 5 min to 50% B at 15 min, 50% B and 50% A from 15 to 25 min, 50% B from 25 min to 98% B at 35 min; flow rate: 0.15 mL/min; particle beam interface desolvation chamber temperature 70°C; source temperature: 200°C; quadrupole temperature: 100°C; mode: PCI, SIM, m/z 73 (400), 113 (400), 131 (200); electron energy: 220 eV; peak identification: (1) L-dehydroascorbic acid (100 ng); (2) L-ascorbic acid (30 ng); (3) thiamine (100 ng); (4) nicotinic acid (10 ng); (5) nicotinamide (10 ng); (6) pantothenic acid (150 ng). (From Ref. 62.)

nized with 0.005 M KOH; the samples of natural products were extracted with hot water. The extracts were centrifuged, purified, and further treated as the plasma samples.

   Aliquots of this solution were analyzed by GC-MF using a wide-bore fused-silica column (DB-17, 15 m × 0.53 mm) with a flow rate of helium of 15 mL/min, and the injection port, column oven, and separator temperatures of 250°C, 200°C, and 250°C, respectively. The MS detector was operating in the electron impact mode at 70 eV, the ionization current was 100 μA and the temperature of the ion source was 200°C. The stable fragment ions $[M-CH_3]^+$ selected for multiple ion detection were at m/z 420, 434, and 448. They were produced from the trimethylsilyl derivatives of pantothenic acid, hopantenic acid, and 5-[2,4-dihydroxy-3,3-dimethyl-1-oxobutyl)amino]pentanoic acid (internal standard), respectively (Fig. 30).

   The calibration curves were linear in the range 5 to 100 ng/mL of plasma and the detection limits were ca. 1 ng/mL plasma samples. The average recoveries of pantothenic acid and hopantenic acid were 92.9 ± 4.6% and 95.5 ± 5.1%, respectively.

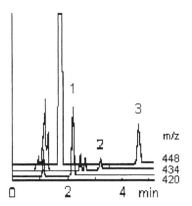

Figure 30 Determination of pantothenic acid and hopantenic acid in green tea by GC-MF. Column: fused silica capillary, 15 m × 0.53 mm; stationary phase: DB-17; mobile phase: helium, 15 mL/min; temperatures: injector 250°C, column 200°C, separator 250°C, ion source 200°C; detector: electron impact mode at 70 eV, ionization current 100 μA; ions: m/z 420, 434, and 448; peak identification: (1) pantothenic acid; (2) hopantenic acid; (3) 5-[2,4-dihydroxy-3,3-dimethyl-1-oxobutyl)amino]pentanoic acid (internal standard). (From Ref. 55.)

2. Optical Isomers of Pantothenic Acid and Hopantenic Acid

GC-MF was also used for the chiral separation and simultaneous determination of pantothenic acid and hopantenic acid enantiomers in rat plasma (56). The method is based on derivatization of the acids and separation of the volatile products using a chiral stationary phase.

The plasma samples were purified on an anion exchange resin MCI GEL CA08P (170 × 10 mm, Cl⁻). DL-Pantothenic acid and DL-hopantenic acid were eluted with 1 M NaCl solution and free acids were extracted with ethyl acetate from the eluate acidified with 6 M HCl in the presence of ammonium sulfate and esterified with 1.5 M HCl in methanol at room temperature for 1 h. The resulting methyl esters were trifluoroacetylated with trifluoroacetic anhydride in dichloromethane at room temperature for 30 min and analyzed by GC-MF. Mass spectrometry with selected-ion monitoring was employed for the simultaneous determination of the enantiomers of pantothenic acid and hopanthenic acid. The stable mass fragments were detected at m/z 257 and 271, being the base peaks of pantothenic acid and hopantenic acid, respectively (Fig. 31).

The calibration curves were linear in the 50 to 2000 ng/mL range. The recoveries of L-pantothenic acid and L-hopantenic acid were 93.5 ± 5.5% and 92.4 ± 8.3%, respectively. The detection limits of pantothenic acid and hopantenic acid were 5 ng/mL and 12 ng/mL, respectively.

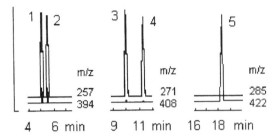

Figure 31 Mass fragmentogram of methyl bis(trifluoroacetates) of DL-pantothenic acid and DL-hopantenic acid. Column: fused silica capillary, 20 m × 0.22 mm, film thickness 0.12 μm; stationary phase: Chirasil-D-Val; coated with L-valine-*tert*-butylamide; mobile phase: helium, 1 kg/cm^2, split ratio 1:30; temperatures: injector 180°C, column 130°C, separator 250°C, ion source 250°C; detector: not specified; ions: *m/z* 257, 271, and 285; peak identification: (1) L-pantothenic acid; (2) D-pantothenic acid; (3) L-hopantenic acid; (4) D-hopantenic acid; (5) D-5-[2,4-dihydroxy-3,3-dimethyl-1-oxobutylamino]pentanoic acid (internal standard). (From Ref. 56.)

Alternatively, the derivatization was conducted using either thionyl chloride or n-butylboronic acid instead of trifluoroacetic anhydride. The resolution factors between the D-pantothenic acid and L-pantothenic acid derivatives were 1.6 (trifluoroacetates), 1.16 (cyclic sulfinates; Fig. 32), and 1.55 (cyclic boronates; Fig. 33), respectively. Although all were sufficiently high, trifluoroacetylation resulted in the best resolution factor. The sulfinates have two chiral centers in their molecules, but the respective chromatogram showed the separation of only one pair of enantiomers (80,81).

3. Coenzyme A Compounds

A sensitive method has been developed for the analysis of long-chain acyl intermediates of fatty acid and lipid biosynthesis that occur covalently bound to either ACP or CoA. The method is based on aminolysis of the sample with n-butylamine

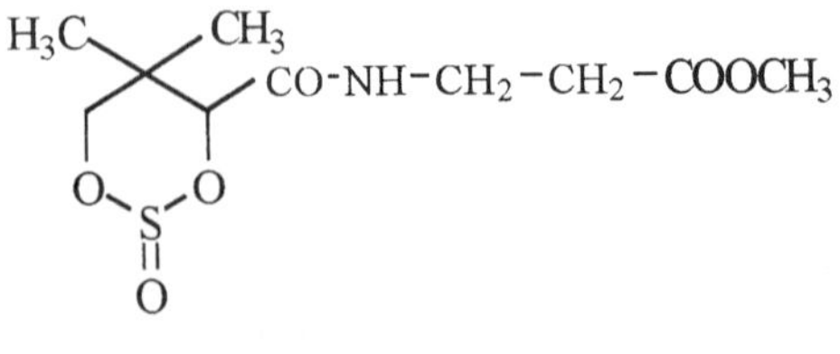

Figure 32 Structure of methyl pantothenate sulfinate.

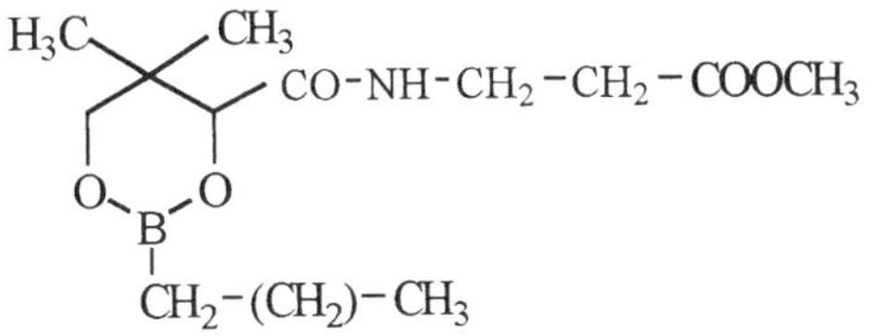

Figure 33   Structure of methyl pantothenate butylboronate.

resulting in selective formation of the corresponding fatty acid butylamides. The acylbutylamides are subsequently analyzed by GC-MS. The reaction was specific for thioester-linked acyl groups and 90% conversion was achieved with ACP and acyl-CoA in aqueous solution. EI mass spectra exhibited two intense diagnostic ions at m/z 115 and 128, common to butylamides of saturated and unsaturated fatty acids. The limit for the quantitative analysis of the long-chain $C_{18:0}$-butyl-amides and $C_{18:1}$-butylamides was 1.5 pmol and the detection limit was <0.5 pmol. The utility of this method was demonstrated by the analysis of a standard solution of acyl-CoA mixtures and spinach leaf acyl-ACP. A purification step based on DE 52 anion exchange chromatography was necessary to separate spin-ach acyl-ACP and acyl-CoA from tissue extracts (82).

VI. RECENT TECHNIQUES

Capillary zone electrophoresis (CE) is a relatively new analytical method currently under investigation for use in research and control laboratories for the analysis of ionic forms of vitamins. Micellar electrokinetic capillary chromatography (MECC) is a modification of CE which allows the separation of both neutral and ionic forms using buffers with micellar additives. Both methods have been used to separate water-soluble vitamins but very rarely pantothenic acid, which has only been analyzed by this technique in model mixtures and pharmaceuticals.

A. Capillary Zone Electrophoresis

A mixture of several water-soluble vitamins including calcium pantothenate was recently evaluated by CE by Jegle (83). The sample was analyzed in a 0.02 M sodium phosphate buffer (pH 7) and separated using a three-dimensional capillary zone electrophoresis system (fused-silica, 50 µm i.d., straight, length to detector 400 mm, total length 485 mm, injection pressure 4.6 sec at 4 kPa, postinjection pressure 4 sec at 40 kPa, polarity positive, voltage 20 kV, capillary temperature

25°C, detection 215 nm). Between runs, the capillary was conditioned for 2 min with 0.1 M NaOH and for 3 min with the buffer. The method seems to be well suited for thiamine, nicotinamide, and nicotinic acid but less so for other water-soluble vitamins including pantothenate (the average migration time of pantothenate was 10.4 min).

B. Micellar Electrokinetic Capillary Chromatography

Dinelli and Bonetti (84) have recently shown that a mixture of vitamin B_1, B_2, B_3, B_6, B_{12}, C, and calcium pantothenate can be completely resolved and eluted within 30 min using MECC. They used a 700×0.1 mm i.d. capillary at 25°C. The system was operated with a 0.05 M $Na_2B_4O_7$, 0.0225 M SDS, 10% v/v methanol, pH 8.0 electrolyte buffer, under applied voltage of 16 kV with a duration of 10 min and current intensity of 33 µA. The samples were injected with a pressure of 3.44 kPa for 5 sec, corresponding to an injection volume of approximately 0.03 µL. The on-column UV detection was affected at 214 nm. The method was applied to the quantitative determination of one multivitamin preparation containing calcium pantothenate. Direct analysis of this preparation dissolved in water was unsatisfactory as the mean overall recovery ranged from 43% to 66%. Using solid-phase enrichment of the vitamins, the recovery improved, ranging from 93% to 103% and the coefficient of variation was <5% (retention time of pantothenate was 24.4 ± 0.3 min).

VII. FUTURE TRENDS

The literature on chromatography of pantothenic acid and its derivatives that has appeared during the last decade has shown that chromatographic techniques became useful analytical tools, especially for the quantitative determination of the vitamin in pharmaceuticals, for which they offer certain advantages over earlier analytical methods. The minor attention that has been paid to the analysis of naturally occurring pantothenic acid in foods can to a certain extent be ascribed to the fact that part of the vitamin is supplied to some extent by the intestinal microflora of animals and humans. Together with the vitamin coming from the diet, the latter provides adequate quantities to cover nutritional requirements. Another reason is the complexity of the matrix which makes the analysis of the vitamin by modern instrumental techniques very difficult; thus, the official microbiological method using *Lactobacillus plantarum* is still in use.

Modern chromatography of pantothenic acid and other compounds in pharmaceuticals that possess the vitamin activity is mainly restricted to HPLC and GLC techniques. The main reason is that HPLC is an attractive alternative to the more time-consuming chemical and microbiological methods as it allows simple sample preparation, fast determinations, and good sensitivity and selectivity.

Currently employed HPLC methods for pantothenic acid and/or pantothenates have been applied solely to pharmaceuticals and simple matrices such as fortified infant formulas, whereas assays of coenzyme A and its acyl analogs have also been successfully performed on animal tissues. In the last few years, chiral stationary phases have been developed for optical resolution of pantothenic acid and related compounds by HPLC, and also HPLC-MS has become a promising technique. However, the newly developed HPLC procedures still require increased sensitivity and selectivity to make them applicable for the analysis of the total vitamin content in complex matrices such as foods and feeds.

Appropriate GLC procedures have also been used for the analysis of pure vitamins, their optical isomers, pharmaceuticals, and, rarely, even foods. The main disadvantage of this group of chromatographic methods seems to be the tedious and relatively time-consuming sample preparation (derivatization and/or cleanup procedures) prior to the GLC analysis. Nevertheless, the newly developed and introduced GC-MS techniques show that GLC is a sufficiently sensitive tool for the trace analysis of pantothenic acid, its higher homolog and other related compounds such as acyl-CoA, even in biological samples (serum, brain, foodstuffs).

Capillary zone electrophoresis and micellar electrokinetic capillary chromatography appeared as relatively new analytical methods during the last few years. These electrophoretic methods seem to be extremely powerful analytical tools because of their speed, selectivity, reproducibility, and possible level of automation comparable to HPLC and GLC methods. The methods are now under investigation for use in research and control laboratories; however, the theoretical approach still prevails with respect to the practical one. Nevertheless, it is evident that these methods have a potential to perform a complete resolution and quantitation of water-soluble vitamins mixtures, possibly even in complex biological samples.

Currently used methods for quantitative estimation of the vitamin in foods and feeds still rely mainly on microbiological assay procedures. Some other techniques, especially the sensitive radioimmunoassay procedures, have also been applied on a routine basis. Other chromatographic techniques, such as thin-layer chromatography and ion exchange chromatography, remain valuable particularly for qualitative analysis and preparative purposes.

REFERENCES

1. AG Moiseenok, ed. Chemistry, Biochemical Function, and Use of a Pantothenic Acid. Minsk: Izd Nauka Tekhnika, 1977.

2. W Friedrich. Vitamins. Berlin: Walter de Gruyter, 1988.
3. HM Fox. Pantothenic acid. In: LJ Machlin, ed. Handbook of Vitamins. 2nd ed. New York: Marcel Dekker, 1991, pp 429–451.
4. GF Combs Jr. The Vitamins: Fundamental Aspects in Nutrition and Health. San Diego: Academic Press, 1992.
5. J Velíšek, J Davídek, T Davídek. Pantothenic acid. In: AP de Leenheer, WE Lambert, HJ Nelis, eds. Modern Chromatographic Analysis of Vitamins. New York: Marcel Dekker, 1992, pp 513–560.
6. R Macrae, RK Robinson, MJ Sadler, eds. Encyclopedia in Food Technology and Nutrition. Vol. 5. London: Academic Press, 1993.
7. DM Sullivan, DE Carpenter, eds. Methods of Analysis for Nutritional Labelling. Arlington, VA: AOAC-International, 1993, pp 327–330.
8. GFM Ball. Water-Soluble Vitamins Assays in Human Nutrition. London: Chapman and Hall, 1994.
9. RJ Williams, EM Bradway. The further fractionation of yeast nutrilites and their relationship of vitamin B and Wilders' "bios." J Am Chem Soc 53:783–789, 1931.
10. RJ Williams, CM Lyman, GM Goodyear, JE Truesdayl, D Holaday. "Pantothenic acid" a growth determinant of universal biological occurrence. J Am Chem Soc 55: 2912–2927, 1933.
11. RJ Williams. A Textbook of Chemistry. New York: Van Nostrand, 1938.
12. ET Stiller, SA Harris, J Finkelstein, JC Keresztesy, K Folkers. Pantothenic acid VIII. The total synthesis of pure pantothenic acid. J Am Chem Soc 62:1785–1790, 1940.
13. F Lipmann, MO Kaplan. A common factor in the enzymic acetylation of sulfanilamide and of choline. J Biol Chem 162:743–744, 1946.
14. F Lipmann, MO Kaplan, GD Novelli, LC Tutte, BM Guylard. Coenzyme for acetylation, a pantothenic acid derivative. J Biol Chem 167:869–870, 1947.
15. F Lynen, E Reichert, L Rueff. Zum biologischen Abbau der Essigsäure VI. "Aktivierte Essigsäure," ihre Isolierung aus Hefe und ihre chemische Natur. Ann Chem 574:1–32, 1951.
16. JG Moffatt, HG Khorana. Nucleoside polyphosphates XII. The total synthesis of coenzyme A. J Am Chem Soc 83:663–675, 1961.
17. AM Michelson. Chemistry of the nucleotides. Annu Rev Biochem 30:133–164, 1961.
18. EL Pugh, SJ Wakil. Mechanism of fatty acid synthesis XIV. The prosthetic group of acyl carrier protein and the mode of its attachment to the protein. J Biol Chem 240:4727–4733, 1965.
19. M Shimizu. Pantothenic acid and coenzyme A. Method Chim 11:73–76, 1977.
20. E Knobloch, J Černá-Heyrovská. Fodder Biofactors. Their Methods of Determination. Prague: Academia, 1979.
21. HM Fox. Pantothenic acid. In: LJ Machlin, ed. Handbook of Vitamins. New York: Marcel Dekker, 1984.
22. EJ Vandamme. Production of vitamins, coenzymes and related biochemicals by biotechnological processes. J Chem Tech Biotech 53:313–327, 1992.
23. P Tarli, S Benocci, P Neri. Gas chromatographic determination of pantothenates and panthenol in pharmaceutical preparations by pantoyl lactone. Anal Biochem 42:8–13, 1971.

24. Y Umeno, K Nakai, E Matsushima, T Marunaka. Gas chromatographic–mass frag-mentographic determination of homopantothenic acid in plasma. J Chromatogr 226: 333–339, 1981.

25. JV Kathman, C Kies. Pantothenic acid status of free living adolescent and young adults. Nutr Res 4:245–250, 1984.

26. AG Moiseenok, VM Kopelevich, MA Izraelit, LM Smuilovich. D-homopantothenic acid, its physicochemical and pharmaceutical properties, metabolism and clinical use. Farmakol Toksikol (Moscow) 36:489–494, 1973.

27. J Davídek, J Velíšek, J Pokorný, eds. Chemical Changes During Food Processing. Amsterdam: Elsevier, 1990.

28. K Hoppner, B Lampi. Pantothenic acid and biotin retention in cooked legumes. J Can Inst Food Sci Technol 22:170–172, 1989.

29. M Kirchgessner, DA Roth-Maier, U Heindl, FJ Schwarz. B-vitamins (thiamin, vita-min B_6, pantothenic acid) in lean tissue of growing cattle of the German Simmental breed under different feeding intensities. Z Lebensm Unters Forsch 201:20–24, 1995.

30. S Lang. Vitamins in Milk. Munich: Volkswirtschaftlicher Verlag, 1990.

31. L Dolfini, R Kueni, P Eberhard, D Fuchs, PU Gallman, W Strahm, R Sieber. Behav-iour of supplemented vitamins during storage of UHT skin milk. Mitt Geb Lensm Hyg 82:187–198, 1991.

32. R Sieber. Behaviour of vitamins during storage of UHT milk. Mitt Geb Lensm Hyg 80:467–489, 1989.

33. AM Berthier. Yogurt: composition and nutritional value. Rev Laitier Fr 493:36–37, 1990.

34. K Hoppner, B Lampi. Total folate, pantothenic acid and biotin content of yoghurt products. J Can Inst Food Sci Technol 23:223–225, 1990.

35. K Hoppner, B Lampi. Pantothenic acid and biotin retention in cooked legumes. J Food Sci 58:1084–1085, 1993.

36. R Strohecker, HM Henning. Vitamin-Bestimmungen. Weinheim: Verlag Chemie, 1963, pp 220–231.

37. MH Hashmi. Assay of Vitamins in Pharmaceutical Preparations. New York: Wiley, 1973.

38. J Velíšek, J Davídek. Gas-liquid chromatography of vitamins in foods: the water soluble vitamins. J Micronutr Anal 2:25–42, 1986.

39. JT Tanner, SA Barnett, MK Moutford. Analysis of milk-based infant formula. Phase V. Vitamin A and E, folic acid, and pantothenic acid: Food and Drug Administration Infant Formula Council: collaborative study. J Assoc Off Anal Chem 76:399–413, 1993.

40. S Dupré, R Chiaraluce, M Nardini, C Cannella, G Ricci, D Cavallini. Continuous spectrophotometric assay of pantothenase activity. Anal Biochem 142:175–181, 1984.

41. G Bertelsen, PM Finglas, J Longhridge, JM Faulks, MRA Morgan. Investigation into the effects of conventional cooking on levels of thiamin (determined by HPLC) and pantothenic acid (determined by ELISA) in chicken. Food Sci Nutr 42F:83–96, 1988.

42. KM Knights, R Drew. A radioisotopic assay of picomolar concentrations of coen-zyme A in liver tissues. Anal Biochem 168:94–99, 1988.

43. HC Morris, PM Finglas, RM Faulks, MRA Morgan. The development of an enzyme-linked immunosorbent assay (ELISA) for the analysis of pantothenic acid and analogues. Part I. Production of antibodies and establishment of ELISA systems. J Micronutr Anal 4:33–45, 1988.

44. HC Morris, PM Finglas, RM Faulks, MRA Morgan. The development of an enzyme-linked immunosorbent assay (ELISA) for the analysis of pantothenic acid and analogues. Part II. Determination of pantothenic acid in foods. J Micronutr Anal 4:47–59, 1988.

45. PM Finglas, MRA Morgan. Application of biospecific methods to the determination of B-group vitamins in food—a review. Food Chem 49:191–201, 1994.

46. T Takeuchi, Y Kabasawa, R Horikawa, T Tanimura. Flow injection determination of drugs by specific detection of carboxylic acid. Analyst 113:1673–1676, 1988.

47. SS Nag, SK Das. Identification and quantification of panthenol and pantothenic acid in pharmaceutical preparations by thin-layer chromatography and densitometry. J Assoc Off Anal Chem 75:898–901, 1992.

48. C Li, J Zhong, J Sun, J Zhang, M Xia, X Gao. Determination of meprobamate and calcium pantothenate with ammonia gas-sensing electrode. Yiyao Gongye 19:308–311, 1988; Chem Abstr 109:176434c, 1988.

49. S Tagami, S Miyajima. Determination of nitrogen in drugs by an ammonia-selective electrode. Buneski Kagaku 36:T129–T131, 1987; Chem Abstr 108:82213g, 1988.

50. JA Timmons, JC Meyer, DJ Steible, SP Assenza. Reverse-phase liquid chromatographic assay for calcium pantothenate in multivitamin preparations and raw materials. J Assoc Off Anal Chem 70:510–513, 1987.

51. MC Gennaro. Separation of water-soluble vitamins by reserved-phase ion-interaction reagent high-performance liquid chromatography: application to multivitamin pharmaceuticals. J Chromatogr Sci 29:410–415, 1991.

52. RK Gharehbagh, S Ebel. Stability analysis of dexpanthenol. I. Determination of dexpanthenol and pantolactone by HPLC. Pharmazie 50:39–40, 1995.

53. JM Romera, M Ramirez, A Gil. Determination of pantothenic acid in infant milk formulas by high-performance liquid chromatography. J Dairy Sci 79:523–526, 1995.

54. T Arai, H Matsuda, H Oizumi. Determination of optical purity by high-performance liquid chromatography on chiral stationary phases: pantothenic acid and related compounds. J Chromatogr 474:405–410, 1989.

55. K Banno, M Matsuoka, S Horimoto, J Kato. Simultaneous determination of pantothenic acid and hopantenic acid in biological samples and natural products by gas chromatography–mass fragmentography. J Chromatogr 525:255–264, 1990.

56. K Banno, S Horimoto, M Matsuoka. Analytical studies on the chiral separation and simultaneous determination of pantothenic acid and hopantenic acid enantiomers in rat plasma by gas chromatography–mass fragmentography. J Chromatogr 564:1–10, 1991.

57. MT King, PD Reiss. Separation and measurement of short-chain coenzyme A compounds in rat liver by reversed-phase high-performance liquid chromatography. Anal Biochem 146:173–179, 1985.

58. Y Hosokawa, Y Shimomura, RA Harris, T Ozawa. Determination of short-chain

acyl-coenzyme A esters by high-performance liquid chromatography. Anal Biochem 153:45–49, 1986.

59. W Huth, C Worm-Breitgoff, U Möller, I Wunderlich. Evidence for an in vivo modification of mitochondrial proteins by coenzyme A. Biochim Biophys Acta 1077:1–10, 1991.

60. A Demoz, A Garras, DK Asiedu, B Netteland, RK Berge. Rapid method for the separation and detection of tissue short-chain coenzyme A esters by reversed-phase high-performance liquid chromatography. J Chromatogr 667:148–152, 1995.

61. A Hermans-Lokkerbol, R van der Heijden, R Verpoorte. Isocratic high-performance liquid chromatography of coenzyme A esters involved in the metabolism of 3S-hydroxy-3-methylglutaryl-coenzyme A. Detection of related enzyme activities in *Catharanthus roseus* plant cell cultures. J Chromatogr A 752:123–130, 1996.

62. M Careri, R Cilloni, MT Lugari, P Manini. Analysis of water-soluble vitamins by high-performance liquid chromatography–particle beam–mass spectrometry. Anal Commun 33:159–162, 1996.

63. TJ Hudson, S Subramanian, RJ Allen. Determination of pantothenic acid, biotin, and vitamin B_{12} in nutritional products. J Assoc Off Anal Chem 67:994–998, 1984.

64. TJ Hudson, RJ Allen. Determination of pantothenic acid in multivitamin pharmaceutical preparations by reverse-phase high-performance liquid chromatography. J Pharm Sci 73:113–115, 1984.

65. P Jonvel, G Andermann, JF Barthelemy. Determination of calcium pantothenate in multivitamin preparations by high-performance liquid chromatography. J Chromatogr 281:371–376, 1983.

66. TJ Franks, JD Stodola. A reverse-phase HPLC assay for the determination of calcium pantothenate utilizing column switching. J Liq Chromatogr 7:823–837, 1984.

67. H Umagat, R Tscherne. High-performance liquid chromatographic determination of panthenol in bulk, premix and multivitamin preparations. Anal Chem 52:1368–1370, 1980.

68. LL Bieber. Quantitation of CoASH and acyl-CoA. Anal Biochem 204:228–230, 1992.

69. MS DeBuysere, MS Olson. The analysis of acyl-coenzyme A derivatives by reverse-phase high-performance liquid chromatography. Anal Biochem 133:373–379, 1983.

70. BE Corkey, JT Deeney, eds. Progress in Clinical and Biological Research, Fatty Acid Oxidation: Clinical, Biological and Molecular Aspects. New York: Alan R. Liss, 1990, pp 217–232.

71. K Bartlett, AG Causey. Radiochemical high-performance liquid chromatography methods for the study of branched-chain amino acid metabolism. Methods Enzymol 166:79–96, 1988.

72. AR Prosser, AJ Sheppard. Gas-liquid chromatographic determination of pantothenates and panthenol. J Pharm Sci 58:718–721, 1969.

73. AR Prosser, AJ Sheppard. GLC of trimethylsilyl derivatives of pantothenyl alcohol and pantothenates. J Pharm Sci 60:909–912, 1971.

74. E Schulze zur Wiesch, C Hesse, D Hötzel. Gas chromatographic method for the determination of pantothenic acid. Z Klin Chem Klin Biochem 12:55–56, 1974.

75. E Tesmer. Gaschromatographische Bestimmung von Pantothensäure in Lebensmittel. PhD dissertation, Reinischen Friedrich-Wilhelms-Universität, Bonn, 1978.

76. E Tesmer, J Leinert, D Hötzel. Gaschromatographische Bestimmung von Pantothen-säure in Lebensmittel. Nahrung 24:697–704, 1980.

77. J Davídek, J Velíšek, J Černá, T Davídek. Gas chromatographic determination of pantothenic acid in foodstuffs. J Micronutr Anal 1:39–46, 1985.

78. WA König, U Sturm. Determination of optical purity by enantioselective capillary gas chromatography of panthenol and related compounds. J Chromatogr 328:357–361, 1985.

79. A Takasu, K Ohya. Separation and determination of the enantiomers of pantolactone by gas-liquid chromatography. J Chromatogr 389:251–255, 1987.

80. K Banno. Analytical studies on chiral separation and simultaneous determination of pantothenic acid and hopantenic acid enantiomers in rat plasma by gas chromatogra-phy–mass fragmentography. J Chromatogr 576:181, 1992.

81. E Küsters, C Spöndlin, C Eder. Analytical studies on chiral separation and simultane-ous determination of pantothenic acid and hopantenic acid enantiomers in rat plasma by gas chromatography–mass fragmentography: a reply. J Chromatogr 576:179–180, 1992.

82. J Kopka, JB Ohlrogge, JG Jaworski. Analysis of in vivo levels of acyl-thioesters with gas chromatography/mass spectrometry of the butylamide derivative. Anal Chem 224:51–60, 1995.

83. U Jegle. Separation of water-soluble vitamins via high-performance capillary elec-trophoresis. J Chromatogr A 652:495–501, 1993.

84. G Dinelli, A Bonetti. Micellar electrokinetic capillary chromatography analysis of water-soluble vitamins and multi-vitamin integrators. Electrophoresis 15:1147–1150, 1994.

Index

RETURN TO ➡ **CHEMISTRY LIBRARY** 2257
100 Hildebrand Hall   642-3753

LOAN PERIOD 1	2	3
	1 MONTH	
4	5	6

ALL BOOKS MAY BE RECALLED AFTER 7 DAYS
Renewable by telephone

DUE AS STAMPED BELOW

NON-CIRCULATING		
UNTIL: 8/18/00		

FORM NO. DD5

UNIVERSITY OF CALIFORNIA, BERKELEY
BERKELEY, CA 94720
℗s